# Netter's Anatomy **Coloring Book**

## Third Edition

**John T. Hansen, PhD**

Emeritus Professor of Neuroscience
Former Schmitt Chair of Neurobiology and Anatomy and
Associate Dean for Admissions
University of Rochester Medical Center
Rochester, New York

**ARTISTS**

Art based on the works of the **Frank H. Netter, MD,** collection
*www.netterimages.com*

Modified for coloring by
Carlos A.G. Machado, MD
and
Dragonfly Media Group

ELSEVIER

1600 John F. Kennedy Blvd.
Ste 1800
Philadelphia, PA 19103-2899

NETTER'S ANATOMY COLORING BOOK, THIRD EDITION ISBN: 978-0-323-82673-0



*Executive Content Strategist:* Elyse O'Grady
*Senior Content Development Specialist:* Marybeth Thiel
*Publishing Services Manager:* Deepthi Unni
*Senior Project Manager:* Beula Christopher
*Book Designer:* Renee Duenow

Printed in China

9 8 7 6 5 4 3 2

For **Amy**, daughter, wife, mother, and physician, who colored her way through medical school and made me a believer...

For **Sean**, son, husband, father, and engineer, who colored outside the lines and showed me his creativity...

And, for **Paula**, wife, mother, grandmother, teacher, and soul mate, who understood the value of coloring and always gave us encouragement.

# About the Author

**John T. Hansen, PhD,** is Emeritus Professor of Neuroscience, the former Killian J. and Caroline F. Schmitt Professor and Chair of Neurobiology and Anatomy, and the Associate Dean for Admissions at the University of Rochester Medical Center, Rochester, New York.

Dr. Hansen is the recipient of numerous teaching awards from students at three different medical schools. From 1995 to 1998, he was Professor and a Robert Wood Johnson Dean's Senior Teaching Scholar. In 1999, he was the recipient of the *Alpha Omega Alpha* Robert J. Glaser Distinguished Teacher Award given annually by the Association of American Medical Colleges to nationally recognized medical educators. From 2004 to 2005, Dr. Hansen was Chair of the Northeast Group on Student Affairs for the Association of American Medical Colleges. In 2013, he was selected as an Honored Member of the American Association of Clinical Anatomists, that organization's highest recognition. In 2018, he was elected to membership in *Alpha Omega Alpha* by the Rochester Medical Class of 2018, and in 2020 he was the recipient of the James S. Armstrong Alumni Service Award from the University of Rochester Medical Center.

Dr. Hansen has served on the USMLE Step Anatomy Test Material Development Committee and given numerous Faculty and Curricular Development Workshops at a number of US and foreign medical schools.

In 2010, Dr. Hansen was the first recipient of the University of Rochester's Presidential Diversity Award in recognition of his "advocacy, support, mentoring, planning, and leading the medical school's initiatives to increase the recruitment, retention, excellence, and graduation of students from diverse backgrounds."

Dr. Hansen's investigative research career encompassed the study of the peripheral and central nervous systems' dopaminergic pathways, neural plasticity, and central nervous system inflammation. He has received a prestigious 5-year NIH Research Career Development Award; a number of foundation and NIH research grants; and has presented his research findings at major US universities, national meetings, and at a number of international meetings. In addition to over 100 full-length research publications, he also is the co-author of the 2002 edition of the *Netter's Atlas of Human Physiology*; the lead consulting editor of the 3rd through the 7th editions of the *Netter's Atlas of Human Anatomy* (from 2003 to 2021); author of *Netter's Anatomy Flash Cards*, the *Essential Anatomy Dissector*, and *Netter's Anatomy Coloring Book*; and is the co-author of the *TNM Staging Atlas with Oncoanatomy*, which was selected from 630 worldwide entries as the *Book of the Year in 2008* by the British Medical Association.

# About the Artists

## Frank H. Netter, MD

Frank H. Netter was born in 1906, in New York City. He studied art at the Art Student's League and the National Academy of Design before entering medical school at New York University, where he received his MD degree in 1931. During his student years, Dr. Netter's notebook sketches attracted the attention of the medical faculty and other physicians, allowing him to augment his income by illustrating articles and textbooks. He continued illustrating as a sideline after establishing a surgical practice in 1933, but he ultimately opted to give up his practice in favor of a full-time commitment to art. After service in the United States Army during World War II, Dr. Netter began his long collaboration with the CIBA Pharmaceutical Company (now Novartis Pharmaceuticals). This 45-year partnership resulted in the production of the extraordinary collection of medical art so familiar to physicians and other medical professionals worldwide.

In 2005, Elsevier, Inc. purchased the Netter collection and all publications from Icon Learning Systems. There are now over 50 publications featuring the art of Dr. Netter available through Elsevier, Inc. (in the US: www.us.elsevierhealth.com/Netter; outside the US: www.elsevierhealth.com).

Dr. Netter's works are among the finest examples of the use of illustration in the teaching of medical concepts. The 14-book *Netter Collection of Medical Illustrations*, which includes the greater part of thousands of paintings created by Dr. Netter, became and remains one of the most famous medical works ever published. Netter's *Atlas of Human Anatomy*, first published in 1989, presents the anatomical paintings from the Netter collection. Now translated into 16 languages, it is the anatomy atlas of choice among medical and health professions students the world over.

The Netter illustrations are appreciated not only for their aesthetic qualities but, more important, for their intellectual content. As Dr. Netter wrote in 1949, "...clarification of a subject is the aim and goal of illustration. No matter how beautifully painted, how delicately and subtly rendered a subject may be, it is of little value as a *medical illustration* if it does not serve to make clear some medical point." Dr. Netter's planning, conception, point of view, and approach are what inform his paintings and what makes them so intellectually valuable.

Frank H. Netter, MD, physician and artist, died in 1991.

Learn more about the physician-artist whose work has inspired the Netter Reference collection:

https://netterimages.com/artist-frank-h-netter.html

## Carlos A.G. Machado, MD

Carlos Machado was chosen by Novartis to be Dr. Netter's successor. He continues to be the main artist who contributes to the Netter collection of medical illustrations.

Self-taught in medical illustration, cardiologist Carlos Machado has contributed meticulous updates to some of Dr. Netter's original plates and has created many paintings of his own in the style of Netter as an extension of the Netter collection. Dr. Machado's photorealistic expertise and his keen insight into the physician–patient relationship inform his vivid and unforgettable visual style. His dedication to researching each topic and subject he paints places him among the premier medical illustrators at work today.

Learn more about his background and see more of his art at:

https://netterimages.com/artist-carlos-a-g-machado.html

# PREFACE: **HOW TO USE THIS BOOK**

Human anatomy is a fascinating and complex subject, and one that is interesting to virtually every one of us. Learning anatomy does not have to be difficult and can actually be enjoyable. Exploring human anatomy in a simple, systematic, and fun way is what the *Netter's Anatomy Coloring Book* is all about. This coloring book is for students of all ages; curiosity is the only prerequisite!

The images in *Netter's Anatomy Coloring Book* are based on the famous beautifully rendered medical illustrations of human anatomy by Frank H. Netter, MD, as compiled in his *Atlas of Human Anatomy*. This is the most widely used anatomy atlas in the world and is translated into 18 different languages, and with good reason. The Netter illustrations have withstood the test of time and have illuminated human anatomy for millions of students around the world.

Why use an anatomy coloring book? The best reason, in my opinion, is because "active learning" always trumps passive learning. Seeing, doing, and learning go hand in hand; said another way, "eye to hand to mind to memory." This is how most of us learn best. Textbooks, flash cards, videos, and anatomy atlases all have their place in learning human anatomy, but those elements that engage us the most and allow us to participate in an active learning experience cement the material into our memory.

The *Netter's Anatomy Coloring Book* approaches human anatomy by body system. In each coloring book plate, the most important structures are emphasized. The coloring exercises, labels, text, bullet points of essential material, and tables are provided to help you understand why the carefully chosen views of the human body are important both anatomically and functionally. I intentionally did not over-label each image because I want you to focus on the most important aspects of the anatomy; however, this is *your* coloring book! Feel free to color everything you wish; add your own labels as desired; cover structures to quiz yourself; in short, use each image as creatively as you wish to enhance your learning experience. In most cases, I let you choose the colors you want but would encourage you to color arteries bright red, veins blue, muscles reddish-brown, nerves yellow, and lymph nodes green, as these are common colors used in most color atlases of anatomy. Finally, I think you probably will find that colored pencils work best, but if crayons, colored pens, highlighters, or markers are your preferred medium, by all means use them! Most of all, have fun learning anatomy—after all, it is your anatomy too!

JOHN T. HANSEN, PHD

Cross-references to Netter's *Atlas of Human Anatomy* available online (see inside front cover)

# Chapter 1 Orientation and Introduction

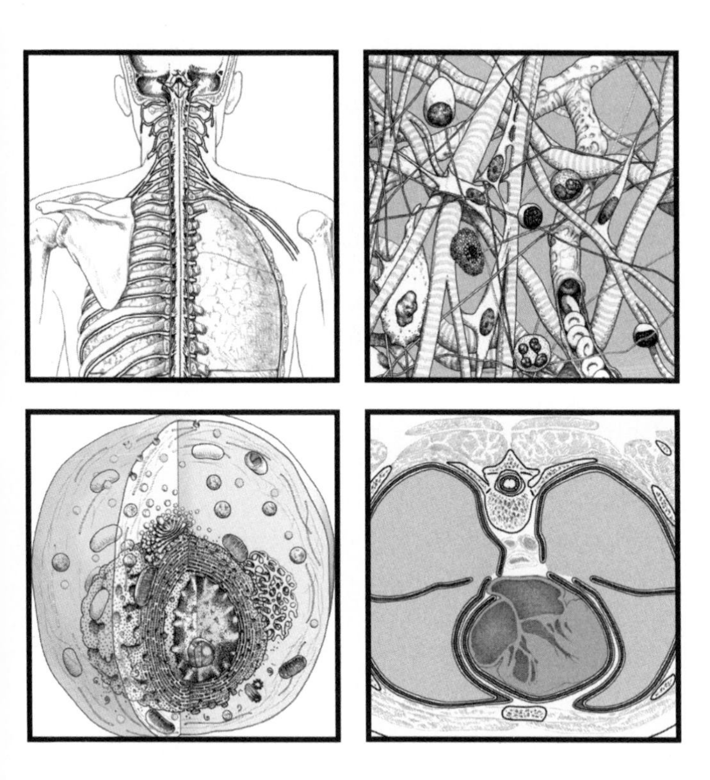

# 1 Terminology

Anatomy requires a clinical vocabulary that defines positions, movements, relationships, and planes of reference. By convention, anatomical descriptions of the human body are based on a person standing in the **anatomical position**. This position is defined as:

- Standing erect with the head and toes facing anteriorly (forward)
- Arms (upper limbs) hanging at the sides, palms facing anteriorly
- Legs (lower limbs) placed together, feet slightly apart and directed anteriorly

**COLOR** the major regions, beginning with the head and working inferiorly to the lower limb, using a different color for each region:

- ☐ 1. **Head (cephalon)**
- ☐ 2. **Neck (cervicis)**
- ☐ 3. **Thorax (chest)**
- ☐ 4. **Abdomen**
- ☐ 5. **Pelvis**
- ☐ 6. **Upper limb**
- ☐ 7. **Lower limb**

Regions of the body were originally named with Latin or Greek terms, and those terms are often still used today, especially in textbooks of anatomy. In countries where English is spoken, anglicized terms may also be used. In Plate 1-1, major regions and specific areas of the human body are designated by terms currently used in anatomical and clinical settings.

The study of anatomy may be by body region (**Regional Anatomy**) or by body systems (**Systemic Anatomy**). Most dissection courses of medical human anatomy in the United States use a regional approach, integrating all applicable body systems into the study of a particular region. This typically involves all applicable systems found in a particular region, e.g., in the forearm this would include the skin, muscles, nerves, bones, and vasculature. Of course, one must also remember that the body regions also may be reliant upon the other systems such as the endocrine system, lymphatic system, the gastrointestinal system for nourishment, etc.

The study of anatomy also may be organized around a systems approach, where the organ system is studied independently but with the understanding that organ systems do not function in isolation from one another. The **Netter's Anatomy Coloring Book** is organized by systems, which is a common approach used in undergraduate human anatomy courses (and Human Anatomy and Physiology courses), some medical school courses, and in anatomy courses where previously dissected anatomical specimens and/or models may be used to teach and learn anatomy.

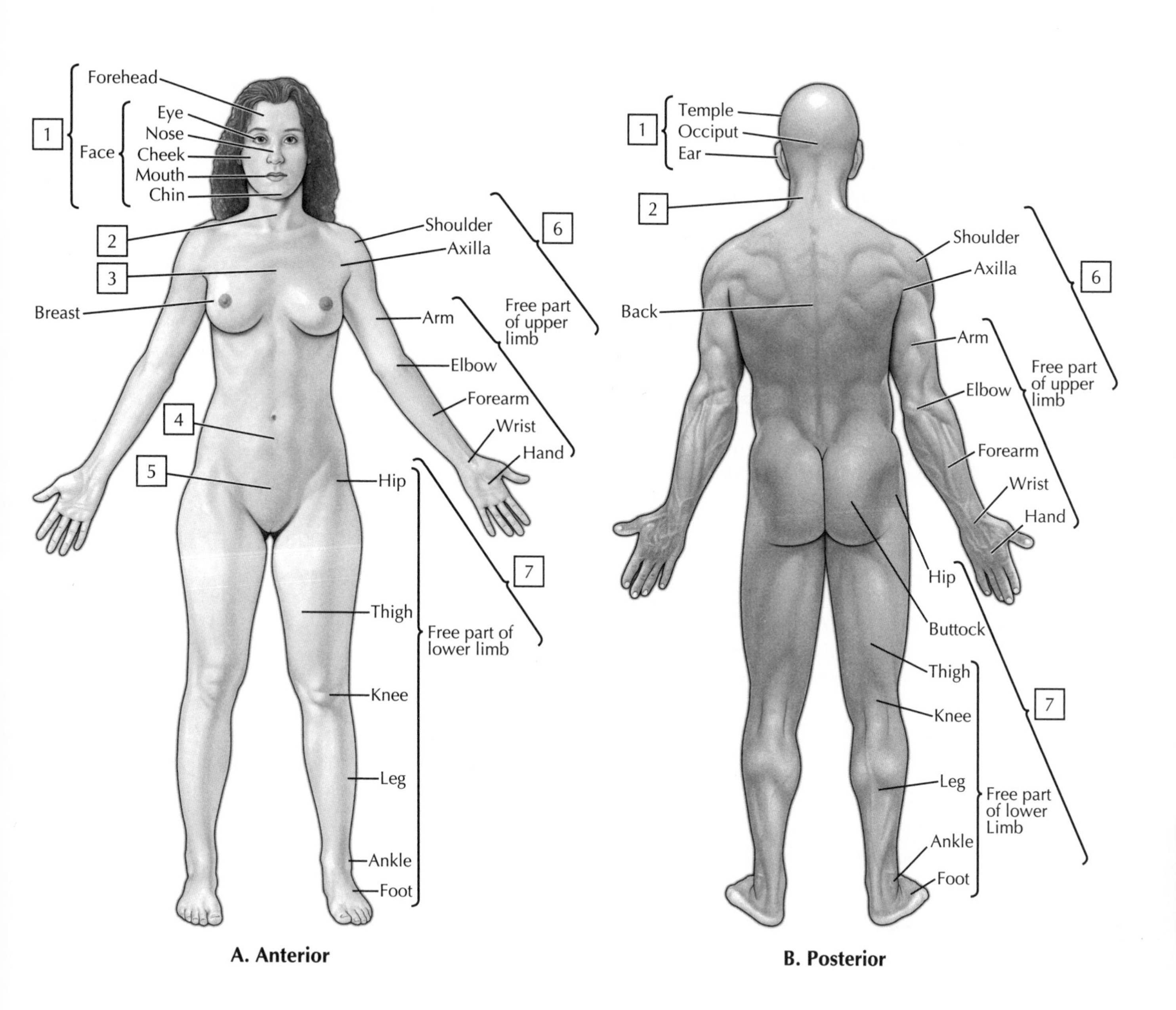


A. Anterior

B. Posterior

# 1 Body Planes and Terms of Relationship

Anatomical descriptions are referenced to one of four planes that pass through the human body in anatomical position. The **four planes** include the following:

- The **median plane**, also known as the **median sagittal** or **midsagittal plane**, is a vertical plane that passes through the center of the body, dividing it into equal right and left halves.
- **Sagittal planes**, other than the median sagittal plane, are vertical planes that are parallel to the median sagittal plane and are often called **parasagittal planes**.
- **Frontal planes**, also known as **coronal planes**, are vertical planes that pass through the body and divide it into anterior (front) and posterior (back) sections.
- **Transverse planes**, also known as cross sections, horizontal planes, or **axial planes**, are planes that are at right angles to the sagittal and frontal planes and divide the body into superior (upper) and inferior (lower) sections. Radiologists often refer to these as axial or transaxial planes.

The table on this page provides you with an overview of the commonly used anatomical terms of relationship and comparison. Additionally, some terms can be used in combination. For example, **superiormedial** means closer to the head and nearer to the median sagittal plane. Additionally, the term **palm** (palmar surface) refers to the palmar surface of the hand, and the term **sole** (plantar surface) refers to the bottom of the foot while standing barefoot.

When anatomists or physicians refer to "right" and "left," it is always the person's or patient's right side and left side that they are referring to, NOT the student's, anatomist's, or physician's right and left side.

Paired right and left side structures are **bilateral**, while those present on only one side of the body are **unilateral** (e.g., gallbladder or spleen). **Ipsilateral** refers to the same side of the body (e.g., the right thumb and right big toe), and **contralateral** refers to the opposite side of the body (e.g., the right foot is contralateral to the left foot).

**COLOR** the three planes shown on the figure using different colors.

- ☐ **1. Sagittal plane (midsagittal or median plane)**
- ☐ **2. Coronal plane (frontal plane)**
- ☐ **3. Transverse plane**

| TERM | DESCRIPTION |
|---|---|
| Anterior (ventral) | Nearer the front |
| Posterior (dorsal) | Nearer the back |
| Superior (cranial) | Upward or nearer the head |
| Inferior (caudal) | Downward or nearer the feet |
| Medial | Toward the midline or median plane |
| Lateral | Farther from the midline or median plane |
| Proximal | Near to a reference point |
| Distal | Away from a reference point |
| Superficial | Closer to the surface |
| Deep | Farther from the surface |
| Median plane | Divides body into equal right and left halves |
| Midsagittal plane | Median plane |
| Sagittal plane | Divides body into unequal right and left halves |
| Coronal plane | Divides body into equal or unequal anterior and posterior parts |
| Transverse plane | Divides body into equal or unequal superior and inferior parts (cross sections or axial sections) |

A. Body planes

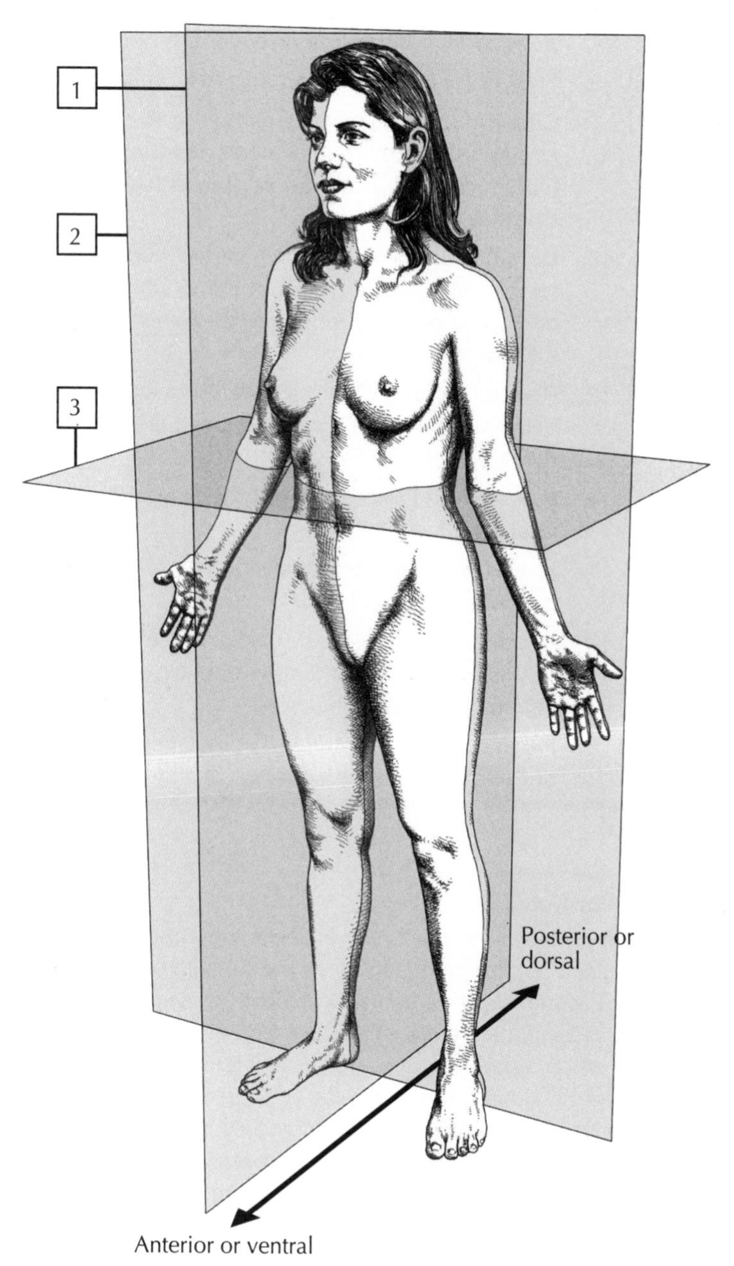


B. Terms of relationship

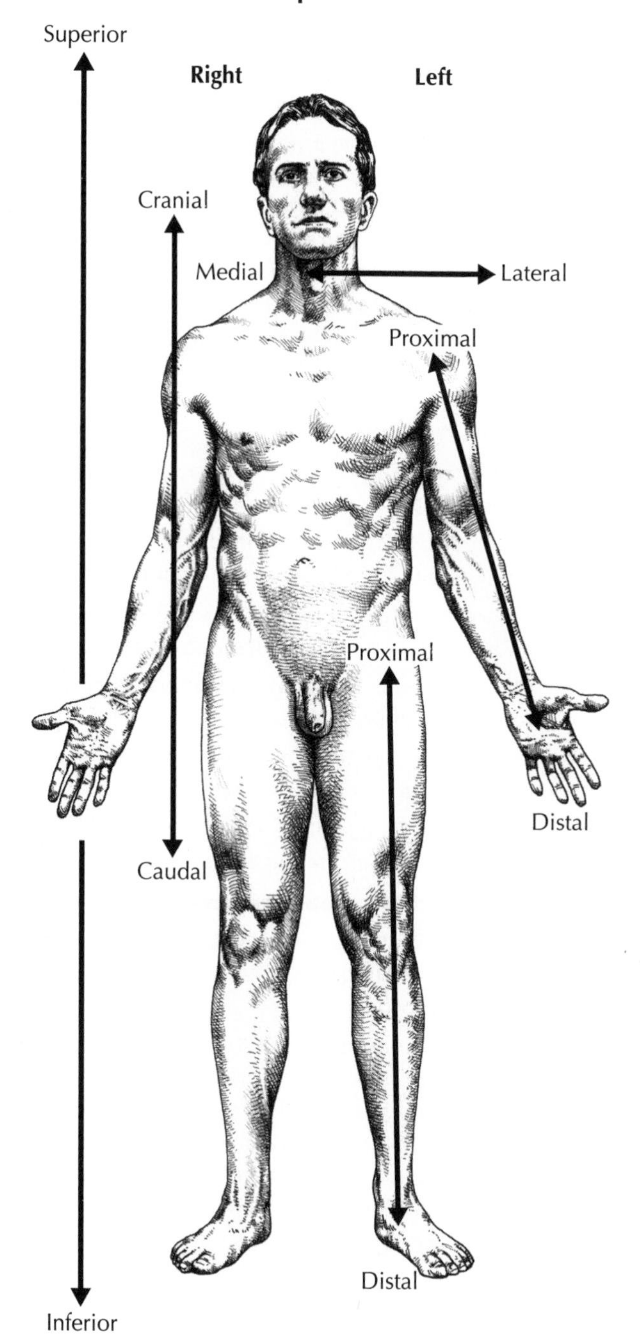

# Movements

Body movements occur at the joints, the points of articulation between two or more adjacent skeletal elements. Generally, when we refer to body movements, we are focusing on movements about a joint that occur from the contraction (physical shortening) of skeletal muscle. These contractions result in the movement of a limb, the bending of the spine, the fine movements of our fingers, or the tensing of our vocal cords for speaking (phonation). Of course, many other types of movements also occur throughout the body; for example, peristaltic movements of the intestinal smooth muscle that move food through the intestines. However, the major movements about the joints that occur because of skeletal muscle contraction are highlighted in the following list and illustrated.

**COLOR** the circle on the images corresponding to the numbered movement in the following list, using a different color for each movement. Note that the letter abbreviation of the movement (e.g., F = flexion) is shown in the circle and corresponds to the key in the list below. For example, you may wish to color all abduction (AB) movements red, all adduction (AD) movements blue, and so on. Also, it helps if you try to mimic the illustrated movements on yourself and take note of where on your body you feel the skeletal muscles contract or relax.

- ☐ 1. **Abduction (AB): movement away from a central reference point**
- ☐ 1. **Adduction (AD): movement toward a central reference point; the opposite of abduction**
- ☐ 1. **Lateral rotation (L): turning a bone or limb around its long axis laterally or away from the midline**
- ☐ 1. **Medial rotation (M): opposite of lateral rotation; turning medially toward the midline**
- ☐ 2. **Flexion (F): usually a movement that decreases the joint's angle**
- ☐ 2. **Extension (E): usually a movement that increases the joint's angle; the opposite of flexion**
- ☐ 3. **Elevation (EL): lifting superiorly, as in shrugging the shoulders**
- ☐ 3. **Depression (D): a movement of a portion of the body inferiorly**
- ☐ 4. **Flexion (F) and extension (E) of the spine: as these terms relate to the spine, flexion decreases the angle between the vertebral bodies and extension increases this angle; when we bend forward we flex the spine, and when we bend backward to arch the back we are extending the spine.**
- ☐ 5. **Flexion (F) and extension (E) at the elbow**
- ☐ 6. **Flexion (F) and extension (E) at the wrist**
- ☐ 7. **Pronation (P): rotation of the radius about the ulna in the forearm, causing the palm to face posteriorly (in anatomical position) or inferiorly (if the hand is held forward with the palm upward)**
- ☐ 7. **Supination (S): opposite of pronation; causes the palm to face anteriorly or superiorly**
- ☐ 8. **Flexion (F) and extension (E) at the knee joint**
- ☐ 9. **Circumduction (C): movement in space that circumscribes a circle or cone about a joint (circumduction of the lower limb at the hip joint is illustrated)**
- ☐ 10. **Dorsiflexion (DF): lifting the foot at the ankle joint (similar to extension at the wrist, but at the ankle it is referred to as dorsiflexion rather than extension)**
- ☐ 10. **Plantarflexion (PF): a downward movement or depression of the foot at the ankle (similar to wrist flexion)**
- ☐ 11. **Eversion (EV): movement of the sole of the foot laterally**
- ☐ 11. **Inversion (I): movement of the sole of the foot medially**
- ☐ 12. **Retraction (R): posterior displacement of a portion of the body, in this example the jaw, without a change in angular movement**
- ☐ 12. **Protraction (PT): anterior displacement of a portion of the body without a change in angular movement**

***Clinical Note:***

Often physicians will test their patient's movements to assess the range of movement as well as the strength of movement (by offering resistance to the movement). This can be important in accessing the **"range of motion"** and the appropriate strength of a particular movement, taking into consideration the patient's build and age. Reflexive movements also are tested at various joints; for example, the "knee jerk" reflex. This not only accesses the muscle reflex but also assesses the integrity of the neuromuscular components (nerve conduction and muscle strength).

AB
AD
L
M
AD
AB
M
L
1
F
E
F
E
2
EL
D
3
E
F
4
F
E
5
E
F
6
S
P
7
F
E
8
C
9
DF
PF
10
EV
I
11
R
PT
12

# 1 The Cell

The cell is the basic unit, structurally and functionally, of all of the body's tissues. Like people, cells come in many different varieties, but, also like people, almost all cells share many basic internal structures that we call **organelles**. Organelles function cooperatively in a variety of ways that allow the cells and tissues to perform their unique functions. Cells may also contain **inclusions**; unlike organelles, inclusions are not surrounded by a membrane. Depending upon the type of cell, some will contain more of one type or another of an organelle or inclusion than others.

**COLOR** each of these 13 cellular components, using different colors, noting their morphology and function as you do so.

☐ 1. **Peroxisomes: small vesicles in cytoplasm that detoxify toxic substances; for example, they may contain enzymes that degrade hydrogen peroxide and fatty acids**

☐ 2. **Golgi apparatus: one or more flattened stacks of membranes that modify and package proteins and lipids for intracellular or extracellular use**

☐ 3. **Plasma membrane: the cell membrane, composed of a lipid bilayer that functions in protection, secretion, uptake, sensitivity (maintains a resting potential essential for excitable cells), adhesion, and support; the membrane can fuse with a secretory vesicle to release its contents in a process called exocytosis, or take up extracellular substances in a process called pinocytosis; the membrane also may possess specialized receptors along its surface, e.g., neurotransmitters and hormones**

☐ 4. **Cytoplasm: the aqueous matrix called the cytosol of the cell outside of the nucleus, containing inorganic ions, organic molecules, intermediate metabolites, carbohydrates, proteins, lipids, and RNA; also contains various organelles and inclusions (see next page)**

☐ 5. **Mitochondria: double-membrane organelles that produce adenosine triphosphate (ATP) via oxidative phosphorylation for energy; mitochondria possess an outer membrane and a folded inner membrane**

☐ 6. **Lysosomes: vesicles containing digestive enzymes**

☐ 7. **Endoplasmic reticulum (ER): membranous network in the cytoplasm, studded with ribosomes for protein synthesis (rough ER, 7A) or lacking ribosomes and involved with lipid and steroid synthesis (smooth ER, 7B)**

☐ 8. **Centrioles: paired bundle-like inclusions essential for chromosome movement in cell division**

☐ 9. **Nucleolus: a small, nonmembrane-bound structure within the nucleus (usually one or two per nucleus), containing condensations of ribosomal RNA and proteins; are larger in actively growing cells**

☐ 10. **Cell nucleus: the control structure of the cell surrounded by inner and outer membranes; the nucleus contains chromosomes, enzymes, and the nucleolus; the nuclear membrane, or envelope, is perforated by small nuclear pores; all three types of RNA (ribosomal, transfer, and messenger) are formed in the nucleus and migrate into the cytoplasm**

☐ 11. **Ribosomes: tiny particles in cytoplasm containing RNA and proteins, both free and attached to rough ER; ribosomes are involved in protein synthesis by translating the amino acid protein coding under the direction of messenger RNA (mRNA)**

☐ 12. **Microfilaments: fine filaments of actin, the contractile protein, that provide strength, support, and/or intracellular movement, and are involved in muscle contraction**

☐ 13. **Microtubules: tubulin protein inclusions that support the cell and assist in intracellular transport**

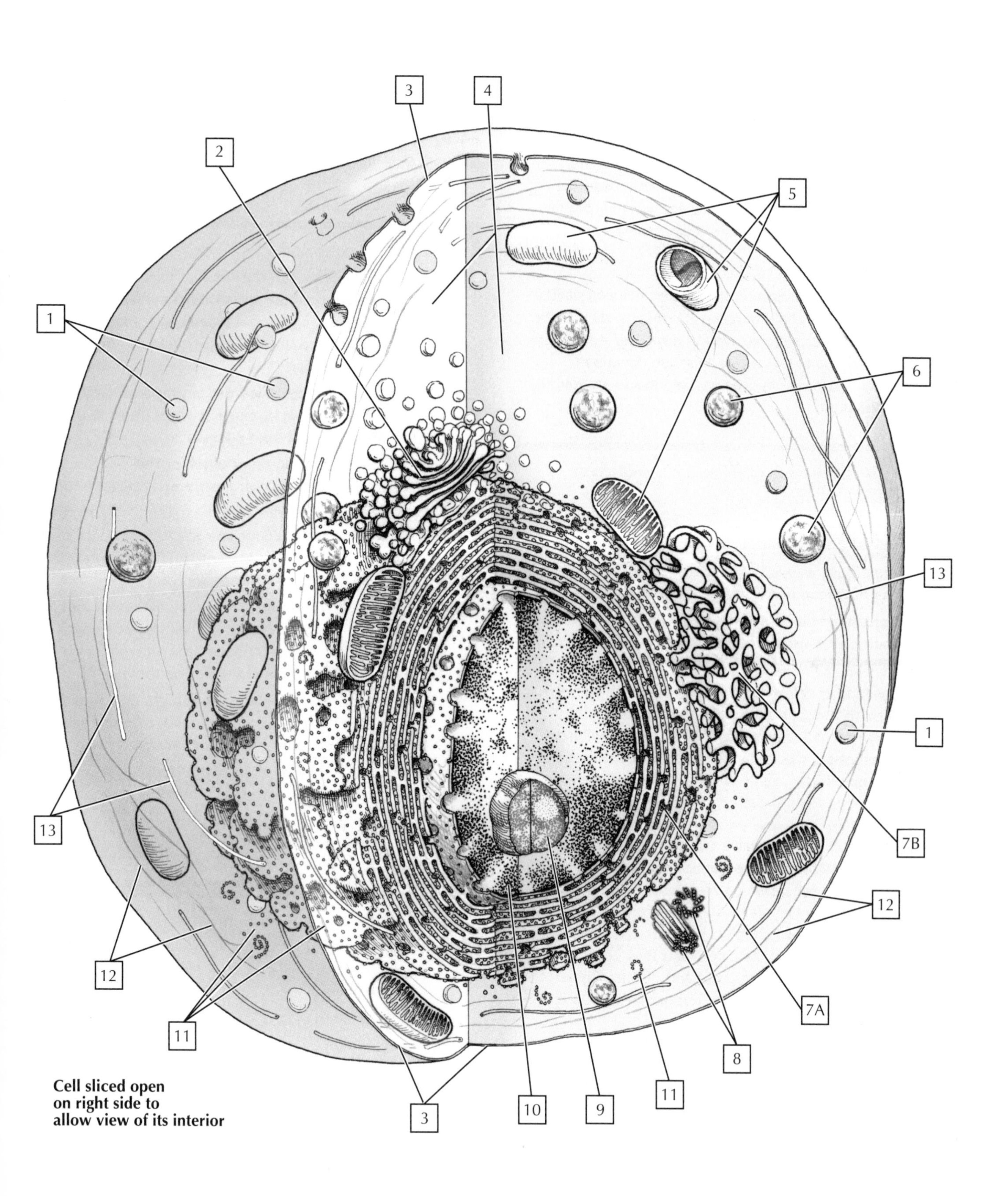


Cell sliced open
on right side to
allow view of its interior

# 1 Epithelial Tissues

The epithelial cells form the epithelium, one of the four basic tissue types found in the human body (the other three are connective tissue, muscle tissue, and nervous tissue). The epithelium covers the body surfaces; lines the body cavities, the ducts of organs and glands, the vasculature, and the organs; and forms the secretory portions of glands. Tight junctions may form between adjacent cells to provide a barrier; the cells may participate in absorption or secretion, and/or they may possess the ability to distend and spread out along an expanded surface (e.g., in the epithelium lining the distended urinary bladder). The epithelium rests on a basement membrane.

Epithelium is classified in several ways. First, it is classified according to the number of cell layers it contains, as **simple epithelium** (which has one cell layer) or **stratified epithelium** (which has two or more cell layers). It is also described according to the shape of its cells, as **squamous**, **cuboidal**, or **columnar epithelium**.

**COLOR** the three types of epithelium based on cell shape:

- ☐ **1. Squamous: thin, flattened cells; the width of each cell is greater than its height**
- ☐ **2. Cuboidal: "cubes" of cells; the width, depth, and height of each cell are approximately equal**
- ☐ **3. Columnar: taller, cylindrical cells; the height of each cell is greater than its width**

The classifications of cell layers and shapes may be combined to describe six different kinds of epithelium, and two specialized types, called **pseudostratified** and **transitional epithelium,** are also included in the classification, for a total of eight types of epithelium.

*Clinical Note:*

In adults, the most common types of tumors (neoplasms) originate in epithelial cells. **Benign neoplasms** occur locally (they do not travel to other sites within the body) and are often encapsulated and slow growing. **Cancers** are **malignant neoplasms** that are not encapsulated, and their cells tend to divide, grow, and travel to adjacent and distant sites where they form secondary cancers. This process is called **metastasis**.

**Cancers** may spread by direct contact with adjacent tissues or, more commonly, via the blood and/or lymphatics. Many cancers grow rapidly and consume a large portion of the body's nutrients, thereby leading to a loss of energy, weight loss, and wasting.

Epithelial cancers are termed **carcinomas**. As discussed above, these tumors may be benign or malignant. Before becoming cancerous, tumors usually undergo a change described as **dysplasia** (abnormal development) or **metaplasia** (abnormal transformation). Most cancers originate in the epithelium of the skin, breast (largely females, although males also may develop breast cancer [about 4-5% of breast cancers]), colon, lung, and prostate (males).

**COLOR** examples of the eight types of epithelium typically seen in tissues and organs:

- ☐ **4. Simple squamous: lines body cavities and the vasculature, offering a barrier to transport or functioning as an exchange system, often by simple diffusion**
- ☐ **5. Simple cuboidal: lines ducts of glands and kidney tubules, offering a passageway with or without the ability to absorb and secrete**
- ☐ **6. Simple columnar: lines much of the gastrointestinal system, offering a surface for absorption and secretion**
- ☐ **7. Pseudostratified: lines the trachea, bronchi of the lungs, and ductus deferens, offering a passageway with or without barrier or secretory functions**
- ☐ **8. Stratified squamous: forms the skin and lines the oral cavity, esophagus, and vagina, offering a protective surface; the skin may have a protective layer of keratin overlying the epithelium**
- ☐ **9. Stratified cuboidal: lines the ducts of sweat glands and other large exocrine glands, offering a conduit and/or serving as a barrier to transport**
- ☐ **10. Stratified columnar: lines the large ducts of exocrine glands, offering a conduit and serving as a barrier**
- ☐ **11. Transitional: lines the urinary system, offering a conduit; has the ability to distend**

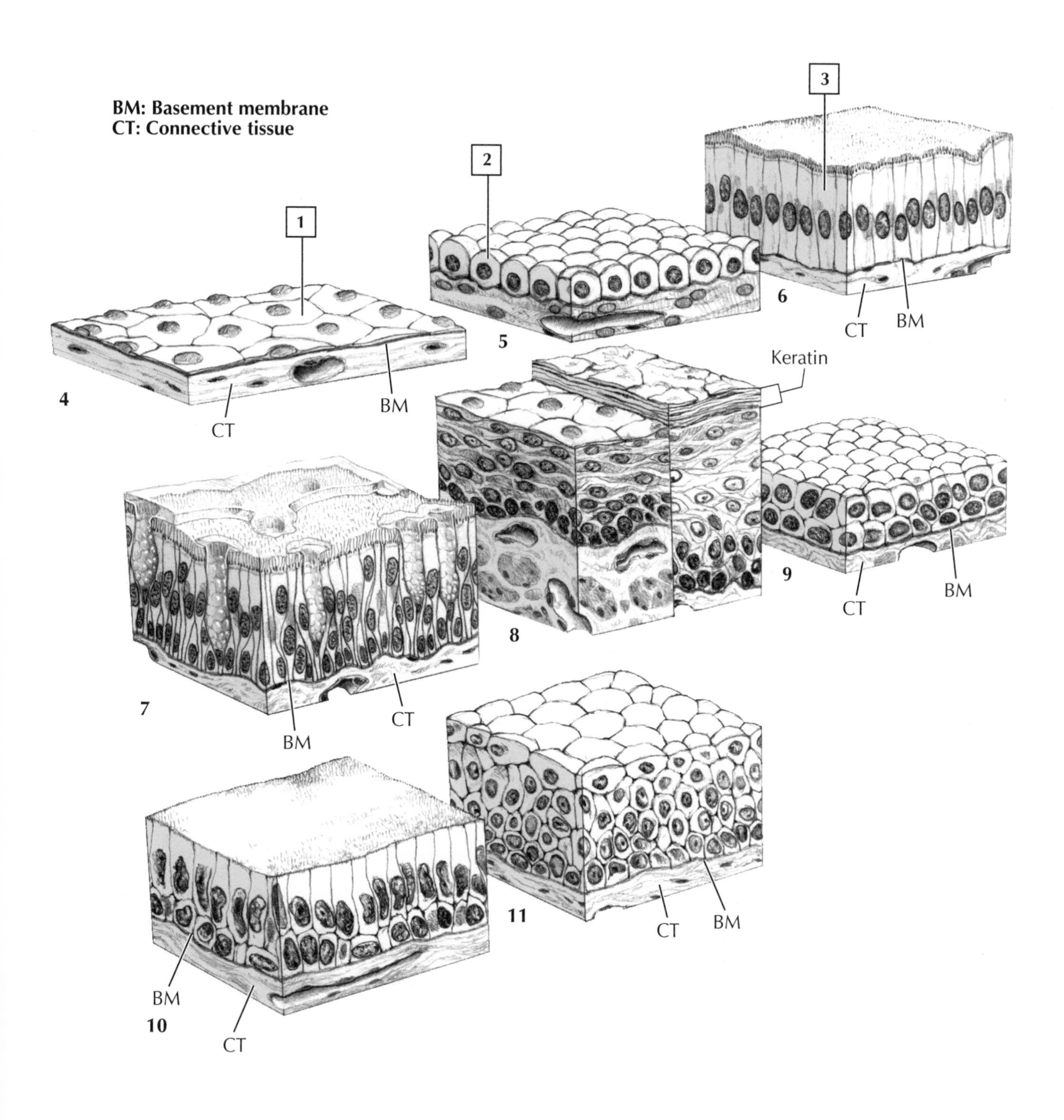
BM: Basement membrane
CT: Connective tissue
1
2
3
4
5
6
7
8
9
10
11
BM
CT
Keratin

# 1 Connective Tissues

Connective tissues are a diverse group of specialized cells and tissues. Connective tissues function in:

- Support
- Transport
- Storage
- Immune defense
- Thermoregulation

Two major groupings of connective tissues are recognized:

- **Connective tissue proper**: includes loose and dense connective tissues (arranged in either an irregular or a regular conformation)
- **Specialized connective tissue**: includes cartilage, bone, adipose tissue (fat), hemopoietic tissue, blood, and lymph

Connective tissue proper includes a variety of cell types and fibers enmeshed in a ground substance that comprises an **extracellular matrix. Loose connective tissue (areolar)** is found largely under the epithelium that forms the body's surface (skin) and the epithelium that lines the internal organ systems. Along with the skin, loose connective tissue is often the first line of defense against infection. **Dense connective tissue** has many fibers but few cells and includes tendons, ligaments, the submucosa, and reticular layers that offer support.

The fibrous elements in connective tissue include:

- **Collagen fibers**: numerous in connective tissues; offer flexibility and strength
- **Elastic fibers**: interwoven fibers that offer flexibility and retain their shape if stretched
- **Reticular fibers**: thinner collagen fibers that provide strength; they are the least common of the fibrous elements

***Clinical Note:***

Malignant tumors of the connective tissues and muscle are called **sarcomas**, which are tumors that arise from mesenchymal tissues. The most common adult soft tissue sarcoma is a malignant fibrous histiocytoma, thought to derive from perivascular mesenchymal cells.

Although there are over 25 different types of collagen, types I through IV are the most common. Type I collagen accounts for 90% of the body's collagen and is common in the skin, muscle tendons, ligaments, and bones. Type II collagen is found in cartilage. Type III collagen is found in loose connective tissue and forms a loose reticular meshwork or supportive scaffold for the tissues and organs. Type IV collagen is found in the basement membrane supporting the epithelium.

**Keloids** occur when scar tissue of the skin grows well beyond the boundary of the initial wound and does not normally regress.

**Fibrosis** is a term used to describe the deposition and overgrowth of fibrous connective tissue that forms scar tissue. This usually occurs because of a cut (laceration), infection, allergy, or long-term inflammation.

Of the many connective tissue diseases that primarily target collagen, **scleroderma** (systemic sclerosis) is a chronic, degenerative disease that results from an excessive production of collagen due to an autoimmune dysfunction. While there currently is no known cure, recent studies point to an upregulation of collagen gene expression in fibroblasts.

**Marfan syndrome** is an inherited connective tissue disorder caused by molecular defects in a gene that encodes an extracellular protein that is a component in microfibrils. These microfibrils serve as scaffolds for the deposition of elastic fibers.

**Chronic inflammation** results in fibrosis and tissue necrosis. The resulting inflammation is linked to many autoimmune disorders, e.g., rheumatoid arthritis and some cancers.

**Lipomas** are the most common mesenchymal soft tissue tumors in adults; they grow slowly and are usually found in proximal limbs, back, shoulder, and neck. They can be removed by liposuction or surgical excision.

**COLOR** each of the commonest cellular elements in connective tissue, using a different color for each type, as they appear in the different varieties of connective tissue:

- ☐ **1. Plasma cells: secrete immunoglobulins and are derived from B lymphocytes**
- ☐ **2. Macrophages: phagocytic cells (which engulf pathogens and cell debris) derived from monocytes in the blood**
- ☐ **3. Lymphocytes: the principal cells of the immune system**
- ☐ **4. Mast cells: respond early to immunologic challenges and secrete powerful vasoactive and chemotactic substances**
- ☐ **5. Adipocytes: store and release triglycerides as needed by the body (fat cells), and produce hormones and growth factors**
- ☐ **6. Fibroblasts: abundant cells that synthesize all the fibrous elements and elaborate the matrix**
- ☐ **7. Eosinophils: phagocytes that respond to allergens and parasitic infections**
- ☐ **8. Myofibroblasts: are capable of contraction and function in ways that are similar to fibroblasts and smooth muscle cells**
- ☐ **9. Neutrophils: respond to injury and immunologic challenges and are capable of phagocytosis**

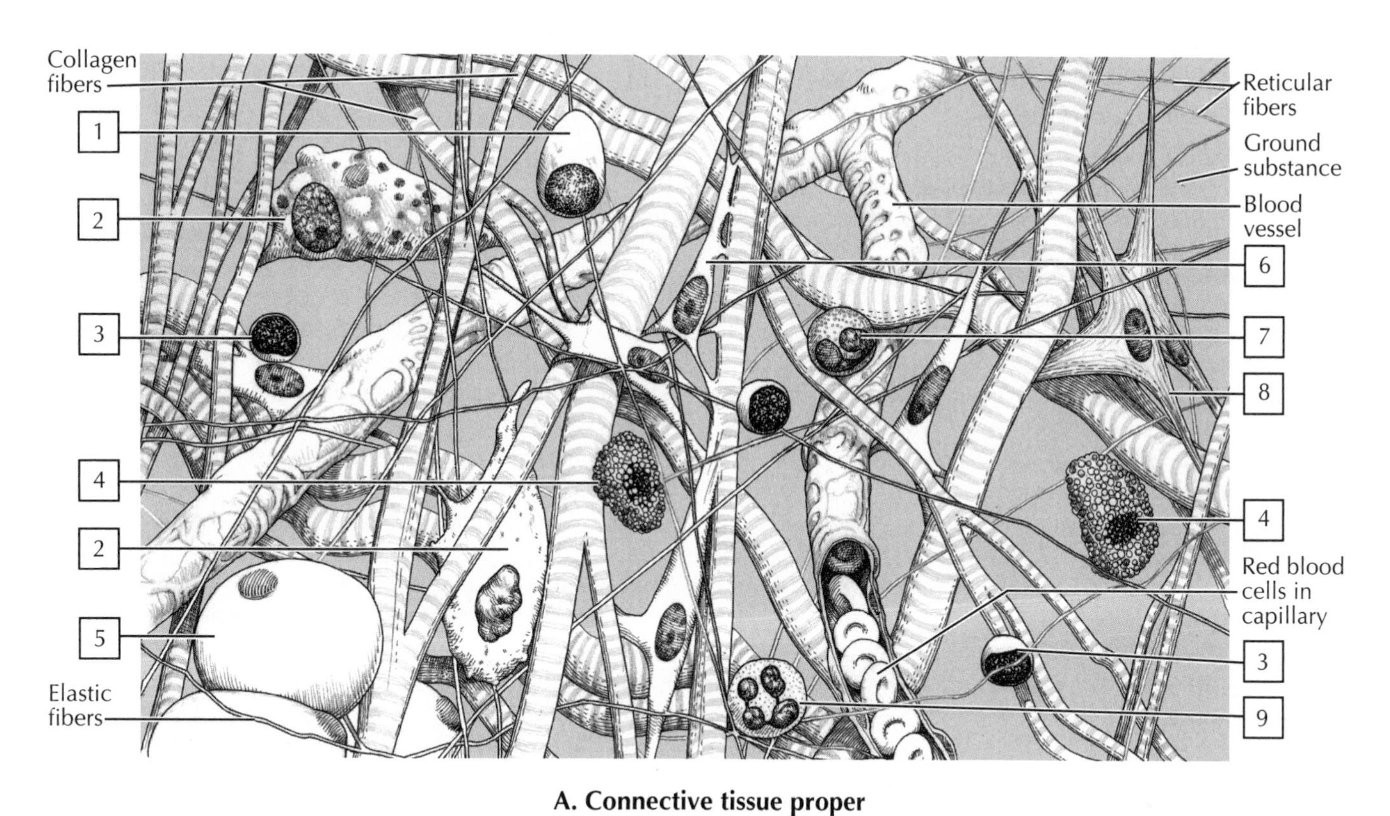


A. Connective tissue proper

5

B. Adipose tissue

6

C. Tendon

Chondrocytes (cartilage cells)

D. Cartilage

# 1 Skeleton

The human skeleton is divided into two descriptive regions: axial and appendicular.

**COLOR** each skeletal region a different color to differentiate them from one another:

- ☐ 1. **Axial skeleton: the bones of the skull, vertebral column (spine), ribs, and sternum (they form the axis, or central line of the body)**
- ☐ 2. **Appendicular skeleton: the bones of the limbs, including the pectoral (shoulder) and pelvic girdles (i.e., the bones of the upper and lower limbs, which attach to the axial skeleton)**

The **axial skeleton** includes 80 bones:

- The skull and associated bones (auditory ossicles and hyoid bone) account for 29 bones
- The thoracic cage (sternum and ribs) accounts for 25 bones
- The vertebral column accounts for 26 bones

The **appendicular skeleton** includes 134 bones:

- The pectoral girdle (paired clavicles and scapulae) accounts for 4 bones
- The upper limbs account for 64 bones
- The pelvic girdle (coxal or hip bone) accounts for 2 bones
- The lower limbs account for 64 bones

In total, the skeleton has 214 bones (including 8 sesamoid bones of the hands and feet). One must add that this number may vary somewhat among people, often because of a variable number of small sesamoid bones.

The bones and cartilages of the human skeletal system are formed from living, dynamic, rigid connective tissue. Cartilage is attached to some bones, especially those for which flexibility is important, and it also covers many of the articular (joint) surfaces of bones. About 99% of the body's calcium is stored in bones. Many bones possess a central cavity that contains bone marrow, a collection of hemopoietic (blood-forming) cells. The bones of the skeleton account for about 20% of a person's body mass.

The **skull** is the most complex of the bones. Not only does it include 29 associated bones, but it also has about 85 named openings (foramen [holes; some regular and some irregular], canals and fissures) which provide passageways for the spinal cord, nerves and blood vessels. The **mandible** (lower jaw) is the largest and strongest of the facial bones of the skull, and it articulates with the temporal bone via its temporomandibular joint. The **hyoid bone** is somewhat unique in that it is included within the bones of the skull but actually does not articulate with any other bones! It lies at about the level of the third cervical vertebra and is the attachment point for several neck muscles used to raise and lower the larynx during speech or when swallowing.

Most individual bones can be classified into one of five shapes: flat, irregular, short, long, or sesamoid.

**COLOR** using a different color for each shape, the five different types of bones:

- ☐ 3. **Flat bone**
- ☐ 4. **Irregular bone**
- ☐ 5. **Short bone**
- ☐ 6. **Long bone**
- ☐ 7. **Sesamoid bone**

Functions of the skeletal system and bones include:

- Support
- Protecting vital tissues or organs
- Providing a mechanism, along with muscles, for movement
- Storing calcium and other salts, growth factors, and cytokines
- Providing a source of blood cells

There are two types of bone:

- **Compact bone:** a relatively solid mass of bone commonly seen as a superficial layer of bone that provides strength
- **Spongy bone (trabecular or cancellous):** a less dense trabeculated network of bone spicules making up the substance of most bones and surrounding an inner marrow cavity

Most articular surfaces of bone are covered by **hyaline cartilage**, the most common type of cartilage. A second type of cartilage is **fibrocartilage**, which is found where more support is needed (e.g., the meniscus of the knee joint and the intervertebral discs between the bodies of the vertebrae). The third type of cartilage is **elastic cartilage**, which is found where flexibility is needed (e.g., the auricle of the ear and the epiglottis).

***Clinical Note:***

**Osteoporosis** (porous bone) is the most common bone disease. It results from an imbalance between bone resorption and bone formation, which then places the bones at great risk for fracture. Approximately 10 million Americans (80% of them women) have osteoporosis.

**Supernumerary (accessory) bones** may occur and are most common in the foot and cranial vault.

**Avascular necrosis** occurs from a loss of blood supply. This often occurs at a fracture site and usually involves only a small portion of the bone.

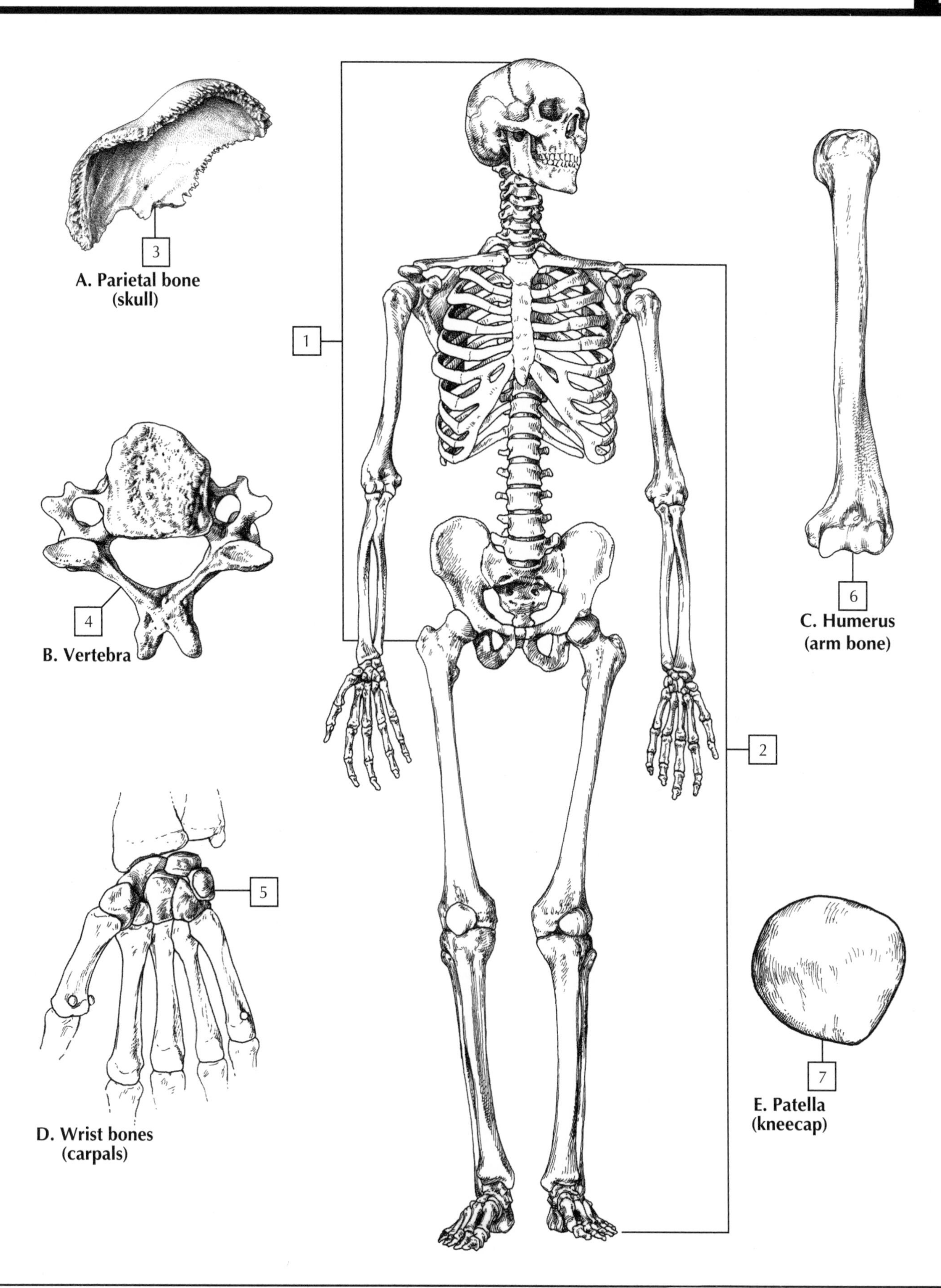
3
A. Parietal bone
(skull)
1
4
B. Vertebra
5
D. Wrist bones
(carpals)
2
6
C. Humerus
(arm bone)
7
E. Patella
(kneecap)

# 1 Joints

Joints are articulations between bones. Three types of joints are identified in humans:

- **Fibrous (synarthroses)**: joints formed by fibrous connective tissue (e.g., sutures of some skull bones, fibrous connections between some long bones **[syndesmosis]**, and gomphosis [teeth in the jaw])
- **Cartilaginous (amphiarthroses)**: joints formed by cartilage or by cartilage and fibrous tissue; they include primary cartilaginous joints, such as the epiphysial plates of growing bones (**synchondrosis**), and secondary cartilaginous joints (**symphysis**), such as the intervertebral discs between adjacent vertebrae of the spine
- **Synovial (diarthroses)**: a joint cavity filled with synovial fluid and surrounded by a capsule, with articular cartilage covering the opposed surfaces, such as the knee joint. This is the most common type of joint and often allows for considerable movement.

**COLOR** the following features of each of the three major types of joints:

☐ 1. **Suture: a type of fibrous joint that allows little movement**

☐ 2. **Interosseous membrane: a type of fibrous joint that permits some movement**

☐ 3. **Epiphysial plate: a cartilaginous joint that is immovable**

☐ 4. **Intervertebral disc: a cartilaginous joint that permits some movement**

☐ 5. **Synovial joint: the most common type of joint, which permits a range of movements (color the fibrous capsule, synovial membrane, articular cartilage, and synovial joint cavity each a different color)**

Generally, the more movement that occurs at a joint, the more vulnerable that joint is to injury or dislocation. Joints that allow little or no movement offer greater support and strength.

***Clinical Note:***

**Osteoarthritis** is characterized by the progressive loss of articular cartilage and failure of repair. It can affect any synovial joint but most often involves the foot, hip, spine, and hand. Once the articular cartilage is degraded and lost, the exposed bony surfaces, called the subchondral (beneath the cartilage) bones, rub against one another and undergo some remodeling; this often causes significant pain.

**Avascular necrosis** is the loss of blood supply to the epiphysis or another portion of a bone that will result in death of the bone.

**Degenerative joint disease** (osteoarthritis) is a common condition in the elderly and most often affects the weight-bearing joints (the hip and knees).

Rheumatoid arthritis is an immune response or infectious process that can damage the articular cartilages.

**Gout** causes severe pain because sharp crystals are deposited in the joint. This irritation also can cause the formation of calcareous deposits that deform the joint, thus limiting the range of motion at that joint.

**Osteoporosis** (porous bone) is the most commonly occurring bone disease. The loss of normal bone deforms the bone's normal microarchitecture.

A deficiency in calcium during growth of a bone can cause **rickets**. Both nutritional and hormonal factors affect bone mineralization.

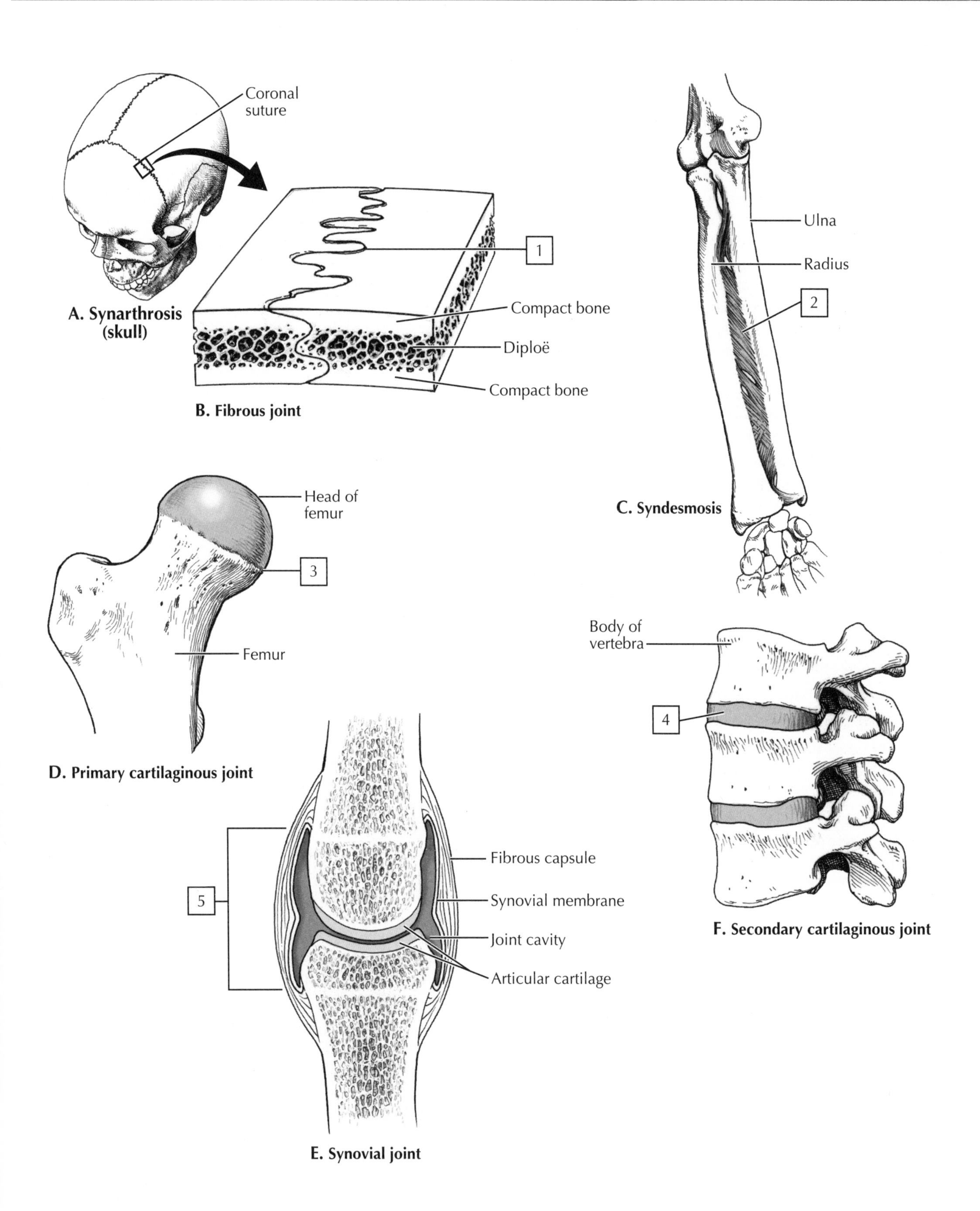


A. Synarthrosis (skull)

B. Fibrous joint

C. Syndesmosis

D. Primary cartilaginous joint

E. Synovial joint

F. Secondary cartilaginous joint

# 1 Synovial Joints

Generally, synovial joints offer considerable movement. They are classified according to their shape and the type of movement that they permit (uniaxial, biaxial, or multiaxial; movements in one, two, or multiple planes, respectively). The six types of synovial joints include:

- **Hinge (ginglymus)**: uniaxial joints that permit flexion and extension, such as the elbow joint
- **Pivot (trachoid)**: uniaxial joints that permit rotation, such as the joint between the atlas and axis (first two cervical vertebrae) of the neck, which pivots from side to side when you shake your head as if to signify "no"
- **Saddle**: biaxial joint for flexion, extension, abduction, adduction, and circumduction, such as the joint at the base of the thumb (carpometacarpal joint)
- **Condyloid (ellipsoid)**: biaxial joint for flexion, extension, abduction, adduction, and circumduction, such as the finger joints
- **Plane (gliding)**: joint that allows a simple gliding movement, such as the joint at the shoulder between the clavicle and scapula (acromioclavicular joint)
- **Ball-and-socket (spheroid)**: multiaxial joint for flexion, extension, abduction, adduction, medial and lateral rotation, and circumduction, such as the hip joint

**COLOR** the distal bone of each joint (the bone that usually undergoes the greatest amount of movement when a synovial joint moves):

- ☐ 1. **Ulna of the elbow's *hinge* joint**
- ☐ 2. **Axis of the atlantoaxial *pivot* joint**
- ☐ 3. **Metacarpal of the thumb's *saddle* joint**
- ☐ 4. **Tibia of the knee's *condyloid* joint**
- ☐ 5. **Femur of the hip's *ball-and-socket* joint (the acetabulum of the pelvis forms the "socket" of this joint)**
- ☐ 6. **Scapula of the acromioclavicular *plane* joint at the shoulder (the plane joint between the acromion of the scapula and the clavicle)**

Within the synovial joint cavity a small amount of **synovial fluid**, a filtrate of blood flowing in the capillaries of the synovial membrane, lubricates the joint. This fluid has the consistency of albumen (egg white).

As muscles pass over a joint, their tendons may be cushioned by a fibrous sac called a **bursa**, which is lined by a synovial membrane (synovium) and contains a small amount of synovial fluid. These fluid-filled "bags" cushion the tendon as it slides over the bone and act like a ball bearing to reduce some of the friction. Humans have over 150 bursae in different locations in the subcutaneous tissues associated with muscle tendons, bones, and joints at sites where cushioning helps to protect the tendon.

***Clinical Note:***

Movement at the joint can lead to inflammation of the tendons surrounding the joint and secondary inflammation of the bursa (**bursitis**) that cushions the joint and tendon. This inflammation is painful and can lead to a significant increase in the amount of synovial fluid in the bursa.

The knee joint (a synovial joint) is the most complex joint in humans and is especially vulnerable in athletes. The most common type of injury is **rupture of the anterior cruciate ligament** (ACL), often related to sharp turns when the knee is twisted while the foot is firmly on the ground or from blows to the lateral aspect of an extended knee. Further damage can involve rupture of the tibial collateral ligament and a tear to the medial meniscus, in addition to the rupture of the anterior cruciate ligament (This tripartite injury is called the "unhappy triad.").

While synovial joints can withstand normal wear and tear, they do tend to display some degenerative changes as one ages. The articular cartilage begins to undergo an irreversible degenerative change and the joint becomes less able to absorb the normal stress of weight-bearing and movement. Considerable pain results in **degenerative joint disease**, resulting in discomfort, stiffness and pain.

Joints receive blood from small articular arteries that arise from the larger arteries that pass around the joint. Fortunately, these articular arteries form **peri-arterial anastomoses** which supply blood to the joint at rest and during active use. Articular veins accompany these arteries and, like the articular arteries, are commonly found in the joint capsule and primarily in the synovial membrane.

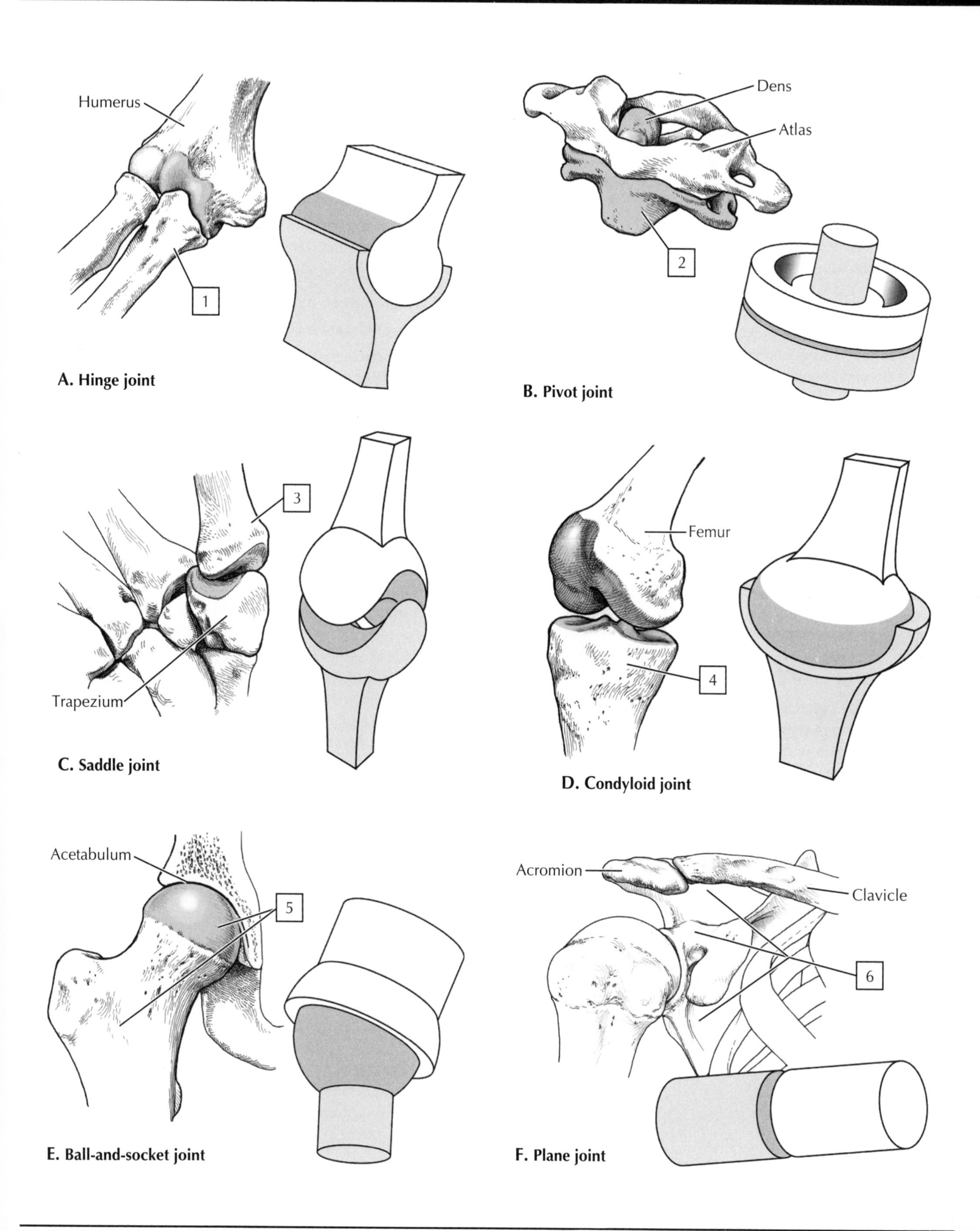
Humerus
1
A. Hinge joint
Dens
Atlas
2
B. Pivot joint
3
Trapezium
C. Saddle joint
Femur
4
D. Condyloid joint
Acetabulum
5
E. Ball-and-socket joint
Acromion
Clavicle
6
F. Plane joint

# 1 Muscle

Muscle cells (fibers) produce contractions (shortenings in length) that result in movements, maintain posture, produce changes in shape, or move fluids through hollow tissues or organs. There are three different types of muscle:

- **Skeletal**: striated fibers that usually are attached to bone and are responsible for movement of the skeleton at its joints. However, some skeletal muscles are attached to the eyeball, the skin (facial muscles), and to mucous membranes (intrinsic tongue muscles). Skeletal muscle is innervated by the somatic nervous system.
- **Cardiac**: striated fibers that make up the walls of the heart (the myocardium). Innervated by the autonomic nervous system.
- **Smooth**: unstriated fibers that line various organs, attach to hair follicles, are present within the eyeball (control pupil size and lens thickness), and line blood vessels. Innervated by the autonomic nervous system.

Muscle contractions occur in response to nerve stimulation at neuromuscular junctions, to paracrine stimulation (by localized release of various stimulating agents) in the local environment of the muscle, and to endocrine (hormonal) stimulation (see Plate 11-1).

**Skeletal muscle** is divided into bundles or fascicles. These fascicles are composed of fibers. The fibers are composed of myofibrils, and myofibrils contain myofilaments.

**COLOR** the elements of skeletal muscle, using a different color for each element:

- ☐ 1. **Muscle fascicles: which are surrounded by a connective tissue sheath known as the perimysium; epimysium is the connective tissue sheath that surrounds multiple fascicles to form a complete muscle "belly"**
- ☐ 2. **Muscle fibers: which are composed of a muscle cell that is a syncytium because it is multinucleated (the muscle fibers are surrounded by the endomysium)**
- ☐ 3. **Muscle myofibrils: which are longitudinally oriented and extend the full length of the muscle fiber cell**
- ☐ 4. **Muscle myofilaments: which are the individual myosin (thick) filaments and actin (thin) filaments that slide over one another during muscle contraction**

Skeletal muscle moves bones at their joints. Each skeletal muscle has an **origin** (the muscle's fixed or proximal attachment) and an **insertion** (the muscle's moveable or distal attachment). At the gross level, the shape of a muscle allows anatomists to classify it (see D in the illustration).

**COLOR** each of the five different conformations that characterize the gross appearance of skeletal muscle.

- ☐ 5. **Fusiform: thick in the center and tapered at the ends (spindle-shaped)**
- ☐ 6. **Quadrate: four-sided muscle**
- ☐ 7. **Flat: formed by parallel fibers**
- ☐ 8. **Circular: sphincters that close off tubes**
- ☐ 9. **Pennate: feathered in appearance (unipennate, bipennate, or multipennate forms)**

**Cardiac muscle** has myofilaments that are arranged similarly to those of skeletal muscle, but it has other structural features that distinguish it from skeletal muscle. Moreover, cardiac muscle has unique contraction properties, including an intrinsic **rhythmic contraction** and specialized conduction features that coordinate its contraction.

**Smooth muscle** usually occurs in bundles or sheets of elongated cells with a fusiform or tapered appearance. Smooth muscle is specialized for slow, prolonged contraction, and it also can contract in a wavelike fashion known as **peristalsis**.

In general, skeletal muscle does not undergo mitosis, and it responds to increased demand by hypertrophy (i.e., cells increase in size but not in number). Cardiac muscle also normally does not undergo mitosis, and it also responds to increased demand by hypertrophy. Smooth muscle can undergo mitosis, and it responds to increased demand by hypertrophy and hyperplasia (i.e., cells increase in both size and number). It also has the ability to regenerate.

---

***Clinical Note:***

Muscle action can be tested by having a patient perform a movement against resistance to test the contraction strength of the muscle. Muscles also can be tested by electrical stimulation (**electromyography,** or **EMG**).

**Muscle atrophy** is "wasting" of muscle tissue that may result from immobilization or a disorder of the muscle itself, e.g., muscle tears, rupture of its tendon, a decrease in blood supply (myocardial infarct in cardiac muscle), and a loss of innervation.

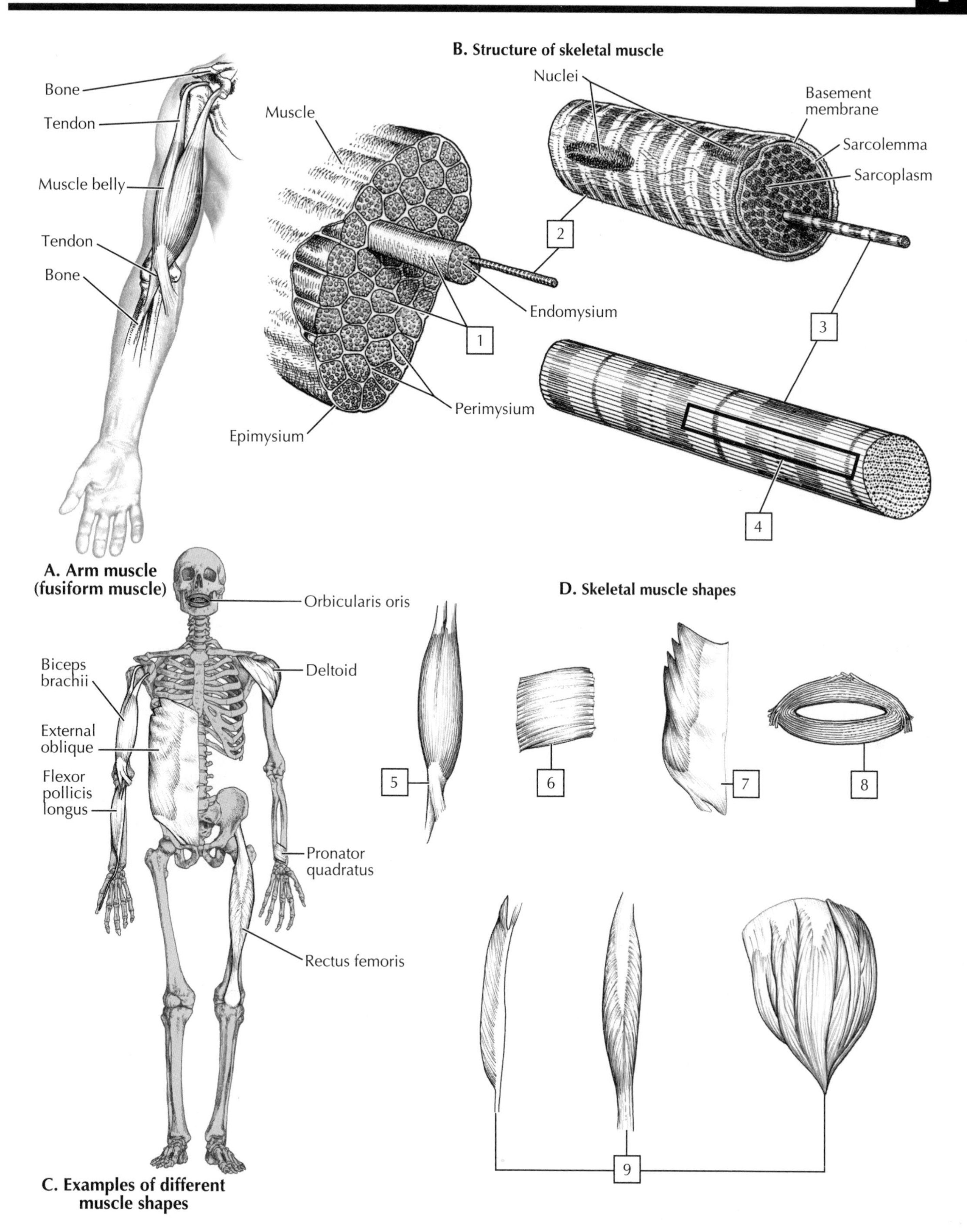
B. Structure of skeletal muscle
Bone
Tendon
Muscle belly
Tendon
Bone
Muscle
Nuclei
Basement membrane
Sarcolemma
Sarcoplasm
2
Endomysium
1
3
Perimysium
Epimysium
4
A. Arm muscle (fusiform muscle)
Orbicularis oris
D. Skeletal muscle shapes
Biceps brachii
Deltoid
External oblique
Flexor pollicis longus
5
6
7
8
Pronator quadratus
Rectus femoris
9
C. Examples of different muscle shapes

# Nervous System

The nervous system integrates and regulates many body activities, sometimes at discrete locations (specific targets) and sometimes more globally. The nervous system usually acts quite rapidly and can also modulate effects of the endocrine and immune systems.

The nervous system has two structural divisions:

- **Central nervous system (CNS)** (includes the brain and spinal cord)
- **Peripheral nervous system (PNS)** (includes the nerve cell bodies and fibers outside of the CNS somatic, autonomic, and enteric nerves in the periphery)

The brain includes the:

- **Cerebral cortex**: highest center for sensory and motor processing, composed of the two cerebral hemispheres separated by a longitudinal fissure, with each hemisphere composed of a frontal, parietal, temporal, and occipital lobe
- **Diencephalon**: includes the thalamus (a relaying and processing function, it serves as the "gateway" to the cerebral cortex, acts as an "executive secretary," and regulates input and output from the cortex) and the hypothalamus (regulates visceral functions, emotions, is involved in autonomic control, and plays a role in hormone production via the anterior and posterior pituitary glands)
- **Cerebellum**: coordinates smooth motor activities and balance, processes muscle position, and may play a role in behavior and cognition
- **Brainstem (midbrain, pons, and medulla)**: conveys motor and sensory information from the body, and mediates important autonomic functions

**COLOR** the subdivisions of the cerebral cortex, using a different color for each lobe:

☐ 1. **Cortex, frontal lobe: processes motor, visual, speech, and personality modalities**

☐ 2. **Cortex, parietal lobe: processes sensory information, subserves analysis of movement and positional relationships of viewed objects (spatial discrimination), taste, and receptive speech**

☐ 3. **Cortex, temporal lobe: processes language, auditory, and memory modalities**

☐ 4. **Cortex, occipital lobe: processes visual input**

**Peripheral nerves** arise from the spinal cord and form networks; each network is called a **plexus**. The 31 pairs of spinal nerves (8 cervical pairs, 12 thoracic pairs, 5 lumbar pairs, 5 sacral pairs, and 1 coccygeal pair of nerves) contribute to four major nerve plexuses.

**COLOR** the four major nerve plexuses formed by the anterior (ventral) pairs of spinal nerves, using a different color for each plexus:

☐ 5. **Cervical plexus: largely innervates muscles of the neck (anterior rami of 4 cervical pairs of nerves, C1-C4)**

☐ 6. **Brachial plexus: largely innervates muscles of the shoulder and upper limb (anterior rami of 5 pairs of nerves, 4 cervical and 1 pair of thoracic nerves, C5-T1)**

☐ 7. **Lumbar plexus: largely innervates muscles of the anterior and medial thigh (anterior rami of 4 lumbar pairs of nerves, L1-L4)**

☐ 8. **Lumbosacral plexus: largely innervates muscles of the buttock, pelvis, perineum, and lower limb (anterior rami of 6 pairs of nerves, 2 lumbar pairs and 4 sacral pairs, L4-S4)**

***Clinical Note:***

Vascular lesions such as **infarcts** and **hemorrhage** may damage various specific cortical regions. More global damage may occur with **ischemia** (lack of sufficient blood flow) or **anoxia** (lack of oxygen), causing more expansive dysfunction, cognitive deficits, and even coma. For example, a stroke resulting in cortical damage in a dominant hemisphere (usually the left hemisphere in right-handed individuals and most left-handed individuals) can result in **expressive aphasia** (impaired motor speech function), **receptive aphasia** (impaired understanding of spoken language), or **global aphasia** (all speech and communication are impaired).

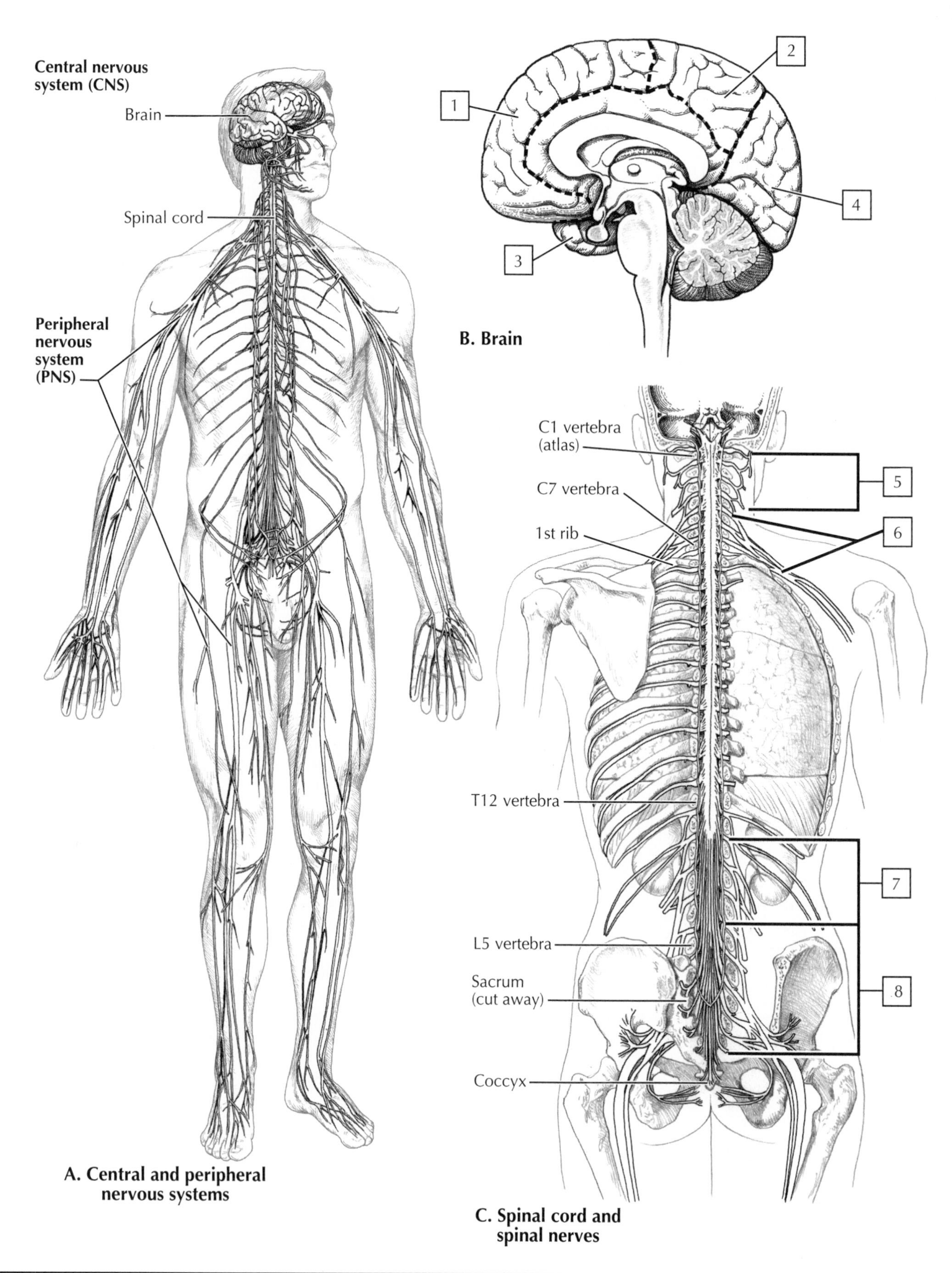


A. Central and peripheral nervous systems

B. Brain

C. Spinal cord and spinal nerves

# 1 Skin (Integument)

The skin is the largest organ in the body, accounting for about 15% of the total body mass. The skin consists of two layers: epidermis and dermis.

**COLOR** the brackets that delineate the two layers of the skin, using two different colors:

- ☐ 1. **Epidermis: an outer protective layer consisting of a keratinized stratified squamous epithelium derived from embryonic ectoderm**
- ☐ 2. **Dermis: a dense connective tissue layer that gives skin most of its thickness and support and is derived from embryonic mesoderm**

The outer epidermal layer itself consists of four layers.

**COLOR** the four layers of the epidermis, listed below from outermost to innermost, using different colors than previously used:

- ☐ 3. **Stratum corneum: an anuclear cell layer that is thick, comprised of 20-30 cells in thickness, and contains flattened cells filled almost entirely with keratin filaments; this layer provides a durable protective covering that resists water loss**
- ☐ 4. **Stratum granulosum: a layer one to three cells thick; its cells contain keratohyalin granules, which contain a protein that will aggregate the keratin filaments of the next layer; these cells begin to flatten, and their nuclei and cell organelles begin to disintegrate**
- ☐ 5. **Stratum spinosum: a layer that is several cell layers thick; its cells contain cytoplasmic processes, but the processes are lost as the cells ascend toward the surface of the skin**
- ☐ 6. **Stratum basale: a single germinal basal cell layer that is mitotically active and provides cells for the layers superficial to it; it is attached to the underlying dermis**

The epidermis is renewed by cells from the basal layer that rise up through the skin to the surface. This outer layer of skin is renewed about every 25 to 45 days, as we lose millions of skin cells every day!

The dermis is divided into a papillary layer and a reticular layer and contains epidermal skin appendages. Dermal papillae extend up into the underside of the epidermis and increase the surface area for the attachment of the epidermis to the underlying dermal layer. This layer contains abundant collagen and elastin fibers, and is heavily vascular. The reticular dermis lies deeper, is thicker (accounts for about 60% to 70% of the dermal thickness), and is less cellular than the papillary layer. Deep within the dermis and subcutaneous tissue lie atriovenous shunts, which participate in thermoregulation, along with a variable number of sweat glands. **Apocrine sweat glands** are odoriferous and found in the axilla (armpit), scrotum, prepuce, labia minora, nipples, and perianal region. **Modified apocrine glands** include ceruminous glands in the external ear (secrete earwax) and Moll glands in the eyelids. **Eccrine sweat glands** serve a thermoregulatory role and can produce 500-750 mL or more of sweat/day.

**COLOR** the epidermal skin appendages found in the dermal layer:

- ☐ 7. **Sebaceous glands; associated with hair follicles and secrete sebum (under hormonal control)**
- ☐ 8. **Hair follicles**
- ☐ 9. **Sweat glands (several types; see description above)**

Additionally, the dermis contains capillaries, specialized receptors and nerves, pigment cells, immune cells, and smooth muscle (arrector pili muscles attached to the hair follicles).

Also, if you wish, color the small arteries red, the small veins blue, and the nerve fibers yellow. **Note that from this point forward, arteries will always be colored red, veins blue, and nerves yellow.**

Beneath the dermis lies a loose connective tissue layer, the **hypodermis** or subcutaneous tissue (superficial fascia), which is of variable thickness and often contains a significant amount of adipose (fat) cells.

Skin functions include:

- Protection, via both mechanical abrasion and immune responses
- Temperature regulation, via vasodilatation or vasoconstriction, and by sweat gland activity (water evaporation as a cooling mechanism)
- Sensation, via touch (mechanoreceptors such as Pacinian and Meissner's corpuscles), pain (nociceptors), and temperature receptors (thermoreceptors)
- Endocrine, via secretion of hormones, cytokines, and growth factors
- Exocrine, via secretion of sweat from sweat glands and oily sebum from sebaceous glands

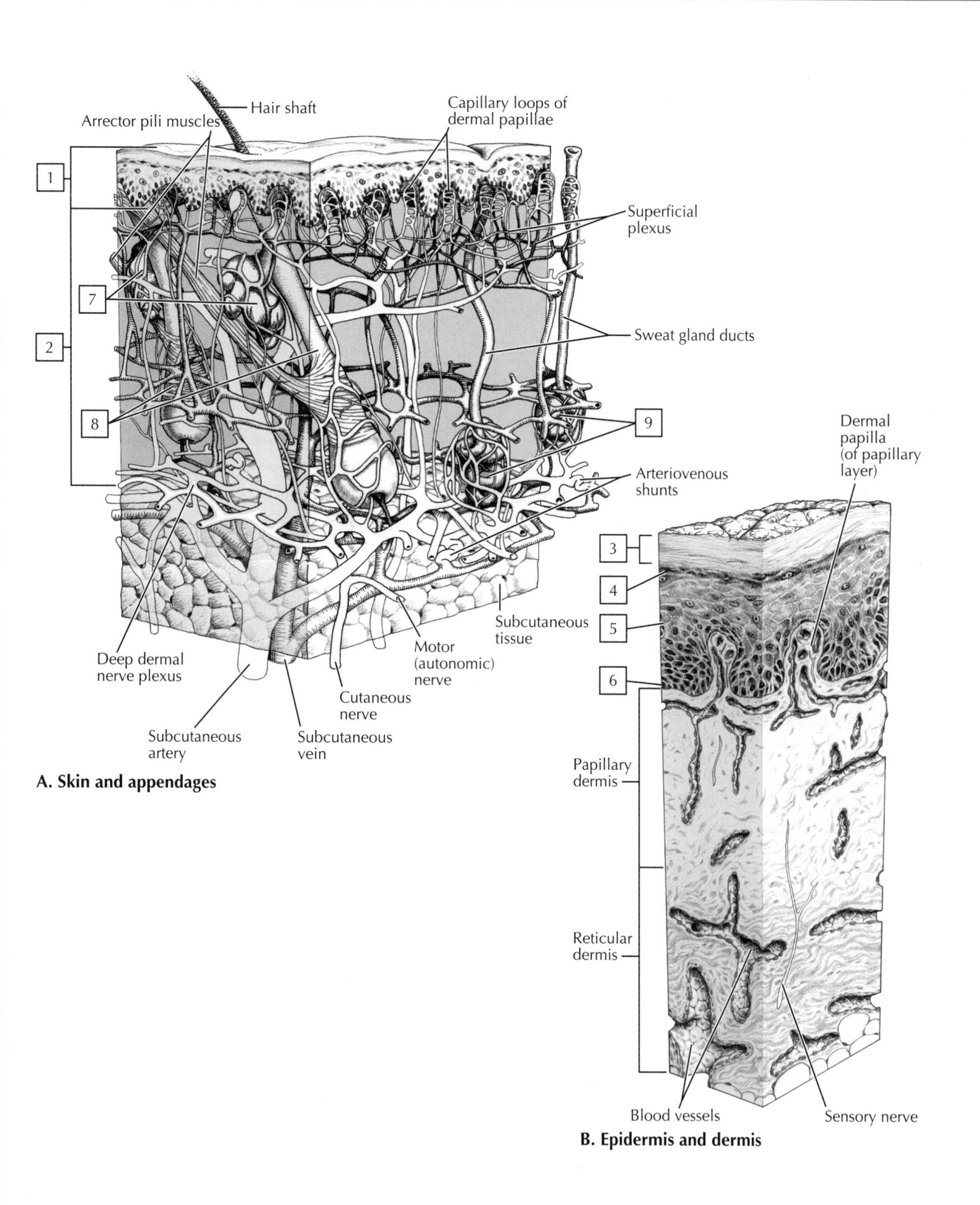


A. Skin and appendages

B. Epidermis and dermis

# 1 Body Cavities

Organ systems and other visceral structures are often segregated into body cavities. These cavities can protect the viscera and also may allow for some expansion and contraction in size. Two major collections of body cavities are recognized:

- **Dorsal cavities**: include the brain, surrounded by the meninges and bony cranium, and the spinal cord, surrounded by the same meninges as the brain and also surrounded by the vertebral column
- **Ventral cavities**: include the **thoracic** and **abdominopelvic cavities,** separated from each other by the respiratory diaphragm (skeletal muscle important in respiration)

The central nervous system (brain and spinal cord) is surrounded by three membranes (see Plate 4-18):

- **Pia mater**: a delicate, transparent inner layer that intimately covers the brain and spinal cord
- **Arachnoid mater**: a fine, weblike membrane beneath the outer dura mater
- **Dura mater**: a thick, tough outermost layer that is vascularized and richly innervated by sensory nerve fibers

**COLOR** the brain and spinal cord, using a different color for each and for their coverings:

- ☐ 1. **Brain and its dural lining (1A)**
- ☐ 2. **Spinal cord and its dural lining (2A)**

The thoracic cavity contains **two pleural cavities** (right and left; see Plate 7-5) and a single midline space called the **mediastinum** (middle space). The heart and structures lying posterior to the heart, including the descending thoracic aorta and esophagus, lie within the thoracic cavity. The heart itself resides in its sac, called the **pericardial sac** (see Plate 5-3), which has a parietal layer and a visceral layer.

**COLOR** the two pleural cavities and the serous membrane lining these cavities:

- ☐ 3. **Parietal pleura: lines the thoracic walls and abuts the mediastinum medially**
- ☐ 4. **Visceral pleura: encases the lungs themselves and reflects off of the lung surfaces to be continuous with the parietal pleura**
- ☐ 5. **Heart and its surrounding pericardium (5A)**

The abdominopelvic cavity also is lined by a serous membrane, called the **peritoneum**, which likewise has a parietal layer and a visceral layer.

**COLOR** the abdominopelvic cavity and its peritoneal membranes (see Plate 8-5):

- ☐ 6. **Parietal peritoneum: lining the body walls**
- ☐ 7. **Visceral peritoneum: reflects off of the body walls and covers the abdominal visceral (organ) structures**

***Clinical Note:***

In a healthy person, the pleural, pericardial, and peritoneal spaces are considered potential spaces, because between the parietal and visceral layers we usually find only a small amount of serous lubricating fluid that is keeping the surfaces of the organs within the spaces moist and slick. This lubrication reduces friction from movements such as respiration, the beating of the heart, or intestinal peristalsis. If, however, inflammation develops or traumatic injury occurs, fluids (pus or blood) can collect in these spaces and restrict movements of the viscera. In such situations, these potential spaces expand to become real spaces. It may become necessary to remove the offending fluid to prevent the compromise of organ function or the exacerbation of an ongoing infection.

Excessive fluid in the pleural space may be removed by a hypodermic needle in a procedure termed **thoracentesis**. Drainage of serous fluid from the pericardial cavity is termed **pericardiocentesis**. Aspiration or drainage of fluid from the peritoneal cavity is termed **paracentesis**.

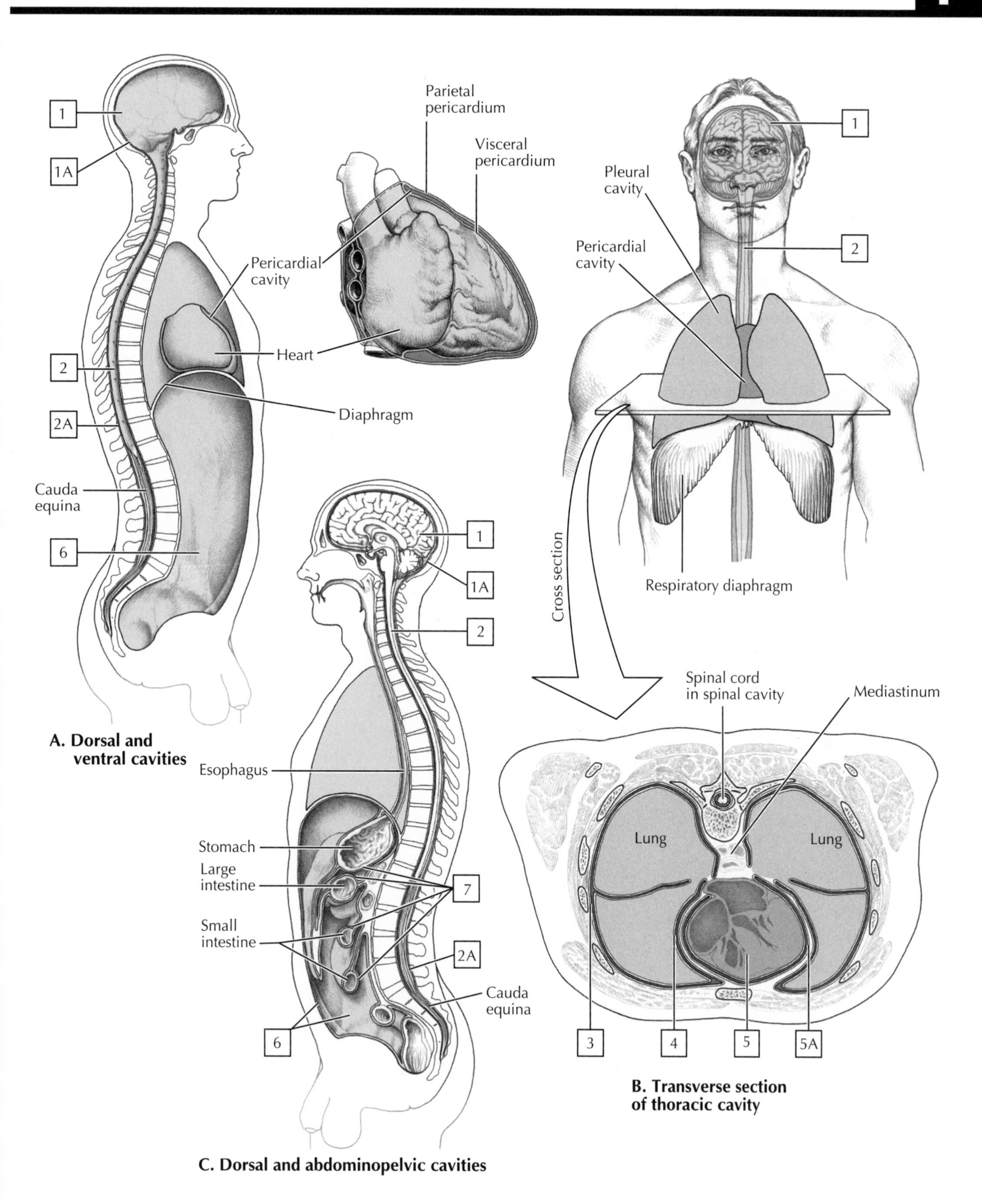


A. Dorsal and ventral cavities

B. Transverse section of thoracic cavity

C. Dorsal and abdominopelvic cavities

## REVIEW QUESTIONS

1. Write the correct term of relationship for each of the following:
   A. Nearer to the head: ______
   B. Closer to the surface: ______
   C. Divides the body into equal right and left halves: ______
2. Which term below best describes the position of the hand when the palm is facing toward the ground?
   A. Abduction
   B. Extension
   C. Plantarflexion
   D. Pronation
3. 
   A. Which intracellular organelle produces ATP? ______
   B. Which intracellular organelle has pores in its membrane? ______
   C. Which intracellular organelle is a condensation of RNA? ______
4. List the three types of epithelium based on cell shape. ______
5. List the three types of joints found in humans. ______
6. List the three types of muscles found in humans. ______
7. What two structures make up the central nervous system in humans? ______
8. The spinal cord is covered by (A) pia mater, (B) arachnoid mater, and (C) dura mater. Using a red pencil, circle the covering (A, B, or C) that lies closest to the spinal cord. With a blue pencil, circle the layer that is richly innervated and vascularized. With a green pencil, circle the layer that lies between the other two layers.

## ANSWER KEY

1A. Superior (cranial)

1B. Superficial

1C. Median plane

2D. Pronation

3A. Mitochondria

3B. Nucleus

3C. Nucleolus

4. Squamous, cuboidal, columnar

5. Fibrous, cartilaginous, synovial

6. Skeletal, cardiac, smooth

7. Brain and spinal cord

8. Red: Pia mater
   Blue: Dura mater
   Green: Arachnoid mater

# Chapter 2 Skeletal System

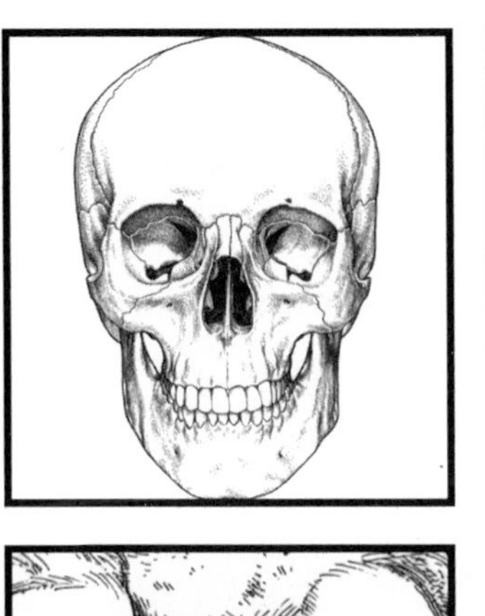
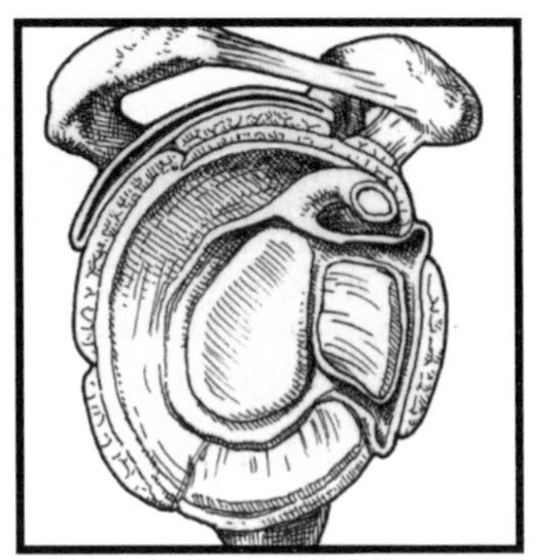
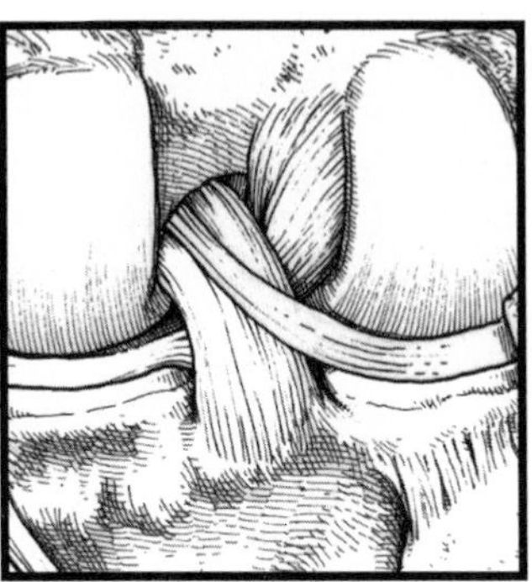
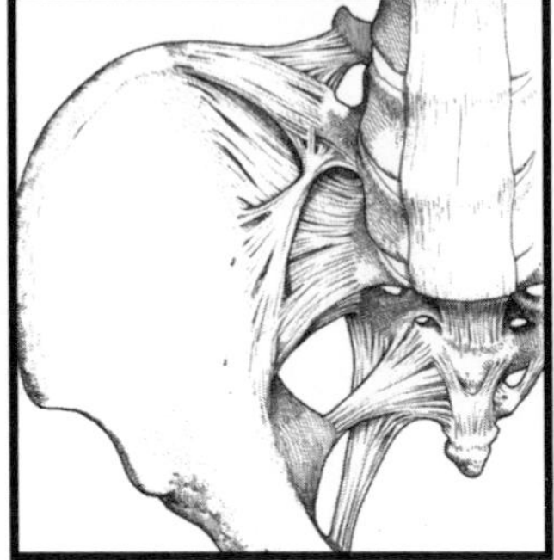

# 2 Bone Structure and Classification

Bone is a specialized form of connective tissue, consisting of cells and a matrix. The matrix is mineralized with calcium phosphate (hydroxyapatite crystals), giving it a hard texture and allowing it to serve as a significant reservoir for calcium. Bone is classified as:

- **Compact:** dense bone that forms the outer layer of a bone
- **Spongy:** cancellous bone that contains a meshwork of thin trabeculae or spicules of bone tissue; it is found in the epiphyses and metaphyses of long bones

A typical long bone has the following structural elements:

- **Diaphysis:** the shaft of the bone
- **Epiphysis:** an expanded area of bone found at either end of a bone that is covered by articular cartilage
- **Metaphysis:** lies between the diaphysis and epiphysis; it is a conical region adjacent to the area where active bone growth will occur
- **Marrow cavity:** the central portion of the shaft of many bones, it contains stem cells that produce blood cells

**COLOR** each of the following features of a long bone, using a different color for each feature:

- ☐ **1. Epiphysis (highlight the bracket)**
- ☐ **2. Metaphysis (highlight the bracket)**
- ☐ **3. Diaphysis (highlight the bracket)**
- ☐ **4. Articular cartilage (hyaline cartilage)**
- ☐ **5. Spongy bone**
- ☐ **6. Periosteum: a thin fibrous connective tissue sheath or capsule that surrounds the shaft of a bone but is not found on the articular surfaces, which are covered by articular cartilage**
- ☐ **7. Marrow cavity**
- ☐ **8. Compact bone**

Bone formation occurs largely by the deposition of matrix (osteoid) that later becomes calcified, and by resorption of bone. Thus it is a dynamic process just like the formation of any other living tissue in the body. Three major types of cells participate in this process:

- **Osteoblasts:** cells that form new bone by laying down osteoid
- **Osteocytes:** mature bone cells, formerly osteoblasts, that become surrounded by bone matrix and are responsible for maintaining the bone matrix
- **Osteoclasts:** large cells that enzymatically dissolve bone matrix and are commonly found at sites of active bone remodeling

**COLOR** the following features of compact bone:

- ☐ **9. Osteon**
- ☐ **10. Vein (color blue)**
- ☐ **11. Artery (color red)**
- ☐ **12. Lamellae of bone matrix: with osteocytes embedded within the lamellae**
- ☐ **13. Osteocytes**

An **osteon** (haversian system) is the cylindrical unit of bone and consists of a central canal (haversian canal), which contains the neurovascular bundle supplying the osteon. This canal is surrounded by concentric lamellae of bone matrix and radially oriented small canaliculi that contain the processes of **osteocytes**, which are the bone cells. Compact bone is organized into these haversian systems, but spongy bone is trabecular and its arrangement is not nearly as concentric or uniformly organized (see left side of image B).

***Clinical Note:***
**Rickets** is a disease process in which calcium deficiency during active growth leads to matrix formation that is not normally mineralized with calcium. It can occur from a lack of dietary calcium or a vitamin D deficiency, or both, because vitamin D is necessary for the normal absorption of calcium by the small intestine.

As living entities, injured bones—e.g., a **fracture**—bleed and remodel themselves based on the stresses that are placed on that injured bone. They have blood vessels, lymphatics, and nerves. If they lose their blood supply, bones will become necrotic (dead).

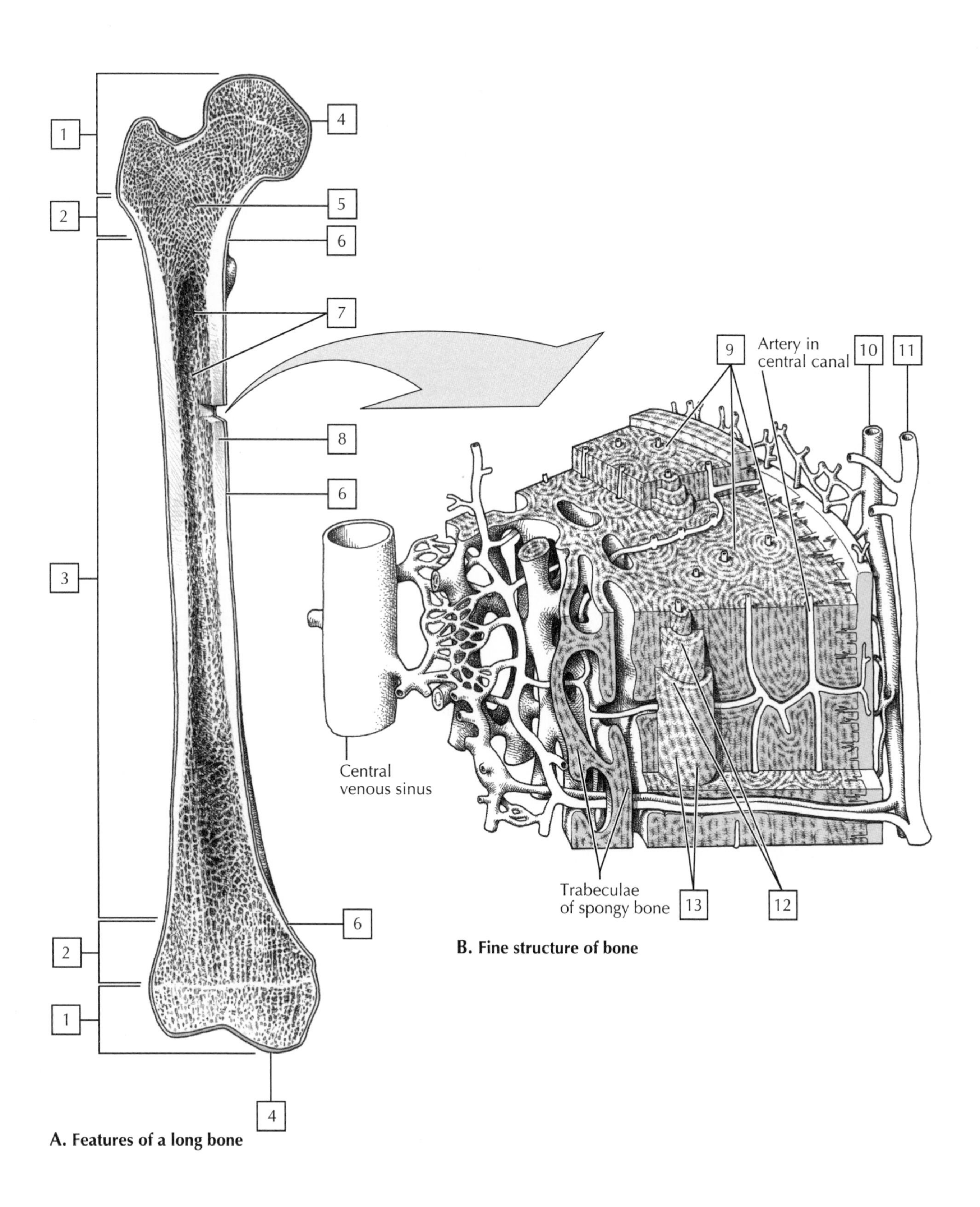


A. Features of a long bone

B. Fine structure of bone

# 2 External Features of the Skull

The skull is divided into the **neurocranium**, or calvaria (contains the brain and its meningeal coverings), and the **viscerocranium** (facial skeleton). The skull is composed of 22 bones (excluding the middle ear ossicles, 3 small bones for each ear), with 8 forming the cranium and 14 forming the face. The orbits (eye sockets) lie between the calvaria (skull cap) and the facial skeleton and are formed by contributions from 7 different bones.

**COLOR** the bones of the calvaria (skullcap), using either solid colors or diagonal lines or stippling of different colors for the larger bones:

- ☐ 1. **Frontal**
- ☐ 2. **Parietal (paired bones)**
- ☐ 3. **Sphenoid**
- ☐ 4. **Temporal (paired bones)**
- ☐ 5. **Occipital**
- ☐ 6. **Ethmoid**

The bones of the calvaria are attached to each other by sutures, a type of fibrous joint that is immobile. The sutures are labeled in the illustrations and include:

- Coronal suture
- Lambdoid suture
- Sagittal suture
- Squamous suture
- Sphenoparietal suture
- Sphenosquamous suture
- Parietomastoid suture
- Occipitomastoid suture

**COLOR** the bones of the **viscerocranium** (facial skeleton) (all paired bones except the vomer and mandible), using different colors or patterns from those used to highlight the bones of the calvaria:

- ☐ 7. **Nasal**
- ☐ 8. **Lacrimal**
- ☐ 9. **Zygomatic**
- ☐ 10. **Maxilla**
- ☐ 11. **Inferior nasal concha**
- ☐ 12. **Vomer**
- ☐ 13. **Mandible**
- ☐ 14. **Palatine**

***Clinical Note:***
The lateral aspect of the skull, where the frontal, parietal, sphenoid, and temporal bones converge, is called the **pterion**. The skull is thin here, and trauma to the side of the head in this region can lead to intracranial bleeding (**epidural hematoma**) from a lacerated middle meningeal artery, which lies between the inner aspect of these bones and the dura mater covering the brain. These fractures can be life-threatening.

Fractures of the skull, especially to some of the thinner bones of the calvaria, may result in a **depressed fracture** where the bone is pushed inward and may compress the underlying brain. If the bone is fractured into several pieces, it is called a **comminuted fracture**. Fractures to the cranial base are **basilar fractures**.

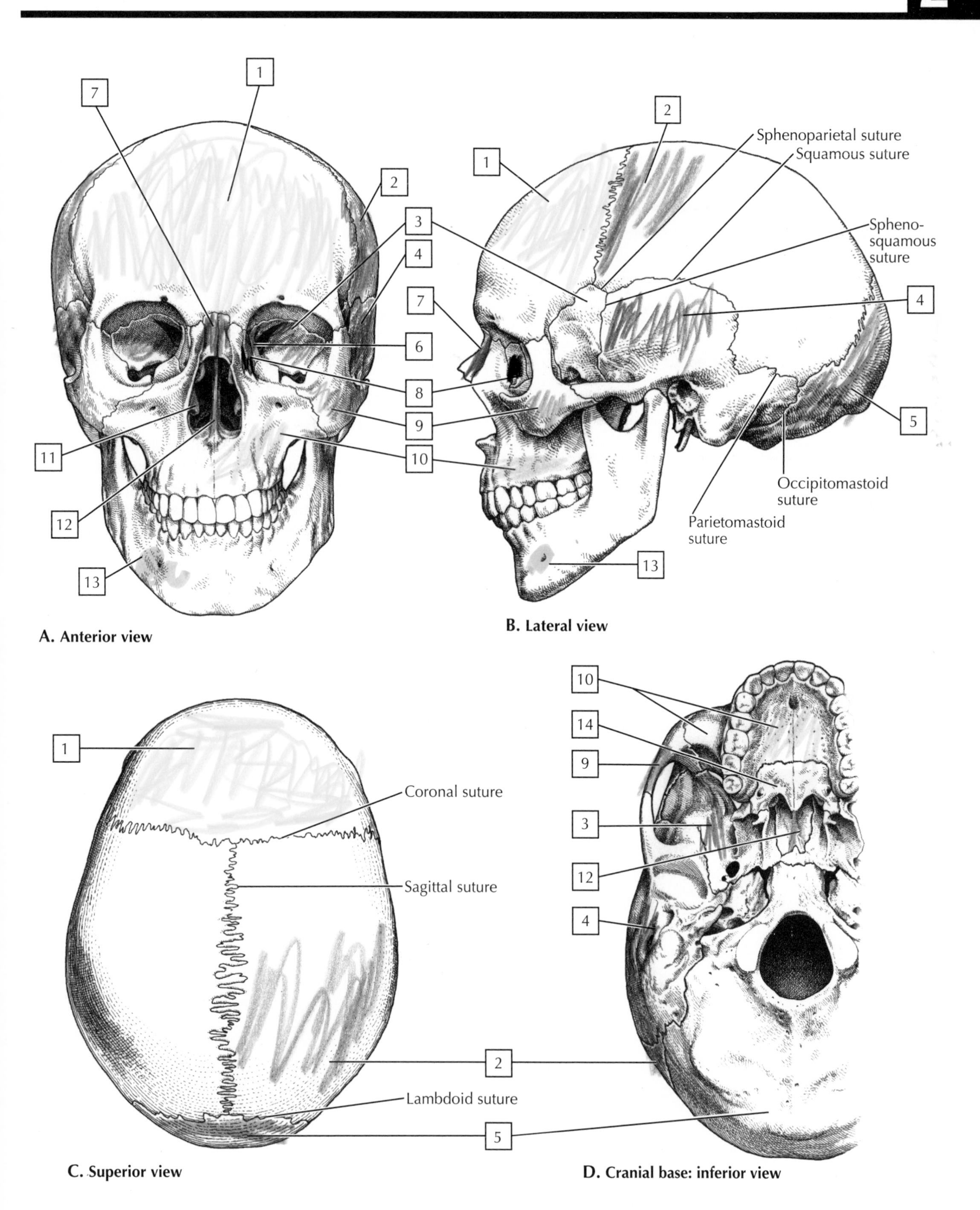


A. Anterior view

B. Lateral view

C. Superior view

D. Cranial base: inferior view

# 2 Internal Features of the Skull

The **nasal septum** is formed by:

- Perpendicular plate of the ethmoid
- Vomer
- Palatine bones
- Septal cartilages

The lateral nasal wall is formed by seven bones (see below).

**COLOR** the bones that make up the **lateral nasal wall**, using a different color for each bone:

- ☐ 1. **Nasal bone**
- ☐ 2. **Ethmoid (superior and middle conchae) (bone)**
- ☐ 3. **Lacrimal bone**
- ☐ 4. **Inferior concha (a separate bone)**
- ☐ 5. **Maxilla**
- ☐ 6. **Palatine bone**
- ☐ 7. **Sphenoid bone**

The inferior aspect of the skull (cranial base or floor) is divided into **three cranial fossae**:

- Anterior: contains the orbital roof and frontal lobes of the brain
- Middle: contains the temporal lobes of the brain
- Posterior: contains the cerebellum, pons, and medulla of the brain

Numerous holes appear in the cranial floor, and they are called **foramina** (singular, **foramen**). Important structures, especially cranial nerves arising from the brain, pass through the foramina to reach the exterior. Often blood vessels may accompany these nerves. These important structures are labeled on the illustration of the cranial base.

**COLOR** the leader line and cranial foramen for each identified foramen and the structures that pass through that foramen. Note that in **Part C Foramina of the cranial base: superior view**, that the base is divided into three cranial fossae. The **anterior cranial fossa** is formed by the frontal bone anteriorly, the ethmoid bone centrally, and portions of the sphenoid bone posteriorly. The **middle cranial fossa** is comparatively small and is formed by the body of the sphenoid bone and part of the temporal bones. The **posterior cranial fossa**, the largest of the cranial fossae, is formed by the occipital and temporal bones, with some contributions from the sphenoid and parietal bones.

A. Skull: sagittal aspect

B. Lateral nasal wall with nasal septum removed

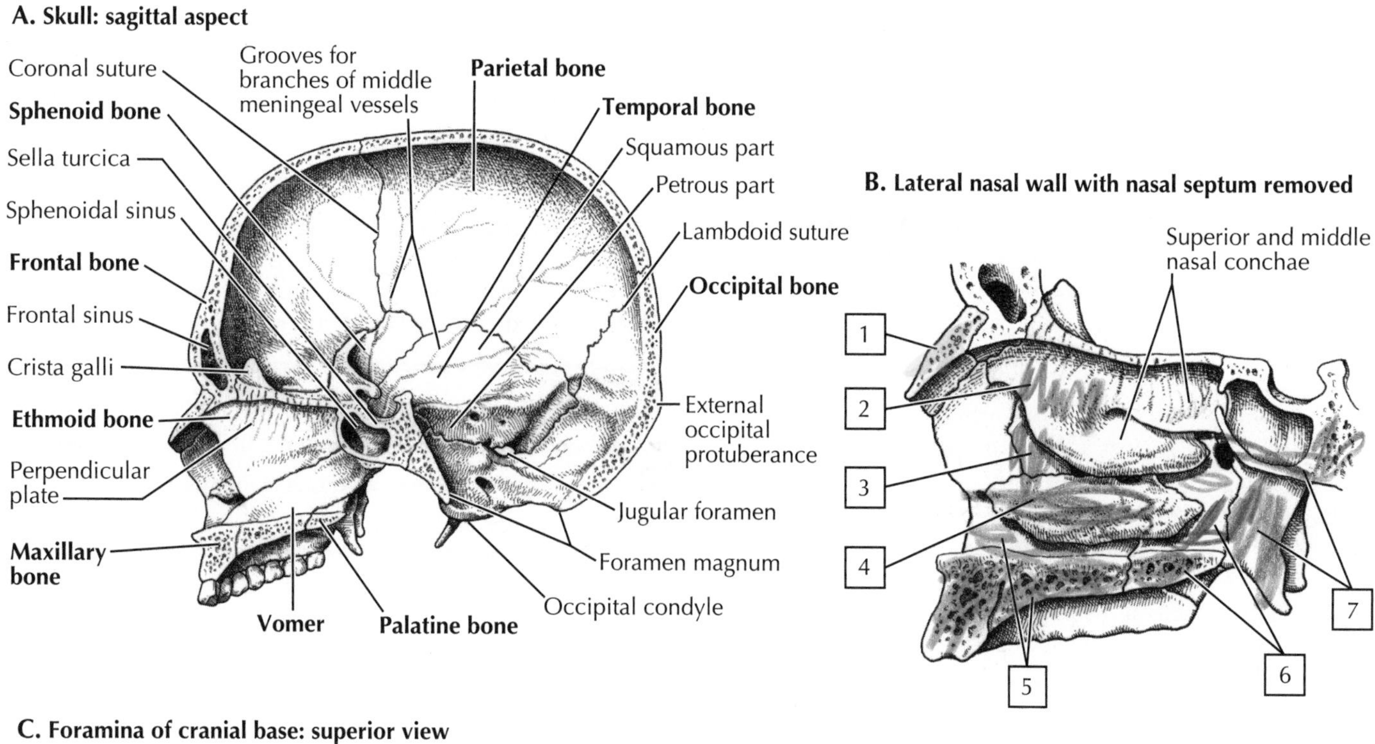


C. Foramina of cranial base: superior view

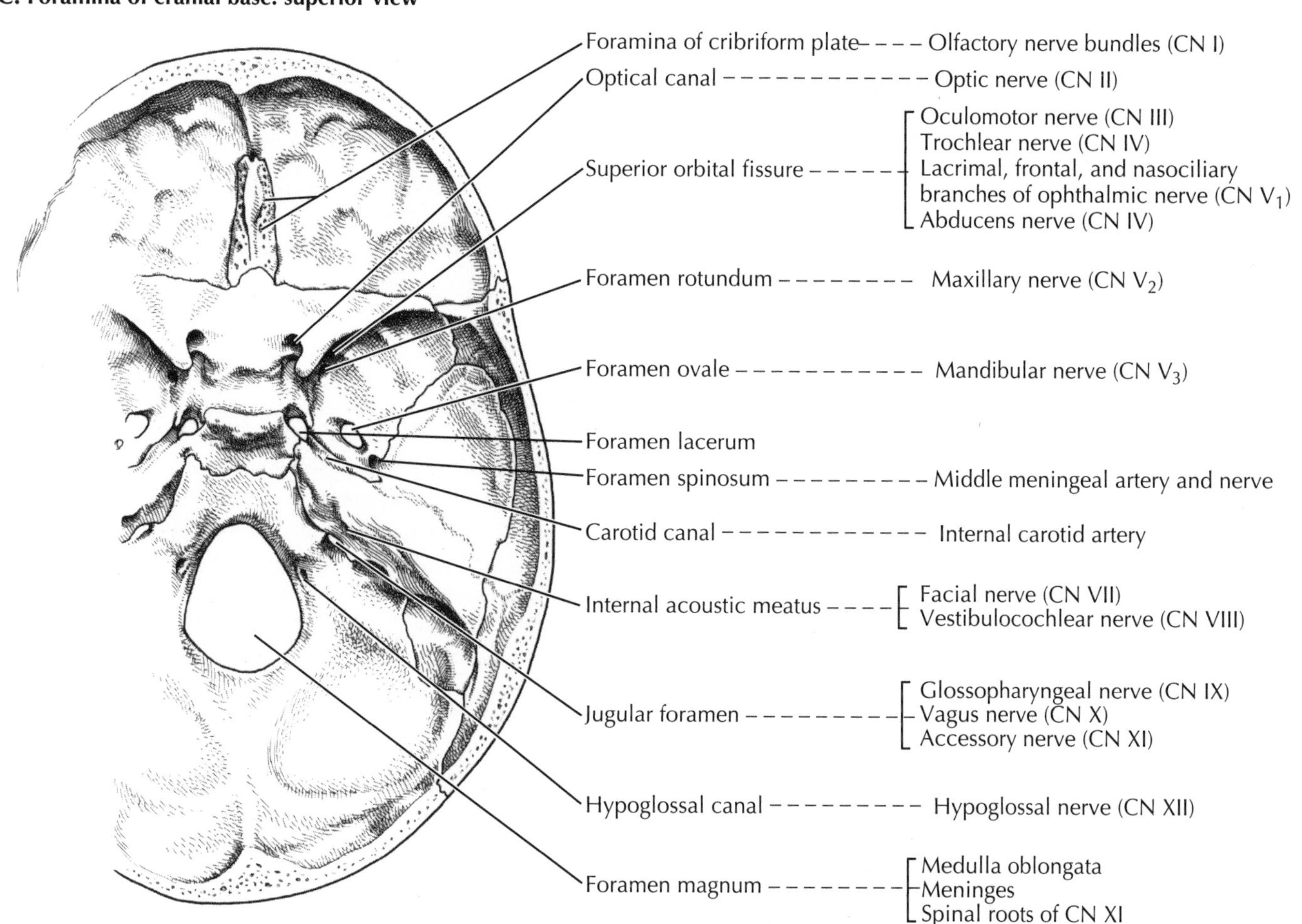

# 2 Mandible and Temporomandibular Joint

Features of the mandible (lower jaw) are summarized in the table below. The mandible articulates with the temporal bone, and in chewing or speaking, it is only the mandible or lower jaw that moves; the upper jaw or maxilla remains stationary. The teeth of the lower jaw are contained in the alveolar portion of the mandible.

| FEATURE | CHARACTERISTICS |
|---|---|
| Mandibular head | Articulates with mandibular fossa of temporal bone |
| Mandibular foramen | Inferior alveolar foramen, artery, and vein enter mandible at this opening |
| Teeth | 16 teeth: 4 incisors, 2 canines, 4 premolars (bicuspid), 6 molars (third molars are called wisdom teeth) |

**COLOR** the mandibular teeth, using a different color for each type. Note that in adults there are 16 teeth in the mandible and 16 teeth in the maxilla, but in children there are a total of 20 deciduous teeth, also called "milk teeth" or "baby teeth." By 24 months of age, most or all of these deciduous teeth have emerged.

- ☐ **1. Molars (the third molars are called wisdom teeth) (6 teeth)**
- ☐ **2. Premolars (bicuspids) (4 teeth)**
- ☐ **3. Canines (2 teeth)**
- ☐ **4. Incisors (4 teeth)**

The **temporomandibular joint (TMJ)** is actually two synovial joints in one, separated by an articular disc. The articular surfaces of most synovial joints are covered by hyaline cartilage, but the TMJ surfaces are covered by fibrocartilage.

The TMJ is a modified hinge type of synovial joint, and its features are summarized in the table below.

| FEATURE | ATTACHMENT | COMMENT |
|---|---|---|
| Capsule | Temporal fossa and tubercle to mandibular head | Permits side-to-side motion, protrusion, and retrusion |
| Lateral (TMJ) ligament | Temporal to mandible | Thickened fibrous band of capsule |
| Articular disc | Between temporal bone and mandible | Divides joint into two synovial compartments |

**COLOR** the following features of the TMJ:

- ☐ **5. Joint capsule**
- ☐ **6. Lateral (temporomandibular) ligament**
- ☐ **7. Articular disc**

***Clinical Note:***

Because of its vulnerable location, the mandible is the second most commonly **fractured** facial bone (the nasal bone is the first). **Dislocation** of the TMJ can occur when the mandibular condyle moves anterior to the articular eminence (just anterior to the "open position" seen in part E). Sometimes, just a wide yawn is enough to cause dislocation, which can be quite painful.

**Dental caries** (cavities) occur as a result of decay of the hard tissues of the tooth. If the decay invades the pulp cavity of the tooth, a painful infection usually results.

Likewise, poor dental hygiene may result in **gingivitis**, an inflammation of the gingivae (gums). Left untreated, the infection may invade the alveolar bone, a condition that is called **periodontitis**.

An **inferior alveolar nerve block** is a common procedure used by dentists to anesthetize the inferior alveolar nerve (a branch of CN V3), which innervates the ipsilateral (same side) mandibular teeth. The anesthetic agent is infiltrated around the mandibular foramen, where the inferior alveolar nerve emerges (along with an inferior alveolar artery and vein). This blocks the innervation of all of the mandibular teeth on the side of the injection. The skin and mucosa of the lower lip, gingiva (gums), and skin of the chin also are anesthetized.

Decay of the tooth results in a **dental caries**, commonly known as a cavity. If the decay invades the pulp cavity of the tooth, a painful infection will result (called **pulpitis**); commonly called a "toothache."

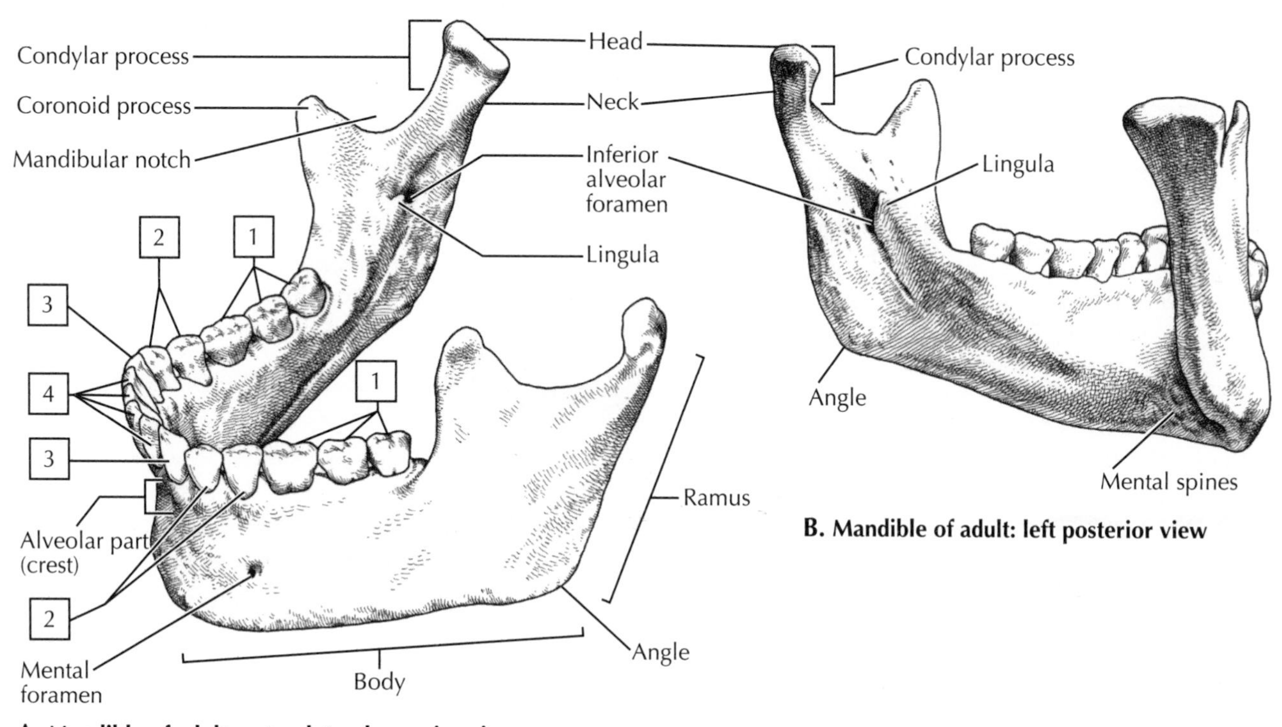


A. Mandible of adult: anterolateral superior view

B. Mandible of adult: left posterior view

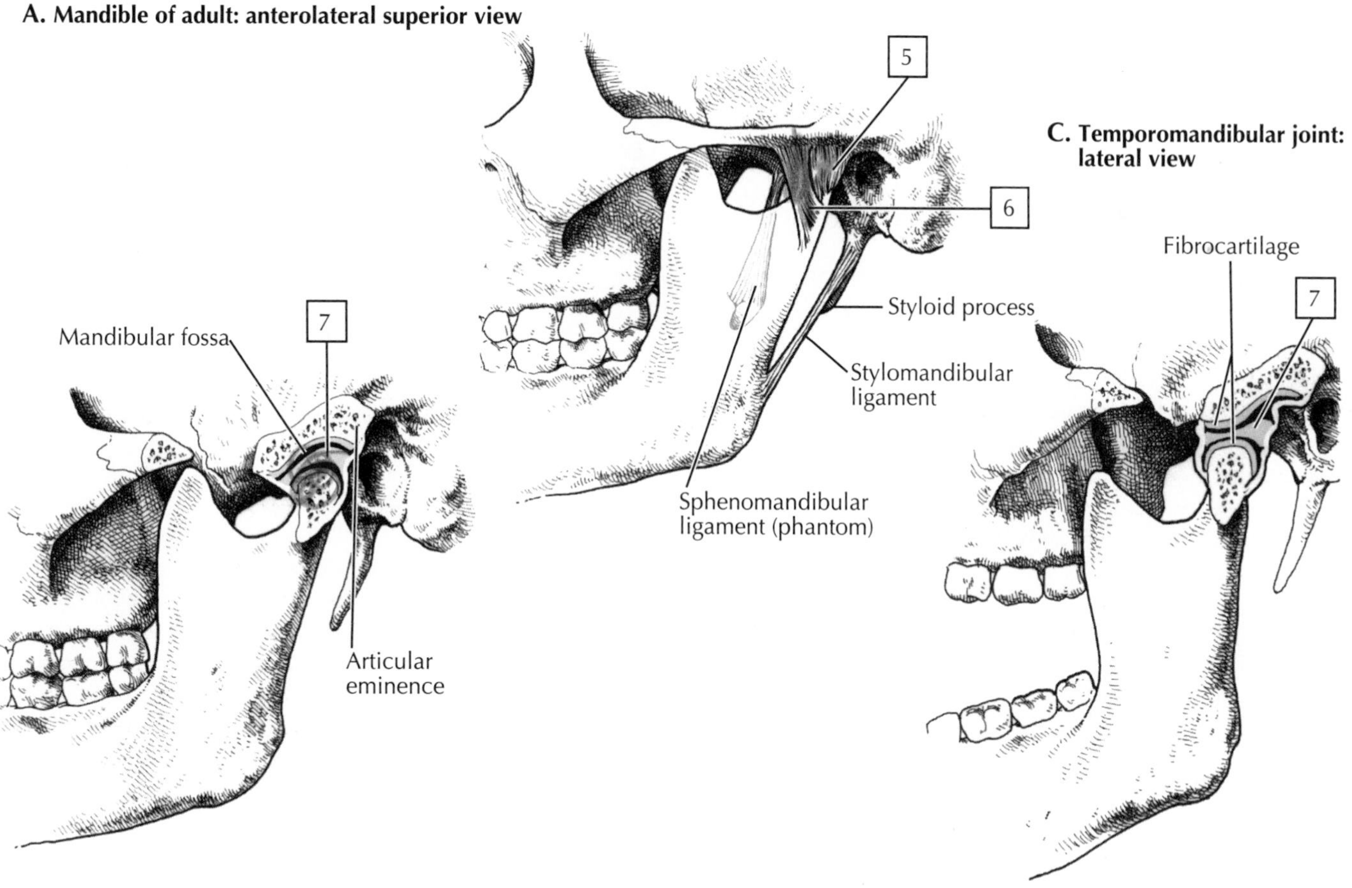


C. Temporomandibular joint: lateral view

D. Temporomandibular joint: jaws closed

E. Temporomandibular joint: jaws widely opened (hinge and gliding actions combined)

# Vertebral Column

The vertebral column (spine) forms the central axis of the human body, highlighting the segmental nature of all vertebrates, and is composed of 33 vertebrae distributed as follows:

- **Cervical vertebrae:** 7 total, with the first two called the atlas (C1) and axis (C2)
- **Thoracic vertebrae:** 12 total, each articulating with a pair of ribs
- **Lumbar vertebrae:** 5 total, large vertebrae to support the body's weight
- **Sacrum:** 5 fused vertebrae
- **Coccyx:** 4 total vertebrae, Co1 often not fused but Co2-Co4 are fused, a remnant of our embryonic tail

Viewed from the lateral aspect, one can identify the:

- **Cervical lordosis:** acquired secondarily when the infant can support the weight of its own head
- **Thoracic kyphosis:** a primary curvature present in the fetus
- **Lumbar lordosis:** acquired secondarily when the infant assumes an upright posture
- **Sacral kyphosis:** a primary curvature present in the fetus

A "typical" vertebra has several consistent features:

- **Body:** weight-bearing portion that tends to increase in size as one descends the spine
- **Arch:** projection formed by paired pedicles and laminae
- **Transverse processes:** lateral extensions from the union of the pedicle and lamina
- **Articular processes (facets):** two superior and two inferior facets for articulation
- **Spinous process:** projection that extends posteriorly from the union of two laminae
- **Vertebral notches:** superior and inferior features that in articulated vertebrae form intervertebral foramina
- **Intervertebral foramina:** traversed by spinal nerve roots and associated vessels
- **Vertebral foramen (canal):** formed from the vertebral arch and body, the foramen contains the spinal cord and its meningeal coverings
- **Transverse foramina:** apertures that exist in transverse processes of cervical vertebrae and transmit vertebral vessels

**COLOR** the following features of a typical vertebra, using a different color for each feature:

- ☐ 1. **Vertebral body**
- ☐ 2. **Transverse process**
- ☐ 3. **Articular facets**
- ☐ 4. **Spinous process**
- ☐ 5. **Arch**

Additionally, adjacent articulated vertebrae are secured by ligaments, and their individual vertebral bodies are separated by **fibrocartilaginous intervertebral discs**. Each intervertebral disc acts as a shock absorber and compresses and expands slightly in response to weight-bearing. The central portion of the disc is a gelatinous **nucleus pulposus** that is surrounded by concentric layers of fibrocartilage called the **anulus fibrosus**. If the anulus is exposed to excessive pressure or the dehydration associated with aging, it can begin to weaken, and the nucleus pulposus can **herniate ("slipped disc")** through the cartilaginous lamellae and impinge on a nerve root as it exits the spinal cord (see Plate 2-7).

**COLOR** the intervertebral discs and the key ligaments observed in a lateral "cutaway" view of several adjacent vertebrae:

- ☐ 6. **Intervertebral discs (IVD): fibrocartilaginous discs between adjacent vertebral bodies**
- ☐ 7. **Anterior longitudinal ligament: connects adjacent vertebral bodies and the IVD along the anterior aspects of the bodies**
- ☐ 8. **Posterior longitudinal ligament: connects adjacent vertebral bodies and IVD along the posterior aspects of the bodies**
- ☐ 9. **Supraspinous ligament: extends between adjacent spinous processes**
- ☐ 10. **Interspinous ligament: extends between adjacent spinous processes**
- ☐ 11. **Ligamenta flavum: connects adjacent laminae; contains pale yellow elastic fibers and is able to resist sudden flexion, which might otherwise cause separation of the vertebral laminae. This ligament also assists in maintaining one's posture and in straightening the spine after flexion.**

***Clinical Note:***

Accentuated curvatures of the spine may occur congenitally or be acquired. **Scoliosis** is an accentuated lateral and rotational curve of the thoracic or lumbar spine; it is more common in adolescent girls. Hunchback is an accentuated **kyphosis** of the thoracic spine, usually resulting from poor posture or osteoporosis. Swayback is an **accentuated lordosis** of the lumbar spine, usually caused by weakened trunk muscles or obesity; it is also commonly seen in women in late pregnancy.

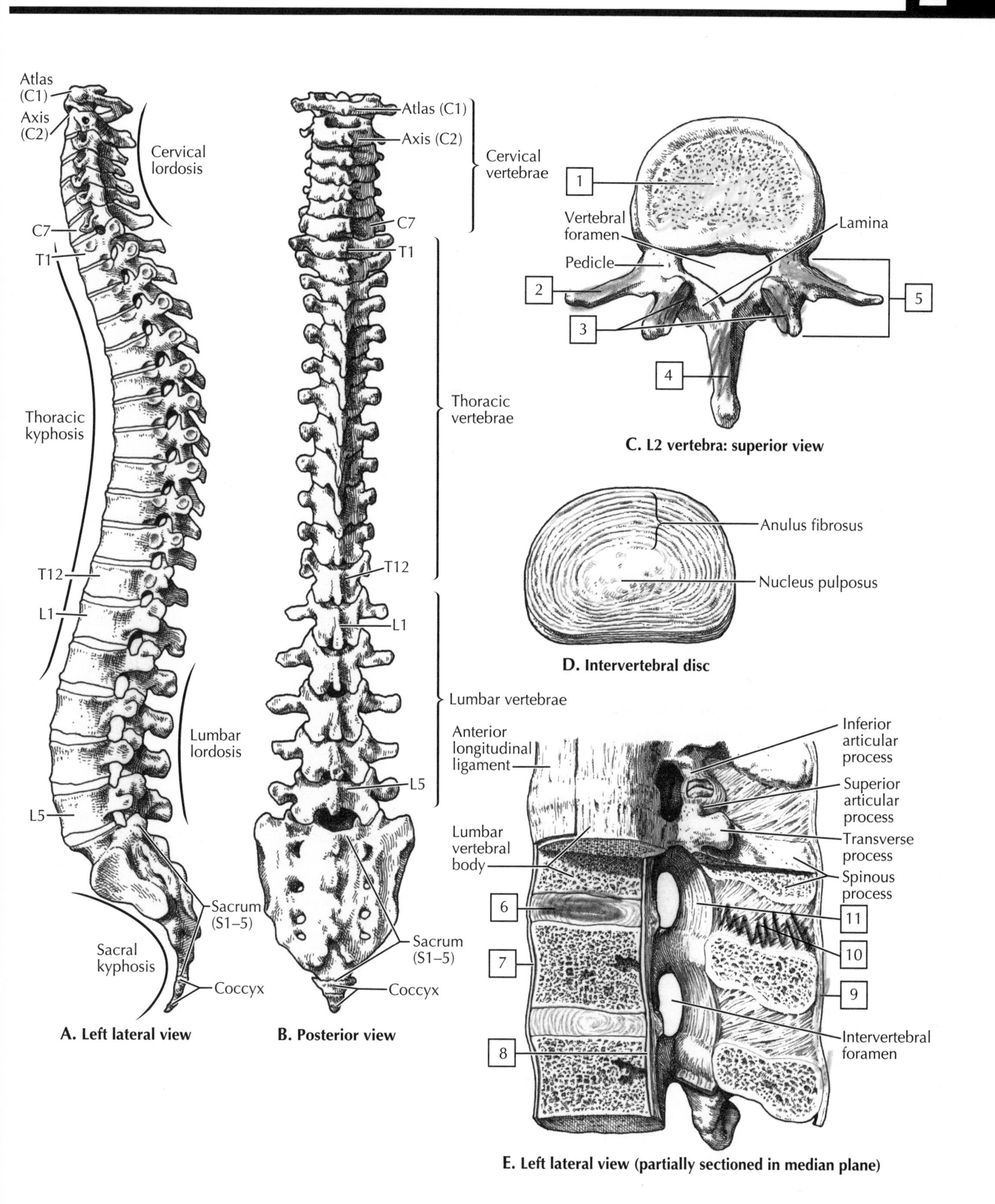


A. Left lateral view

B. Posterior view

C. L2 vertebra: superior view

D. Intervertebral disc

E. Left lateral view (partially sectioned in median plane)

# 2 Cervical and Thoracic Vertebrae

The **cervical spine** is composed of seven cervical vertebrae. The first two cervical vertebrae are unique and are termed the **atlas** (C1) and **axis** (C2). The atlas (C1) holds the head on the neck and gets its name from the Titan of Greek mythology, Atlas, who held the heavens on his shoulders. The axis (C2) is the point of articulation where the head turns on the neck, providing an "axis of rotation." The cervical region is a fairly mobile portion of the spine, allowing for flexion and extension as well as rotation and lateral bending. Features of the seven cervical vertebrae are summarized in the table below.

| ATLAS (C1) | CERVICAL VERTEBRAE C3 TO C7 |
|---|---|
| Ringlike bone; superior facet articulates with occipital bone | Large triangular vertebral foramen |
| Two lateral masses with facets | Foramen transversarium, through which the vertebral artery passes |
| No body or spinous process | C3 to C5: short bifid spinous process |
| C1 rotates on articular facets of C2 | C6 to C7: long spinous process |
| Vertebral artery runs in groove on posterior arch | C7 called vertebra prominens; usually the most superior vertebral spine you can see on the lower neck |
| | Narrow intervertebral foramina |
| | Nerve roots at risk of compression |
| **AXIS (C2)** | |
| Dens projects superiorly | |
| Strongest cervical vertebra | |

**COLOR** the following features of the cervical vertebrae (parts ***A*** to ***C***), using a different color for each feature:

- ☐ **1. Posterior arch of the atlas**
- ☐ **2. Vertebral canal (the spinal cord passes through the vertebral canal)**
- ☐ **3. Dens axis**
- ☐ **4. Foramen transversarium**
- ☐ **5. Intervertebral discs (note that no disc exists between the atlas and axis)**
- ☐ **6. Vertebral body (note that the atlas does not possess a body)**
- ☐ **7. Transverse process**
- ☐ **8. Bifid spine**
- ☐ **9. Lamina**

The **thoracic spine** is composed of 12 thoracic vertebrae. The 12 pairs of ribs articulate with the thoracic vertebrae, and this region of the spine is more rigid and less flexible than the cervical spine. Key features of the thoracic vertebrae include:

- Heart-shaped body, with facets for rib articulation
- Small circular vertebral foramen (the spinal cord passes through the vertebral foramen)
- Long transverse processes, which have costal facets for rib articulation (T1 to T10 only)
- Long spinous processes, which slope posteriorly and overlap the next vertebra below

**COLOR** the following features of the thoracic vertebrae (parts ***D*** and ***E***):

- ☐ **10. Body**
- ☐ **11. Superior costal facet**
- ☐ **12. Vertebral foramen**
- ☐ **13. Spinous process**
- ☐ **14. Transverse costal facet**
- ☐ **15. Inferior costal facet**

***Clinical Note:***

A strong blow to the top of the head, e.g., diving into a swimming pool and hitting the pool bottom head-first, can compress the lateral masses of the atlas and **fracture** one or both of the anterior or posterior arches.

**Dislocation**, rather than an outright fracture of the cervical vertebral bodies, if significant, can damage the spinal cord.

Additionally, sudden **cervical hyperextension** can result in a "whiplash" injury that used to be associated with a rear-end vehicular crash. Headrests on the car seats have greatly reduced this type of hyperextension-hyperflexion injury.

The thoracic vertebrae may be counted and examined by having a patient flex their neck and back, thus exposing the spinous process of C7. The thoracic spines can then be counted, but realize that thoracic spines do not overlap the thoracic vertebral bodies, but the thoracic vertebra below (the T3 spinous process overlaps the T4 vertebral body). The transverse processes of the thoracic vertebrae usually can be palpated as well, and, in especially thin individuals, one may be able to palpate the tubercle and angle of the ribs in the lower thoracic spine.

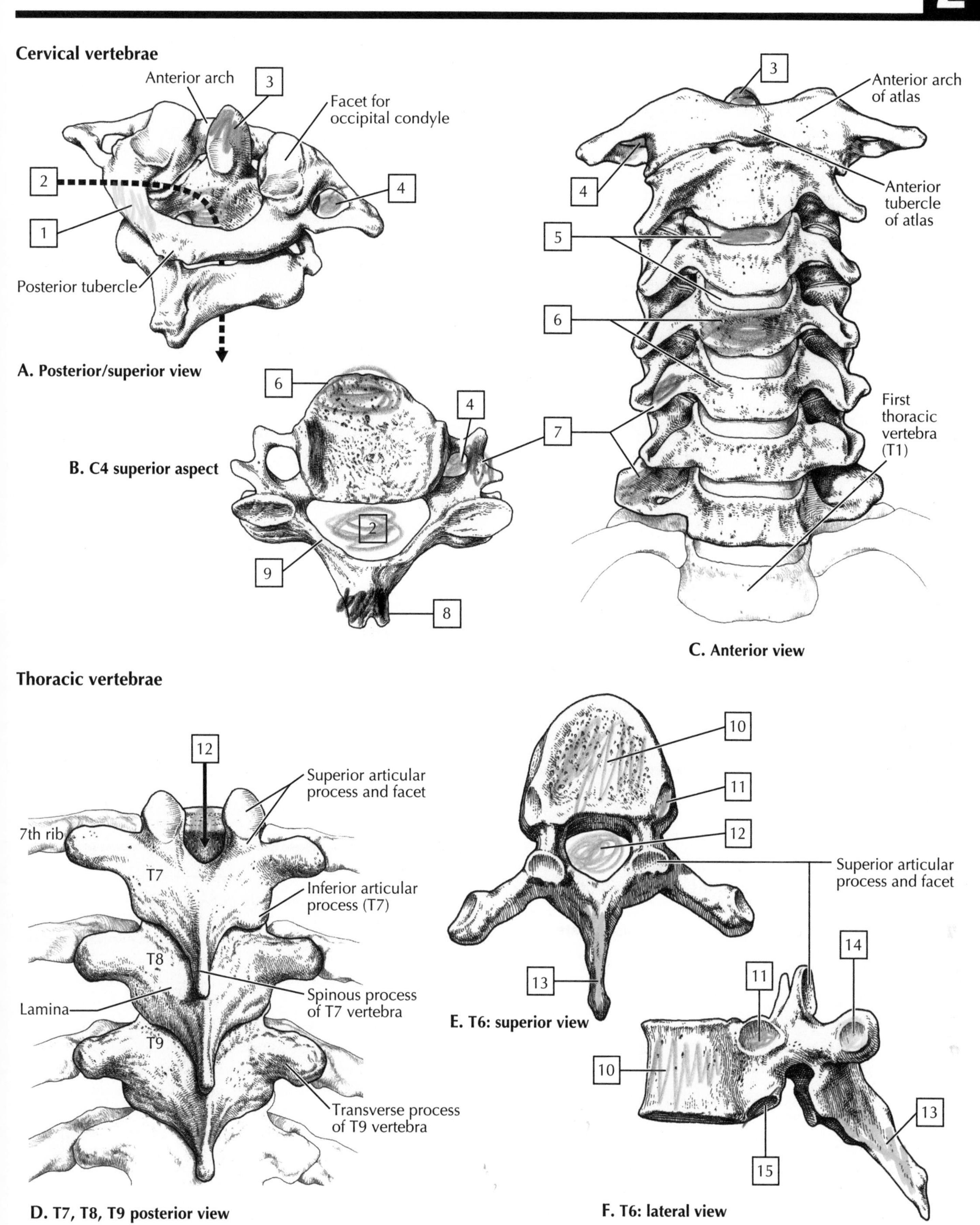
Cervical vertebrae
Anterior arch
3
Facet for occipital condyle
2
4
1
Posterior tubercle
A. Posterior/superior view
6
4
7
B. C4 superior aspect
2
9
8
3
Anterior arch of atlas
4
Anterior tubercle of atlas
5
6
First thoracic vertebra (T1)
C. Anterior view
Thoracic vertebrae
12
Superior articular process and facet
7th rib
T7
Inferior articular process (T7)
T8
Spinous process of T7 vertebra
Lamina
T9
Transverse process of T9 vertebra
D. T7, T8, T9 posterior view
10
11
12
Superior articular process and facet
13
E. T6: superior view
14
11
10
13
15
F. T6: lateral view

# 2 Lumbar, Sacral, and Coccygeal Vertebrae

The **lumbar spine** is composed of five lumbar vertebrae. They are comparatively large, so that they can bear the weight of the trunk, and are also fairly mobile, but they are not nearly as mobile as the cervical spine. The **sacrum** is composed of five fused vertebrae that form a single wedge-shaped bone. The sacrum provides support for the pelvis. The **coccyx** is a remnant of our embryonic tail and usually consists of four vertebrae, with the last three being fused into a single bone. The coccyx lacks vertebral arches and has no vertebral canal. The general features of all of these vertebrae are summarized in the table below.

| THORACIC VERTEBRAE | LUMBAR VERTEBRAE |
|---|---|
| Heart-shaped body, with facets for rib articulation | Kidney-shaped body, massive for support |
| Small circular vertebral foramen | Midsized triangular vertebral foramen |
| Long transverse processes, which have facets for rib articulation in T1-T10 | Facets face in a medial or lateral direction, which permits good flexion and extension |
| Long spinous processes, which slope posteriorly and overlap next vertebra | Spinous process is short and strong |
| | L5 is largest vertebra |
| **SACRAL VERTEBRAE** | **COCCYGEAL VERTEBRAE** |
| Large, wedge-shaped bone, which transmits body weight to pelvis | Co1 often not fused |
| Five fused vertebrae, with fusion complete by puberty | Co2 to Co4 fused |
| Four pairs of sacral foramina on posterior and anterior (pelvic) sides | No pedicles, laminae, spines |
| Sacral hiatus, the opening of sacral vertebral foramen | Remnant of our embryonic tail |

**COLOR** the following features of the lumbar (part ***A***), sacral (parts ***B***, ***C***, ***D***, and ***E***), and coccygeal (parts ***B***, ***C***, and ***E***) vertebrae, using a different color for each feature:

- ☐ **1. Intervertebral foramen: traversed by a spinal nerve as it leaves the spinal cord and passes out to the periphery**
- ☐ **2. Intervertebral disc**
- ☐ **3. Vertebral body**
- ☐ **4. Superior articular process**
- ☐ **5. Spinous process**
- ☐ **6. Lumbosacral articular surface: articulates with the body of the L5 vertebra**
- ☐ **7. Anterior (pelvic) sacral foramina: for the passage of spinal nerves**
- ☐ **8. Coccyx**
- ☐ **9. Median sacral crest: equivalent of vertebral spinous processes elsewhere along the vertebral column**

**COLOR** the following features of the image (part ***D***) of the articulated lower spine (lumbar, sacral, and coccygeal vertebrae):

- ☐ **10. Anterior longitudinal ligament**
- ☐ **11. Intervertebral discs**
- ☐ **12. Spinal nerves (color yellow)**
- ☐ **13. Interspinous ligament**
- ☐ **14. Supraspinous ligament**

***Clinical Note:***
Stress- or age-related changes can lead to dehydration of the intervertebral discs. In this process, the central nucleus pulposus **herniates** through the anulus fibrosus, and if the herniation is posterolateral, which is most common, it can compress the spinal nerve or its root as it exits the intervertebral foramen.

This can lead to chronic pain from the compression of the spinal nerve. Most disc herniations occur in the L4-L5 or L5-S1 vertebral levels, but herniation of a cervical disc also may occur.

**Sciatica** is the pain in the lower back and radiating down the posterior thigh and leg, often from a herniated disc, although other causes must be considered and eliminated.

Generalized back pain is common and may occur from several sources, including:

- Nervous elements exiting the intervertebral foramina
- Meninges covering the spinal cord
- Joint pain from the synovial joints of the spine (osteoarthritis)
- Fibroskeletal elements of the spine
- Intrinsic back muscle pain muscle cramping (spasms)

**COLOR**

- ☐ **15. The herniating nucleus pulposus as it compresses a spinal nerve**

***Clinical Note:***
**Osteoarthritis** is the most common form of arthritis and often involves erosion of the articular cartilage of weight-bearing joints, such as those of the vertebral column. Progressive erosion of cartilage in joints of the spine, fingers, knee, and hip are most commonly involved. It is usually significant after the age of 65. Risk factors include one's age, sex (females affected more than males), joint trauma, repetitive stress, obesity, genetic risk, and previous inflammatory joint disease. If the spine is involved, it can lead to spinal nerve impingement.

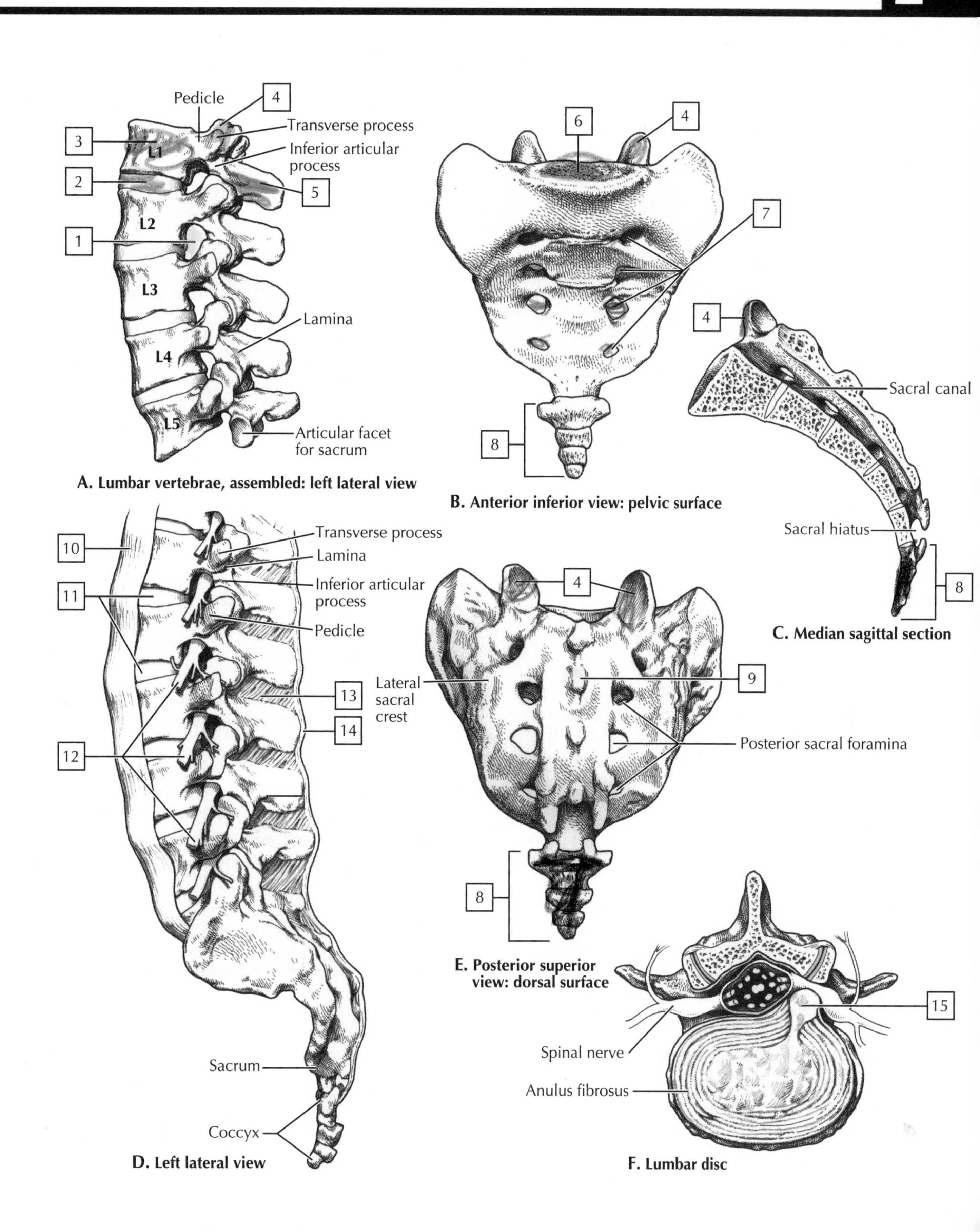


A. Lumbar vertebrae, assembled: left lateral view

B. Anterior inferior view: pelvic surface

C. Median sagittal section

D. Left lateral view

E. Posterior superior view: dorsal surface

F. Lumbar disc

# 2 Thoracic Cage

The thoracic cage is part of the axial skeleton and includes the midline sternum and 12 pairs of ribs, each with a(n):

- **Head:** it articulates with the inferior costal facet of the vertebral body above and with the superior costal facet of the body of its own vertebra (e.g., rib 3 with T3 vertebra)
- **Neck**
- **Tubercle:** it articulates with the transverse process of its own vertebra
- **Angle**
- **Body**

Ribs 1 to 7 articulate anteriorly with the sternum directly and are called "true ribs."

Ribs 8 to 10 articulate with costal cartilages of the ribs above them and are called "false ribs." Their connection with the sternum is indirect.

Ribs 11 to 12 articulate with vertebrae only and are called "floating ribs." Their anterior portions end in muscles of the posterior abdominal wall.

| FEATURE | ATTACHMENT | COMMENT |
|---|---|---|
| | **Sternoclavicular (Saddle-type Synovial) Joint With an Articular Disc** | |
| Capsule | Clavicle and manubrium | Allows elevation, depression, protraction, retraction, circumduction |
| Sternoclavicular ligament | Clavicle and manubrium | Consists of anterior and posterior ligaments |
| Interclavicular ligament | Between both clavicles | Connects two sternoclavicular joints |
| Costoclavicular ligament | Clavicle to 1st rib | Anchors clavicle to 1st rib |
| | **Sternocostal Joints (Primary Cartilaginous Joints, or Synchondroses)** | |
| First sternocostal ligament | First rib to manubrium | Allows no movement at this joint |
| Radiate sternocostal ligament | Ribs 2-7 with sternum | Permits some gliding or sliding movement at these synovial plane joints |
| | **Costochondral (Primary Cartilaginous) Joints** | |
| Cartilage | Costal cartilage to rib | Normally no movement at these joints |
| | **Interchondral (Synovial Plane) Joints** | |
| Interchondral ligament | Between costal cartilages | Allows some gliding movement |

Functionally, the thoracic cage participates in breathing, via its muscle attachments, protection of the vital thoracic organs, including the heart and lungs, and as a conduit for the passage of important structures to and from the head and also the abdomen. The opening at the top of the thoracic cage is the **superior thoracic aperture,** and that at the bottom is called the **inferior thoracic aperture.** The inferior aperture is largely covered by the abdominal diaphragm, an important skeletal muscle used in breathing.

The upper limb attaches to the thoracic cage at the pectoral girdle, which includes the:

- **Clavicle:** acts as a strut to keep the limb at the side of the body wall
- **Scapula:** a flat triangular bone that has 16 different muscles attached to it that largely act on the shoulder joint

**COLOR** the following features of the thoracic cage, using a different color for each feature:

- ☐ 1. **Costal cartilages**
- ☐ 2. **Clavicle**
- ☐ 3. **Sternum and its three parts:**
  - **3A.** ***Manubrium***
  - **3B.** ***Body***
  - **3C.** ***Xiphoid process***
- ☐ 4. **Superior articular facet of head of rib: articulation for head of rib of same number as the vertebra**
- ☐ 5. **Inferior articular facet of head of rib: articulation for head of rib one number greater than the vertebra number**
- ☐ 6. **Parts of a typical rib (6A, head; 6B, neck; 6C, tubercle; and 6D, angle and remainder of the rib, called the body)**

***Clinical Note:***

Thoracic trauma often includes **rib fractures** (the 1st, 11th, and 12th ribs are usually spared), crush injuries (commonly with rib fractures), and penetrating chest wounds (stab and gunshot wounds). The pain associated with rib fractures is often intense because of the expansion and contraction of the rib cage during respiration.

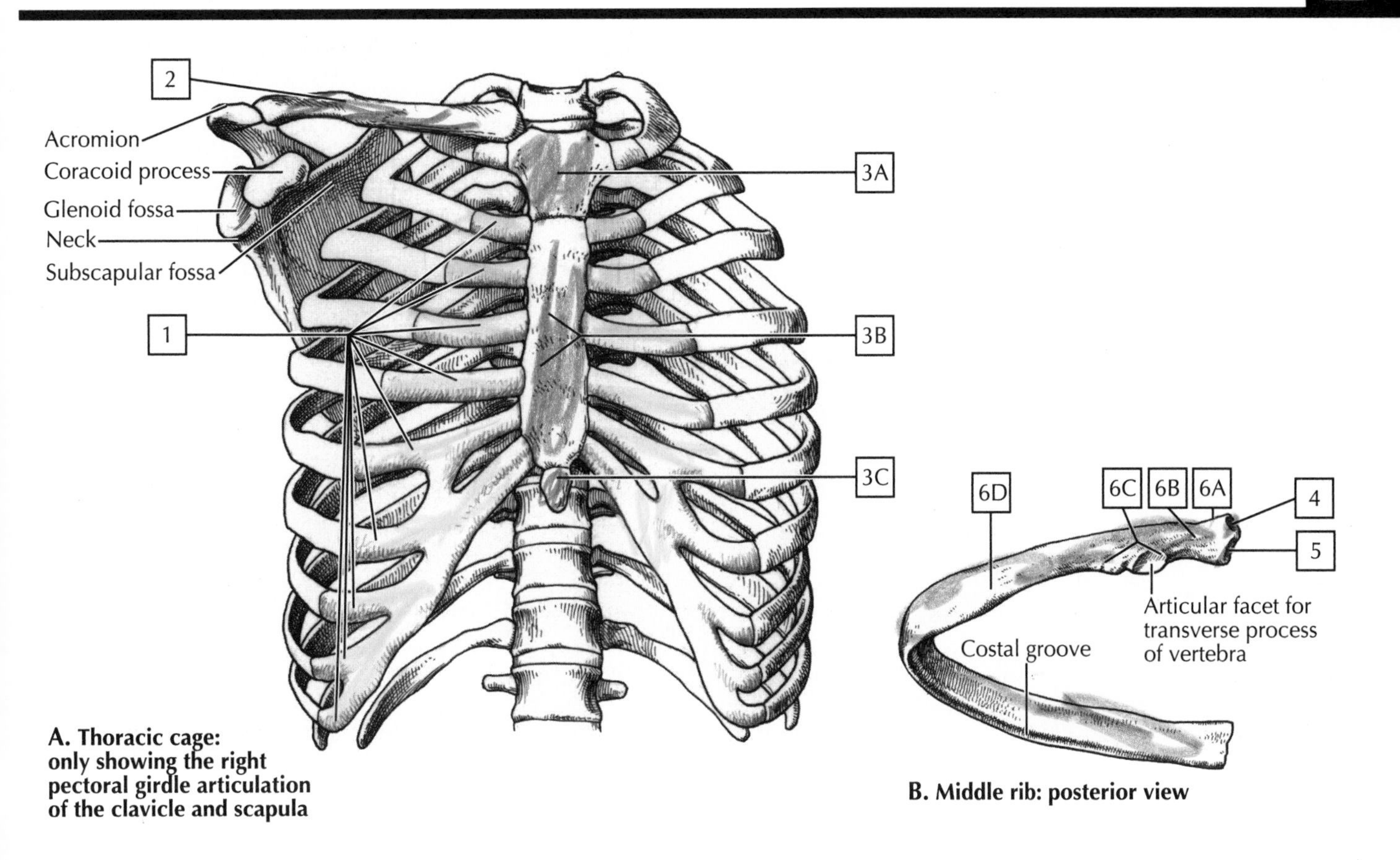


A. Thoracic cage: only showing the right pectoral girdle articulation of the clavicle and scapula

B. Middle rib: posterior view

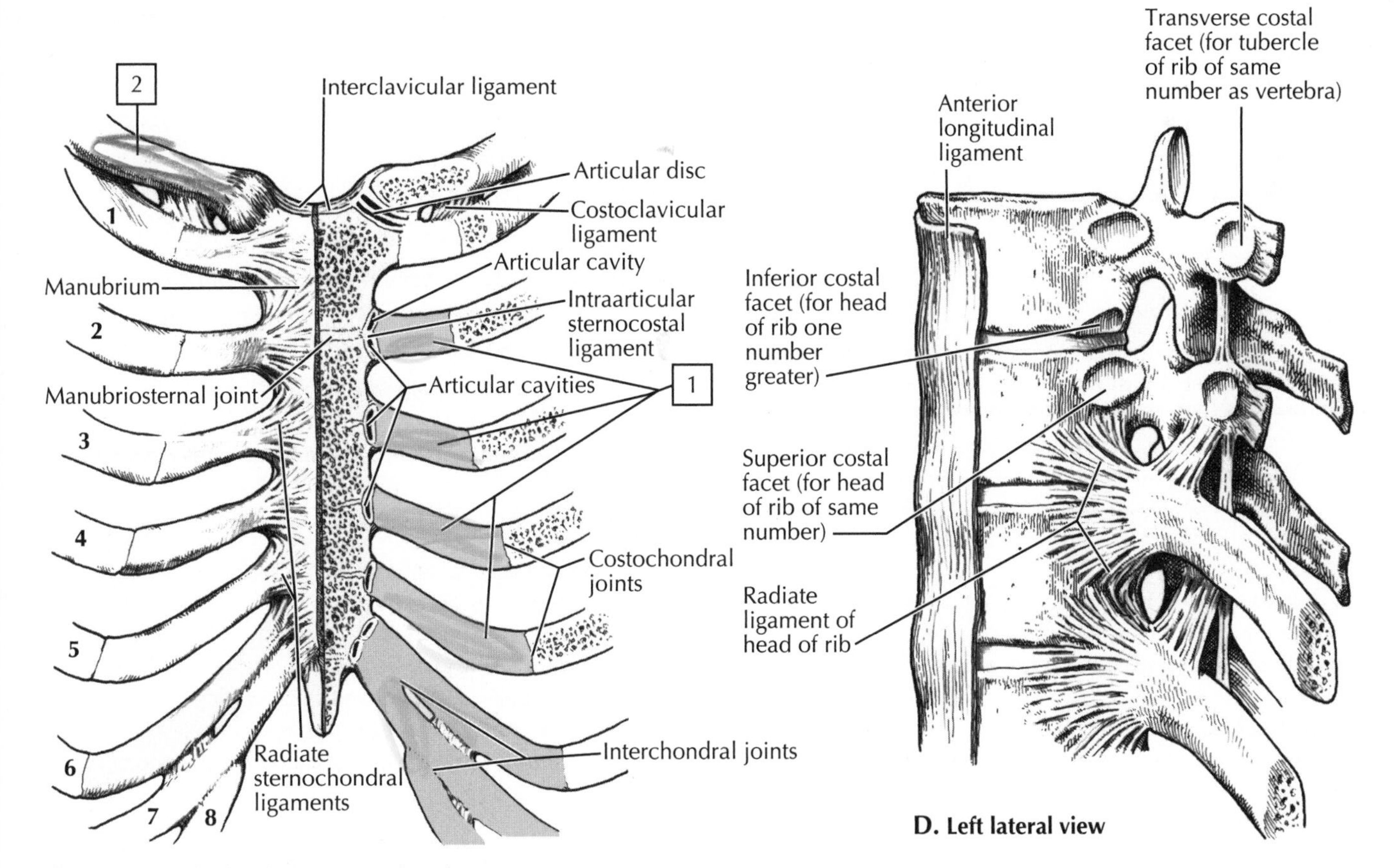


C. Sternocostal articulations: anterior view

D. Left lateral view

# 2 Joints and Ligaments of the Spine

The **craniovertebral joints** are synovial joints that offer a relatively wide range of motion compared with most joints of the spine and include the:

- **Atlantooccipital joint:** between the **atlas** (C1) and the occipital bone of the skull; allows for flexion and extension, as in nodding the head to signify "yes"
- **Atlantoaxial joint:** between the atlas and the **axis** (C2); allows for rotational movement, as in shaking the head to signify "no"

| FEATURE | ATTACHMENT | COMMENT |
|---|---|---|
| **Atlantooccipital (Biaxial Condyloid Synovial) Joint** | | |
| Articular capsule | Surrounds facets and occipital condyles | Allows flexion and extension |
| Anterior and posterior membranes | Anterior and posterior arches of C1 to foramen magnum | Limit movement of joint |
| **Atlantoaxial (Uniaxial Synovial) Joint** | | |
| Tectorial membrane | Axis body to margin of foramen magnum | Is continuation of posterior longitudinal ligament |
| Apical ligament | Dens to occipital bone | Is very small |
| Alar ligament | Dens to occipital condyles | Limits rotation |
| Cruciate ligament | Dens to lateral masses | Resembles a cross; allows rotation |

**COLOR** the following ligaments of the craniovertebral joints (parts ***A*** to ***D***), using a different color for each ligament:

- ☐ 1. **Capsule of the atlantooccipital joint**
- ☐ 2. **Capsule of the atlantoaxial joint**
- ☐ 3. **Posterior longitudinal ligament**
- ☐ 4. **Alar ligaments**
- ☐ 5. **Cruciate ligament: superior and inferior bands and transverse ligament of the atlas**

Joints of the **vertebral arches** are plane synovial joints between the superior and inferior articular facets that allow some gliding or sliding movement.

Joints of the **vertebral bodies** are secondary cartilaginous joints between adjacent vertebral bodies. These stable, weight-bearing joints also serve as shock absorbers.

The **intervertebral discs** consist of an outer fibrocartilaginous **anulus fibrosus** and an inner gelatinous **nucleus pulposus.** The lumbar discs are the thickest and the upper thoracic spine discs the thinnest. Anterior and posterior longitudinal ligaments help to stabilize these joints.

| FEATURE | ATTACHMENT | COMMENT |
|---|---|---|
| **Zygapophysial (Plane Synovial Joints)** | | |
| Articular capsule | Surrounds facets | Allows gliding motion<br>C5-C6 is most mobile<br>L4-L5 permits most flexion |
| **Intervertebral (Secondary Cartilaginous [Symphyses]) Joints** | | |
| Anterior longitudinal (AL) ligament | Anterior bodies and intervertebral discs | Is strong and prevents hyperextension |
| Posterior longitudinal (PL) ligament | Posterior bodies and intervertebral discs | Is weaker than AL and prevents hyperflexion |
| Ligamenta flava | Connect adjacent laminae of vertebrae | Limit flexion and are more elastic |
| Interspinous ligament | Connect spines | Are weak |
| Supraspinous ligament | Connect spinous tips | Are stronger and limit flexion |
| Ligamentum nuchae | C7 to occipital bone | Is cervical extension of supraspinous ligament and is strong |
| Intertransverse ligament | Connect transverse processes | Are weak ligaments |
| Intervertebral discs | Between adjacent bodies | Are secured by AL and PL ligaments |

**COLOR** the following ligaments of the vertebral arches and bodies (parts ***E*** and ***F***), using a different color for each ligament:

- ☐ 6. **Intervertebral disc**
- ☐ 7. **Anterior longitudinal ligament**
- ☐ 8. **Posterior longitudinal ligament**
- ☐ 9. **Ligamentum flavum (appears yellow because it contains elastic fibers)**
- ☐ 10. **Interspinous ligament**
- ☐ 11. **Supraspinous ligament**
- ☐ 12. **Radiate ligament of the head of a rib**

***Clinical Note:***
**Whiplash** is a nonmedical term for a cervical hyperextension injury (muscular, ligament, and/or bone damage), which is usually associated with a rear-end vehicular accident. The relaxed neck is thrown backward, or hyperextended, as the vehicle accelerates rapidly forward. Rapid recoil of the neck into extreme flexion occurs next. Properly adjusted headrests can significantly reduce the occurrence of hyperextension injury.

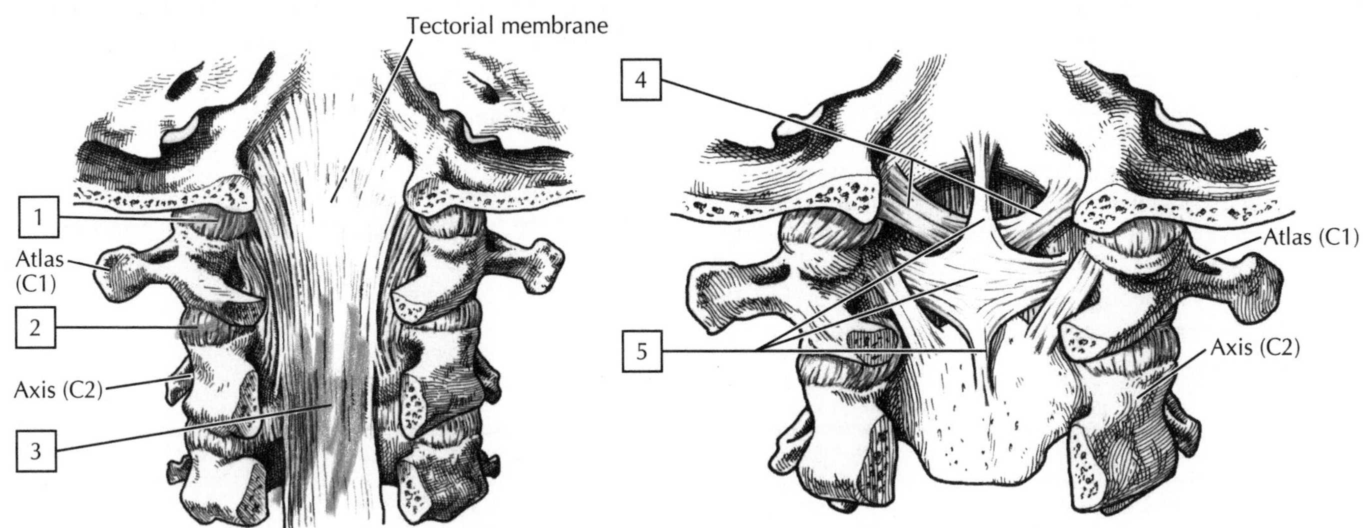


**A.** Upper part of vertebral canal with spinous processes and parts of vertebral arches removed to expose ligaments on posterior vertebral bodies: posterior view

**B.** Principal part of tectorial membrane removed to expose deeper ligament: posterior view

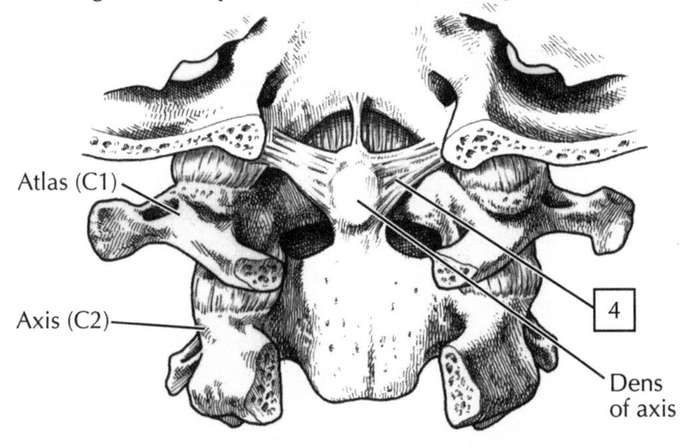


**C.** Cruciate ligament removed to show deepest ligaments: posterior view

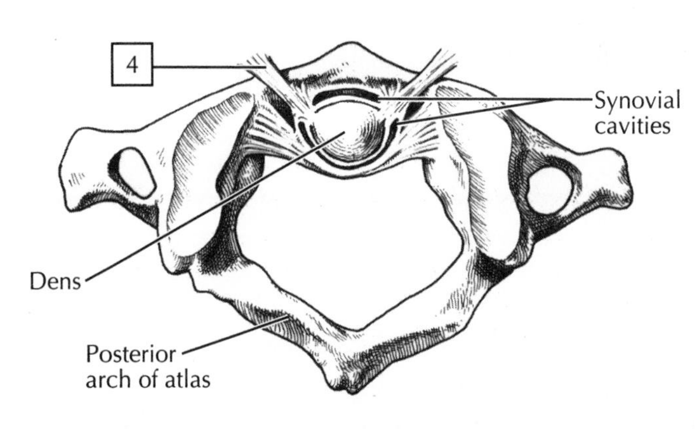


**D.** Median atlantoaxial joint: superior view

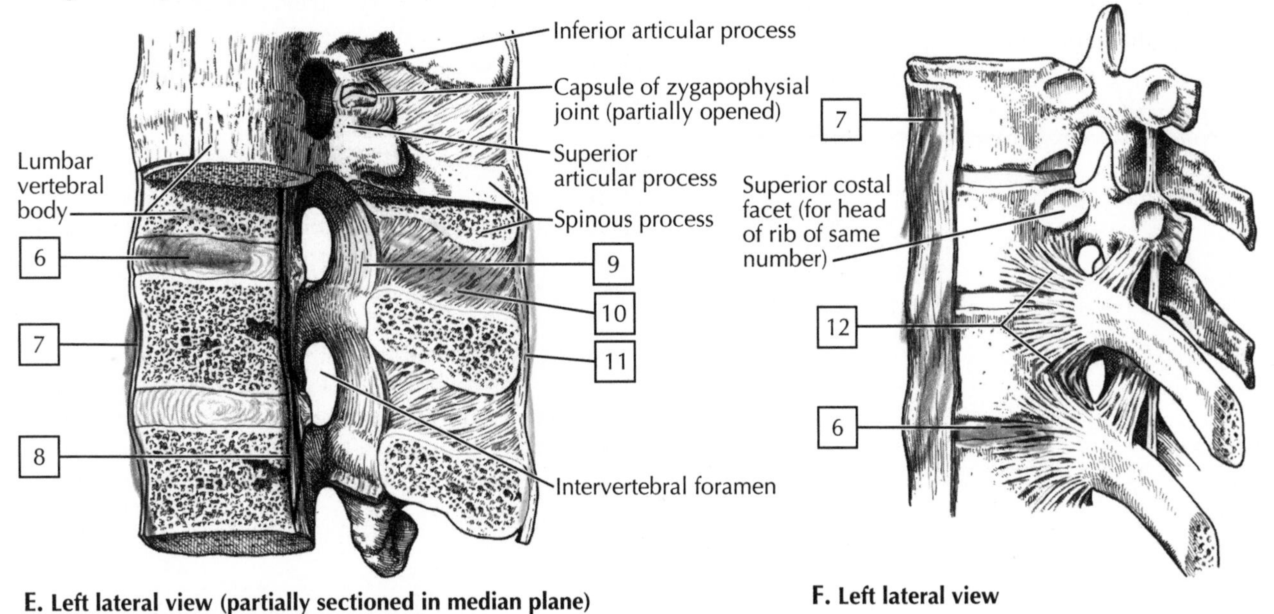


**E.** Left lateral view (partially sectioned in median plane)

**F.** Left lateral view

# 2 Pectoral Girdle and Arm

The **pectoral girdle** is the attachment point of the upper limb to the thoracic wall. The only direct articulation is between the clavicle and the sternum, with the other end of the clavicle articulating with the **scapula** at the acromion. The bone of the arm, called the **humerus,** articulates with the scapula at the glenoid cavity, forming the shoulder or glenohumeral joint. The distal end of the humerus contributes to the elbow joint. Numerous muscles act on the shoulder joint, giving this joint tremendous mobility. The triangular-shaped scapula, for instance, is the site of attachment for 16 different muscles! The features of the clavicle, scapula, and humerus are summarized in the table below.

**COLOR** each of the following bones of the pectoral girdle (part ***A***), using a different color for each bone:

☐ **1. Clavicle**

☐ **2. Scapula**

☐ **3. Humerus**

| CLAVICLE | SCAPULA | HUMERUS |
|---|---|---|
| Cylindrical bone with slight S-shaped curve | Flat triangular bone | Long bone |
| Middle third: narrowest portion | Shallow glenoid cavity | Proximal head: articulates with glenoid cavity of scapula |
| First bone to ossify but last to fuse | Attachment locations for 16 muscles | Distal medial and lateral condyles: articulate at elbow with ulna and radius |
| Formed by intramembranous ossification | Fractures are relatively uncommon | Surgical neck is a common fracture site, which endangers axillary nerve |
| Most commonly fractured bone | | |
| Acts as strut to keep limb away from trunk | | |

**COLOR** each of the following features of the pectoral girdle bones (parts ***B*** and ***C***), using a different color for each feature:

 **4. Coracoid process of the scapula**

 **5. Spine of the scapula**

 **6. Trochlea of the distal humerus: for articulation with the ulna at the elbow**

 **7. Acromial facet of the clavicle: articulates with the scapula at the acromion**

 **8. Sternal facet of the clavicle: articulates with the manubrium of the sternum**

***Clinical Note:***

The **clavicle** is the most commonly fractured bone in the body, especially in children. The fractures usually occur from a fall on an outstretched hand or from direct trauma to the shoulder. Fractures of the clavicle usually occur in the middle third of the bone.

While the clavicle is the first long bone to ossify (by intramembranous ossification) during the second month of embryonic development, it is the last long bone to fully ossify and fuse, which usually occurs between 25 to about 30 years of age.

The acromioclavicular joint (see Plate 2-11) is weak, even though its coracoclavicular ligament is strong, and can be **dislocated (shoulder separation)** by a direct blow, such as might occur in contact sports (football, hockey). It also can be injured by falling on an outstretched arm.

**Fracture of the scapula** usually occurs from severe direct trauma, as may occur in an automobile accident or by direct contact with a high-velocity blunt object. Fortunately, because the scapula is covered by thick muscles, the fracture can usually be treated easily unless it is a displaced or compound (open) fracture (breaks through the skin).

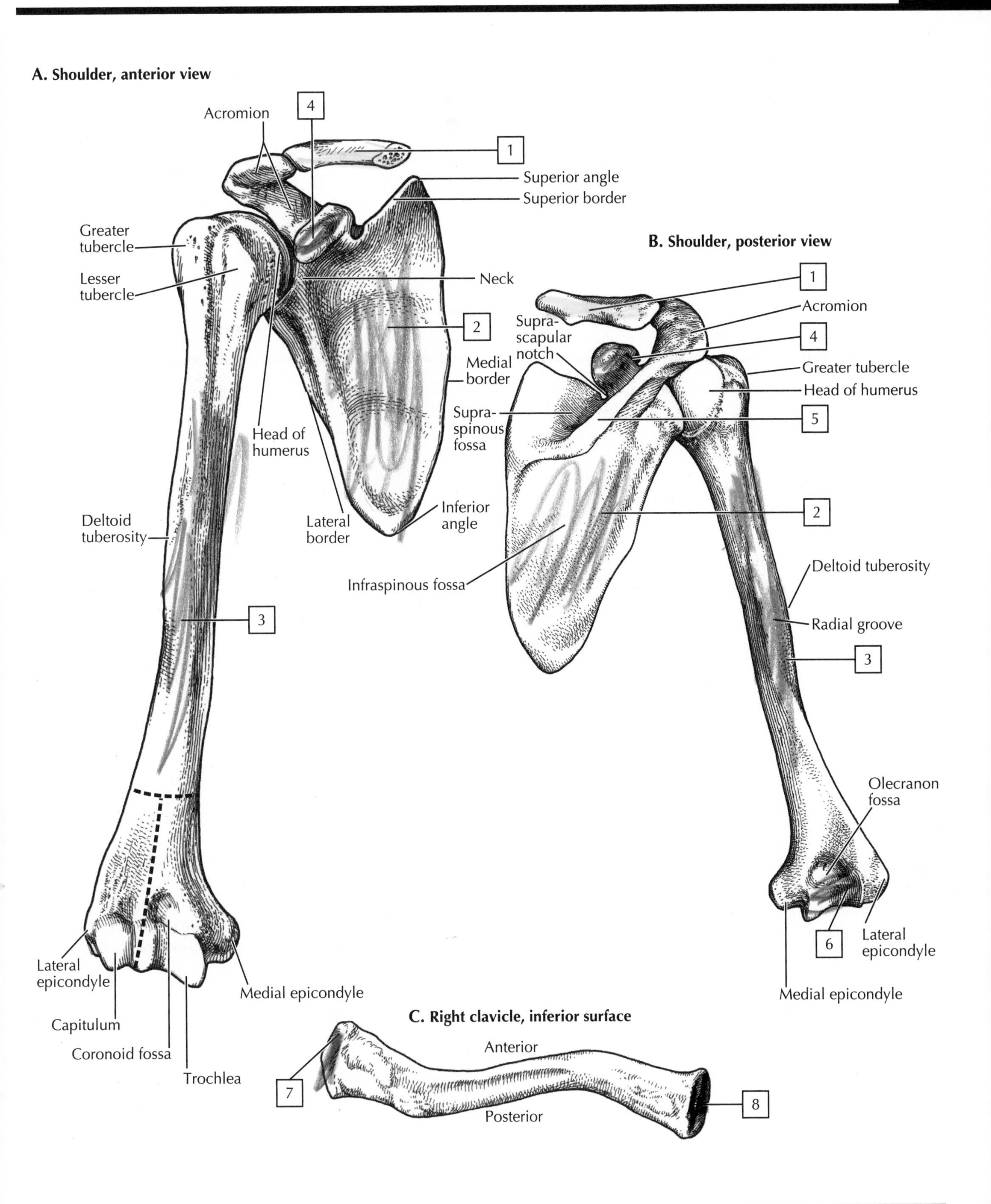
A. Shoulder, anterior view
Acromion
4
1
Superior angle
Superior border
Greater tubercle
Lesser tubercle
Neck
2
Medial border
Head of humerus
Deltoid tuberosity
Lateral border
Inferior angle
3
Lateral epicondyle
Capitulum
Coronoid fossa
Trochlea
Medial epicondyle
B. Shoulder, posterior view
1
Acromion
Supra-scapular notch
4
Greater tubercle
Head of humerus
Supra-spinous fossa
5
2
Infraspinous fossa
Deltoid tuberosity
Radial groove
3
Olecranon fossa
6
Lateral epicondyle
Medial epicondyle
C. Right clavicle, inferior surface
Anterior
7
Posterior
8

# Shoulder Joint

The **shoulder,** or **glenohumeral joint,** is a multiaxial synovial ball-and-socket joint that allows tremendous mobility of the upper limb. Because of the shallow nature of this ball-and-socket joint and its relatively loose capsule, the shoulder joint is one of the most commonly dislocated joints in the body. The **acromioclavicular joint** is a plane synovial joint that permits some gliding movement when the arm is raised and the scapula rotates. The shoulder joint is reinforced by four **rotator cuff muscles,** whose tendons help stabilize the joint (also see Plate 3-17 on rotator cuff muscles):

- Supraspinatus
- Infraspinatus
- Teres minor
- Subscapularis

Bursae help to reduce friction by separating the muscle tendons from the fibrous capsule of the glenohumeral joint. Additionally, although the **glenoid cavity of the scapula** is shallow, a rim of fibrocartilage, called the glenoid labrum ("lip"), lines the peripheral margin of the cavity like a collar and deepens the "socket." Note also that the tendon of the long head of the biceps muscle passes deep to the joint capsule to insert on the supraglenoid tubercle of the scapula. Features of the shoulder joint ligaments and bursae are summarized in the table below.

| LIGAMENT OR BURSA | ATTACHMENT | COMMENT |
|---|---|---|
| | **Acromioclavicular (Synovial Plane) Joint** | |
| Capsule and articular disc | Surrounds joint | Allows gliding movement as arm is raised and scapula rotates |
| Acromioclavicular | Acromion to clavicle | Supports the joint superiorly |
| Coracoclavicular (conoid and trapezoid ligaments) | Coracoid process to clavicle | Reinforces the joint by stabilizing the clavicle |
| | **Glenohumeral (Multiaxial Synovial Ball-and-Socket) Joint** | |
| Fibrous capsule | Surrounds joint | Permits flexion, extension, abduction, adduction, protraction, circumduction; most frequently dislocated joint |
| Coracohumeral | Coracoid process to greater tubercle of humerus | Strengthens the capsule superiorly |
| Glenohumeral | Supraglenoid tubercle to lesser tubercle of humerus | Composed of superior, middle, and inferior thickenings |
| Transverse humeral | Spans greater and lesser tubercles of humerus | Holds long head of biceps tendon in intertubercular groove |
| Glenoid labrum | Margin of glenoid cavity of scapula | Is fibrocartilaginous ligament that deepens glenoid cavity |
| | **Bursae** | |
| Subacromial | | Between coracoacromial arch and suprascapular muscle |
| Subdeltoid | | Between deltoid muscle and capsule |
| Subscapular | | Between subscapularis tendon and scapular neck |

**COLOR** the following ligaments, tendons, and the bursae labeled in ***C*** and ***D*** (color these blue) associated with the shoulder joint, using a different color for each:

☐ **1. Supraspinatus tendon**
☐ **2. Subscapularis tendon**
☐ **3. Biceps brachii tendon (long head)**
☐ **4. Capsular ligaments of the shoulder**
☐ **5. Infraspinatus tendon**
☐ **6. Teres minor tendon**

***Clinical Note:***

Movement at the shoulder joint, or almost any joint, can lead to **inflammation of the tendons** surrounding that joint and secondary inflammation of the bursa that cushions the joint from the overlying muscle or tendon. At the shoulder, the supraspinatus muscle tendon is especially vulnerable because it can become pinched by the greater tubercle of the humerus, the acromion, and the coracoacromial ligament.

About 95% of **shoulder joint dislocations** occur in an anterior-inferior direction. Often this can happen with a throwing motion, which places stress on the capsule and anterior elements of the rotator cuff (especially the subscapularis tendon).

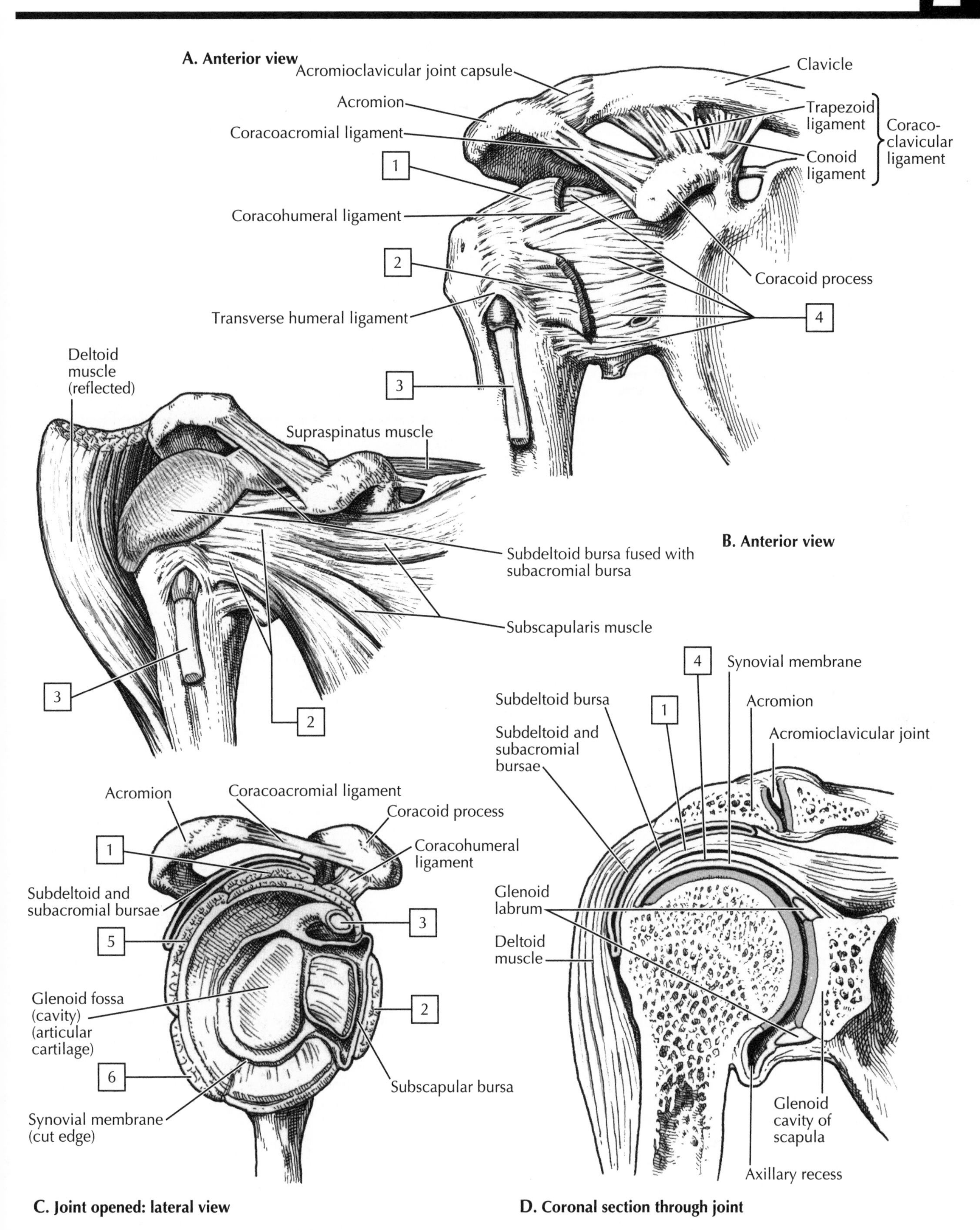
A. Anterior view
Acromioclavicular joint capsule
Clavicle
Acromion
Trapezoid ligament
Coracoacromial ligament
Coraco-clavicular ligament
Conoid ligament
1
Coracohumeral ligament
2
Coracoid process
Transverse humeral ligament
4
3
Deltoid muscle (reflected)
Supraspinatus muscle
Subdeltoid bursa fused with subacromial bursa
B. Anterior view
Subscapularis muscle
3
2
4
Synovial membrane
Subdeltoid bursa
1
Acromion
Subdeltoid and subacromial bursae
Acromioclavicular joint
Acromion
Coracoacromial ligament
Coracoid process
1
Coracohumeral ligament
Subdeltoid and subacromial bursae
Glenoid labrum
3
5
Deltoid muscle
Glenoid fossa (cavity) (articular cartilage)
2
6
Subscapular bursa
Synovial membrane (cut edge)
Glenoid cavity of scapula
Axillary recess
C. Joint opened: lateral view
D. Coronal section through joint

# 2 Forearm and Elbow Joint

The forearm extends from the elbow proximally to the wrist distally and is composed of two bones, the **radius** laterally and the **ulna** medially. The radius is the shorter of the two bones. The region just anterior to the elbow is known as the **cubital fossa** (a cubit is an ancient term for linear measurement and was the length from the elbow to the tip of the middle finger; whoever was king decided how long the cubit was, so it varied from ruler to ruler) and is a common site for venipuncture (access to a vein to withdraw blood or administer fluids).

An interosseous membrane connects the radius and ulna and is a type of fibrous joint. The movements of **supination** (palm facing forward [anteriorly] in anatomical position) and **pronation** (palm facing backward [posteriorly] in anatomical position) are unique movements of the wrist and hand but occur exclusively in the forearm with the radius crossing over the ulna (pronation) or back alongside the ulna (supination) (see parts ***A*** and ***B***).

**COLOR** each bone and note each bone's labeled features:

- ☐ **1. Radius**
- ☐ **2. Ulna**

The elbow joint actually contains several joints, and its ligaments and features are summarized in the table below:

- **Humeroulnar:** for flexion and extension; the ulnar trochlear notch articulates with the trochlea of the humerus
- **Humeroradial:** for flexion and extension; the head of the radius articulates with the capitulum of the humerus
- **Proximal radioulnar:** for supination and pronation; the radial head articulates with the radial notch of the ulna

| LIGAMENT | ATTACHMENT | COMMENT |
|---|---|---|
| | **Humeroulnar Joint (Uniaxial Synovial Hinge [Ginglymus] Joint)** | |
| Capsule | Surrounds joint | Provides flexion and extension |
| Ulnar (medial) collateral ligament | Medial epicondyle of humerus to coronoid process and olecranon of ulna | Is triangular ligament with anterior, posterior, and oblique bands |
| | **Humeroradial Joint** | |
| Capsule | Surrounds joint | Capitulum of humerus to head of radius |
| Radial (lateral) collateral ligament | Lateral epicondyle of humerus to radial notch of ulna and annular ligament | Is weaker than ulnar collateral ligament but provides posterolateral stability |
| | **Proximal Radioulnar (Uniaxial Synovial Pivot) Joint (see Plate 2-14 for distal radioulnar joint)** | |
| Annular ligament | Surrounds radial head and radial notch of ulna | Keeps radial head in radial notch; allows pronation and supination |

**COLOR** the following key ligaments of the elbow joint (parts ***D*** to ***F***), using a different color for each ligament:

☐ **3. Radial collateral ligament: on the lateral side of the elbow**

☐ **4. Annular ligament: surrounds the radial head in the proximal radioulnar articulation**

☐ **5. Ulnar collateral ligament: on the medial side of the elbow**

***Clinical Note:***

**Elbow dislocations** are third in frequency after shoulder and finger dislocations. Dislocation often occurs from a fall on an outstretched hand. A dislocation in the posterior direction is the most common type. Anterior dislocations are rare and may lacerate the brachial artery. Lateral dislocations are uncommon and medial dislocations rare. With elbow dislocations, one must determine if the ulnar nerve (most common) and/or median nerve have been injured.

**Distal fractures** of the ulnar shaft often occur from a direct blow to or forced pronation of the forearm. Fractures of the distal radius account for about 80% of forearm fractures in all age groups and often occur from a fall on an outstretched hand (a Colles' fracture).

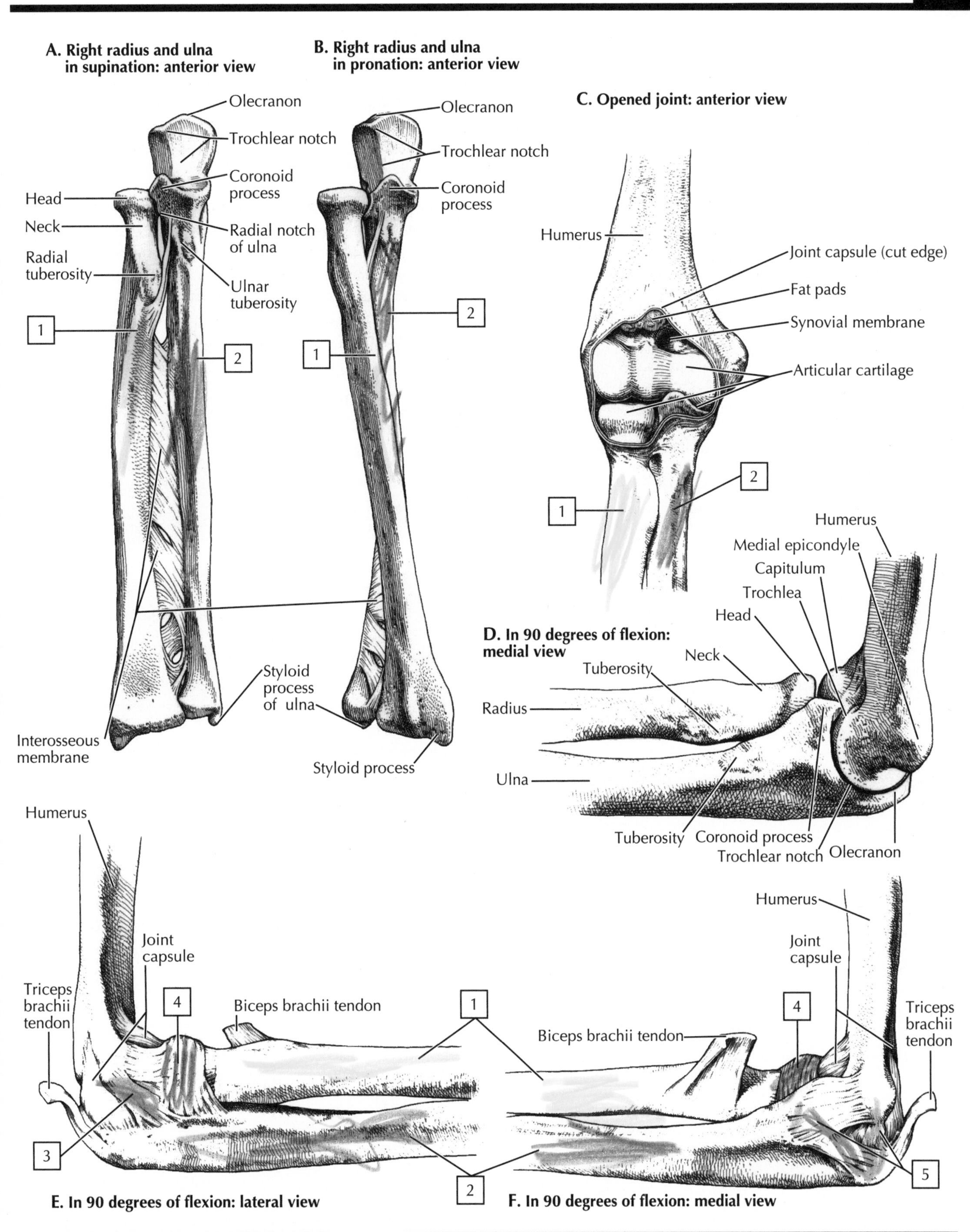
A. Right radius and ulna in supination: anterior view
Olecranon
Trochlear notch
Coronoid process
Head
Neck
Radial notch of ulna
Radial tuberosity
Ulnar tuberosity
1
2
Styloid process of ulna
Interosseous membrane
B. Right radius and ulna in pronation: anterior view
Olecranon
Trochlear notch
Coronoid process
2
1
Styloid process
C. Opened joint: anterior view
Humerus
Joint capsule (cut edge)
Fat pads
Synovial membrane
Articular cartilage
2
1
D. In 90 degrees of flexion: medial view
Humerus
Medial epicondyle
Capitulum
Trochlea
Head
Neck
Tuberosity
Radius
Ulna
Tuberosity
Coronoid process
Trochlear notch
Olecranon
Humerus
Joint capsule
Triceps brachii tendon
4
Biceps brachii tendon
1
3
2
Humerus
Joint capsule
Biceps brachii tendon
4
Triceps brachii tendon
5
E. In 90 degrees of flexion: lateral view
F. In 90 degrees of flexion: medial view

# 2 Wrist and Hand

The wrist and hand are composed of the following 29 bones:

- 8 carpal (wrist) bones, arranged in proximal and distal rows of 4 bones each
- 5 metacarpals, which span the palm of the hand
- 14 phalanges, 2 for the thumb (1st digit) and 3 each for the remaining 4 digits
- 2 sesamoid bones, situated at the distal end of the thumb metacarpal

These bones and their characteristics are summarized in the table below.

| BONE | CHARACTERISTICS |
|---|---|
| **Proximal Row of Carpals** | |
| Scaphoid (boat shaped) | Lies beneath anatomical snuffbox; is most commonly fractured carpal |
| Lunate (moon or crescent shaped) | Broader anteriorly than posteriorly |
| Triquetrum (triangular) | All three bones (scaphoid, lunate, triquetrum) articulate with distal radius |
| Pisiform (pea shaped) | Lies on the palmar surface of the triquetrum |
| **Distal Row of Carpals** | |
| Trapezium (four sided) | Distal row articulates with proximal row of carpals and with metacarpals 1-5; lies on lateral side of carpus |
| Trapezoid | A wedge-shaped bone lying between trapezium and capitate |
| Capitate (round bone) | Largest of the carpal bones |
| Hamate (hooked bone) | Has a hooked process extending anteriorly |
| **Metacarpals** | |
| Numbered 1-5 (thumb to little finger) | Possess a base, shaft, and head |
| | Are triangular in cross section |
| | Fifth metacarpal most commonly fractured |
| Two sesamoid bones | Are associated with head of first metacarpal |
| **Phalanges** | |
| Three for each digit except thumb | Possess a base, shaft, and head |
| | Termed proximal, middle, and distal |
| | Distal phalanx of middle finger commonly fractured |

The carpal bones are not aligned in a flat plane but form an arch, the **carpal arch,** with its concave aspect facing anteriorly. Tendons from forearm muscles, vessels, and nerves pass through or across this arch to gain access to the hand. A tight band of connective tissue, the flexor retinaculum, spans the carpal arch, forming a "carpal tunnel" for the structures passing through this archway.

**COLOR** the following bones of the wrist and hand, using different colors for each carpal bone, a uniform color for the metacarpals, another uniform color for all the phalanges of the digits, and a new color for the sesamoid bones:

- ☐ **1. Scaphoid: some clinicians refer to this bone as the navicular bone ("little ship")**
- ☐ **2. Trapezium**
- ☐ **3. Trapezoid**
- ☐ **4. Lunate**
- ☐ **5. Triquetrum**
- ☐ **6. Pisiform**
- ☐ **7. Hamate**
- ☐ **8. Capitate**
- ☐ **9. Metacarpals**
- ☐ **10. Phalanges of each digit**
- ☐ **11. Sesamoid bones (two at the distal end of the thumb metacarpal)**

***Clinical Note:***

**Fractures** of the scaphoid bone are the most frequent fractures of the carpal bones. This may occur by falling on an extended wrist. Fracture of the middle third of the bone is most common. Loss of the blood supply can lead to nonunion or **avascular osteonecrosis.**

**Finger injuries** are common. It is important to determine if the muscle flexor and extensor tendons or ligaments are disrupted (see Plate 3-23). Fractures of the metacarpal neck commonly result from an end-on blow (punch).

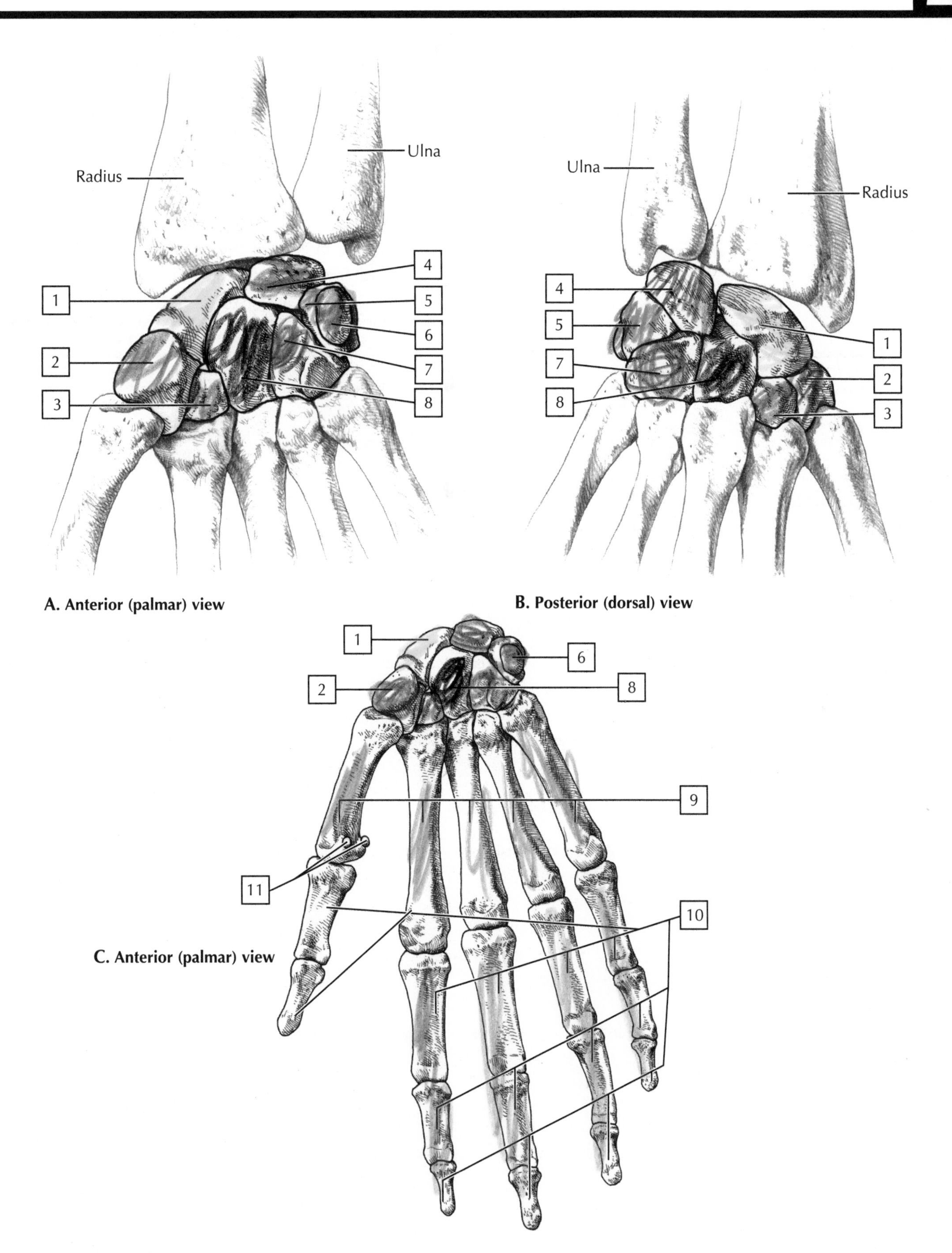


A. Anterior (palmar) view

B. Posterior (dorsal) view

C. Anterior (palmar) view

# Wrist and Finger Joints and Movements

The classification and ligaments of the wrist and finger joints are summarized in the following table. The wrist joint is a radiocarpal (biaxial synovial ellipsoid) joint between the distal radius of the forearm and the scaphoid, lunate, and triquetrum carpals, and the articular disc at the distal ulna. On the facing page, note the finger movements associated with these joints.

**COLOR** the following major ligaments, using a different color for each ligament:

- ☐ **1. Palmar radiocarpal ligaments**
- ☐ **2. Dorsal radiocarpal ligament**
- ☐ **3. Articular disc of the wrist joint**
- ☐ **4. Capsule of a metacarpophalangeal joint**
- ☐ **5. Capsule of a proximal interphalangeal joint**
- ☐ **6. Capsule of a distal interphalangeal joint**
- ☐ **7. Collateral ligament of a metacarpophalangeal joint**
- ☐ **8. Palmar ligament (plate)**

| FEATURE | ATTACHMENT | COMMENT |
|---|---|---|
| | **Radiocarpal (Biaxial Synovial Ellipsoid) Joint** | |
| Capsule and disc | Surrounds joint; radius to scaphoid, lunate, and triquetrum | Provides little support; allows flexion, extension, abduction, adduction, circumduction |
| Palmar (volar) radiocarpal ligaments | Radius to scaphoid, lunate, and triquetrum | Are strong and stabilizing |
| Dorsal radiocarpal ligament | Radius to scaphoid, lunate, and triquetrum | Is weaker ligament |
| Radial collateral ligament | Radius to scaphoid and triquetrum | Stabilizes proximal row of carpals |
| | **Distal Radioulnar (Uniaxial Synovial Pivot) Joint** | |
| Capsule | Surrounds joint; ulnar head to ulnar notch of radius | Is thin superiorly; allows pronation, supination |
| Palmar and dorsal radioulnar ligaments | Extends transversely between the two bones | Articular disc binds bones together |
| | **Intercarpal (Synovial Plane) Joints** | |
| Proximal row of carpal ligaments | Adjacent carpals | Permits gliding and sliding movements |
| Distal row of carpal ligaments | Adjacent carpals | Are united by anterior, posterior, and interosseous ligaments |
| | **Midcarpal (Synovial Plane) Joints** | |
| Palmar (volar) intercarpal ligaments | Proximal and distal rows of carpals | Is the location for one-third of wrist extension and two-thirds of flexion; permits gliding and sliding movements |
| Carpal collateral ligaments | Scaphoid, lunate, and triquetrum to capitate and hamate | Stabilize distal row (ellipsoid synovial joint) |
| | **Carpometacarpal (CMC) (Plane Synovial) Joints (Except Thumb)** | |
| Capsule | Carpals to metacarpals of digits 2-5 | Surrounds joints; allows some gliding movement |
| Palmar and dorsal CMC ligaments | Carpals to metacarpals of digits 2-5 | Dorsal ligament strongest |
| Interosseous CMC ligaments | Carpals to metacarpals of digits 2-5 | |
| | **Thumb (Biaxial Saddle) Joint** | |
| Same ligaments as CMC | Trapezium to first metacarpal | Allows flexion, extension, abduction, adduction, circumduction |
| | | Is common site for arthritis |
| | **Metacarpophalangeal (Biaxial Condyloid Synovial) Joint** | |
| Capsule | Metacarpal to proximal phalanx | Surrounds joint; allows flexion, extension, abduction, adduction, circumduction |
| Radial and ulnar collateral ligaments | Metacarpal to proximal phalanx | Are tight in flexion and loose in extension |
| Palmar (volar) plate | Metacarpal to proximal phalanx | If broken digit, cast in flexion or ligament will shorten during healing |
| | **Interphalangeal (Uniaxial Synovial Hinge) Joints** | |
| Capsule | Adjacent phalanges | Surrounds joints; allows flexion and extension |
| Two collateral ligaments | Adjacent phalanges | Are oriented obliquely |
| Palmar (volar) plate | Adjacent phalanges | Prevents hyperextension |

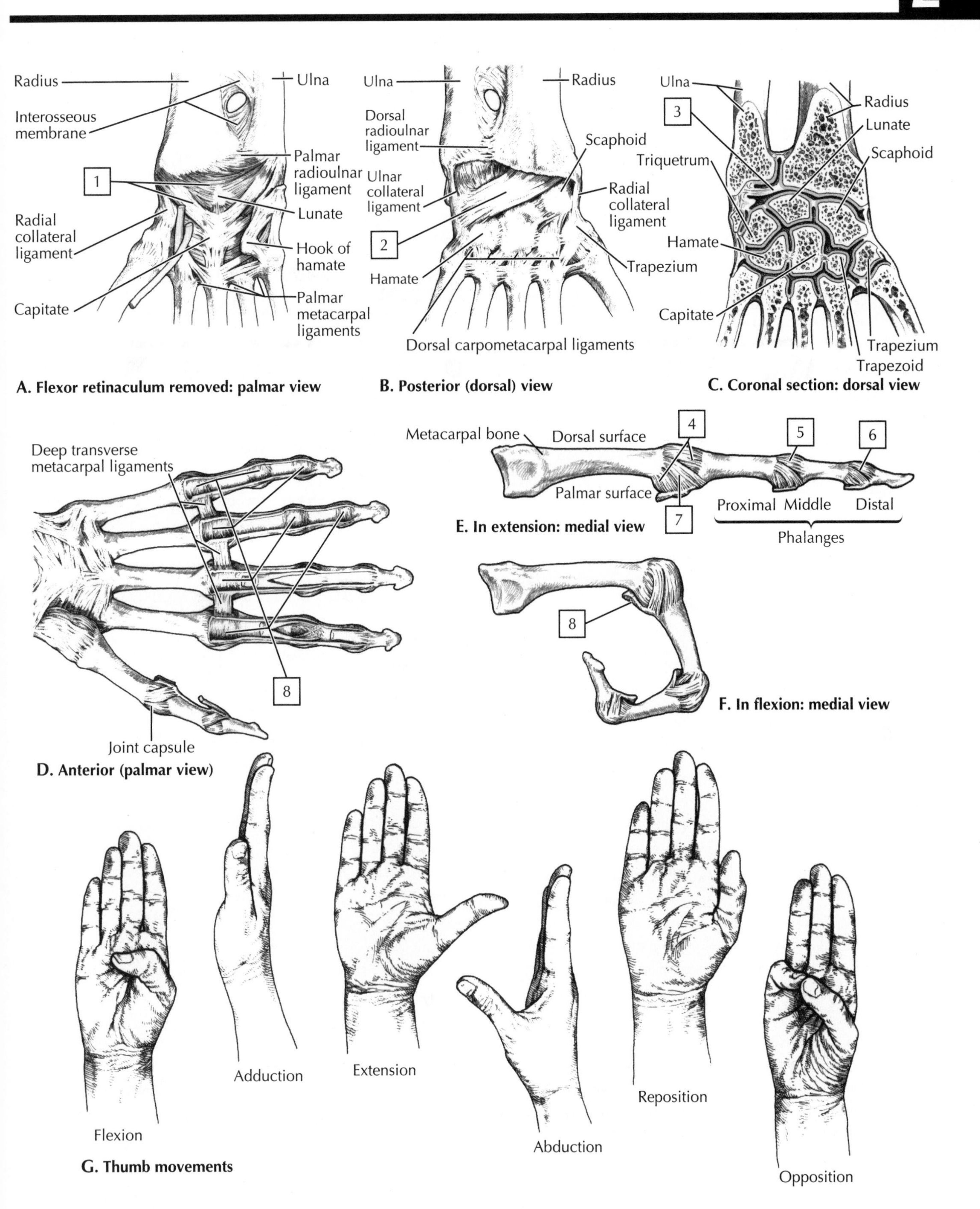


A. Flexor retinaculum removed: palmar view

B. Posterior (dorsal) view

C. Coronal section: dorsal view

D. Anterior (palmar view)

E. In extension: medial view

F. In flexion: medial view

G. Thumb movements

# 2 Pelvic Girdle

The **pelvic girdle** is the point where the lower limb attaches to the trunk of the body. The bony pelvis includes the:

- **Pelvic bone:** a fusion of three separate bones called the **ilium, ischium,** and **pubis;** the three bones join each other in the acetabulum (a cup-shaped feature for articulation with the head of the femur, our thigh bone); the two pelvic bones (right and left) articulate with the sacrum posteriorly and at the pubic symphysis anteriorly
- **Sacrum:** a fusion of five sacral vertebrae of the spine
- **Coccyx:** the terminal end of the spine and a remnant of our embryonic tail

The three pelvic bones fuse into a single bone during late adolescence. Gender differences exist in the structure of the pelvis; in women adaptations have been made for childbirth. The female pelvis has wider iliac crests, the pelvic cavity is wider and shallower, and it has a broader pubic arch than the male pelvis. The pelvis articulates with the sacrum at the sacroiliac (plane synovial) joint, which is reinforced by strong ligaments that provide stability and support. The joints and ligaments of the pelvic girdle are summarized in the table below.

**COLOR** the pelvic girdle, using a different color for each of the following bones (parts ***A*** and ***B***):

- ☐ **1. Ischium**
- ☐ **2. Ilium**
- ☐ **3. Pubis**

**COLOR** the following key ligaments and cartilages of the pelvic articulations (parts ***C*** and ***D***), using a different color for each structure:

- ☐ **4. Posterior sacroiliac ligaments**
- ☐ **5. Sacrospinous ligament: divides the sciatic notch into the greater and lesser sciatic foramina**
- ☐ **6. Sacrotuberous ligament**
- ☐ **7. Anterior sacroiliac ligaments**
- ☐ **8. Pubic symphysis: fibrocartilage that permits some expansion during childbirth**

| BONE | CHARACTERISTICS |
|---|---|
| | **Coxal (Hip) Bone** |
| | Fusion of three bones on each side to form the pelvis, which articulates with the sacrum to form the pelvic girdle |
| Ilium | Body fused to ischium and pubis, all meeting in the acetabulum (socket for articulation with femoral head) |
| | Ala (wing): weak spot of ilium |
| Ischium | Body fused with other two bones; ramus fused with pubis |
| Pubis | Body fused with other two bones; ramus fused with ischium |
| | **Femur (Proximal)** |
| Long bone | Longest bone in the body and very strong |
| Head | Point of articulation with acetabulum of coxal bone |
| Neck | Common fracture site |
| Greater trochanter | Point of the hip; attachment site for several gluteal muscles |
| Lesser trochanter | Attachment site of iliopsoas tendon (strong hip flexor) |

***Clinical Note:***
**Pelvic fractures** may be high or low impact; high-impact fractures (from falls or automobile crashes) often involve significant bleeding and may be life threatening.

| LIGAMENT | ATTACHMENT | COMMENT |
|---|---|---|
| | **Lumbosacral Joint*** | |
| Intervertebral disc | Between L5 and sacrum | Allows little movement |
| Iliolumbar | Transverse process of L5 to crest of ilium | Can be involved in avulsion fracture |
| | **Sacroiliac (Plane Synovial) Joint** | |
| Sacroiliac | Sacrum to ilium | Allows little movement; consists of posterior (strong), anterior (provides rotational stability), and interosseous (strongest) ligaments |
| | **Sacrococcygeal Joint (Symphysis)** | |
| Sacrococcygeal | Between coccyx and sacrum | Allows some movement; consists of anterior, posterior, and lateral ligaments; contains an intervertebral disc between S5 and C1 |
| | **Pubic Symphysis** | |
| Pubic | Between pubic bones | Allows some movement, fibrocartilage disc |
| | **Accessory Ligaments** | |
| Sacrotuberous | Iliac spines and sacrum to ischial tuberosity | Provides vertical stability |
| Sacrospinous | Ischial spine to sacrum and coccyx | Divides sciatic notch into greater and lesser sciatic foramina |

*Other ligaments include those binding any two vertebrae and facet joints.

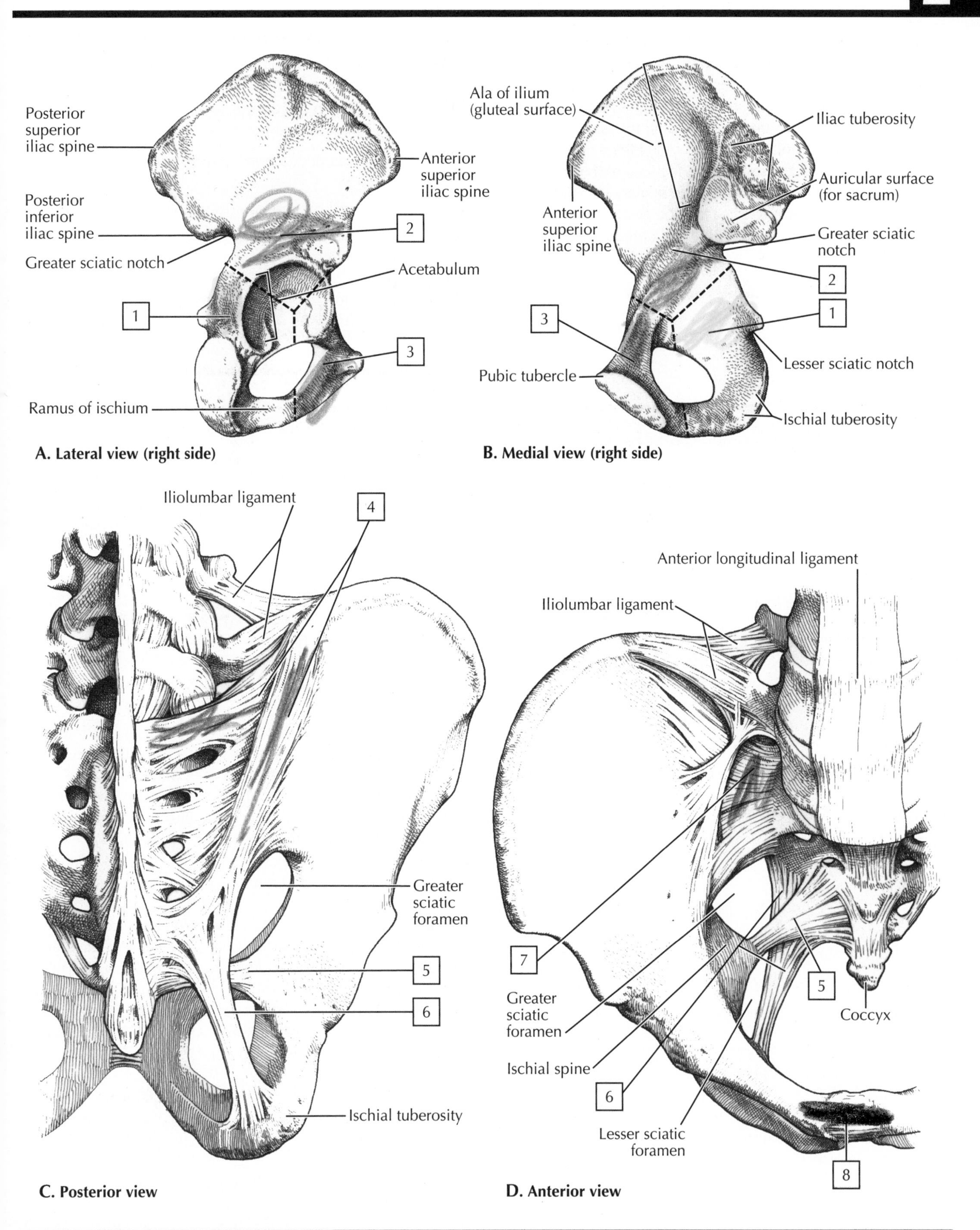


A. Lateral view (right side)

B. Medial view (right side)

C. Posterior view

D. Anterior view

# Hip Joint

The hip joint is a multiaxial synovial ball-and-socket joint between the head of the femur and the acetabulum of the pelvic bone. Unlike the ball-and-socket shoulder joint, the hip joint is designed for stability and support at the expense of some mobility. Similar to the shoulder joint, the acetabulum is rimmed by a fibrocartilaginous "lip" called the **acetabular labrum** that deepens the socket. The features of the hip joint are summarized in the table below. The primary hip joint ligaments include three major ligaments that surround the hip joint and one internal ligament to the head of the femur.

**COLOR** the following ligaments and features of the hip joint, using a different color for each ligament or feature:

- ☐ **1. Iliofemoral ligament (Y ligament of Bigelow): positioned anteriorly**
- ☐ **2. Pubofemoral ligament: positioned anteriorly and inferiorly**
- ☐ **3. Ischiofemoral ligament: positioned posteriorly**
- ☐ **4. Acetabular labrum: fibrocartilage around the rim of the socket**
- ☐ **5. Articular cartilage on the head of the femur**
- ☐ **6. Ligament of the head of the femur: attaches to the acetabular notch and transverse acetabular ligament**

| LIGAMENT | ATTACHMENT | COMMENT |
|---|---|---|
| Capsular | Acetabular margin to femoral neck | Encloses femoral head and part of neck; acts in flexion, extension, abduction, adduction, circumduction |
| Iliofemoral | Iliac spine and acetabulum to intertrochanteric line | Is strongest ligament; forms inverted Y (of Bigelow); limits hyperextension and lateral rotation |
| Ischiofemoral | Acetabulum to femoral neck posteriorly | Limits extension and medial rotation; is weaker ligament |
| Pubofemoral | Pubic ramus to lower femoral neck | Limits extension and abduction |
| Labrum | Acetabulum | Fibrocartilage, deepens socket |
| Transverse acetabular | Acetabular notch interiorly | Cups acetabulum to form a socket for femoral head |
| Ligament of head of femur | Acetabular notch and transverse ligament to femoral head | Artery to femoral head runs in ligament |

***Clinical Note:***

**Hip fractures** are common injuries. In the young, the fracture often results from trauma, whereas in the elderly the cause is often related to osteoporosis and associated with a fall. The neck of the femur is a common site for such fractures. In the United States, about 10 in 1000 infants are born with **developmental dislocation of the hip**. With early diagnosis and treatment, about 96% of affected children have normal hip function. Girls are affected more often than boys.

A "**hip pointer**" injury refers to a contusion (capillary bleeding which infiltrates the tendons, muscles, and surrounding soft tissues) and is a common injury in contact sports such as football and hockey. The term also may be used incorrectly to describe an avulsion of a muscle such as the sartorius to the anterior superior iliac spine; however, this is really an **avulsion fracture** of the hip and not a hip pointer.

A "**charley horse**" is a term given to an acute muscle cramp in the thigh and may result from leg cramps, ischemia, or rupture of blood vessels.

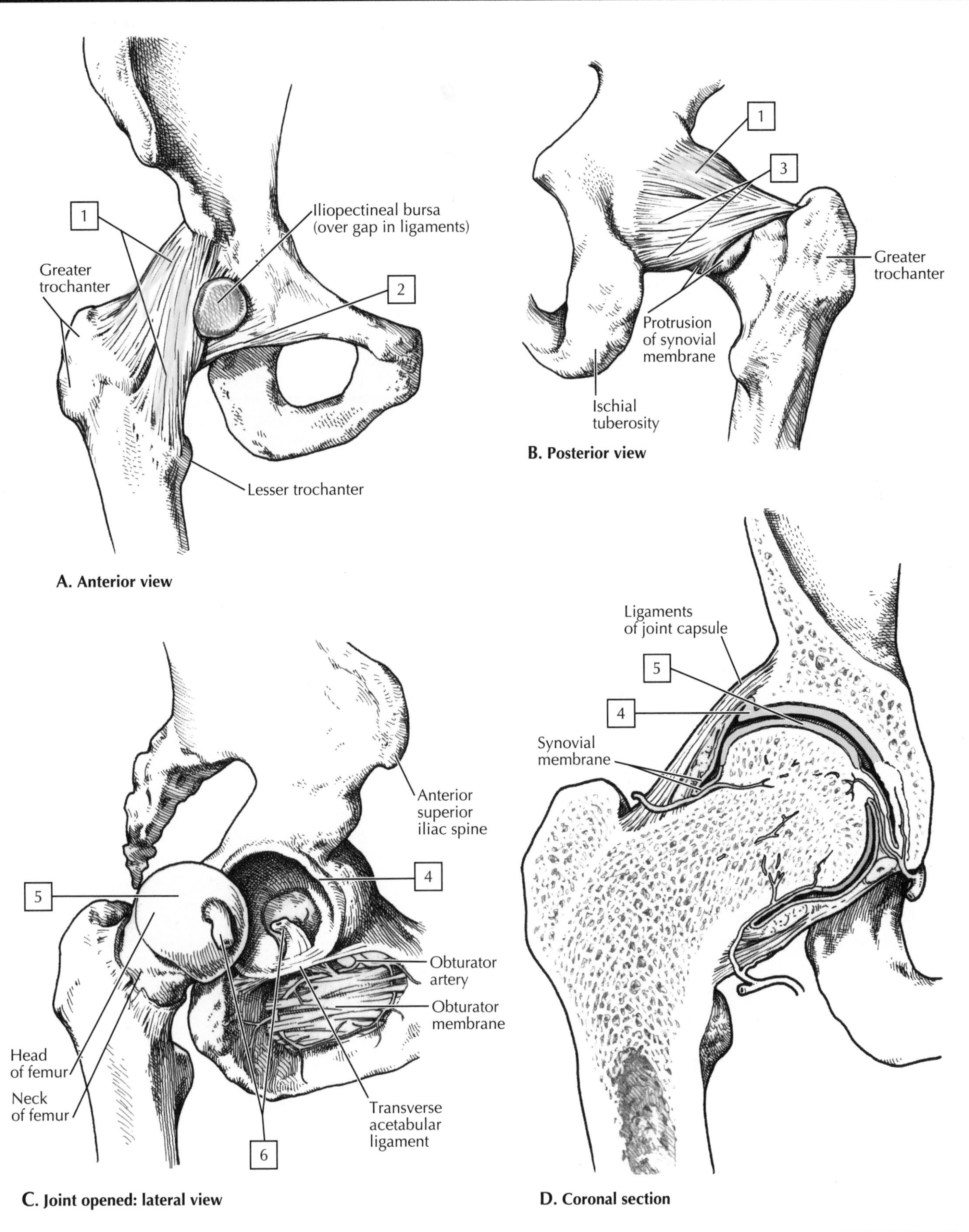


A. Anterior view

B. Posterior view

C. Joint opened: lateral view

D. Coronal section

# 2 Thigh and Leg Bones

The **femur** is the bone of the thigh (anatomically, the thigh is the region between the hip and knee and the leg is the region between the knee and ankle). The femur is the longest bone in the body and transmits the weight of the body from the knee to the pelvis. The major features of the femur are summarized in the table below.

The bones of the leg are the **tibia** and **fibula**. The tibia is the larger of the two leg bones and is medially placed in the leg. Its shaft can be palpated just beneath the skin from the base of the knee to the ankle joint. The articulation of the distal femur and proximal tibia forms the knee joint, and a large sesamoid bone called the **patella** lies anterior to this joint and is embedded in the tendon of the quadriceps femoris muscle. The fibula is not a weight-bearing bone, is found laterally in the leg, and is primarily a bone for muscle attachment. Features of the tibia and fibula are summarized in the table below.

| FEATURE | CHARACTERISTICS |
|---|---|
| | **Femur** |
| Long bone | Longest bone in the body; very strong |
| Head | Point of articulation with acetabulum of coxal bone |
| Neck | Common fracture site |
| Greater trochanter | Point of hip; attachment site for several gluteal muscles |
| Lesser trochanter | Attachment site of iliopsoas tendon (strong hip flexor) |
| Distal condyles | Medial and lateral (smaller) sites that articulate with tibial condyles |
| | **Patella** |
| | Sesamoid bone (largest one in the body) embedded in quadriceps femoris muscle tendon |
| | **Tibia** |
| Long bone | Large, weight-bearing bone |
| Proximal facets | Large plateau for articulation with femoral condyles |
| Tibial tuberosity | Insertion site for patellar ligament |
| Inferior articular surface | Surface for cupping talus at the ankle joint |
| Medial malleolus | Prominence on medial aspect of ankle |
| | **Fibula** |
| Long bone | Slender bone, primarily for muscle attachment |
| Neck | Possible damage to common fibular nerve if fracture occurs here |

**COLOR** the following bones of the thigh and leg, using a different color for each bone:

- ☐ 1. **Femur**
- ☐ 2. **Patella**
- ☐ 3. **Tibia**
- ☐ 4. **Fibula**

***Clinical Note:***

Most **fractures** of the femur occur across the neck of the femur within the articular capsule. **Tibial fractures** occur most frequently where the tibial shaft is narrowest, which is about one-third of the way down the shaft. Because the tibia is largely subcutaneous along its medial border, many of these fractures are open injuries. Tibial fractures may be transverse fractures, spiral fractures, comminuted fractures (multiple small fractures at the site of injury), or segmental fractures, e.g., at two sites, one-third and two-thirds down the length of the tibia.

**Fibular fractures** are most common just proximal to the lateral malleolus, just above the ankle joint on the lateral side.

**Multiple myeloma** is a tumor of plasma cells and is the most malignant type of primary bone tumor. This painful tumor is sensitive to radiation therapy. Additionally, newer chemotherapeutic agents and bone marrow transplantation offer hope for improved survival. This tumor may affect many of the bones in the body; in the lower limb, it is more common in the proximal and distal ends of the femur and tibia.

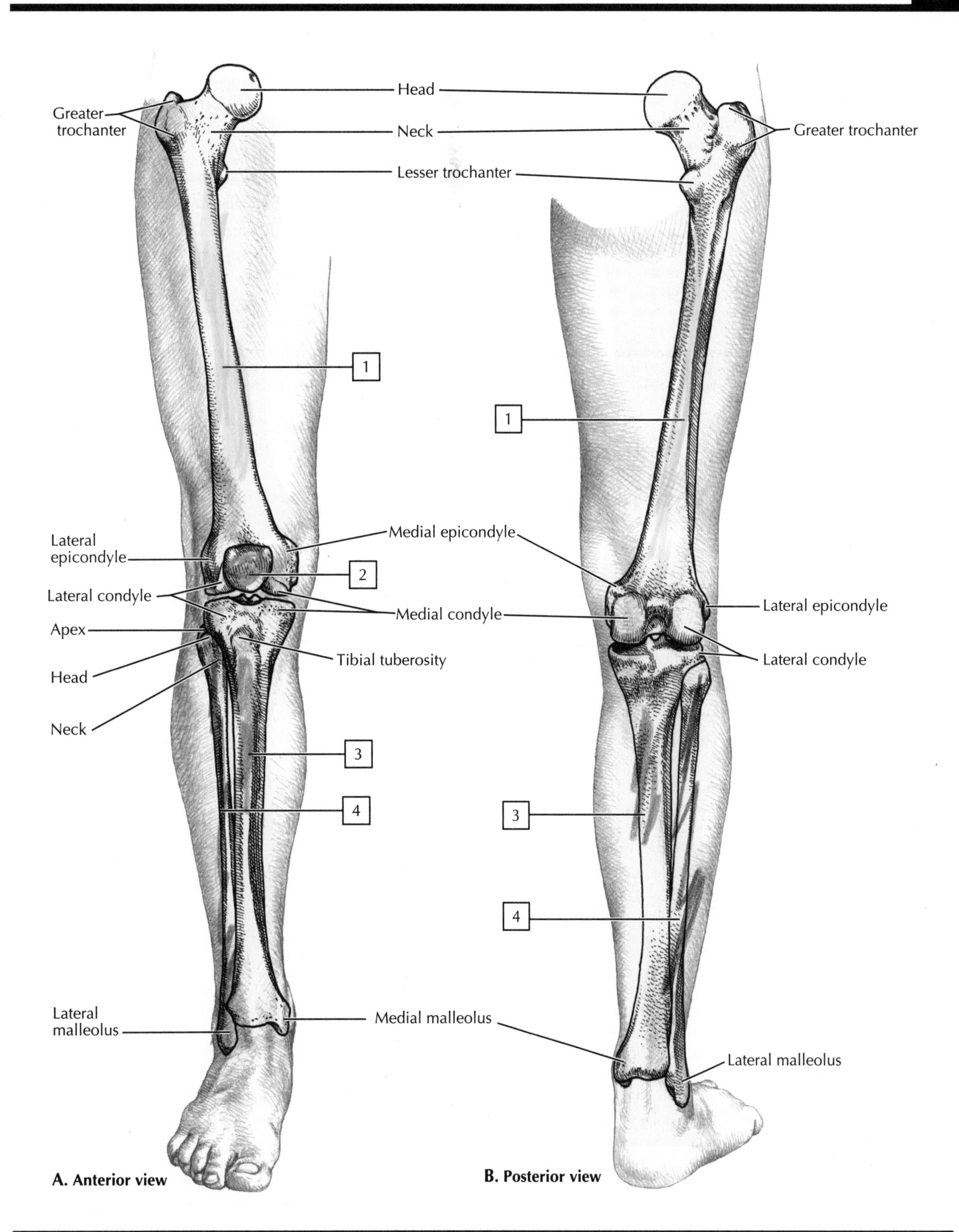


A. Anterior view

B. Posterior view

# 2 Knee Joint

The knee is a biaxial condylar synovial joint and is the most sophisticated joint in the body. It participates in flexion, extension, and some gliding and medial rotation when it is flexed. When the leg is in full extension, the femur rotates medially on the tibia, and the ligaments tighten to "lock" the knee. This patellofemoral biaxial synovial saddle joint accounts for part of the extension mechanisms of the knee by the quadriceps muscles. Features of the knee joint are summarized in the table below. Only the major ligaments are shown in the illustrations.

**COLOR** the following extracapsular and intracapsular ligaments of the knee joint, using a different color for each ligament:

- ☐ 1. **Medial meniscus: fibrocartilaginous disc on the tibia that deepens the articular surface and acts as a shock absorber or cushion**
- ☐ 2. **Tibial (medial) collateral ligament**
- ☐ 3. **Posterior cruciate ligament**
- ☐ 4. **Anterior cruciate ligament**
- ☐ 5. **Lateral meniscus: disc of fibrocartilage on the lateral side of the tibia**
- ☐ 6. **Fibular (lateral) collateral ligament**

| LIGAMENT | ATTACHMENT | COMMENT |
|---|---|---|
| | **Knee (Biaxial Condylar Synovial) Joint** | |
| Capsule | Surrounds femoral and tibial condyles, and patella | Is fibrous, weak (offers little support); allows for flexion, extension, some gliding, and medial rotation |
| | **Extracapsular Ligaments** | |
| Tibial collateral | Medial femoral epicondyle to medial tibial condyle | Limits extension and abduction of leg; attached to medial meniscus |
| Fibular collateral | Lateral femoral epicondyle to fibular head | Limits extension and adduction of leg; overlies popliteus muscle tendon |
| Patellar | Patella to tibial tuberosity | Acts in extension of quadriceps tendon |
| Arcuate popliteal | Fibular head to capsule | Passes over popliteus muscle |
| Oblique popliteal | Semimembranosus tendon to posterior knee | Limits hyperextension and lateral rotation |
| | **Intracapsular Ligaments** | |
| Medial meniscus | Interarticular area of tibia, lies over medial facet, attached to tibial collateral ligament | Is semicircular (C-shaped); acts as cushion; often torn |
| Lateral meniscus | Interarticular area of tibia, lies over lateral facet | Is more circular and smaller than medial meniscus; acts as cushion |
| Anterior cruciate | Anterior intercondylar tibia to lateral femoral condyle | Prevents posterior slipping of femur on tibia; torn in hyperextension |
| Posterior cruciate | Posterior intercondylar tibia to medial femoral condyle | Prevents anterior slipping of femur on tibia; shorter and stronger than anterior cruciate |
| Transverse | Anterior aspect of menisci | Binds and stabilizes menisci |
| Posterior meniscofemoral (of Wrisberg) | Posterior lateral meniscus to medial femoral condyle | Is strong |
| | **Patellofemoral (Biaxial Synovial Saddle) Joint** | |
| Quadriceps tendon | Muscles to superior patella | Is part of extension mechanism |
| Patellar ligament | Patella to tibial tuberosity | Acts in extension of quadriceps muscle tendon; patella stabilized by medial and lateral ligamentous (retinaculum) attachment to tibia and femur |

***Clinical Note:***

**Rupture of the weaker anterior cruciate ligament (ACL)** is a common athletic injury, usually related to twisting of the knee while the foot is firmly on the ground. Because the ACL prevents hyperextension of the knee, movement of the tibia forward on the femur while keeping the foot stable (anterior drawer sign) is used to assess ACL integrity. Often, **ACL injuries** may also be accompanied by a tear of the tibial collateral ligament or the medial meniscus. The medial meniscus attaches to the tibial collateral ligament. The combination of these three ligament tears—ACL, tibial collateral ligament, and medial meniscus—is known as the **"unhappy triad."**

**Osteoarthritis** of the knee, as with the hip, is a painful condition associated with activity, although other causes may also precipitate painful episodes, including changes in the weather.

**Patellar subluxation** (partial dislocation) of the patella, which usually occurs in a lateral direction, is a fairly common occurrence, especially in adolescent girls and young women.

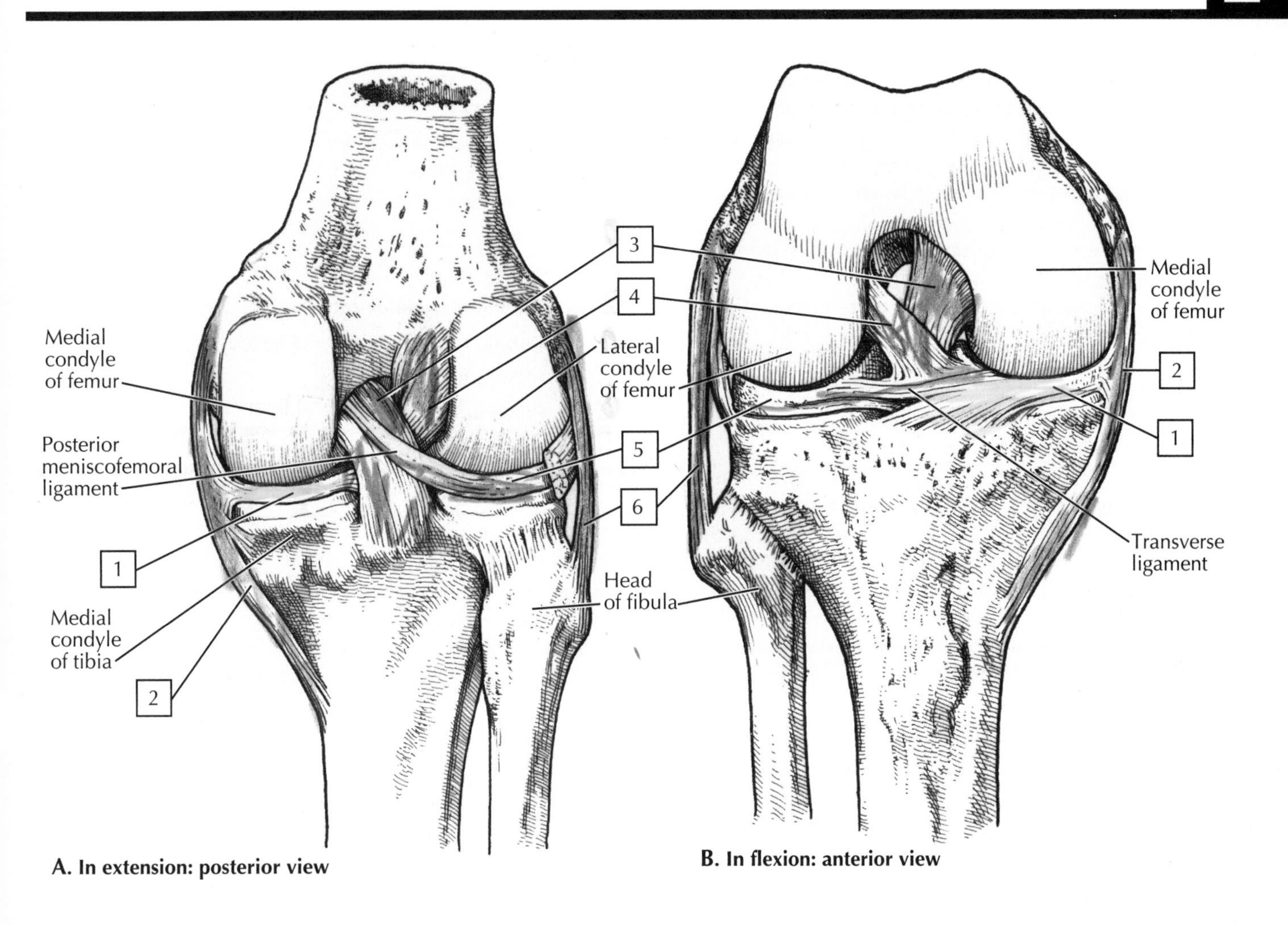


A. In extension: posterior view

B. In flexion: anterior view

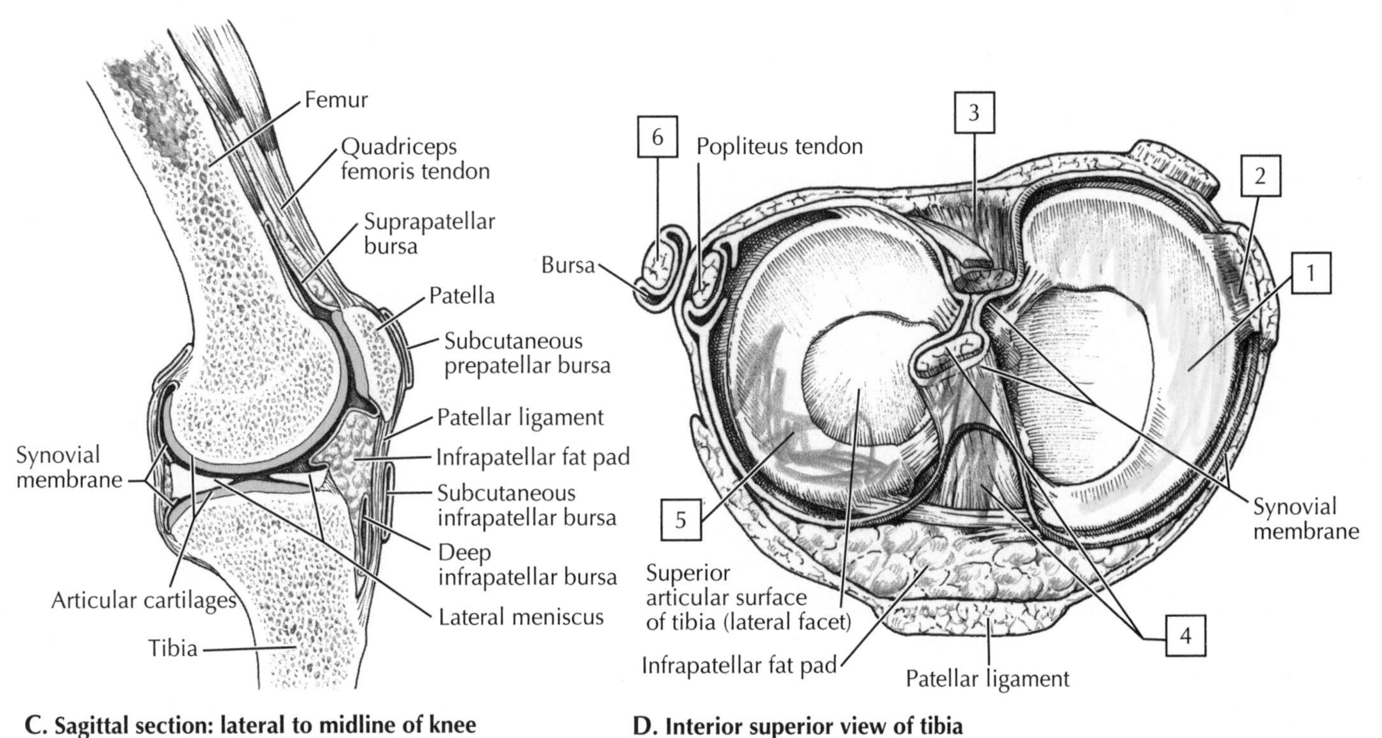


C. Sagittal section: lateral to midline of knee

D. Interior superior view of tibia

# 2 Bones of the Ankle and Foot

The ankle and foot are composed of the following 28 bones:

- 7 tarsal (ankle) bones, arranged in a proximal group of 2 tarsals (talus and calcaneus), a distal row of 4 tarsals (a cuboid and 3 cuneiforms), and a single intermediate tarsal (navicular) bone between these groups
- 5 metatarsals, which span the middle portion of the sole of the foot
- 14 phalanges, 2 for the big toe (hallucis) and 3 each for the other 4 toes
- 2 sesamoid bones, situated on the plantar surface of the distal first metatarsal

The bones of the foot are not aligned in a single flat plane with each bone in contact with the ground. Rather, the foot has two arches, each supported by ligaments and muscles:

- Longitudinal arch, formed by the posterior portion of the calcaneus (heel) and the heads of the five metatarsals; this arch is highest on the medial side of the foot (see part ***E***)
- Transverse arch, formed by the cuboid, the cuneiforms, and the bases of the metatarsals; this arch runs from side to side (see part ***F***)

| BONE | CHARACTERISTICS |
|---|---|
| **Talus (ankle bone) (a tarsal bone)** | Transfers weight from tibia to foot; no muscle attachment |
| Trochlea | Articulates with tibia and fibula |
| Head | Articulates with navicular bone |
| **Calcaneus (heel bone) (a tarsal bone)** | Articulates with talus superiorly and cuboid anteriorly |
| Sustentaculum tali | Medial shelf that supports talar head |
| **Navicular bone (a tarsal bone)** | "Boat shaped," between talar head and three cuneiforms |
| Tuberosity | If large, can cause medial pain in tight-fitting shoe |
| **Cuboid bone (a tarsal bone)** | Most lateral tarsal bone |
| Groove | For fibularis (peroneus) longus tendon |
| **Cuneiform bone (a tarsal bone)** | Three wedge-shaped bones |
| **Metatarsals** | |
| Numbered 1 to 5, from great toe (big toe) to little toe | Possess base, shaft, and head<br>Fibularis brevis tendon inserts on 5th metatarsal |
| Two sesamoid bones | Associated with flexor hallucis brevis tendons |
| **Phalanges** | |
| Three for each digit except great toe | Possess base, shaft, and head<br>Termed proximal, middle, and distal<br>Stubbed 5th toe common injury |

**COLOR** the following bones of the ankle and foot, using a different color for each tarsal, a uniform color for the metatarsals, another uniform color for the phalanges, and a new color for the sesamoid bones.

- ☐ **1. Calcaneus**
- ☐ **2. Talus**
- ☐ **3. Navicular**
- ☐ **4. Cuneiforms (color all three the same color)**
- ☐ **5. Cuboid**
- ☐ **6. Metatarsals**
- ☐ **7. Phalanges**
- ☐ **8. Sesamoid bones**

***Clinical Note:***

**Calcaneal fractures** are the most common tarsal bone fractures and may be extraarticular or intraarticular. Most calcaneal fractures are intraarticular, often caused by a forceful landing on the heel. The talus is "driven" down into the calcaneus, which cannot withstand the force because it is cancellous (spongy) bone.

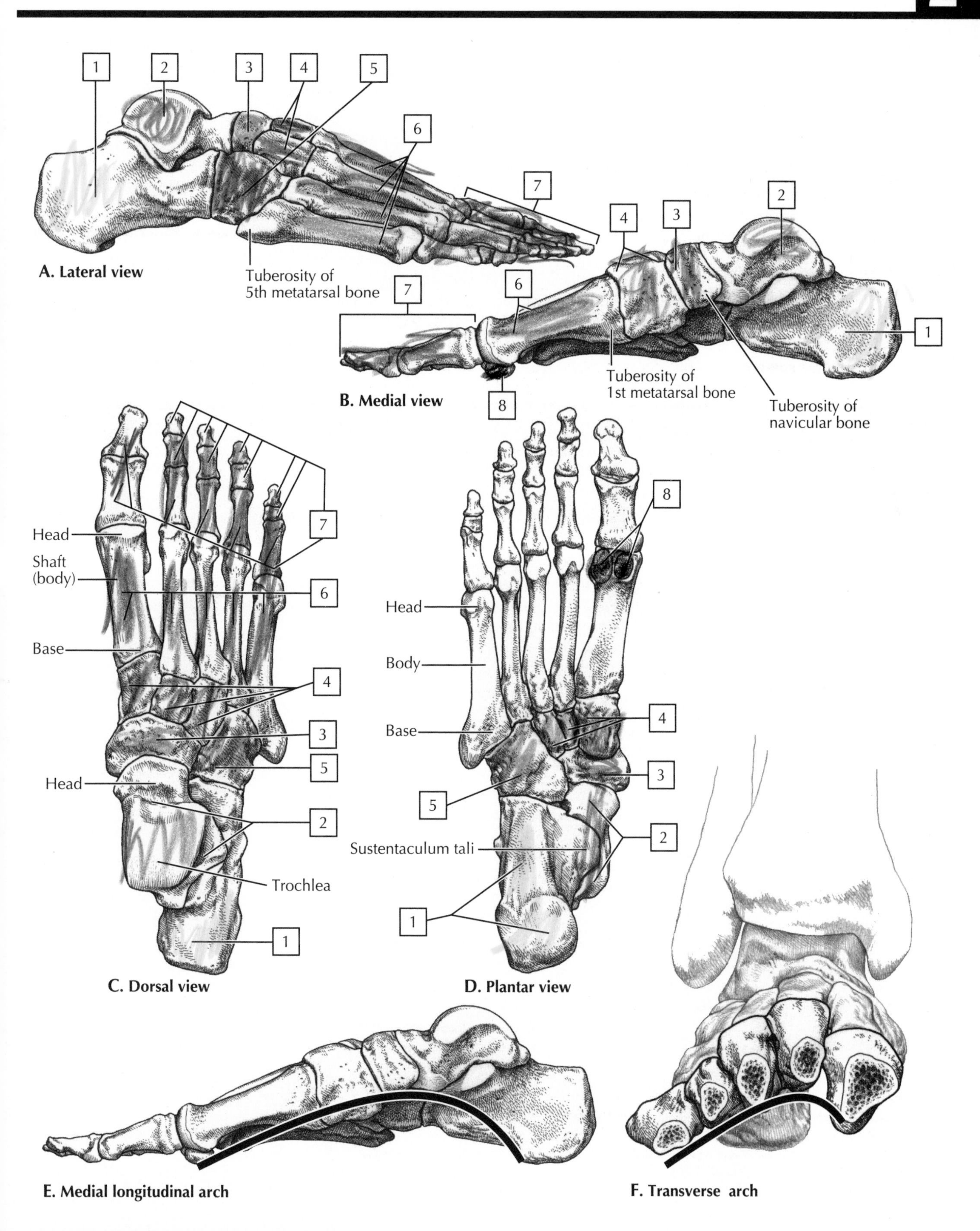
1
2
3
4
5
6
7
Tuberosity of
5th metatarsal bone
A. Lateral view
7
6
4
3
2
1
8
Tuberosity of
1st metatarsal bone
Tuberosity of
navicular bone
B. Medial view
Head
Shaft
(body)
Base
Head
7
6
4
3
5
2
Trochlea
1
C. Dorsal view
Head
Body
Base
8
4
3
5
2
Sustentaculum tali
1
D. Plantar view
E. Medial longitudinal arch
F. Transverse arch

# Ankle and Foot Joints

The classification and ligaments of the ankle and foot joints are summarized in the table below. The ankle joint is primarily a talocrural joint (a weight-bearing joint in which the talus articulates with the distal tibia of the leg) and, laterally, a talofibular joint (in which the talus articulates with the distal fibula of the leg).

**COLOR** the following major ligaments and capsules, using a different color for each ligament or capsule:

- ☐ 1. **Anterior talofibular ligament**
- ☐ 2. **Posterior talofibular ligament**
- ☐ 3. **Calcaneofibular ligament: these first three ligaments together form the "lateral collateral" ligament of the ankle**
- ☐ 4. **Long plantar ligament**
- ☐ 5. **Medial (deltoid) ligament: composed of four separate ligaments extending from the tibia to the talus or calcaneus**
- ☐ 6. **Plantar calcaneonavicular ligament: called the "spring" ligament, it helps support the medial arch of the foot**
- ☐ 7. **Capsule of a proximal interphalangeal joint**
- ☐ 8. **Capsule of a metatarsophalangeal joint**

***Clinical Note:***
Most **ankle sprains** are inversion injuries, in which one lands on the lateral aspect of the foot, the sole is turned medially, and the components of the lateral collateral ligament are stretched or torn.

| LIGAMENT | ATTACHMENT | COMMENT |
|---|---|---|
| | **Distal Tibiofibular Joint (Fibrous Joint, or Syndesmosis)** | |
| Anterior tibiofibular | Anterior distal tibia and fibula | Runs obliquely |
| Posterior tibiofibular | Posterior distal tibia and fibula | Is weaker than anterior ligament |
| Inferior transverse | Medial malleolus to fibula | Is deep continuation of posterior ligament |
| | **Talocrural Joint (Uniaxial Synovial Hinge Joint, or Ginglymus)** | |
| Capsule | Tibia and fibula to talus | Functions in plantarflexion and dorsiflexion |
| Medial (deltoid) | Medial malleolus to talus, calcaneus, and navicular bone | Limits eversion of foot; maintains medial long arch; has four parts |
| Lateral (collateral) | Lateral malleolus to talus and calcaneus | Is weak and often sprained; resists inversion of foot; has three parts |
| | **INTERTARSAL JOINTS (Next three joints)** | |
| | **Talocalcaneal (Subtalar Plane Synovial) Joints** | |
| Capsule | Margins of articulation | Functions in inversion and eversion |
| Talocalcaneal | Talus to calcaneus | Has medial, lateral, and posterior parts |
| Interosseous | Talus to calcaneus | Is strong; binds bones together |
| | **Talocalcaneonavicular (Partial Ball-and-Socket Synovial) Joint** | |
| Capsule | Encloses part of joint | Functions in gliding and rotational movements |
| Plantar calcaneonavicular | Sustentaculum tali to navicular bone | Is strong plantar support for head of talus (called spring ligament) |
| Dorsal talonavicular | Talus to navicular | Is dorsal support to talus |
| | **Calcaneocuboid (Plane Synovial) Joint** | |
| Capsule | Encloses joint | Functions in inversion and eversion |
| Calcaneocuboid | Calcaneus to cuboid | Are dorsal, plantar (short plantar, strong), and long plantar ligaments |
| | **Tarsometatarsal (Plane Synovial) Joints** | |
| Capsule | Encloses joint | Functions in gliding or sliding movements |
| Tarsometatarsal | Tarsals to metatarsals | Are dorsal, plantar, interosseous ligaments |
| | **Intermetatarsal (Plane Synovial) Joints** | |
| Capsule | Base of metatarsals | Provides little movement, supports transverse arch |
| Intermetatarsal | Adjacent metatarsals | Are dorsal, plantar, interosseous ligaments |
| Deep transverse | Adjacent metatarsals | Connect adjacent heads |
| | **Metatarsophalangeal (Multiaxial Condyloid Synovial) Joints** | |
| Capsule | Encloses joint | Functions in flexion, extension, some abduction and adduction, and circumduction |
| Collateral | Metatarsal heads to base of proximal phalanges | Are strong ligaments |
| Plantar (plates) | Plantar side of capsule | Are part of weight-bearing surface |
| | **Interphalangeal (Uniaxial Hinge Synovial) Joints** | |
| Capsule | Encloses each joint | Functions in flexion and extension |
| Collateral | Head of one to base of other | Support the capsule |
| Plantar (plates) | Plantar side of capsule | Support the capsule |

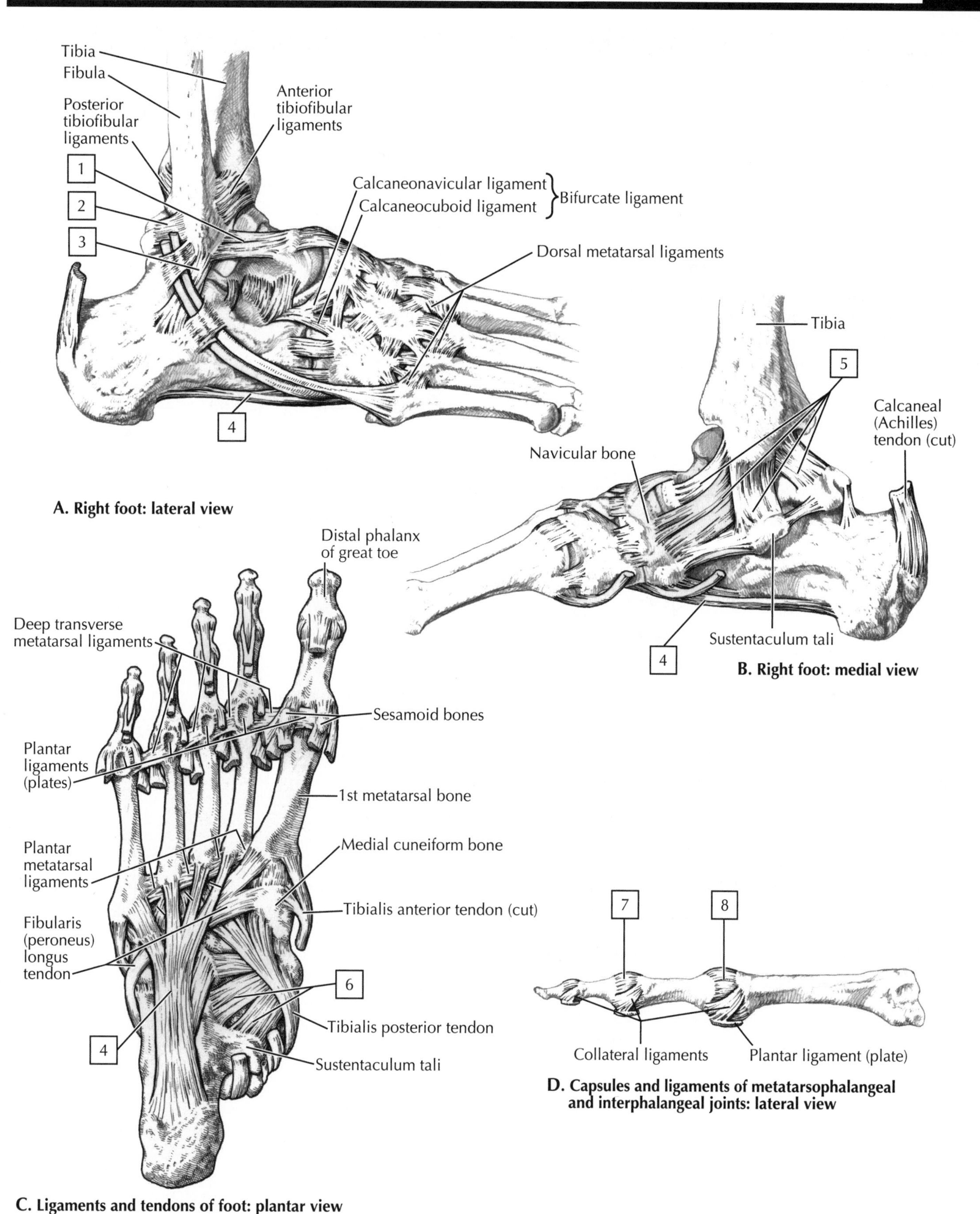
Tibia
Fibula
Posterior tibiofibular ligaments
1
2
3
Anterior tibiofibular ligaments
Calcaneonavicular ligament
Calcaneocuboid ligament
Bifurcate ligament
Dorsal metatarsal ligaments
4
A. Right foot: lateral view
Tibia
5
Calcaneal (Achilles) tendon (cut)
Navicular bone
Sustentaculum tali
4
B. Right foot: medial view
Distal phalanx of great toe
Deep transverse metatarsal ligaments
Sesamoid bones
Plantar ligaments (plates)
1st metatarsal bone
Plantar metatarsal ligaments
Medial cuneiform bone
Tibialis anterior tendon (cut)
Fibularis (peroneus) longus tendon
6
Tibialis posterior tendon
4
Sustentaculum tali
C. Ligaments and tendons of foot: plantar view
7
8
Collateral ligaments
Plantar ligament (plate)
D. Capsules and ligaments of metatarsophalangeal and interphalangeal joints: lateral view

## REVIEW QUESTIONS

1. Color the bones of the human skull indicated by the letters on the image:
   Frontal bone (color green)
   Sphenoid bone (color yellow)
   Zygomatic bone (color brown)
   Mandible (color blue)
   Occipital bone (color red)
   Temporal bone (color orange)

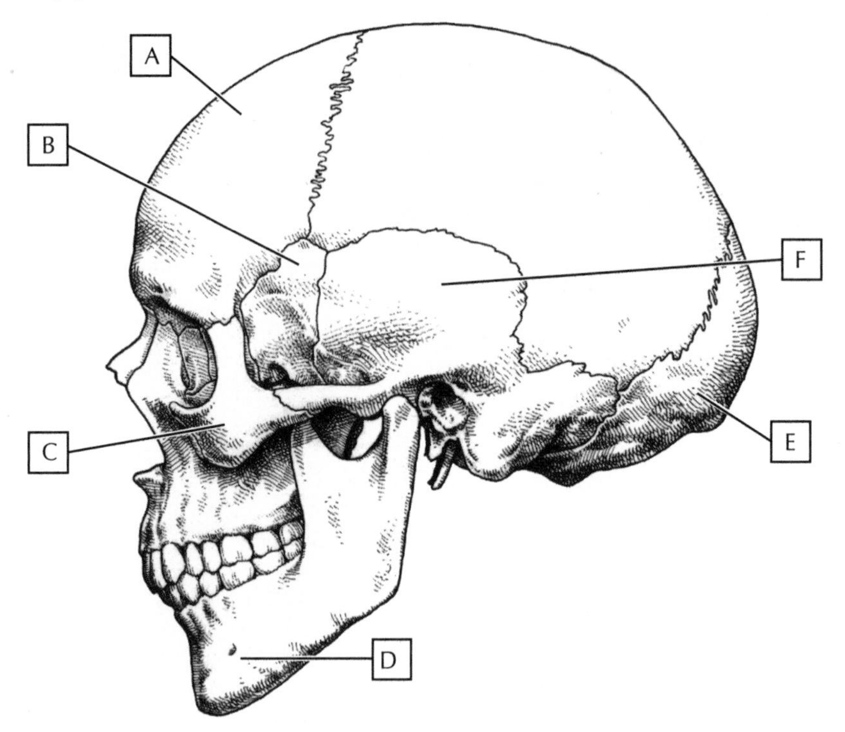


2. What is the name of the four teeth at the front of each jaw? ____________________
3. The arch of a thoracic vertebra is formed by which two paired elements? ____________________
4. What artery passes through the foramen transversarium of the cervical vertebra? ____________________
5. Which three bones form the pectoral girdle and arm of the upper limb? ____________________
6. Which carpal bone articulates with the metacarpal of the thumb? ____________________
7. What are the three bones that fuse to form the bony pelvis? ____________________
8. Most fractures of the femur involve which portion of the bone? ____________________
9. Which ligament of the knee, if torn, will result in excessive extension at the joint? ____________________
10. Which pair(s) of ribs is/are considered "floating ribs"? ____________________

## ANSWER KEY

1.

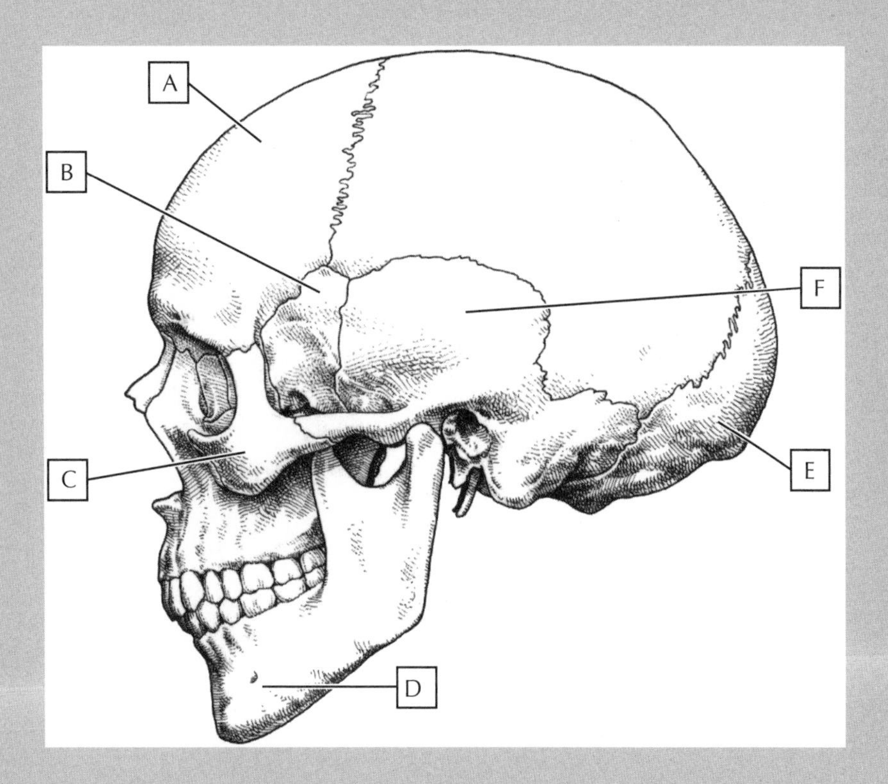


(A) Frontal bone
(B) Sphenoid bone
(C) Zygomatic bone
(D) Mandible
(E) Occipital bone
(F) Temporal bone

2. Incisors

3. Pedicles and laminae

4. Vertebral artery

5. Clavicle, scapula, and humerus

6. Trapezium

7. Ilium, ischium, and pubis

8. Femoral neck

9. Anterior cruciate ligament (ACL)

10. 11th and 12th pairs of ribs are floating ribs

# Chapter 3 Muscular System

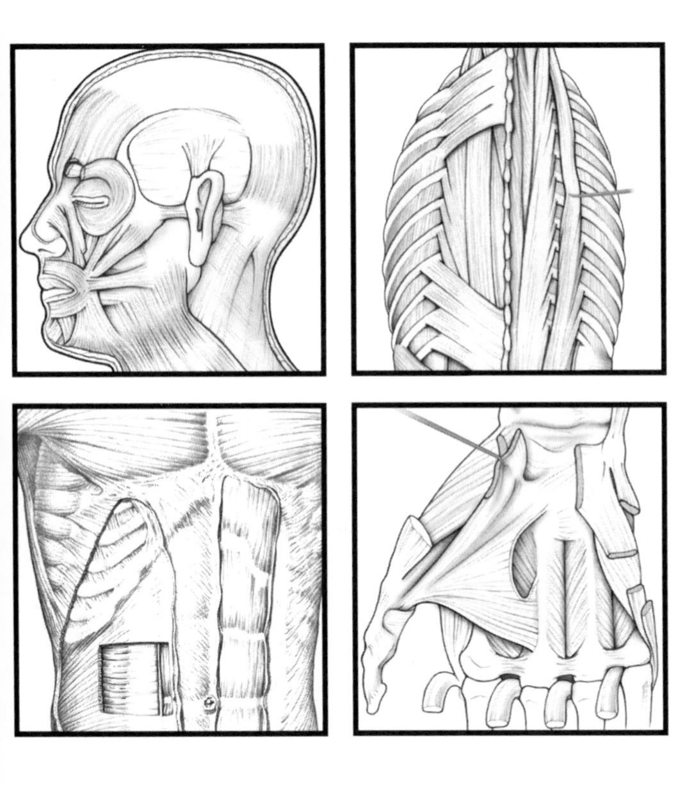

# 3 Muscles of Facial Expression

The muscles of facial expression are in several ways unique among the skeletal muscles of the body. They all originate embryologically from the second pharyngeal arch and are all innervated by terminal branches of the **facial nerve (CN VII)**. Additionally, most arise from the bones of the face or fascia, and insert into the dermis of the skin overlying the scalp, face, and anterolateral neck. Some of the more important muscles of facial expression are summarized in the table below and may be colored on the images on the facing page.

All of these muscles are supplied by the facial nerve (CN VII); while these muscles are skeletal muscles, they are unique in that they are innervated by the five terminal branches of the facial nerve rather than by spinal nerves.

| MUSCLE | ORIGIN ATTACHMENT | INSERTION ATTACHMENT | MAIN ACTIONS |
|---|---|---|---|
| Frontal belly of occipitofrontalis | Epicranial aponeurosis | Skin of forehead | Elevates eyebrows and forehead, and wrinkles forehead |
| Orbicularis oculi | Medial orbital margin, medial palpebral ligament, and lacrimal bone | Skin around margin of orbit; tarsal plate, eyelids | Closes eyelids; orbital part forcefully and palpebral part for blinking |
| Nasalis | Superior part of canine ridge of maxilla | Nasal cartilages | Draws ala of nose toward septum to compress opening |
| Orbicularis oris | Median plane of maxilla superiorly and mandible inferiorly; other fibers from deep surface of skin | Mucous membrane of lips | Closes and protrudes lips (e.g., purses them during whistling) |
| Levator labii superioris | Frontal process of maxilla and infraorbital region | Skin of upper lip and alar cartilage | Elevates lip, dilates nostril, raises angle of mouth |
| Platysma | Skin, superficial fascia of superior deltoid and pectoral regions | Mandible, skin of cheek, angle of mouth, and orbicularis oris | Tenses skin of lower face and neck, depresses mandible (against resistance) |
| Mentalis | Incisive fossa of mandible | Skin of chin | Elevates and protrudes lower lip and wrinkles chin |
| Buccinator | Mandible, pterygomandibular raphe, and alveolar processes of maxilla and mandible | Angle of mouth | Presses cheek against molar teeth, thereby aiding chewing, expels air between lips |

**COLOR** some of the more important muscles of facial expression listed below, using a different color for each muscle:

☐ 1. **Epicranius (frontalis and occipitalis): these two muscles are connected to one another by the galea aponeurotica (a broad, flat tendon)**

☐ 2. **Orbicularis oculi: a sphincter muscle that closes the eyelids (has a palpebral part in the eyelids and an orbital part attached to the bony orbital rim)**

☐ 3. **Levator labii superioris: elevates the lip and flares the nostrils**

☐ 4. **Nasalis: has a transverse and an alar part**

☐ 5. **Orbicularis oris: a sphincter muscle that purses our lips (the "kissing" muscle)**

☐ 6. **Depressor anguli oris: depresses our lips (known as the "sad" muscle, as it turns the corners of our lips downward)**

☐ 7. **Platysma: a broad, thin muscle that covers the anterolateral neck and tenses the skin of the lower face and neck**

☐ 8. **Buccinator: allows us to draw in our cheeks, thereby keeping food between our molars during chewing (sometimes we "bite" this muscle or "bite our cheek" when it contracts too vigorously)**

 9. **Risorius: our "smiling" muscle (helped by the zygomaticus muscles)**

***Clinical Note:***

Unilateral paralysis of the facial nerve (CN VII) (often from inflammation), called **Bell's palsy**, can lead to an asymmetry of the facial features, because the facial muscles are flaccid on the affected side of the face. People with Bell's palsy may not be able to frown or wrinkle the forehead, close their eyelids tightly, smile, purse their lips, or tense the skin of the neck. The acute, idiopathic form of Bell's palsy is most common, but facial nerve palsy also may be caused by herpes simplex virus (HSV) infection. Patients may experience **hyperacusis** (painful sensitivity to sound) and loss of taste on the affected side. The patient often cannot wrinkle their forehead, their eyelid droops slightly, they cannot show their teeth on the affected side when they attempt to smile, and their lower lip droops slightly.

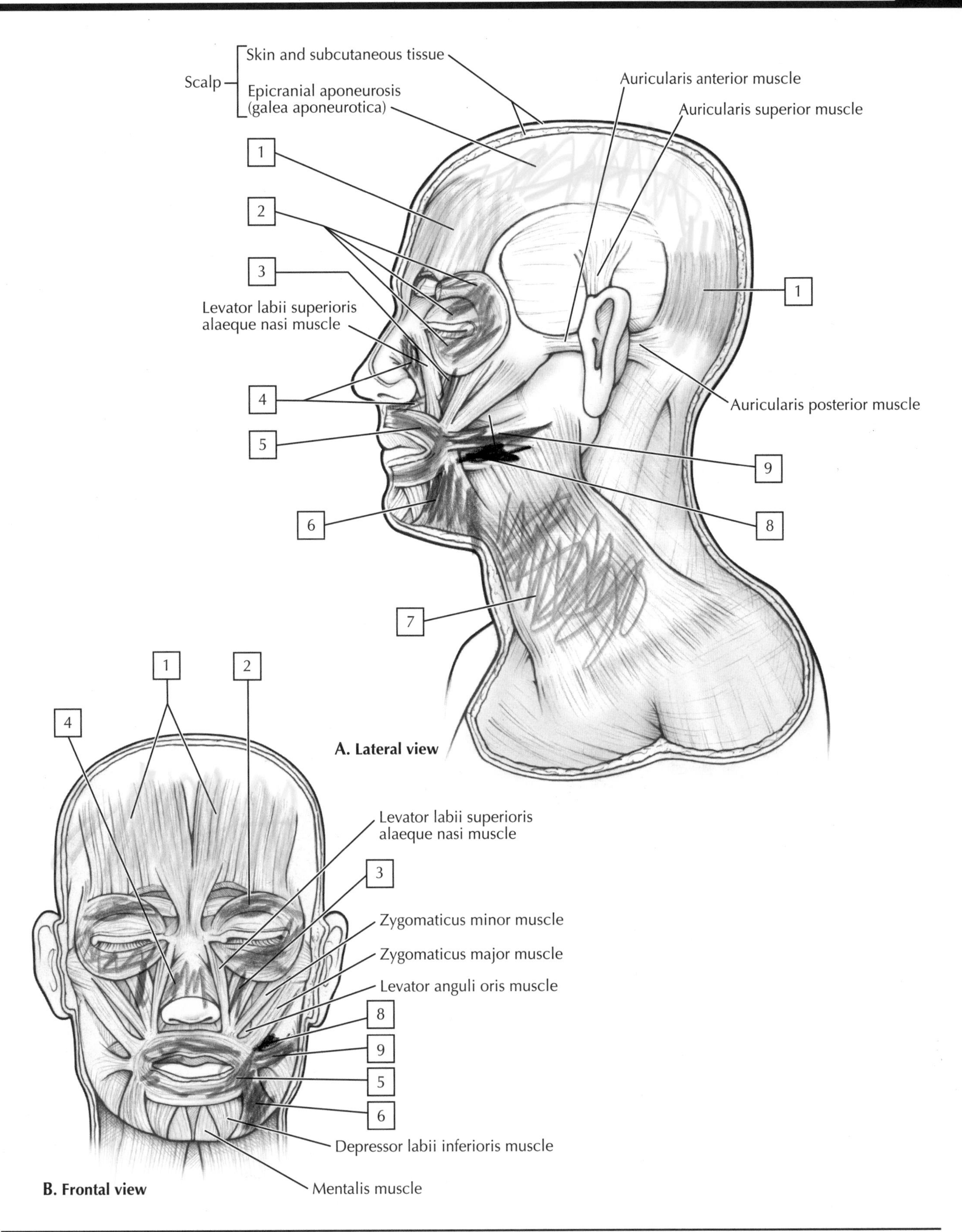
Scalp
Skin and subcutaneous tissue
Epicranial aponeurosis (galea aponeurotica)
Auricularis anterior muscle
Auricularis superior muscle
1
2
3
Levator labii superioris alaeque nasi muscle
4
5
Auricularis posterior muscle
9
8
6
7
A. Lateral view
1
2
4
Levator labii superioris alaeque nasi muscle
3
Zygomaticus minor muscle
Zygomaticus major muscle
Levator anguli oris muscle
8
9
5
6
Depressor labii inferioris muscle
Mentalis muscle
B. Frontal view

# 3 Muscles of Mastication

The muscles of mastication include four pairs of muscles (left and right sides) that attach to the mandible. They are embryological derivatives of the first pharyngeal arch and are all innervated by the mandibular division of the **trigeminal nerve (CN $V_3$)**. These muscles are important in biting and chewing food.

**COLOR** each of the following muscles of mastication, using a different color for each:

☐ 1. **Temporalis: a broad muscle arising from the temporal fossa and overlying the fascia that elevates (closes) the mandible; you see this muscle contract on the side of your head when you are chewing**

☐ 2. **Masseter: a powerful muscle that elevates the mandible and is evident in people who chew a lot of gum, because you can see the muscle contract; chronic gum chewers tend to have chubby cheeks because their masseter muscles are enlarged from chronic use. It arises from the ramus of the mandible and coronoid process**

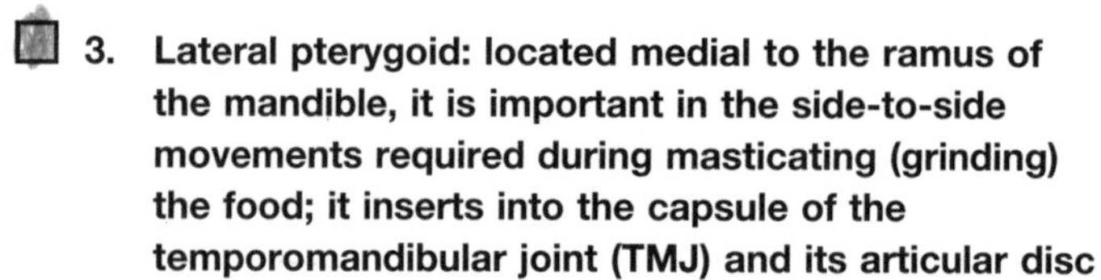

☐ 3. **Lateral pterygoid: located medial to the ramus of the mandible, it is important in the side-to-side movements required during masticating (grinding) the food; it inserts into the capsule of the temporomandibular joint (TMJ) and its articular disc**

☐ 4. **Medial pterygoid: located medial to the ramus of the mandible, it too participates in masticating the food, and because its muscle fibers run in the same direction as the masseter muscle, it also assists this muscle in closing the jaw; acting alternatively, the medial pterygoid muscles produce a grinding motion**

These muscles are summarized in the table below: all are innervated by the mandibular nerve (CN $V_3$), the third division of the trigeminal nerve.

| MUSCLE | ORIGIN ATTACHMENT | INSERTION ATTACHMENT | MAIN ACTIONS |
|---|---|---|---|
| Temporalis | Floor of temporal fossa and deep temporal fascia | Coronoid process and ramus of mandible | Elevates mandible; posterior fibers retract mandible |
| Masseter | Zygomatic arch | Ramus of mandible and coronoid process | Elevates and protrudes mandible; deep fibers retract it |
| Lateral pterygoid | *Superior head*: infratemporal surface of greater wing of sphenoid<br>*Inferior head*: lateral pterygoid plate | Pterygoid fovea, articular disc, and capsule of TMJ | Acting together, protrude mandible and depress chin; acting alone and alternately, each muscle head produces side-to-side grinding movements |
| Medial pterygoid | *Deep head*: medial surface of lateral pterygoid plate and palatine bone<br>*Superficial head*: tuberosity of maxilla | Medial surface of ramus and angle of mandible inferior to mandibular foramen | Elevates mandible; acting together, protrude mandible; acting alone, each muscle head protrudes side of jaw; acting alternately, each muscle head produces grinding motion |

***Clinical Note:***

**Tetanus** is a disease caused by a neurotropic toxin of *Clostridium tetani* that can affect the central nervous system and cause a painful tonic contraction of muscles, especially the masseter muscle, leading to a condition called "lockjaw." This pathogen is often found in soil, dust, and feces, and can enter the body through wounds, blisters, burns, skin ulcers, insect bites, and surgical procedures. Symptoms include restlessness, low-grade fever, and stiffness or soreness. Eventually, nuchal rigidity (back of the neck), trismus (lockjaw), dysphagia (difficulty swallowing), laryngospasm, and acute, massive muscle spasms can occur. There is a vaccination to prevent this disease, so it is important to always keep your immunizations up to date.

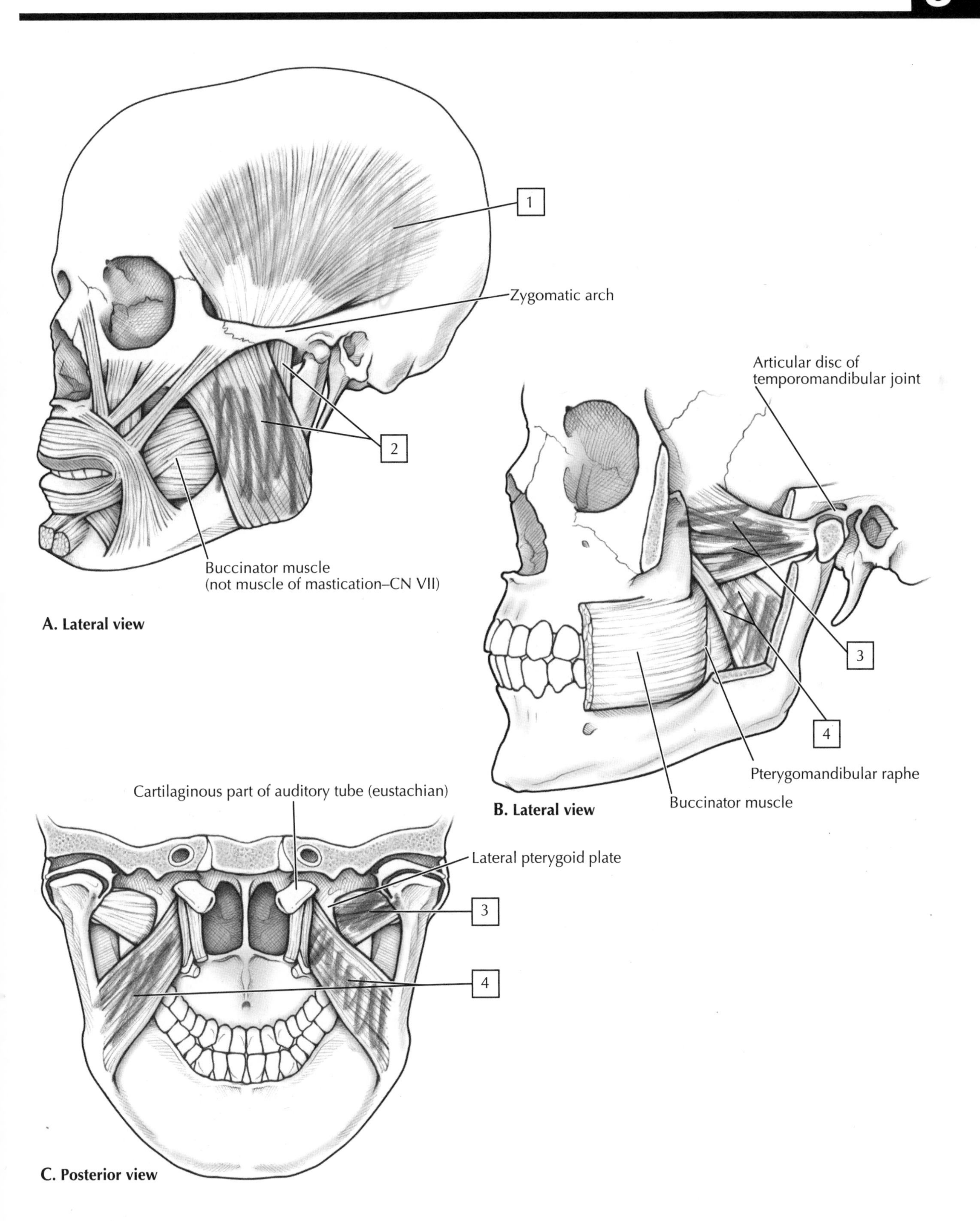


A. Lateral view

B. Lateral view

C. Posterior view

# 3 Extraocular Muscles

The eyeball has two sets of muscles associated with its movements:

- **Extrinsic:** extraocular muscles, six skeletal muscles that move the globe or eyeball proper within the orbit
- **Intrinsic:** smooth muscles that affect the size of the pupil (dilate or constrict the pupil) or that affect the shape of the lens for accommodation (near vision) or distance vision (these smooth muscles will be discussed in Chapter 4, Plate 4-23).

**COLOR** the following extrinsic muscles, using a different color for each muscle:

- ☐ 1. **Superior oblique**
- ☐ 2. **Superior rectus**
- ☐ 3. **Lateral rectus**
- ☐ 4. **Inferior rectus**
- ☐ 5. **Inferior oblique**
- ☐ 6. **Medial rectus**

In addition to the six extraocular muscles, there is another skeletal muscle that works in concert with these muscles to elevate the upper eyelid, called the levator palpebrae superioris muscle (its antagonist would be the orbicularis oculi muscle, a muscle of facial expression that closes the eyelids).

**COLOR** the following muscle:

- ☐ 7. **Levator palpebrae superioris**

Together, the extraocular muscles and the levator palpebrae superioris muscle are innervated by three different cranial nerves, the oculomotor (CN III), trochlear (CN IV), and abducens (CN VI) nerves. These muscles and their innervation are summarized in the table below. The actions of the extraocular muscles are complex and involve multiple subtle movements (including rotational movements), so the movements described in the table are those described anatomically. The movements tested clinically by a physician, where the isolated primary movement of each muscle is observed (elevation, depression, abduction, or adduction), are shown in part ***D*** (also see Clinical Note).

| MUSCLE | ORIGIN ATTACHMENT | INSERTION ATTACHMENT | INNERVATION | MAIN ACTIONS |
|---|---|---|---|---|
| Levator palpebrae superioris | Lesser wing of sphenoid bone, anterosuperior optic canal | Tarsal plate and skin of upper eyelid | Oculomotor nerve (CN III) | Elevates upper eyelid |
| Superior rectus | Common tendinous ring | Superior aspect of sclera just posterior to cornea | Oculomotor nerve (CN III) | Elevates, adducts, and rotates eyeball medially |
| Inferior rectus | Common tendinous ring | Inferior aspect of eyeball, posterior to sclera | Oculomotor nerve (CN III) | Depresses, adducts, and rotates eyeball laterally |
| Medial rectus | Common tendinous ring | Medial aspect of eyeball, posterior to sclera | Oculomotor nerve (CN III) | Adducts eyeball |
| Lateral rectus | Common tendinous ring | Lateral aspect of eyeball, just posterior to sclera | Abducens nerve (CN VI) | Abducts eyeball |
| Superior oblique | Body of sphenoid bone above optic canal | Passes through a trochlea and inserts into sclera | Trochlear nerve (CN VI) | Medially rotates, depresses, and abducts eyeball |
| Inferior oblique | Anterior floor of orbit | Lateral sclera deep to lateral rectus muscle | Oculomotor nerve (CN III) | Laterally rotates, elevates, and abducts eyeball |

***Clinical Note:***

Because the extraocular muscles act as synergists and antagonists and may be responsible for multiple movements, the physician tests the isolated action of each muscle by tracking eye movement while moving her finger in an H pattern. The image at the bottom of the facing page illustrates which muscle is being tested as this happens. For example, when the finger is held up and to the right of the patient's eyes, the patient must primarily use the superior rectus (SR) muscle of his right eye and the inferior oblique (IO) muscle of his left eye to focus on the finger. "Pure" abduction is performed by the lateral rectus (LR) muscle and "pure" adduction by the medial rectus (MR) muscle. In all other cases, three muscles together can abduct (SR, LR, and IR) or adduct (IO, MR, and SO) the eyeball, and two muscles together can elevate (SR and IO) or depress (IR and SO) the globe. If weakness of a muscle is observed, then the physician must determine if it is a muscle problem and/or a nerve problem (damage to the nerve innervating the muscle).

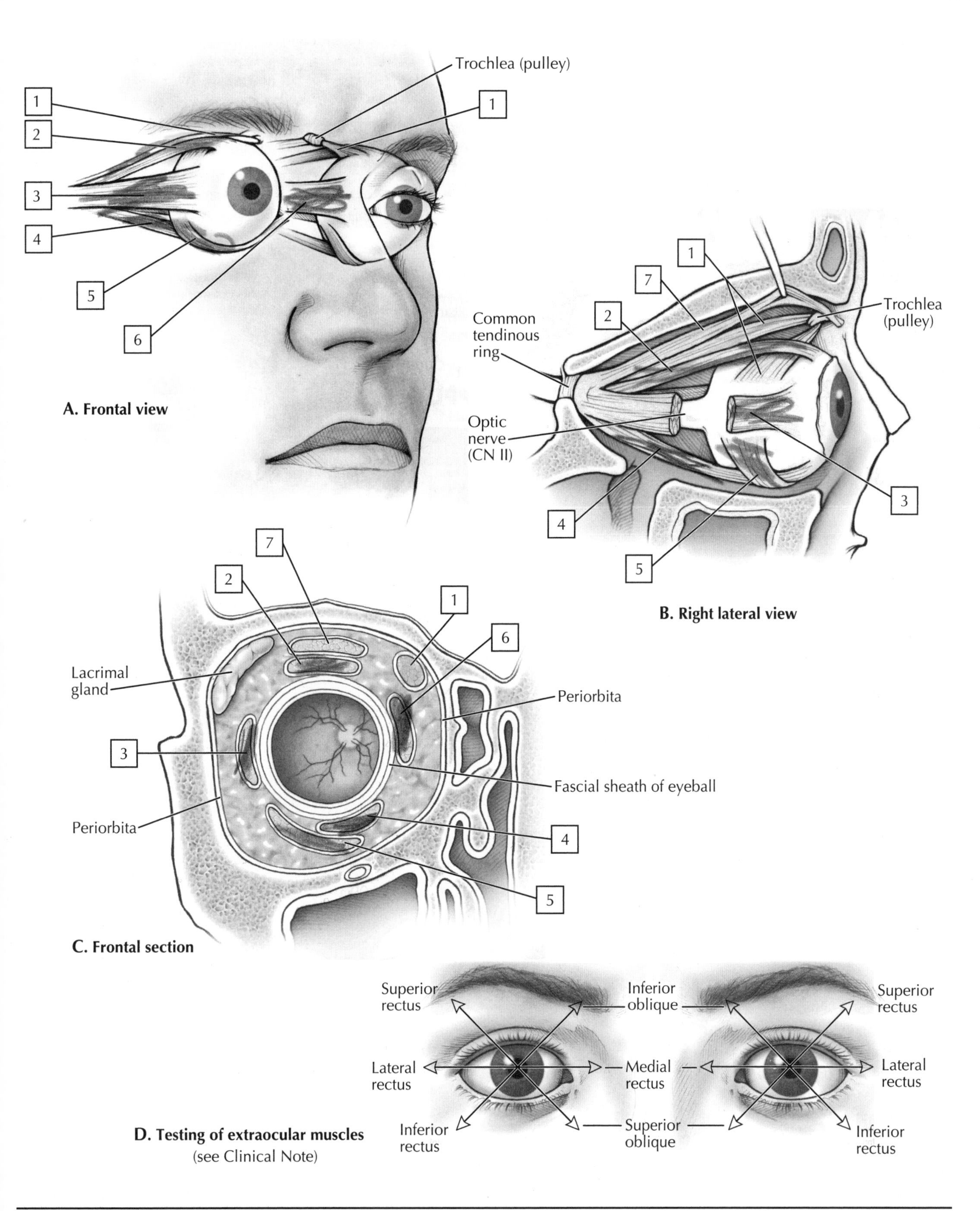


A. Frontal view

B. Right lateral view

C. Frontal section

D. Testing of extraocular muscles
(see Clinical Note)

# 3 Muscles of the Tongue and Palate

The muscles of the tongue are all skeletal muscles and include:

- **Intrinsic muscles:** composed of longitudinal, transverse, and vertical bundles of skeletal muscle that allow one to curl, elongate, and flatten the tongue
- **Extrinsic muscles:** four muscles that move the tongue (protrude, elevate, depress, or retract); all have the suffix "glossus" in their name, referring to the tongue

All of the tongue muscles are innervated by the **hypoglossal nerve** (CN XII) except the palatoglossus muscle, which is innervated by the vagus nerve (CN X). The principal muscle of the tongue is the **genioglossus muscle**, which blends with the intrinsic longitudinal muscle fibers to anchor the tongue to the floor of the mouth. Ounce for ounce, the genioglossus muscle (and its intrinsic muscle component) is the strongest muscle in the body!

**COLOR** the following muscles of the tongue, using a different color for each muscle:

- ☐ 1. **Genioglossus**
- ☐ 2. **Hyoglossus**
- ☐ 3. **Palatoglossus**
- ☐ 4. **Styloglossus**

The muscles of the palate include four muscles, which all act on the soft palate (the anterior two-thirds of the palate is "hard" [bone covered with mucosa], whereas the posterior palate is "soft" [fibromuscular]).

**COLOR** the following muscles of the palate, using a different color for each muscle:

- ☐ 5. **Tensor veli palatini**
- ☐ 6. **Levator veli palatini**
- ☐ 7. **Palatopharyngeus**
- ☐ 8. **Musculus uvulae (uvular muscle)**

The palatoglossus muscle, although grouped with the extrinsic tongue muscles, also acts on the soft palate, so it can be considered a palate muscle as well. The tongue and palate muscles are summarized in the table below.

| MUSCLE | ORIGIN ATTACHMENT | INSERTION ATTACHMENT | INNERVATION | MAIN ACTIONS |
|---|---|---|---|---|
| Genioglossus | Mental spine of mandible | Dorsum of tongue and hyoid bone | Hypoglossal nerve (CN XII) | Depresses and protrudes tongue |
| Hyoglossus | Body and greater horn of hyoid bone | Lateral and inferior aspect of tongue | Hypoglossal nerve (CN XII) | Depresses and retracts tongue |
| Styloglossus | Styloid process and stylohyoid ligament | Lateral and inferior aspect of tongue | Hypoglossal nerve (CN XII) | Retracts tongue and draws it up for swallowing |
| Palatoglossus | Palatine aponeurosis of soft palate | Lateral aspect of tongue | Vagus nerve (CN X) and pharyngeal plexus | Elevates posterior tongue, depresses palate |
| Levator veli palatini | Temporal (petrous portion) bone and auditory tube | Palatine aponeurosis | Vagus nerve (CN X) via pharyngeal plexus | Elevates soft palate during swallowing |
| Tensor veli palatini | Scaphoid fossa of medial pterygoid plate, spine of sphenoid, and auditory tube | Palatine aponeurosis | Mandibular nerve (CN $V_3$) | Tenses soft palate and opens auditory tube during swallowing and yawning |
| Palatopharyngeus | Hard palate and superior palatine aponeurosis | Lateral pharyngeal wall | Vagus nerve (CN X) via pharyngeal plexus | Tenses soft palate; pulls walls of pharynx superiorly, anteriorly, and medially during swallowing |
| Musculus uvulae | Nasal spine and palatine aponeurosis | Mucosa of uvula | Vagus nerve (CN X) via pharyngeal plexus | Shortens, elevates, and retracts uvula |

The oral surface of the tongue is covered with a stratified squamous epithelium that contains many papillae, including the:

- **Filiform papillae:** most numerous mucosal projections that increase the surface area of the tongue but do not contain taste buds
- **Fungiform papillae:** larger than filiform papillae, are rounded and cone-shaped, and contain taste buds
- **Foliate papillae:** rudimentary in humans, found largely along the lateral sides of the tongue near the terminal sulcus, but do contain taste buds
- **Circumvallate papillae:** large, capped papillae found in a single row just anterior to the terminal sulcus, and contain taste buds

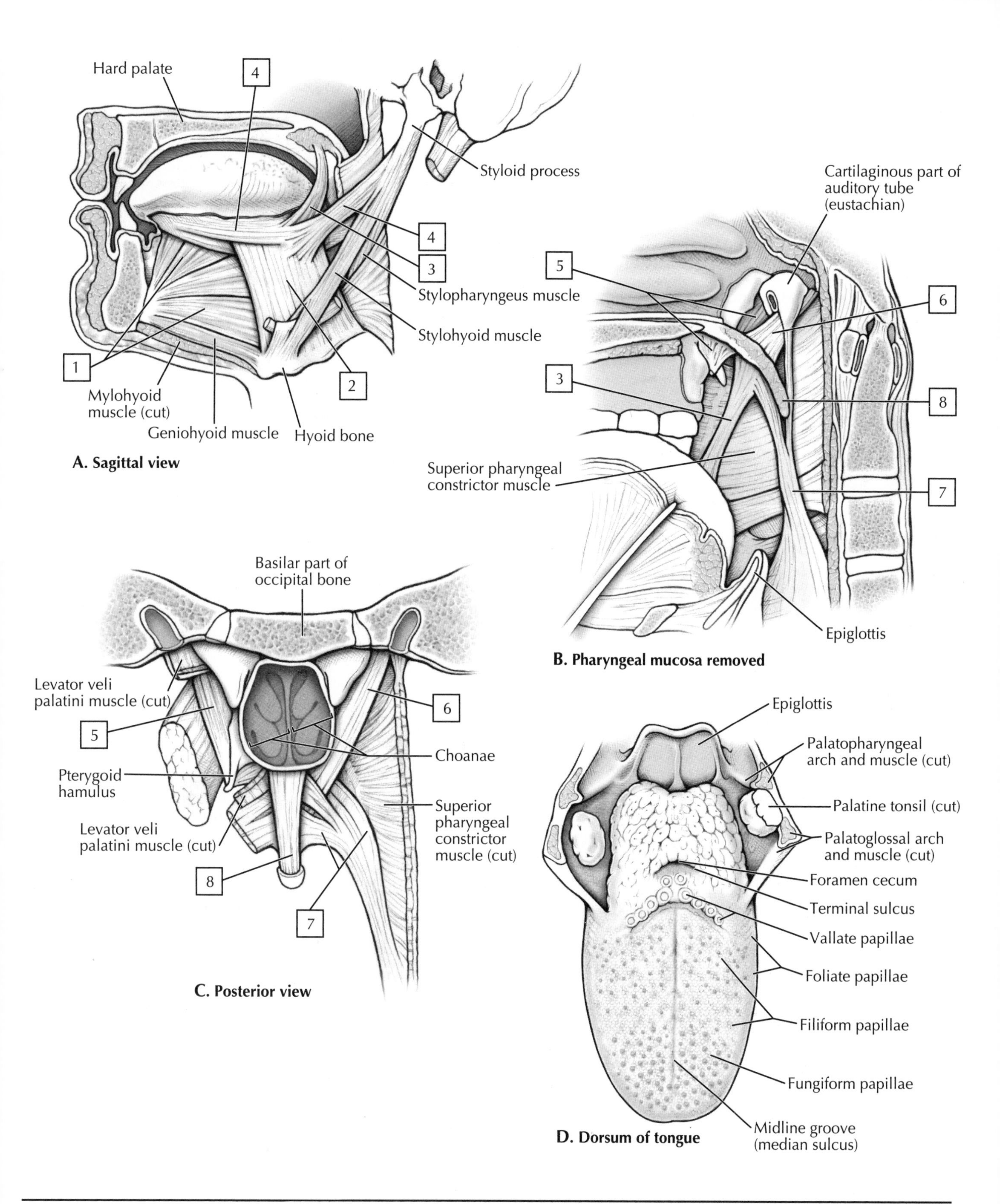


**A. Sagittal view**

**B. Pharyngeal mucosa removed**

**C. Posterior view**

**D. Dorsum of tongue**

# 3 Muscles of the Pharynx and Swallowing

The pharynx (throat) is a muscular tube found just posterior to the nasal and oral cavities that extends inferiorly to become continuous with the esophagus at about the level of the intervertebral disc between the C6 and C7 vertebral bodies. Muscles of the pharynx include the:

- **Superior pharyngeal constrictor:** located behind the nasal and oral cavities
- **Middle pharyngeal constrictor:** located behind the mandible and hyoid bone
- **Inferior pharyngeal constrictor:** located behind the thyroid and cricoid cartilages
- **Stylopharyngeus:** extends from the styloid process into the lateral wall of the pharynx
- **Salpingopharyngeus:** a small interior muscle of the pharynx

**COLOR** the following pharyngeal muscles, using a different color for each muscle:

- ☐ **1. Stylopharyngeus**
- ☐ **2. Superior pharyngeal constrictor**
- ☐ **3. Inferior pharyngeal constrictor**
- ☐ **4. Middle pharyngeal constrictor**

| MUSCLE | ORIGIN ATTACHMENT | INSERTION ATTACHMENT | INNERVATION | MAIN ACTIONS |
|---|---|---|---|---|
| Superior pharyngeal constrictor | Hamulus, pterygomandibular raphe, mylohyoid line of mandible | Median raphe of pharynx | Vagus nerve (CN X) via pharyngeal plexus | Constricts wall of pharynx during swallowing |
| Middle pharyngeal constrictor | Stylohyoid ligament and horns of hyoid bone | Median raphe of pharynx | Vagus nerve (CN X) via pharyngeal plexus | Constricts wall of pharynx during swallowing |
| Inferior pharyngeal constrictor | Oblique line of thyroid cartilage, and cricoid cartilage | Median raphe of pharynx | Vagus nerve (CN X) via pharyngeal plexus | Constricts wall of pharynx during swallowing |
| Salpingopharyngeus | Auditory (pharyngotympanic) tube | Side of pharynx wall | Vagus nerve (CN X) via pharyngeal plexus | Elevates pharynx and larynx during swallowing and speaking |
| Stylopharyngeus | Medial aspect of styloid process | Pharyngeal wall and posterior border of thyroid cartilage | Glossopharyngeal nerve (CN IX) | Elevates pharynx and larynx during swallowing and speaking |

Viewing the interior mucosal-lined wall of the pharynx reveals the three regions of the pharynx (see part ***C***):

- **Nasopharynx:** the region posterior to the choanae, or openings of the nasal cavities, and soft palate
- **Oropharynx:** the region between the soft palate and the posterior third of the tongue
- **Laryngopharynx** (hypopharynx): the region from the epiglottis to the beginning of the esophagus

The pharyngeal muscles contract sequentially, beginning superiorly and moving inferiorly, to "squeeze" a bolus of chewed food down the pharynx and into the upper esophagus. This process of swallowing is called **deglutition** and involves the interplay and coordinated movements of the tongue, soft palate, pharynx, and larynx to work properly. Deglutition includes the following steps:

- The bolus of food is pushed up against the hard palate by the tongue
- The soft palate is elevated to close off the nasopharynx
- The bolus is pushed back into the oropharynx by the action of the tongue
- As the bolus reaches the epiglottis, the larynx is elevated and the tip of the epiglottis is tipped downward over the laryngeal opening (aditus)
- Contractions of the pharyngeal constrictors squeeze the bolus into two streams that pass on either side of the epiglottis and into the upper esophagus, and the soft palate is pulled downward to assist in moving the bolus
- The soft palate is pulled down, the rima glottidis (space between the vocal folds) closes, and once the bolus is safely in the esophagus, all structures return to their starting positions

***Clinical Note:***

The **"gag reflex"** can be elicited by touching the posterior portion of the tongue. Sensation is conveyed by the sensory branches of the glossopharyngeal nerve (CN IX), and the soft palate is elevated by the motor action of the vagus nerve (CN X). During swallowing, this reflex may be elicited to protect the vocal folds and avoid aspiration of food or a foreign object into the trachea.

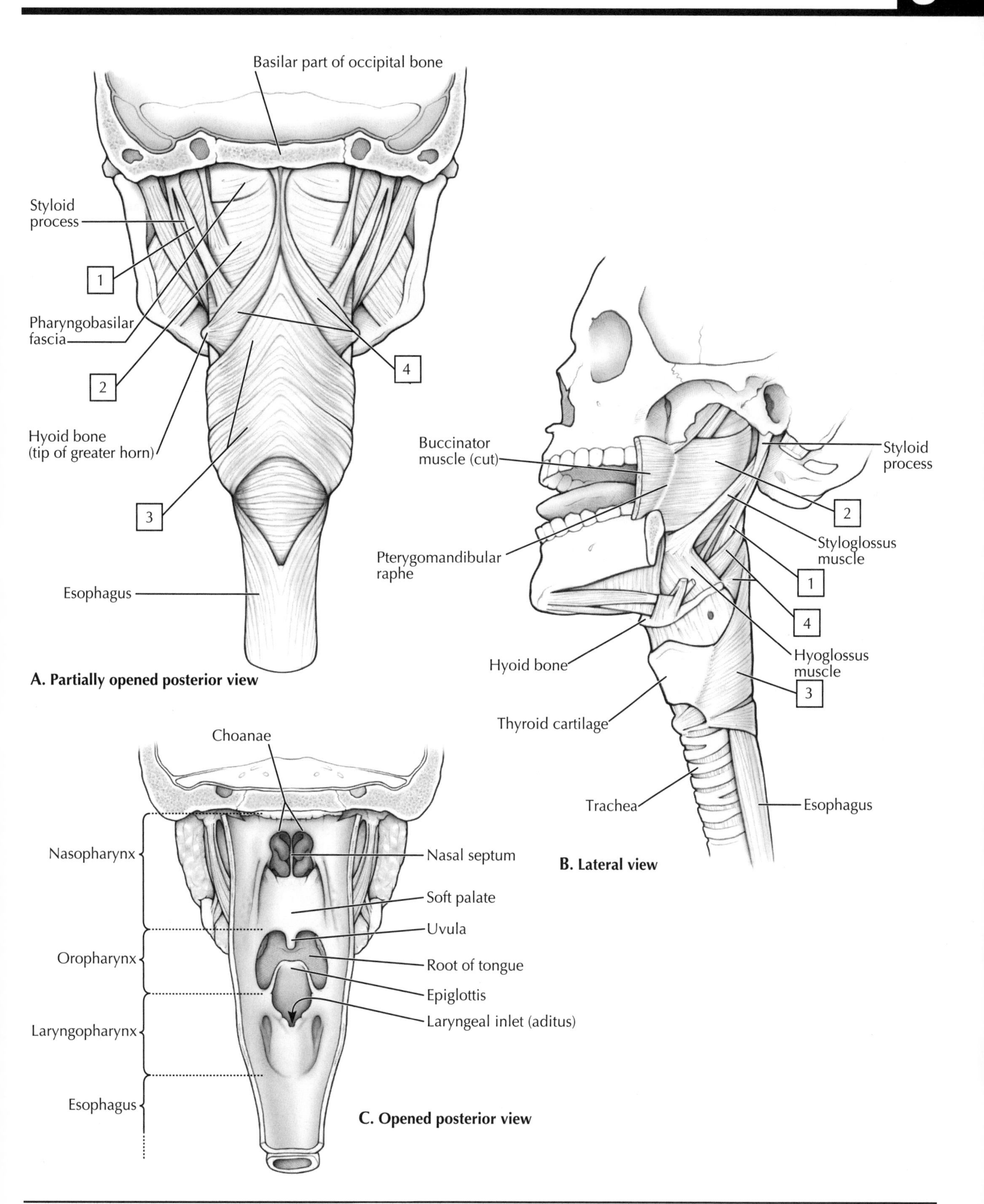


**A. Partially opened posterior view**

**B. Lateral view**

**C. Opened posterior view**

# 3 Intrinsic Muscles of the Larynx and Phonation

The intrinsic muscles of the larynx attach to the cartilages of the larynx, so these will be reviewed first. The larynx (our voice box) is a musculoligamentous and cartilaginous structure that lies at the C3 to C6 vertebral level, just superior to the trachea. It consists of nine cartilages joined by ligaments and membranes. The nine cartilages are summarized in the table below.

| CARTILAGE | DESCRIPTION |
|---|---|
| Thyroid | Two hyaline laminae and the laryngeal prominence (Adam's apple) |
| Cricoid | Signet ring–shaped hyaline cartilage just inferior to thyroid |
| Epiglottis | Spoon-shaped elastic cartilage plate attached to thyroid |
| Arytenoid | Paired pyramidal cartilages that rotate on cricoid cartilage |
| Corniculate | Paired cartilages that lie on apex of arytenoid cartilages |
| Cuneiform | Paired cartilages in aryepiglottic folds that have no articulations |

**COLOR** the following cartilages of the larynx, using a different color for each cartilage:

- ☐ **1. Epiglottis**
- ☐ **2. Thyroid**
- ☐ **3. Cricoid**
- ☐ **4. Arytenoid**

The intrinsic muscles of the larynx act largely to adjust the tension on the vocal cords (ligaments), opening or closing the **rima glottidis** (space between the vocal cords) and opening and closing the **rima vestibuli**, the opening above the **vestibular folds** (false folds). This action is important during swallowing to prevent aspiration into the trachea, but also adjusts the size of the vestibule during phonation to add quality to the sound. All of these intrinsic muscles are innervated by the vagus nerve (CN X), via its recurrent laryngeal branch.

The **vocal folds** (vocal ligaments covered by mucosa) control phonation much like a reed controls sound in a reed instrument (e.g., saxophone, clarinet). Vibrations of the folds produce sounds as air passes through the rima glottidis (the space between the vocal folds). The posterior cricoarytenoid muscles are important because they are the only laryngeal muscles that abduct the vocal folds and maintain the opening between the vocal cords. The vestibular folds are protective in function.

**COLOR** the following intrinsic muscles of the larynx, using a different color for each muscle:

- ☐ **5. Posterior cricoarytenoid muscles: the only pair of muscles that abduct the vocal folds**
- ☐ **6. Arytenoid muscle: composed of transverse and oblique fibers, this muscle adducts the vocal folds and narrows the rima vestibuli**
- ☐ **7. Cricothyroid muscle: pulls the thyroid cartilage anteroinferiorly on the cricoid cartilage and tenses the vocal folds by stretching them**

***Clinical Note:***

**Hoarseness** can be due to any condition that results in improper vibration or coaptation of the vocal folds. **Acute laryngitis** results in inflammation and edema that may be caused by smoking, gastroesophageal reflux disease, chronic rhinosinusitis, cough, cysts, surgical scarring, cancer, overuse of the voice, and infection. The larynx may be examined visually by a procedure called **indirect laryngoscopy**, using a mirror, or by direct visualization using a laryngoscope. The **Heimlich maneuver** may be used in an emergency when someone aspirates a foreign object that is lodged superior to the vestibular folds (laryngeal spasm tenses the vocal folds, closing the rima glottidis) so no air can enter the trachea.

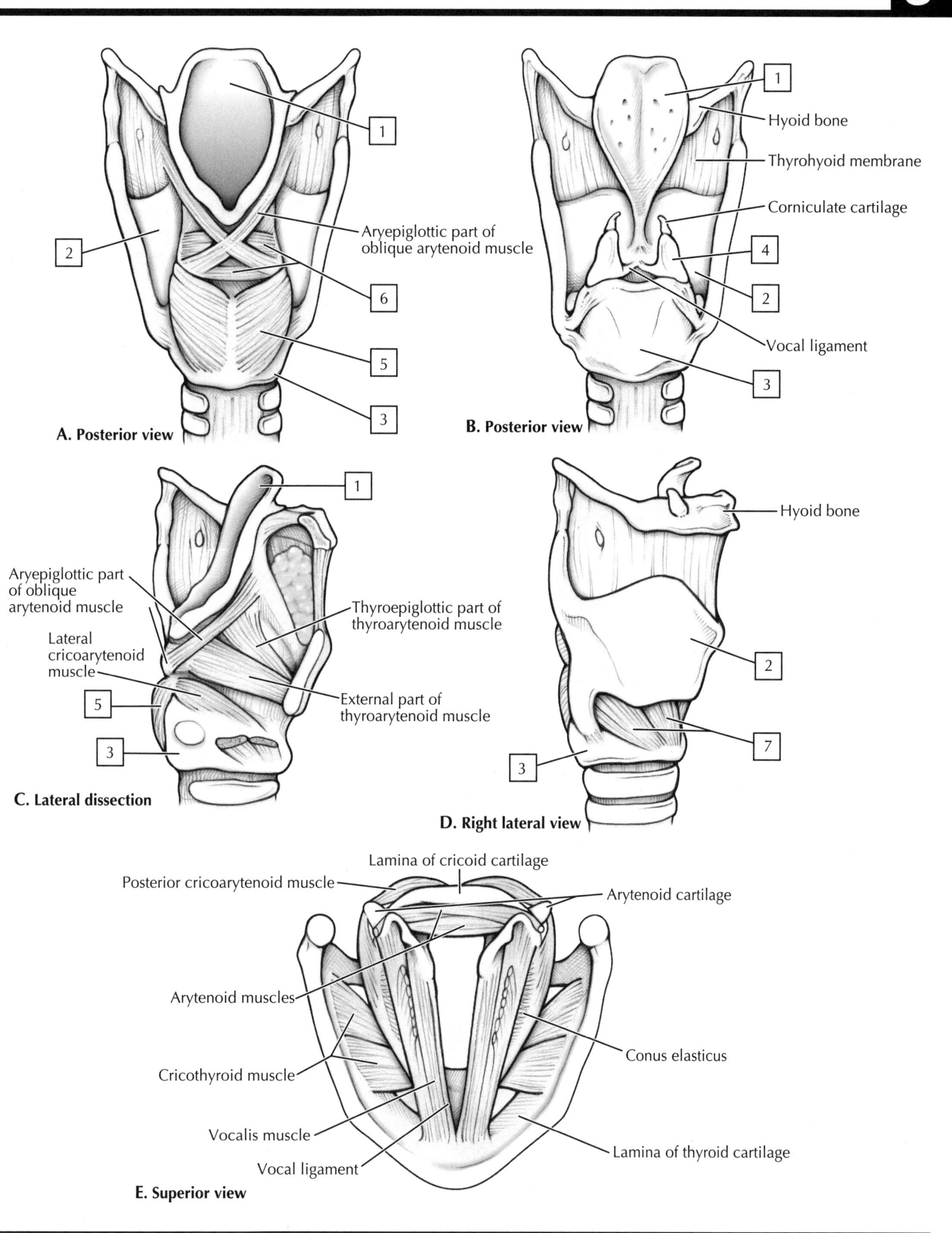
1
2
Aryepiglottic part of oblique arytenoid muscle
6
5
3
A. Posterior view
1
Hyoid bone
Thyrohyoid membrane
Corniculate cartilage
4
2
Vocal ligament
3
B. Posterior view
1
Aryepiglottic part of oblique arytenoid muscle
Lateral cricoarytenoid muscle
5
3
Thyroepiglottic part of thyroarytenoid muscle
External part of thyroarytenoid muscle
C. Lateral dissection
Hyoid bone
2
7
3
D. Right lateral view
Lamina of cricoid cartilage
Posterior cricoarytenoid muscle
Arytenoid cartilage
Arytenoid muscles
Conus elasticus
Cricothyroid muscle
Vocalis muscle
Lamina of thyroid cartilage
Vocal ligament
E. Superior view

# 3 Muscles of the Neck

Muscles of the neck divide the neck into several descriptive "triangles" that are used by surgeons to identify key structures within these regions.

**COLOR** each of these triangles, using a different color to outline the boundaries of each triangle (color over the demarcated outline):

- ☐ 1. **Posterior: between the trapezius and sternocleidomastoid muscles, this triangle is not subdivided further**

  **Anterior, which is further subdivided into the triangles listed below:**
- ☐ 2. **Submandibular: contains the submandibular salivary gland**
- ☐ 3. **Submental: lies beneath the chin**
- ☐ 4. **Muscular: lies anteriorly in the neck below the hyoid bone**
- ☐ 5. **Carotid: contains the carotid artery**

In general, the muscles of the neck position the larynx during swallowing, stabilize the hyoid bone, move the head and upper limb, or are postural muscles attached to the head and/or vertebrae. The key muscles are summarized in the table below. The muscles below the hyoid bone are called "infrahyoid" or "strap" muscles, whereas those above the hyoid bone are called "suprahyoid" muscles.

**COLOR** each of the following muscles, using a different color for each muscle:

- ☐ 6. **Stylohyoid**
- ☐ 7. **Posterior belly of the digastric**
- ☐ 8. **Sternocleidomastoid**
- ☐ 9. **Anterior belly of the digastric**
- ☐ 10. **Thyrohyoid**
- ☐ 11. **Sternohyoid**
- ☐ 12. **Sternothyroid**
- ☐ 13. **Omohyoid**

***Clinical Note:***

The neck provides a conduit that connects the head to the thorax. The muscles, vessels, and visceral structures (trachea and esophagus) are all tightly bound within **three fascial layers** that create compartments within the neck. Infections or masses (tumors) in one or another of these tight spaces can compress softer structures and cause significant pain. The fascial layers themselves also can limit the spread of infection between compartments. On the labeled diagram of the neck in transverse (see part ***E***), **COLOR the three fascial layers** to highlight their extent. The three fascial layers include the:

- **Superficial investing cervical fascia:** surrounds the neck and invests the trapezius and sternocleidomastoid muscles
- **Pretracheal fascia:** limited to the anterior neck, it invests the infrahyoid muscles, thyroid gland, trachea, and esophagus
- **Deep investing cervical fascia:** a tubular sheath, it invests the prevertebral muscles and vertebral column

The **carotid sheath** blends with these fascial layers but is distinct and contains the common carotid artery, internal jugular vein, and vagus nerve (CN X).

| MUSCLE | ORIGIN ATTACHMENT | INSERTION ATTACHMENT | INNERVATION | MAIN ACTIONS |
|---|---|---|---|---|
| Sternocleido-mastoid | *Sternal head:* manubrium<br>*Clavicular head:* medial third of clavicle | Mastoid process and lateral half of superior nuchal line of occipital bone | Spinal root of cranial nerve (CN XI) and C2-C3 (sensory) | Tilts head to one side, i.e., laterally flexes and rotates head so face is turned superiorly toward opposite side; acting together, muscles flex neck |
| Digastric | *Anterior belly*: digastric fossa of mandible<br>*Posterior belly*: mastoid notch | Intermediate tendon to hyoid bone | *Anterior belly*: mylohyoid nerve (CN $V_3$), a branch of inferior alveolar nerve<br>*Posterior belly*: facial nerve (CN VII) | Depresses mandible; raises hyoid bone and steadies it during swallowing and speaking |
| Sternohyoid | Manubrium of sternum and medial end of clavicle | Body of hyoid bone | C1-C3 from ansa cervicalis | Depresses hyoid bone and larynx after swallowing |
| Sternothyroid | Posterior surface of manubrium, 1st costal cartilage | Oblique line of thyroid lamina | C2 and C3 from ansa cervicalis | Depresses larynx and thyroid cartilage after swallowing |
| Thyrohyoid | Oblique line of thyroid cartilage | Body and greater horn of hyoid bone | C1 via hypoglossal nerve | Depresses hyoid bone and elevates larynx when hyoid bone is fixed |
| Omohyoid | Superior border of scapula near suprascapular notch | Inferior border of hyoid bone | C1-C3 from ansa cervicalis | Depresses and fixes hyoid bone |
| Mylohyoid | Mylohyoid line of mandible | Raphe and body of hyoid bone | Mylohyoid nerve, a branch of inferior alveolar nerve of mylohyoid nerve (CN $V_3$) | Elevates hyoid bone, floor of mouth, and tongue during swallowing and depresses mandible |
| Stylohyoid | Styloid process | Body of hyoid bone | Facial nerve (CN VII) | Elevates and retracts hyoid bone |

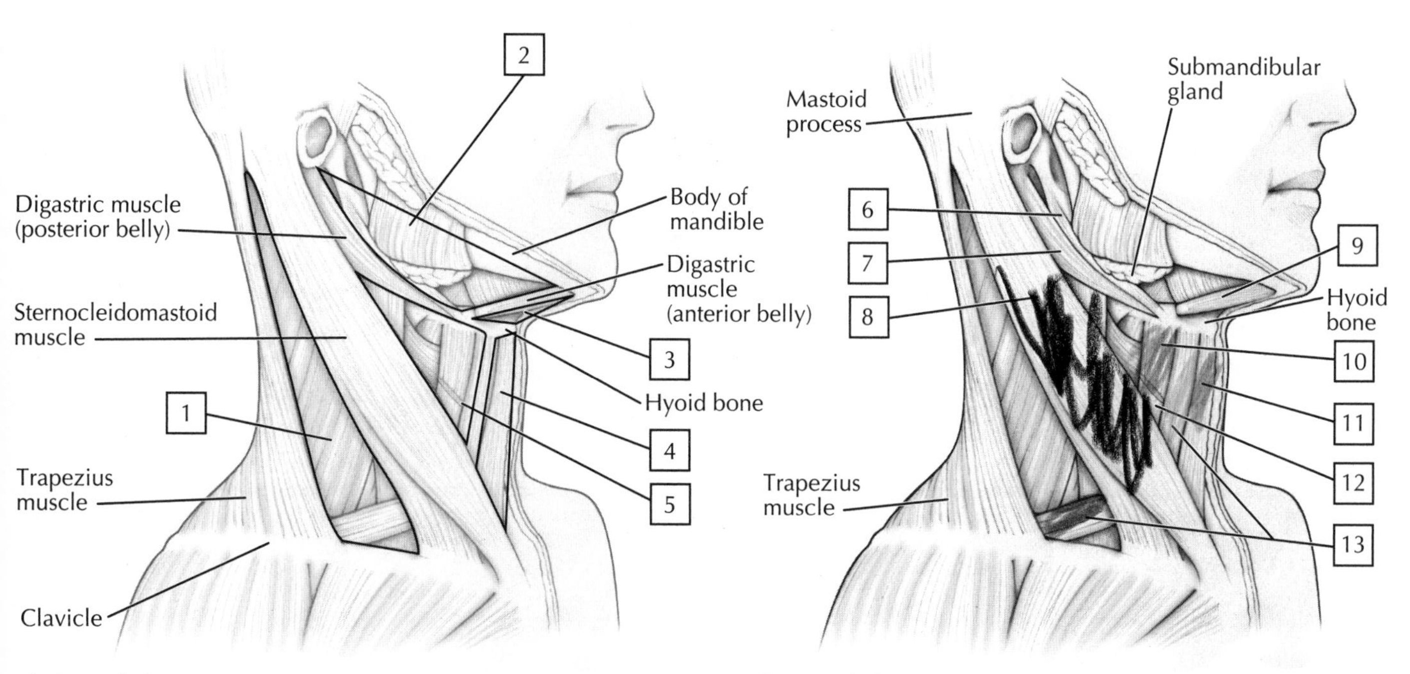


A. Lateral view

B. Lateral view

Hyoid bone
9
6
10
7
Thyroid cartilage
10
13
12
11
Thyroid gland
13
Trachea
Clavicle
Sternohyoid muscle (cut)

C. Anterior view

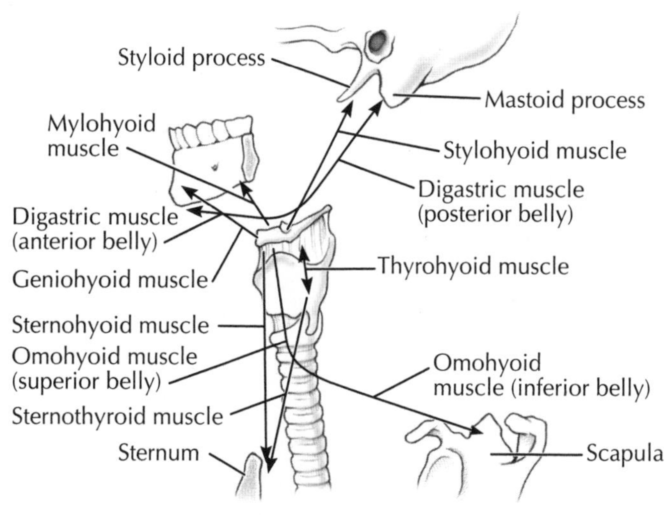


D. Infrahyoid and suprahyoid muscles and their actions

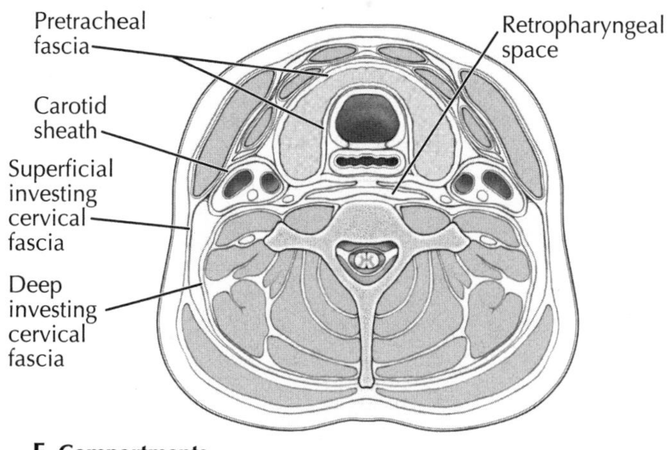


E. Compartments

# 3 Prevertebral Muscles

The prevertebral fascia of the neck encloses many of the prevertebral muscles, which lie anterior to the vertebral column and are muscles that move the head and/or act as postural muscles supporting the head and neck. This group of muscles includes the scalenus muscles (anterior, medius, and posterior) that attach to the upper ribs and also are accessory muscles of respiration. They help raise the thoracic cage during deep inspiration. The prevertebral muscles are summarized in the table below.

**COLOR** the following prevertebral muscles, using a different color for each muscle:

- ☐ **1. Longus capitis (capitis refers to the head)**
- ☐ **2. Longus colli (colli refers to the neck)**
- ☐ **3. Scalenus anterior muscle (note that the subclavian vein passes anterior to this muscle)**
- ☐ **4. Scalenus medius muscle (note that the subclavian artery passes between this muscle and the scalenus anterior muscle)**
- ☐ **5. Scalenus posterior muscle**

| MUSCLE | ORIGIN ATTACHMENT | INSERTION ATTACHMENT | INNERVATION | MAIN ACTIONS |
|---|---|---|---|---|
| Longus colli | Body of T1-T3 with attachments to bodies of C4-C7 and transverse processes of C3-C6 | Anterior tubercle of C1 (atlas), transverse processes of C4-C6, and bodies of C2-C6 | C2-C6 spinal nerves | Flexes cervical vertebrae; allows slight rotation |
| Longus capitis | Anterior tubercles of C3-C6 transverse processes | Basilar part of occipital bone | C2-C3 spinal nerves | Flexes head |
| Rectus anterior capitis | Lateral mass of C1 (atlas) | Base of occipital bone, anterior to occipital condyle | C1-C2 spinal nerves | Flexes head |
| Rectus lateralis capitis | Transverse process of C1 (atlas) | Jugular process of occipital bone | C1-C2 spinal nerves | Flexes laterally and helps stabilize head |
| Scalenus posterior | Posterior tubercles of transverse processes of C4-C6 | Second rib | C5-C8 | Flexes neck laterally; elevates second rib |
| Scalenus medius | Posterior tubercles of transverse processes of C2-C7 | First rib | C3-C7 | Flexes neck laterally; elevates first rib |
| Scalenus anterior | Anterior tubercles of transverse processes of C3-C6 | First rib | C5-C8 | Flexes neck laterally; elevates first rib |

***Clinical Note:***

Looking at the cross section of the neck and the fascial layers in the illustration on the previous page (Plate 3-7), note that there is a space between the pretracheal and deep investing cervical fascia called the **retropharyngeal space**. Infections and abscesses can gain access to this space and spread anywhere from the base of the skull to the upper portion of the thoracic cavity (superior mediastinum). For this reason, clinicians sometimes refer to this space as the "danger" space. A regional nerve block may be used for surgical procedures in the neck. A **cervical plexus block** can be performed by injecting an anesthetic agent along several points along the posterior border of the middle third of the sternocleidomastoid muscle (see Plates 3-7 and 4-28).

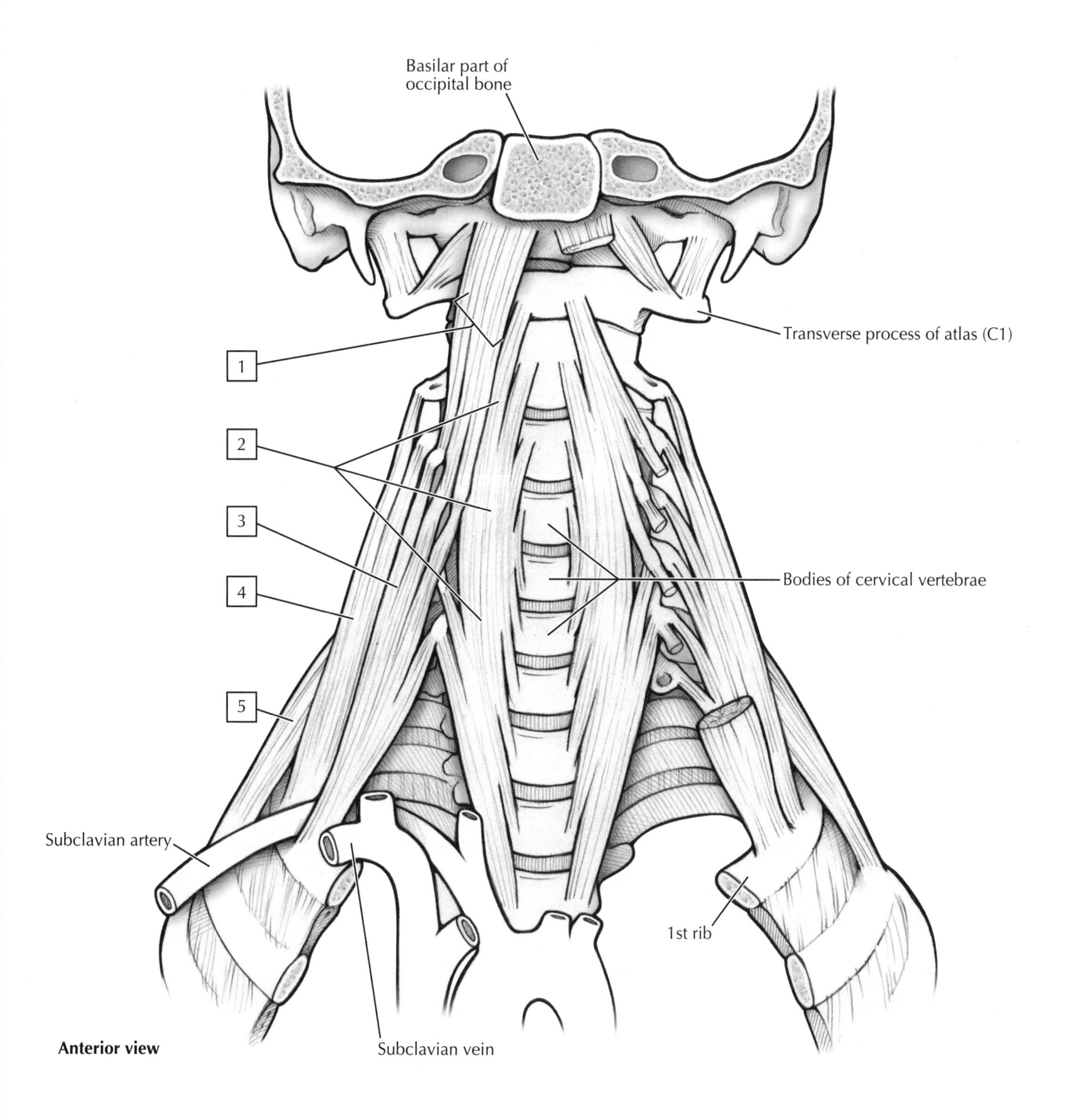
Basilar part of occipital bone
Transverse process of atlas (C1)
1
2
3
4
5
Bodies of cervical vertebrae
Subclavian artery
1st rib
Anterior view
Subclavian vein

# Superficial and Intermediate Back Muscles

The muscles of the back are divided functionally into three groups: superficial, intermediate, and deep.

**Superficial muscles**, which are superficially located, control movements of the upper limbs, largely by acting on the scapulae.

**COLOR** the following superficial muscles, using a different color for each:

☐ 1. **Trapezius: this muscle and the sternocleidomastoid are the only two muscles innervated by the accessory nerve (CN XI)**

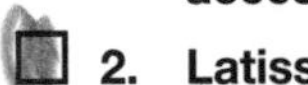

☐ 2. **Latissimus dorsi**

Intermediate muscles, just deep to the superficial layer, are accessory muscles of respiration and have attachments to ribs. The trapezius and latissimus dorsi are removed from the right side of the plate so that you can see this group of muscles.

**COLOR** the following intermediate muscles, using a different color for each:

☐ 3. **Levator scapulae**

☐ 4. **Serratus posterior superior: intermediate group of muscles; have respiratory function**

☐ 5. **Rhomboid major (muscle cut to reveal deeper muscles)**

☐ 6. **Serratus posterior inferior: intermediate group of muscles; have respiratory function**

These groups of back muscles are summarized in the table below.

| MUSCLE | SUPERIOR ATTACHMENT | INFERIOR ATTACHMENT | INNERVATION | MAIN ACTIONS |
|---|---|---|---|---|
| Trapezius | Superior nuchal line, external occipital protuberance, nuchal ligament, and spinous processes of C7-T12 | Lateral third of clavicle, acromion, and spine of scapula | Accessory nerve (CN XI) and C3-C4 (proprioception) | Elevates, retracts, and rotates scapula; lower fibers depress scapula |
| Latissimus dorsi | Spinous processes of T7-T12, thoracolumbar fascia, iliac crest, and last 3-4 ribs | Humerus (intertubercular sulcus) | Thoracodorsal nerve (C6-C8) | Extends, adducts, and medially rotates humerus |
| Levator scapulae | Posterior tubercles of transverse processes of C1-C4 | Medial border of scapula from superior angle to spine | C3-C4 and dorsal scapular (C5) nerve | Elevates scapula medially and tilts glenoid cavity inferiorly |
| Rhomboid minor and major | *Minor*: nuchal ligament and spinous processes of C7-T1 *Major*: spinous processes of T2-T5 | *Minor:* Medial border of scapula at spine of scapula *Major:* Medial border of scapula below base of spine | Dorsal scapular nerve (C4-C5) | Retract scapula, rotate it to depress glenoid cavity, and fix scapula to thoracic wall |
| Serratus posterior superior | Ligamentum nuchae, spinous processes of C7-T3 | Superior aspect of ribs 2-5 | T1-T4 | Elevates ribs |
| Serratus posterior inferior | Spinous processes of T11-L2 | Inferior aspect of ribs 9-12 | T9-T12 | Depresses ribs |

The superficial and intermediate groups of back muscles are segmentally innervated by anterior rami of spinal nerves (except the trapezius muscle, which is innervated by CN XI). The superficial group of muscles migrates onto the back during development of the embryo, although they function as muscles that act on the upper limb.

***Clinical Note:***

Should the serratus anterior muscle become paralyzed (see Plate 3-18), due to injury to its innervation by the long thoracic nerve, the medial border of the scapula will protrude posteriolaterally, giving the appearance of a wing ("winged scapula").

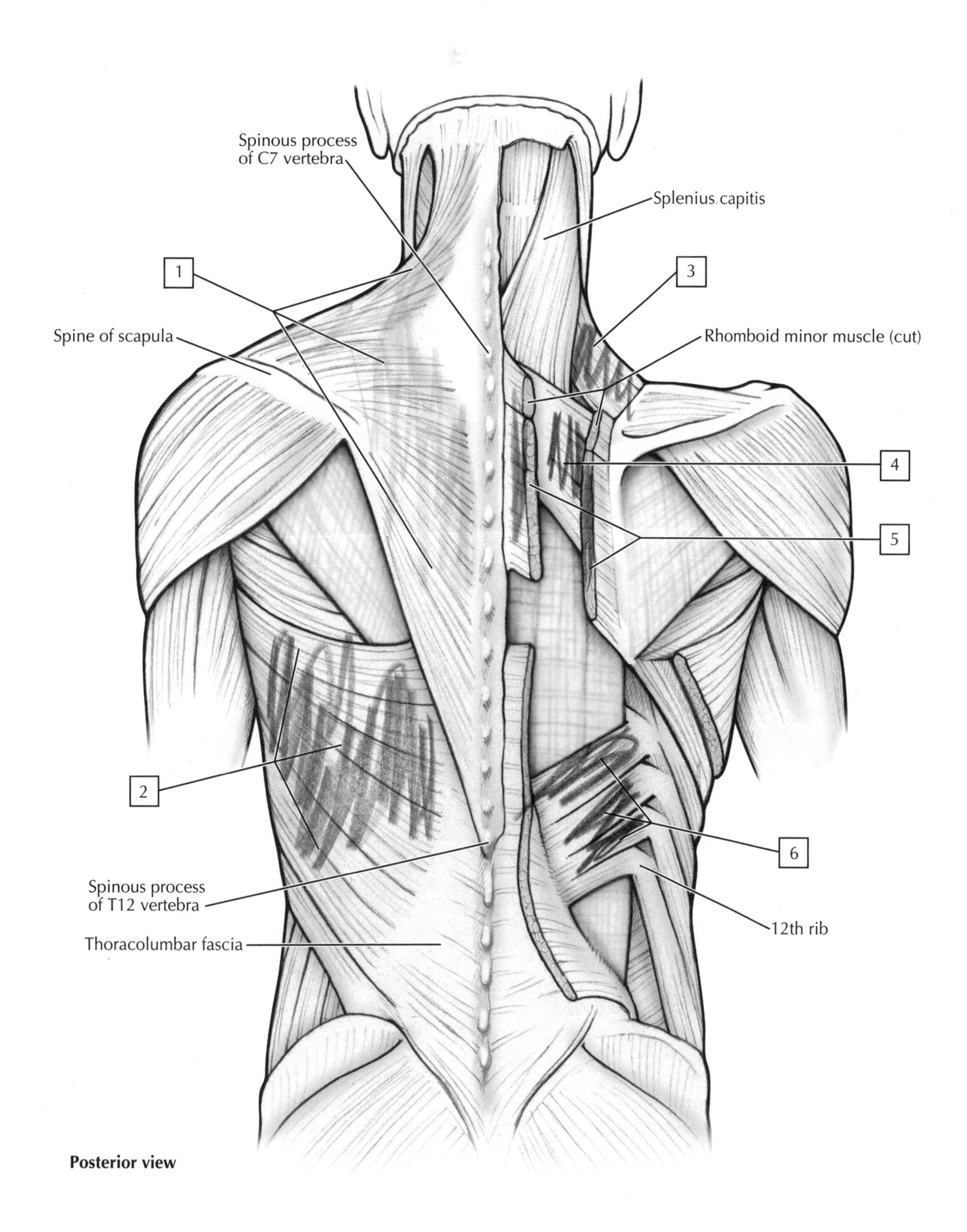


Posterior view

# 3 Deep (Intrinsic) Back Muscles

The deep, or intrinsic, back muscles are beneath the intermediate layer. They participate in movement of the head and neck or postural control of the vertebral column. They are composed of **superficial** (splenius muscles), **intermediate** (erector spinae muscles), and **deep layers** (transversospinal muscles). They support the spine and permit movements of the spine, and they are innervated by posterior rami of spinal nerves. Additionally, the muscles of the back of the neck are transversospinal muscles that are found in the suboccipital region. The muscles are summarized in the table below.

**COLOR** each of the following intrinsic muscles, using a different color for each muscle:

- ☐ 1. **Splenius capitis**
- ☐ 2. **Iliocostalis (erector spinae group, just lateral to the longissimus muscles)**
- ☐ 3. **Longissimus (erector spinae group, just lateral to the spinalis muscles)**
- ☐ 4. **Spinalis (erector spinae group, found most medially in the back)**
- ☐ 5. **Rectus capitis posterior major (suboccipital region)**
- ☐ 6. **Obliquus capitis inferior (suboccipital region; muscles 5 to 7 in this list form the "suboccipital triangle" (see part *C*))**
- ☐ 7. **Obliquus capitis superior (suboccipital region)**

| MUSCLE | SUPERIOR ATTACHMENT | INFERIOR ATTACHMENT | INNERVATION* | MAIN ACTIONS |
|---|---|---|---|---|
| | | **Superficial Layer** | | |
| Splenius capitis | Nuchal ligament, spinous processes of C7-T4 | Mastoid process of temporal bone and lateral third of superior nuchal line | Middle cervical nerves | *Bilaterally*: extends head *Unilaterally*: laterally bends (flexes) and rotates face to same side |
| Splenius cervicis | Spinous processes of T3-T6 | Transverse processes (C1-C3) | Lower cervical nerves | *Bilaterally*: extends neck *Unilaterally*: laterally bends (flexes) and rotates neck toward same side |
| | | **Intermediate Layer** | | |
| Erector spinae | Posterior sacrum, iliac crest, sacrospinous ligament, supraspinous ligament, and spinous processes of lower lumbar and sacral vertebrae | *Iliocostalis*: angles of lower ribs and cervical transverse processes *Longissimus*: between tubercles and angles of ribs, transverse processes of thoracic and cervical vertebrae, and mastoid process *Spinalis*: spinous processes of upper thoracic and midcervical vertebrae | Respective spinal nerves of each region | Extends and laterally bends vertebral column and head |
| Semispinalis | Transverse processes of C4-T12 | Spinous processes of cervical and thoracic regions | Respective spinal nerves of each region | Extends head, neck, and thorax and rotates them to opposite side |
| Multifidi | Sacrum, ilium, and transverse processes of T1-T12, and articular processes of C4-C7 | Spinous processes of vertebrae above, spanning two to four segments | Respective spinal nerves of each region | Stabilize spine |
| Rotatores | Transverse processes of cervical, thoracic, and lumbar regions | Lamina and transverse process of spine above, spanning one or two segments | Respective spinal nerves of each region | Stabilize, extend, and rotate spine |
| | | **Deep Layer** | | |
| Rectus capitis posterior major | Spine of axis | Lateral inferior nuchal line | Suboccipital nerve (C1) | Extends head and rotates to same side |
| Rectus capitis posterior minor | Tubercle of posterior arch of atlas | Median inferior nuchal line | Suboccipital nerve (C1) | Extends head |
| Obliquus capitis superior | Transverse process of atlas | Occipital bone | Suboccipital nerve (C1) | Extends head and bends it laterally |
| Obliquus capitis inferior | Spine of axis | Atlas transverse process | Suboccipital nerve (C1) | Rotates atlas to turn face to same side |

*Dorsal rami of spinal nerves.

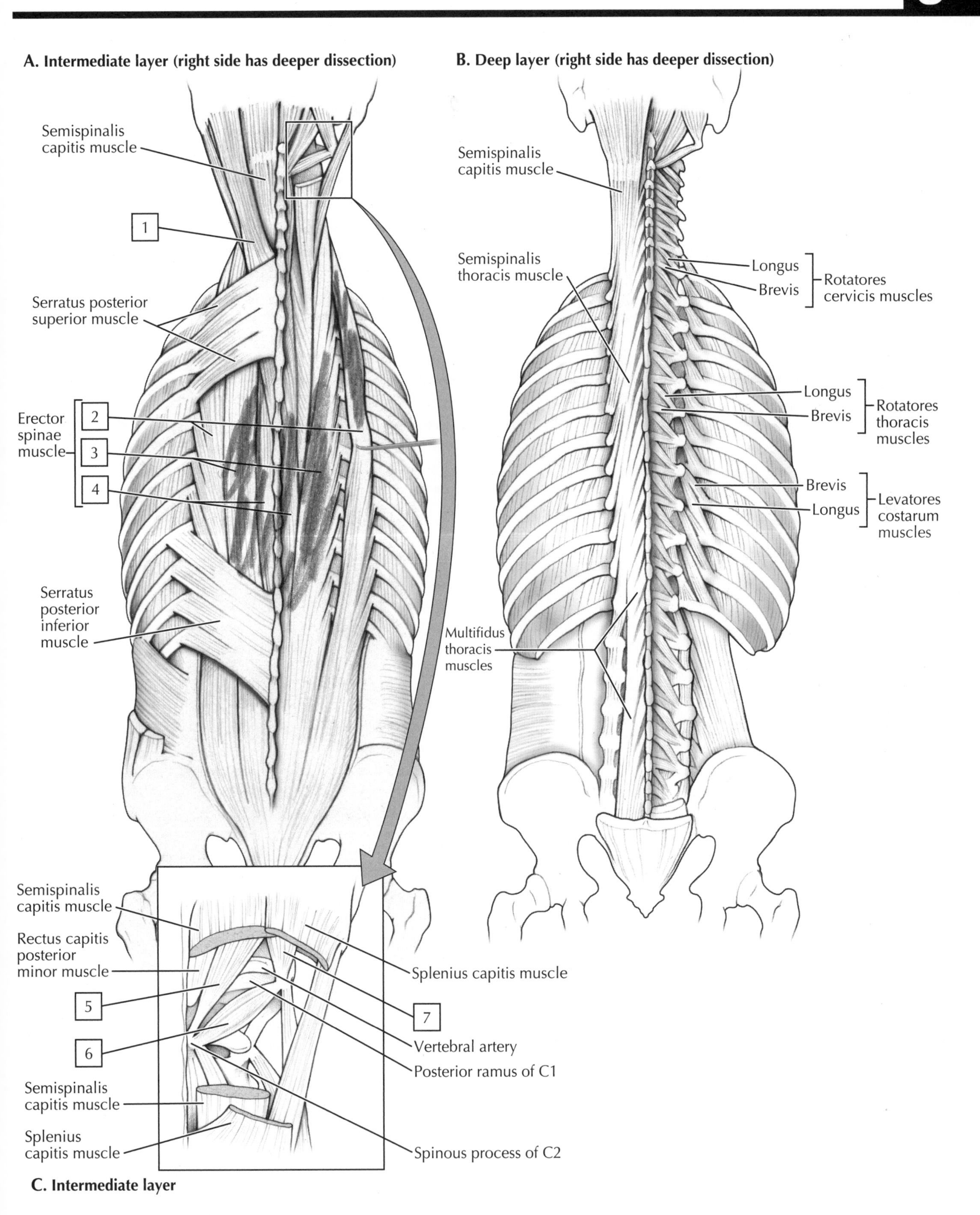
A. Intermediate layer (right side has deeper dissection)
Semispinalis capitis muscle
1
Serratus posterior superior muscle
Erector spinae muscle
2
3
4
Serratus posterior inferior muscle
B. Deep layer (right side has deeper dissection)
Semispinalis capitis muscle
Semispinalis thoracis muscle
Longus
Brevis
Rotatores cervicis muscles
Longus
Brevis
Rotatores thoracis muscles
Brevis
Longus
Levatores costarum muscles
Multifidus thoracis muscles
Semispinalis capitis muscle
Rectus capitis posterior minor muscle
5
6
Semispinalis capitis muscle
Splenius capitis muscle
Splenius capitis muscle
7
Vertebral artery
Posterior ramus of C1
Spinous process of C2
C. Intermediate layer

# 3 Thoracic Wall Muscles

Muscles of the thoracic wall fill the spaces between adjacent ribs, or have attachments to the sternum or vertebrae and then attach to ribs or costal cartilages. Functionally, the muscles of the thoracic wall keep the intercostal spaces rigid, thereby preventing them from bulging out during expiration or being sucked in during inspiration. The exact role of individual intercostal muscles on the movements of the ribs is difficult to interpret despite many electromyographic studies.

On the anterior chest wall, the pectoralis major and minor muscles overlie the intercostal muscles, but these two muscles really act on the upper limb and will be discussed later (see Plate 3-18). Segmental intercostal nerves and vessels travel between the internal and innermost intercostal muscles, as seen in the cross section of the thoracic wall.

**COLOR** each of the following muscles, using a different color for each muscle:

☐ **1. External intercostals: outermost layer of the three intercostal muscles; fibers run from superolateral to inferomedial**

☐ **2. Internal intercostals: middle layer of intercostals; fibers tend to run from superomedial to inferolateral**

☐ **3. Innermost intercostals: fibers almost parallel those of the internal intercostals and may sometimes be fused to these muscles**

☐ **4. Transversus thoracis**

| MUSCLE | SUPERIOR ATTACHMENT | INFERIOR ATTACHMENT | INNERVATION | MAIN ACTIONS |
|---|---|---|---|---|
| External intercostal | Inferior border of rib above | Superior border of rib below | Intercostal nerves | Elevates ribs, supports intercostal space |
| Internal intercostal | Inferior border of rib above | Superior border of rib below | Intercostal nerves | Prevents pushing out or drawing in of intercostal spaces in inspiration, lowers ribs in forced expiration |
| Innermost intercostal | Lower border of ribs | Upper border of rib below rib of origin | Intercostal nerves | Elevates ribs |
| Transversus thoracis | Internal surfaces of costal cartilages 2-6 | Posterior surface of lower sternum | Intercostal nerves | Depresses ribs and costal cartilages |
| Subcostal | Internal surfaces of lower ribs near their angles | Superior borders of second or third ribs below | Intercostal nerves 2nd-5th | Depresses ribs |
| Levator costarum | Transverse processes of C7 and T1-T11 | Subjacent ribs between tubercle and angle | Posterior rami of lower thoracic nerve | Elevates ribs |

***Clinical Note:***

Sometimes it is necessary to introduce a needle or catheter through the chest wall into the underlying pleural cavity, usually to drain off fluids (blood or extracellular fluid and pus) or air that accumulates in this space and could potentially collapse the lung. Careful positioning of the needle or catheter is necessary to avoid impaling the intercostal nerve and vessels, which pass inferior to each rib in the costal groove.

**Asthma and emphysema** are fairly common respiratory problems making it difficult to breathe. Patients tend to use accessory respiratory muscles to help expand their thoracic cavities. They may use neck muscles, e.g., the sternocleidomastoid and scalene muscles to assist with inspiration.

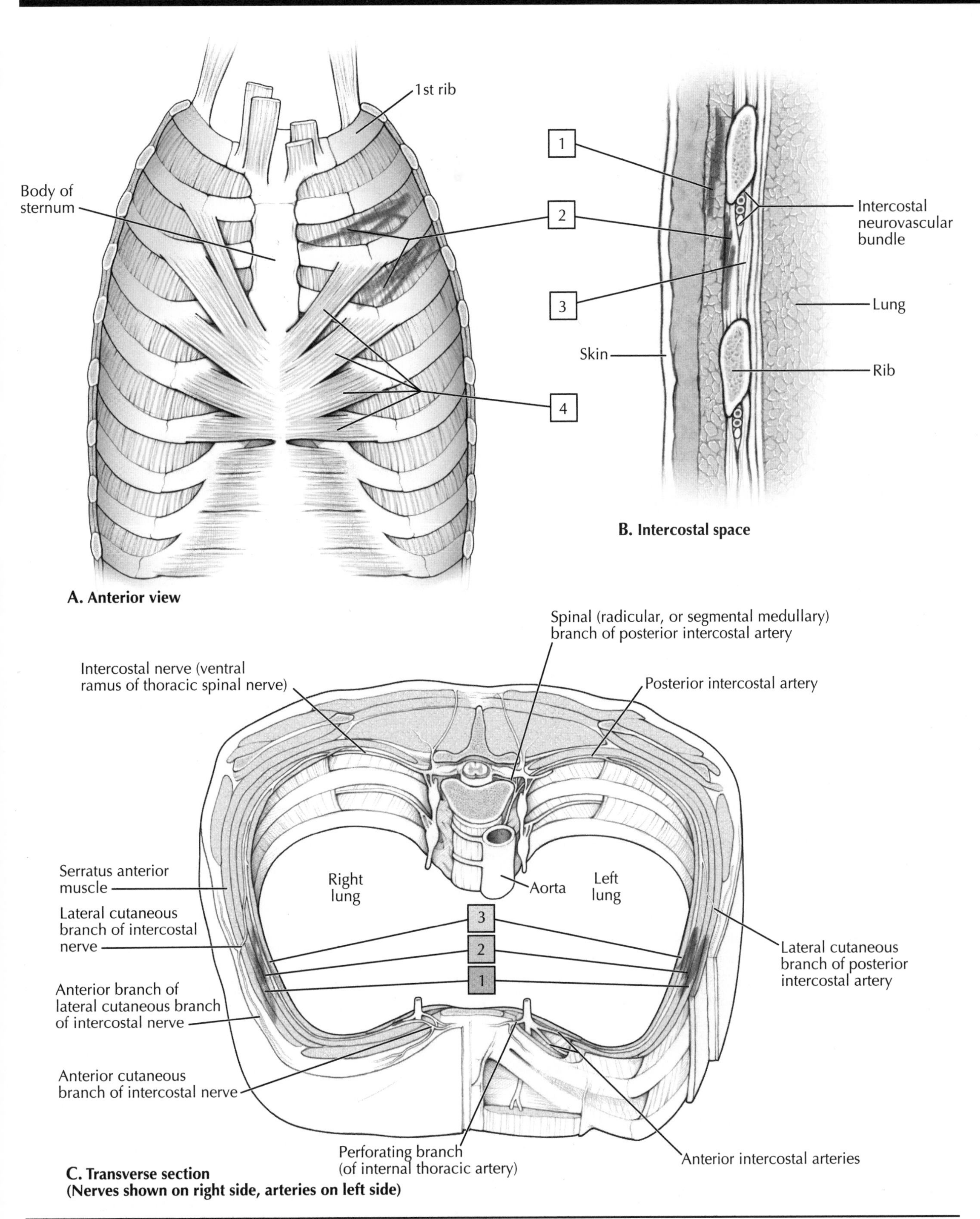


**A. Anterior view**

**B. Intercostal space**

**C. Transverse section (Nerves shown on right side, arteries on left side)**

# 3 Anterior Abdominal Wall Muscles

Three muscles (external abdominal oblique, internal abdominal oblique, and transversus abdominis) wrap around the abdominal wall and are direct continuations of the three muscle layers found in the thoracic wall, where they lie between the ribs and form the intercostal muscles.

The functions of these anterior abdominal muscles include:

- Compressing the abdominal wall and increasing the intra-abdominal pressure, especially when one is lifting an object and during urination, defecation, and childbirth
- Assisting the diaphragm during forced expiration (this occurs unexpectedly when a blow is administered to the anterior abdominal wall and the "wind is knocked out of you")
- Helping flex and rotate the trunk
- Tensing the abdominal wall

**COLOR** these three labeled muscles using a different color for each. Work from the superficial to the deeper layer and note the direction of the muscle fibers as you color:

- ☐ **1. External abdominal oblique**
- ☐ **2. Internal abdominal oblique**
- ☐ **3. Transversus abdominis**

| MUSCLE | ORIGIN ATTACHMENT | INSERTION ATTACHMENT | INNERVATION | MAIN ACTIONS |
|---|---|---|---|---|
| External abdominal oblique | External surfaces of 5th to 12th ribs | Linea alba, pubic tubercle, and anterior half of iliac crest | Inferior six thoracic nerves and subcostal nerve | Compresses and supports abdominal viscera; flexes and rotates trunk |
| Internal abdominal oblique | Thoracolumbar fascia, anterior two-thirds of iliac crest, and lateral two-thirds of inguinal ligament | Inferior borders of 10th to 12th ribs, linea alba, and pubis via conjoint tendon | Anterior rami of inferior six thoracic and 1st lumbar nerves | Compresses and supports abdominal viscera; flexes and rotates trunk |
| Transversus abdominis | Internal surfaces of 7-12 costal cartilages, thoracolumbar fascia, iliac crest, and lateral third of inguinal ligament | Linea alba with aponeurosis of internal oblique, pubic crest, and pecten pubis via conjoint tendon | Anterior rami of inferior six thoracic and 1st lumbar nerves | Compresses and supports abdominal viscera |
| Rectus abdominis | Pubic symphysis and pubic crest | Xiphoid process and costal cartilages 5-7 | Anterior rami of inferior six thoracic nerves | Flexes trunk and compresses abdominal viscera |

Two midline muscles (rectus abdominis and pyramidalis) lie within the rectus sheath, a tendinous sheath composed of the aponeurotic layers of the three abdominal muscles colored (1 to 3). The layers (lamina) that compose the sheath are deficient below the arcuate line (in the lower quarter) of the rectus sheath, where only the transversalis fascia lies in contact with the rectus abdominis muscle (see part ***B***, Below arcuate line).

**COLOR** the midline muscles of the anterior abdominal wall, using a different color from those used previously:

- ☐ **4. Rectus abdominis (note the three tendinous intersections—the infamous "six-pack abs")**
- ☐ **5. Pyramidalis**

**COLOR** the aponeurosis extending from the muscle to form the layers of the rectus sheath. Use a color different from the muscle colors, but note the relationship to the muscles.

- ☐ **1A. Aponeurosis of external abdominal oblique muscle**
- ☐ **2A. Aponeurosis of internal abdominal oblique muscle**
- ☐ **3A. Aponeurosis of transversus abdominis muscle**

| LAYER | COMMENT |
|---|---|
| Anterior lamina above arcuate line | Formed by fused aponeuroses of external and internal abdominal oblique muscles |
| Posterior lamina above arcuate line | Formed by fused aponeuroses of internal abdominal oblique and transversus abdominis muscles |
| Below arcuate line | All three muscle aponeuroses fuse to form anterior lamina, with rectus abdominis in contact only with transversalis fascia posteriorly |

***Clinical Note:***

**Hernias**, abnormal outpouchings of underlying structures due to a weakness of the wall, can occur on the anterior abdominal wall. The most common types include:

- Umbilical hernias—usually seen up to age 3 years or after the age of 40
- Linea alba hernias—often occur in the epigastric region along the midline linea alba
- Incisional hernias—occur at sites of previous abdominal surgical scars
- Inguinal hernias—related to the inguinal canal in the inguinal region (where abdomen and thigh meet)

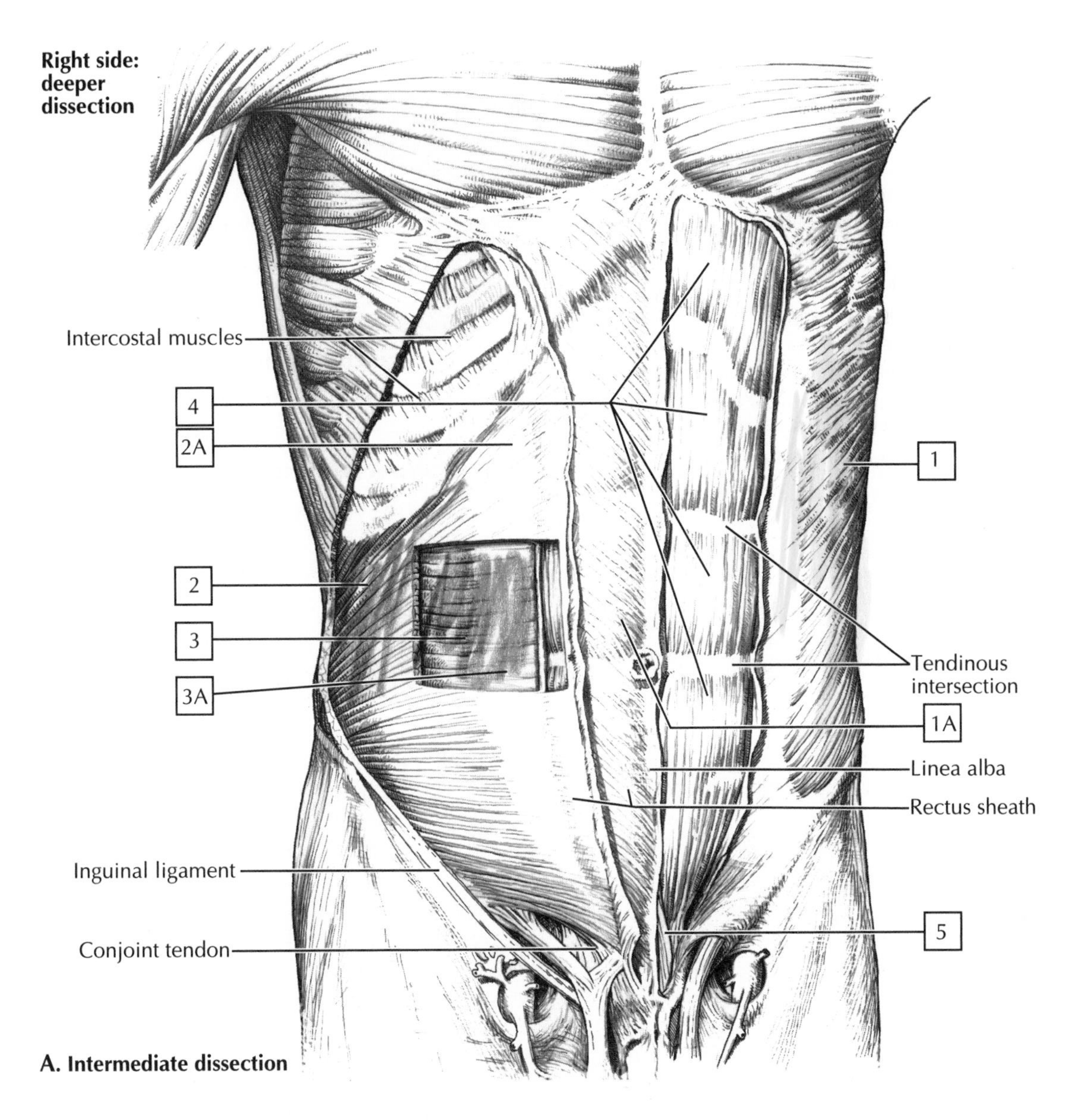


A. Intermediate dissection

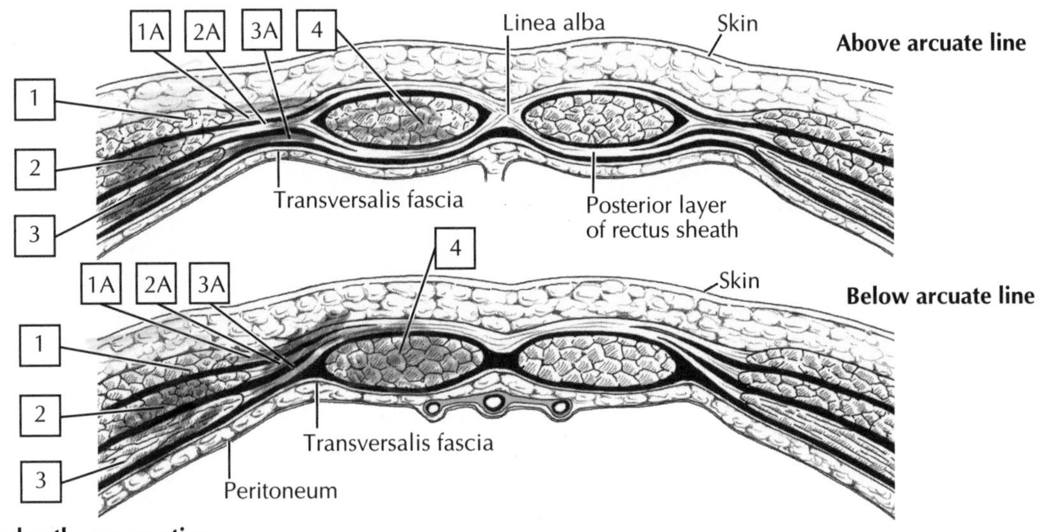


B. Rectus sheath cross section

# Muscles of the Male Inguinal Region

The muscles of the male and female inguinal regions are similar. However, the presence of the spermatic cord in the inguinal canal and the descent of the testis during fetal development render this region clinically unique in males and predispose males to inguinal hernias.

During development, the testis descends from its site of embryological origin in the posterior abdominal region through the inguinal canal (an oblique, lateral-to-medial passageway through the lower anterior abdominal wall) and into the scrotum. Each testis is tethered by its **spermatic cord**, which among other structures contains the ductus (vas) deferens, which will provide a passageway for sperm to re-enter the body cavity and join with the ducts of the seminal vesicles (right and left) and form the ejaculatory ducts which course through the prostatic gland and empty into the prostatic portion of the urethra during sexual arousal (see Plate 10-6).

As the spermatic cord runs in the inguinal canal, it picks up spermatic fascial layers derived from the abdominal wall structures as the testis descends. These derivatives include the:

- **External spermatic fascia**: derived from the external abdominal oblique aponeurosis
- **Middle (cremasteric) fascia**: derived from the internal abdominal oblique muscle, this fascia really includes small skeletal muscle fibers of the cremaster muscle
- **Internal spermatic fascia**: derived from the transversalis fascia

The spermatic cord contains the following structures:

- Ductus (vas) deferens
- Testicular and cremasteric arteries, and the artery of the ductus deferens
- Pampiniform plexus of veins
- Autonomic sympathetic nerve fibers
- Genital branch of the genitofemoral nerve (innervates the cremasteric muscle)
- Lymphatics

The **inguinal canal** itself is a small passageway through the abdominal musculature that is demarcated at both ends by inguinal rings, the deep ring opening in the abdomen and the superficial ring opening externally just lateral to the pubic tubercle. The features of the inguinal canal are noted in the table below.

**COLOR** the following features of the inguinal region and spermatic cord, using a different color for each feature:

- ☐ 1. **Ductus deferens**
- ☐ 2. **External abdominal oblique muscle and aponeurosis**
- ☐ 3. **Internal abdominal oblique muscle**
- ☐ 4. **Transversus abdominis muscle**
- ☐ 5. **Transversalis fascia**
- ☐ 6. **External spermatic fascia (covering the spermatic cord)**
- ☐ 7. **Cremasteric fascia (muscle)**
- ☐ 8. **Internal spermatic fascia**

***Clinical Note:***

**Inguinal hernias** are of two types:

- **Indirect:** 75% of inguinal hernias, they occur lateral to the inferior epigastric vessels and pass through the deep inguinal ring and inguinal canal in a protrusion of peritoneum within the spermatic cord (covered by all three layers of the spermatic cord)
- **Direct:** 25% of hernias, they occur medial to the inferior epigastric vessels and pass through the posterior wall of the inguinal canal; they are separate from the spermatic cord
- Inguinal hernias are much more frequent in males than females, probably related to the descent of the testes in males

The testes begin their descent from the dorsal abdominal wall and arrive in the superior lumbar region around the 9th-12th weeks of fetal development. This "repositioning" of the testes is largely due to the expansive growth of the vertebral column and pelvis. Descent into the scrotum usually occurs shortly before birth or normally soon after birth.

The fetal ovaries, on the other hand, descend from the dorsal abdominal wall and around the 12th week of development relocate into the pelvis, where they remain.

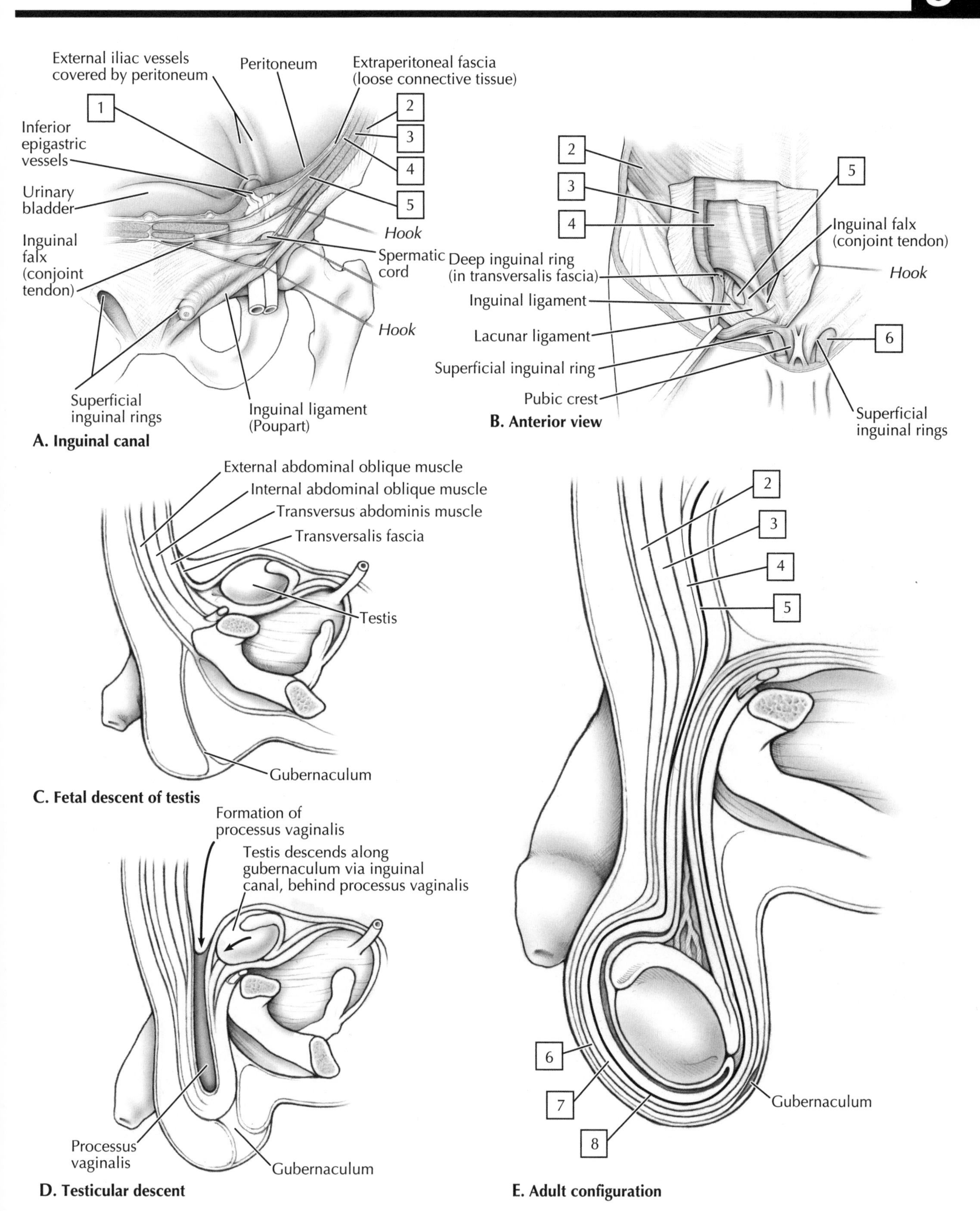
External iliac vessels covered by peritoneum
Peritoneum
Extraperitoneal fascia (loose connective tissue)
1
2
3
4
5
Inferior epigastric vessels
Urinary bladder
Inguinal falx (conjoint tendon)
Hook
Spermatic cord
Hook
Superficial inguinal rings
Inguinal ligament (Poupart)
A. Inguinal canal
2
3
4
5
Inguinal falx (conjoint tendon)
Hook
Deep inguinal ring (in transversalis fascia)
Inguinal ligament
Lacunar ligament
6
Superficial inguinal ring
Pubic crest
Superficial inguinal rings
B. Anterior view
External abdominal oblique muscle
Internal abdominal oblique muscle
Transversus abdominis muscle
Transversalis fascia
Testis
Gubernaculum
C. Fetal descent of testis
Formation of processus vaginalis
Testis descends along gubernaculum via inguinal canal, behind processus vaginalis
Processus vaginalis
Gubernaculum
D. Testicular descent
2
3
4
5
6
7
8
Gubernaculum
E. Adult configuration

# 3 Muscles of the Posterior Abdominal Wall

The muscles of the posterior abdominal wall lie behind the peritoneal cavity and their anterior surface is separated from this cavity by the following:

- Transversalis fascia
- A layer of extraperitoneal fat of variable thickness
- Parietal peritoneum lining the peritoneal cavity

These muscles fill in the space between the lower edge of the rib cage and line the abdominopelvic cavity to the level of the true pelvis. Often the **abdominal diaphragm** is included with these muscles, and its superior extent rises almost to the level of the 8th thoracic vertebral body. Contraction of the diaphragm pulls the central tendon inferiorly and this action increases the volume of the thoracic cavity, causing a drop in pressure slightly below that of the ambient pressure outside of the body. As a result, the air passively passes into the trachea and lungs. Relaxation of the diaphragm and the elastic recoil of the lungs expels the air during normal expiration. These muscles are summarized in the table below.

**COLOR** the following muscles of the posterior abdominal wall, using a different color for each muscle:

☐ 1. **Respiratory diaphragm (leave the central tendon uncolored)**

☐ 2. **Quadratus lumborum**

☐ 3. **Psoas major**

☐ 4. **Iliacus: this muscle and the psoas fuse to function as one muscle, the iliopsoas**

The psoas minor muscle is not always present but does act as a weak flexor of the lumbar vertebral column.

The diaphragm has a central tendinous portion and is attached to the lumbar vertebrae by a right crus and a left crus (leg), which are joined centrally by the median arcuate ligament that passes over the emerging abdominal aorta. The **inferior vena cava** passes through the diaphragm at the T8 vertebral level to enter the right atrium of the heart. The **esophagus** passes through the diaphragm at the T10 vertebral level, along with the anterior and posterior vagal trunks, and the aorta passes through the diaphragm at the T12 vertebral level.

| MUSCLE | ORIGIN ATTACHMENT | INSERTION ATTACHMENT | INNERVATION | ACTIONS |
|---|---|---|---|---|
| Psoas major | Transverse processes of lumbar vertebrae; sides of bodies of T12-L5 vertebrae, and intervening intervertebral discs | Lesser trochanter of femur | Lumbar plexus via anterior branches of L1-L3 nerves | Acting superiorly with iliacus, flexes hip; acting inferiorly, flexes vertebral column laterally; used to balance trunk in sitting position; acting inferiorly with iliacus, flexes trunk |
| Iliacus | Superior two-thirds of iliac fossa, ala of sacrum, and anterior sacroiliac ligaments | Lesser trochanter of femur and shaft inferior to it, and to psoas major tendon | Femoral nerve (L2-L4) | Flexes thigh at hip and stabilizes hip joint; acts with psoas major |
| Quadratus lumborum | Medial half of inferior border of 12th rib and tips of lumbar transverse processes | Iliolumbar ligament and internal lip of iliac crest | Anterior rami of T12 and L1-L4 nerves | Extends and laterally flexes vertebral column; fixes 12th rib during inspiration |
| Respiratory diaphragm | Xiphoid process, lower six costal cartilages, L1-L3 vertebrae | Converge into central tendon | Phrenic nerve (C3-C5) | Draws central tendon down and forward during inspiration |

***Clinical Note:***

An infection of an intervertebral disc at the level of the psoas major muscle can lead to a **psoas abscess**, which first appears at the superior origin of the muscle. This infection can spread beneath the psoas fascial sheath that covers this muscle and even extend inferior to the inguinal ligament.

The **iliopsoas muscle**, because of its position in the posterior abdominal wall, has a relationship with a variety of abdominal organs (ureters, kidneys, pancreas, colon, appendix, lymphatics, and nerves), so when inflammation occurs in this region a physician can perform the iliopsoas test. A patient lying on their unaffected side is asked to extend their thigh, against resistance, on the affected (painful) side. The resulting pain is known as a "positive psoas sign." For example, if the pain is from an inflamed appendix, the patient lying on their left side and extending their right leg against resistance will produce pain and this is a **positive psoas sign**.

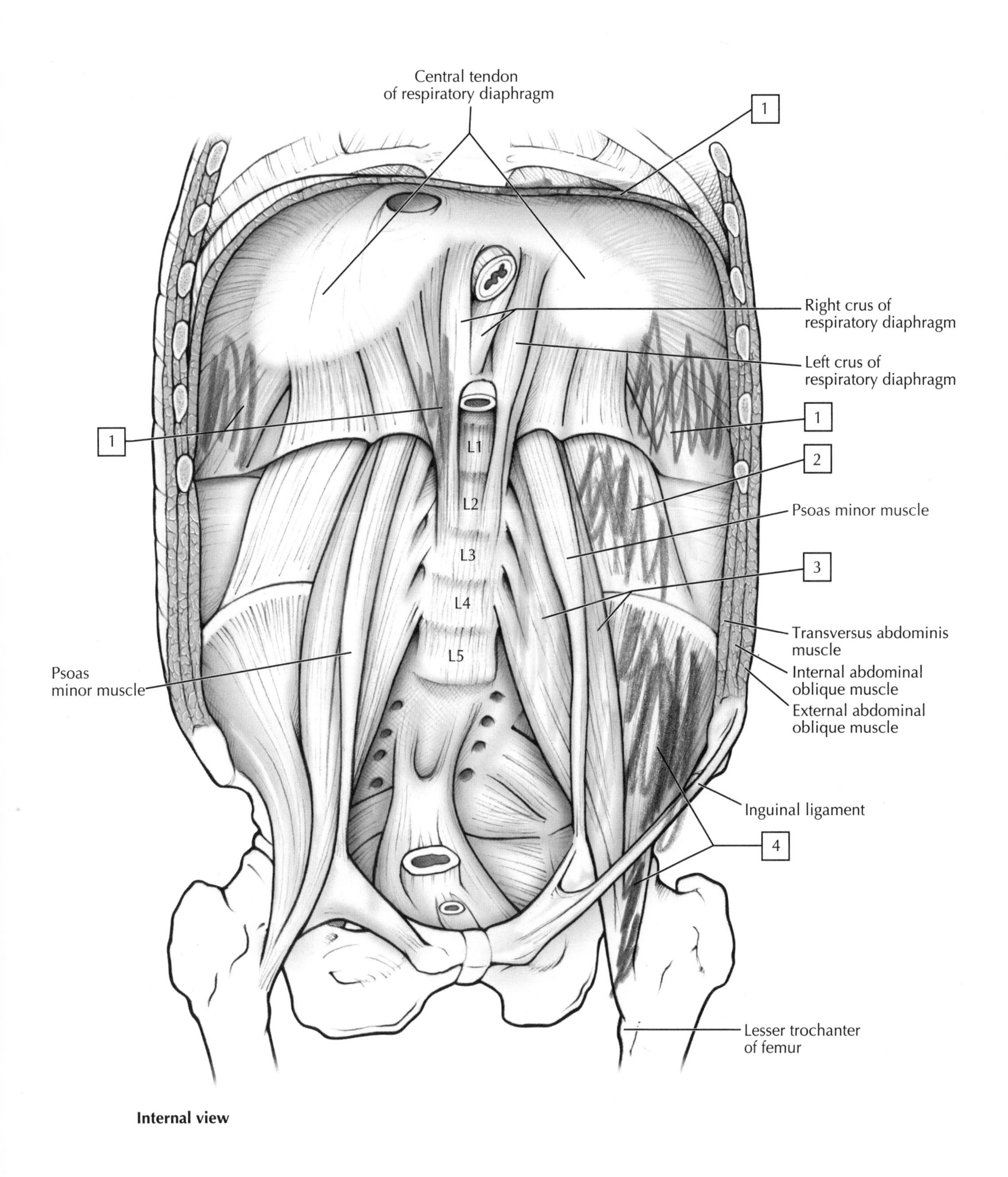


**Internal view**

# 3 Muscles of the Pelvis

The muscles of the pelvis line the lateral pelvic walls (obturator internus and piriformis) and attach to the femur (thigh bone) or cover the floor of the pelvis (levator ani and coccygeus) and form a "**pelvic diaphragm**." We have co-opted the two muscles that form our pelvic diaphragm for a different use than the one for which they were originally intended in most land-dwelling vertebrates. Most land-dwelling mammals, for example, are quadrupeds, whereas we are bipeds and exhibit an upright posture. Bipedalism places greater pressure on our lower pelvic floor as it supports our abdominopelvic viscera. Thus, muscles that are used in animals to tuck the tail between their hind legs (coccygeus) or to wag their tail (levator ani) have become, in humans, muscles with a supportive function because humans no longer have a tail! The levator ani muscle is really a fusion of three separate muscles: the pubococcygeus, puborectalis, and iliococcygeus muscles. The pelvic muscles are summarized in the table below.

**COLOR** the following muscles of the pelvis, using a different color for each muscle:

- ☐ 1. **Levator ani: really composed of three fused muscles, it was a "tail-wagging" muscle in our earliest ancestors**
- ☐ 2. **Obturator internus**
- ☐ 3. **Coccygeus: often partially fibrous, it was a "tail-tucking" muscle in our earliest ancestors**
- ☐ 4. **Piriformis: a pear-shaped muscle; wider on one end than the other, like a pear**

| MUSCLE | ORIGIN ATTACHMENT | INSERTION ATTACHMENT | INNERVATION | MAIN ACTIONS |
|---|---|---|---|---|
| Obturator internus | Pelvic aspect of obturator membrane and pelvic bones | Medial surface of the greater trochanter of femur | Nerve to obturator internus (L5-S1) | Rotates extended thigh laterally; abducts flexed thigh at hip |
| Piriformis | Anterior surface of 2nd to 4th sacral segments and sacrotuberous ligament | Superior border of greater trochanter of femur | Anterior rami of L5, S1-S2 | Rotates extended thigh laterally; abducts flexed thigh; stabilizes hip joint |
| Levator ani | Body of pubis, tendinous arch of obturator fascia, and ischial spine | Perineal body, coccyx, anococcygeal raphe, walls of prostate gland or vagina, rectum, and anal canal | Anterior rami of S3-S4, perineal nerve of the pudendal nerve | Supports pelvic viscera; raises pelvic floor |
| Coccygeus (ischiococcygeus) | Ischial spine and sacrospinous ligament | Inferior sacrum and coccyx | Anterior rami of S4-S5 | Supports pelvic viscera; draws coccyx forward |

***Clinical Note:***

During **defecation**, the levator ani muscle fibers, especially the fibers around the rectum, relax to allow the anorectal (rectum and anal canal) region to straighten and facilitate evacuation. The normal angle between the rectum above and the anal canal below is about 90 degrees (this helps to close off the anorectal junction), but during defecation this angle increases about 40 to 50 degrees (the anal canal swings forward). This relaxation, along with relaxation of the anal sphincters (not shown), opens the anal canal.

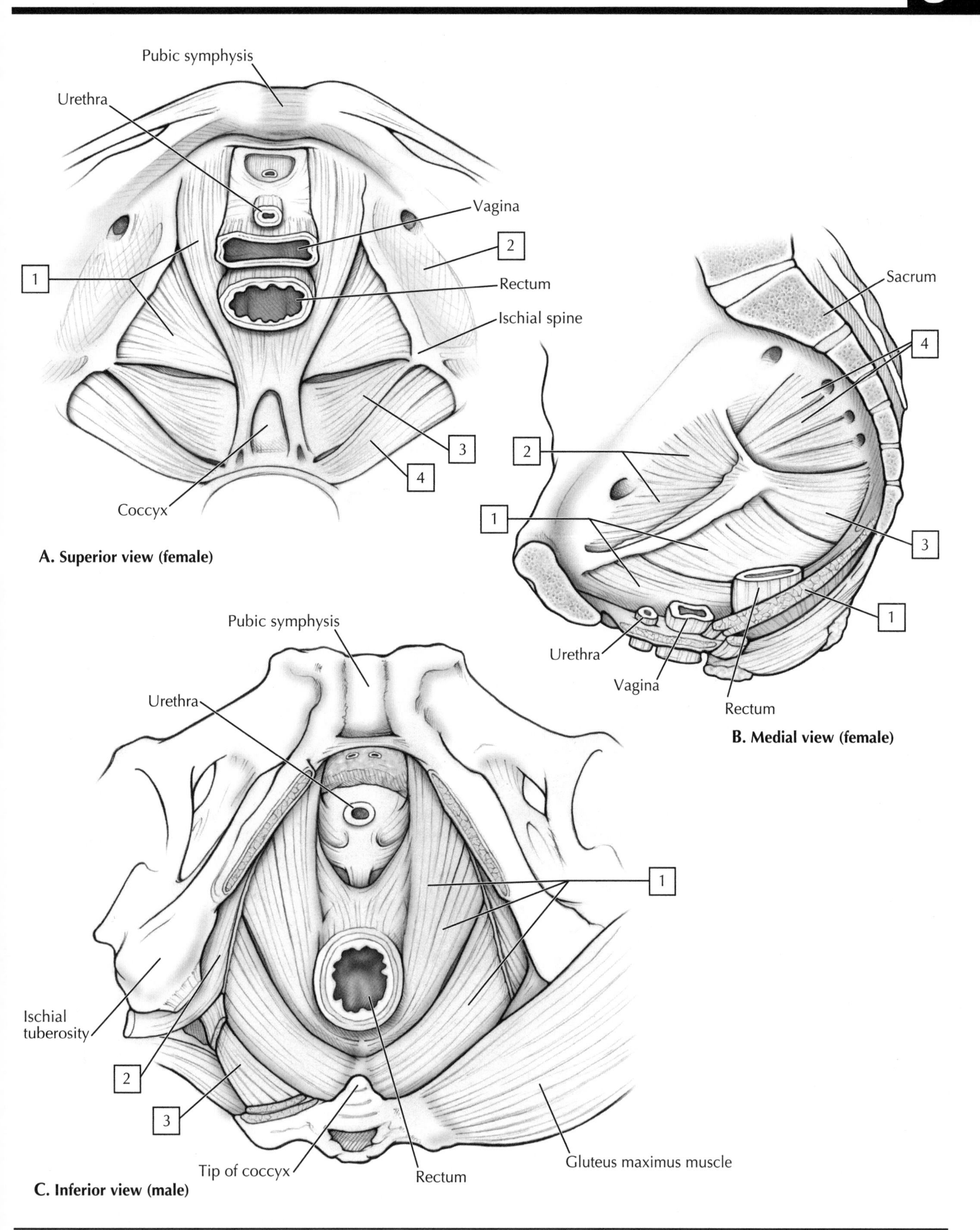


A. Superior view (female)

B. Medial view (female)

C. Inferior view (male)

# Muscles of the Perineum

The perineum is a diamond-shaped region between the thighs. It can be divided into an anterior **urogenital** (UG) **triangle** and a posterior **anal triangle** by an imaginary horizontal line connecting the two ischial tuberosities. Boundaries of the perineum include the:

- Pubic symphysis anteriorly
- Ischial tuberosities laterally
- Coccyx posteriorly

The muscles of the superficial perineal space are skeletal muscles and include the:

- **Ischiocavernosus**: paired muscles that surround the corpus cavernosum (erectile tissue) in males or the crus of the clitoris (also erectile tissue) in females
- **Bulbospongiosus**: a midline muscle that surrounds the bulb of the penis in males or splits to surround the bulbs of the vestibule in females; these also are erectile tissue structures
- **Superficial transverse perineal**: paired muscle that stabilizes the central tendon of the perineum (this muscle is often very small and difficult to identify)
- **External anal sphincter**: closes off the anal canal and rests upon the underlying levator ani muscle

The **central tendon of the perineum** is an important anchoring structure for the perineum. The bulbospongiosus, superficial transverse perineal, levator ani muscles, and the external anal sphincter all have attachments to the central tendon. The UG triangle contains the external genitalia of both sexes, whereas the **anal triangle** (the space is called the ischioanal fossa) is largely filled with fat and fibrous tissue.

Deep to the muscles of the UG triangle lies the external urethral sphincter in males (closes the membranous urethra except when passing urine, or during orgasm and ejaculation of semen). In females, the urethral sphincter blends with the compressor urethrae and sphincter urethrovaginalis muscles in the deep perineal space. All of these muscles, in both sexes, are under voluntary control and innervated by the pudendal (means "shameful") nerve (S2-S4) from the sacral plexus (ventral rami).

**COLOR** the muscles of the perineum, using a different color for each muscle:

- ☐ **1. Bulbospongiosus**
- ☐ **2. Ischiocavernosus**
- ☐ **3. External urethral sphincter (in male)**
- ☐ **4. Urethral sphincter (in female)**
- ☐ **5. Compressor urethrae (in female)**
- ☐ **6. External anal sphincter**

***Clinical Note:***

During childbirth, it may become necessary to enlarge the birth opening to prevent extensive stretching or tearing of the perineum. An incision, called an **episiotomy**, can be made in the posterior midline (median episiotomy) or posterolaterally to the vaginal opening to facilitate delivery of the child. It is important to suture the episiotomy carefully so that the integrity of the central tendon of the perineum is preserved, because this is an important support structure for the muscles of the perineum. Episiotomies now are performed less frequently than in the past.

**Stress incontinence in women** is the involuntary loss of urine after an increase in intra-abdominal pressure and is often associated with a weakening of the support structures of the pelvic floor, including the following:

- Medial and lateral pubovesical ligaments (ligaments supporting the urinary bladder)
- Levator ani muscle (provides support at the urethrovesical junction)
- Functional integrity of the urethral sphincter muscle

The area surrounding the opening of the rectum (the ischioanal fossae) may be prone to an infection, leading to an abscess or collection of pus (an **ischioanal abscess**), which can be painful and can sometimes rupture into the adjacent anal canal.

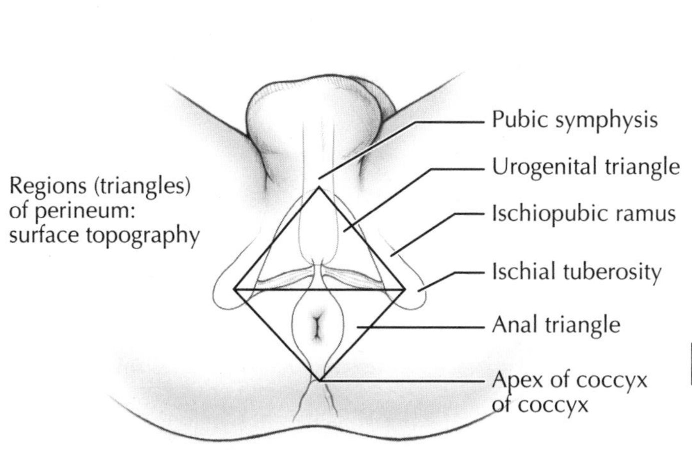


A. Regions (triangles) of perineum: surface anatomy

Penis
Inguinal ligament
Superficial inguinal ring
Spermatic cord
1
2
6
Superficial transverse perineal muscle
Ischial tuberosity
Levator ani muscle

B. Deep dissection

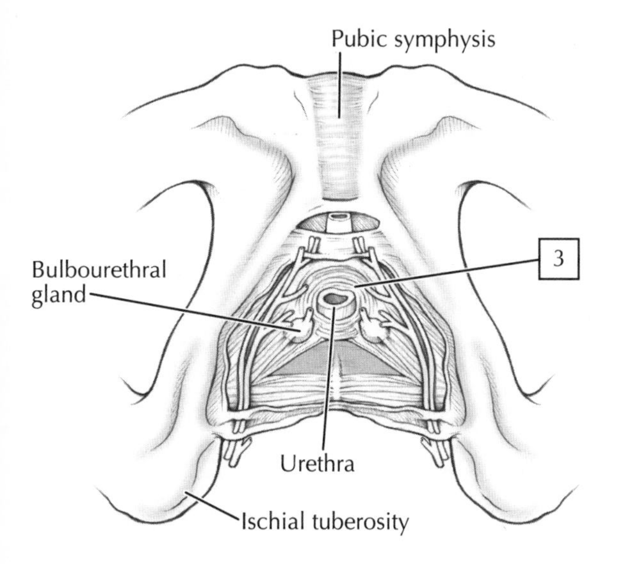


C. Male: inferior view

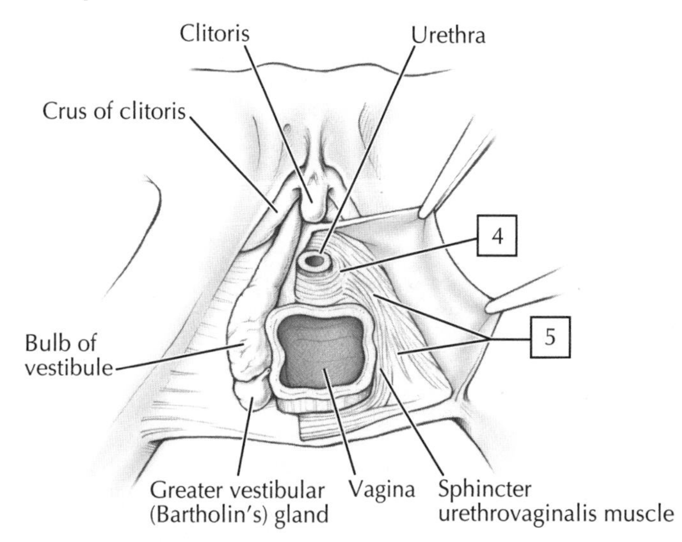


D. Female: deep dissection

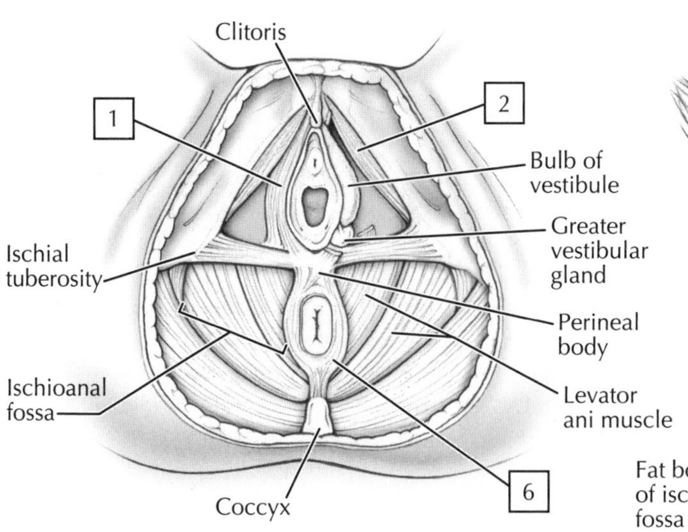


E. Female: deep perineum

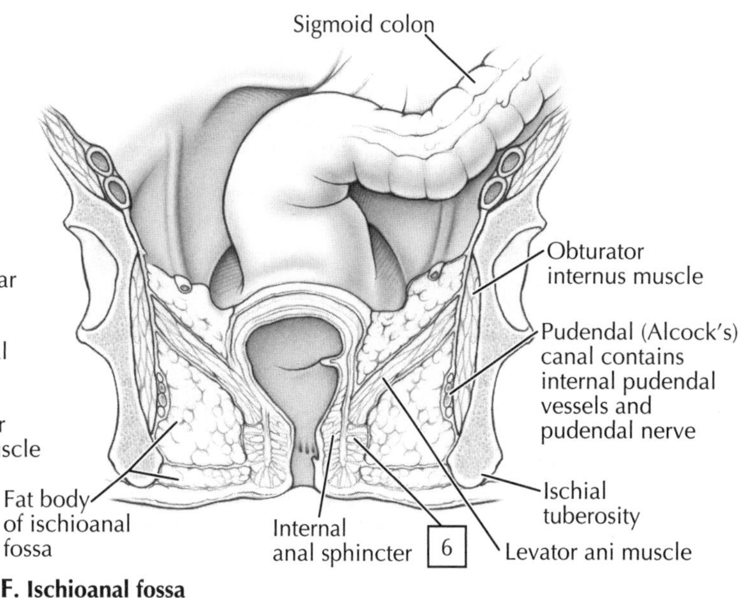


F. Ischioanal fossa

# 3 Posterior Shoulder Muscles

Muscles of the posterior shoulder have attachments to the scapula (the latissimus dorsi may or may not have a slight attachment to the inferior angle) and help in movements of the scapula and shoulder joint. Realize that when your arm is abducted above 20 degrees (the angle between your armpit and your body as your arm is abducted), your scapula begins to rotate with the inferior angle swinging laterally (this tilts the glenoid fossa upward). These muscles largely elevate the scapula, facilitate its rotation, or bring it back into its resting position (arm adducted against the body). These muscles are summarized in the table below.

Among these muscles, four play a unique role in stabilizing the shallow ball-and-socket joint of the shoulder (shallow to provide for extensive mobility) and are called the **rotator cuff muscles**. They include the:

- Supraspinatus
- Infraspinatus
- Teres minor
- Subscapularis: lies on the anterior aspect of the scapula in the subscapular fossa

| MUSCLE | PROXIMAL ATTACHMENT | DISTAL ATTACHMENT | INNERVATION | MAIN ACTIONS |
|---|---|---|---|---|
| Trapezius | Medial third of superior nuchal line; external occipital protuberance, ligamentum nuchae, and spinous processes of C7-T12 | Lateral third of clavicle, acromion, and spine of scapula | Accessory nerve (CN XI) | Elevates, retracts, and rotates scapula; superior fibers elevate, middle fibers retract, and inferior fibers depress scapula |
| Latissimus dorsi | Spinous processes of T7-T12, thoracolumbar fascia, iliac crest, and inferior three ribs | Intertubercular groove of humerus | Thoracodorsal nerve (C6-C8) | Extends, adducts, and medially rotates humerus at shoulder |
| Levator scapulae | Transverse processes of C1-C4 | Superior part of medial border of scapula | Dorsal scapular (C4-C5) and cervical (C3-C4) nerves | Elevates scapula and tilts its glenoid cavity inferiorly by rotating scapula |
| Rhomboid minor and major | *Minor*: ligamentum nuchae and spinous processes of C7 and T1 *Major*: spinous processes of T2-T5 | Medial border of scapula from level of spine to inferior angle | Dorsal scapular nerve (C4-C5) | Retracts scapula and rotates it to depress glenoid cavity; fixes scapula to thoracic wall |
| Supraspinatus (rotator cuff muscle) | Supraspinous fossa of scapula and deep fascia | Superior facet on greater tubercle of humerus | Suprascapular nerve (C5-C6) | Initiates arm abduction and acts with rotator cuff muscles |
| Infraspinatus (rotator cuff muscle) | Infraspinous fossa of scapula and deep fascia | Greater tubercle of humerus | Suprascapular nerve (C5-C6) | Laterally rotates arm at shoulder; helps hold head in glenoid cavity |
| Teres minor (rotator cuff muscle) | Lateral border of scapula | Greater tubercle of humerus | Axillary nerve (C5-C6) | Laterally rotates arm at shoulder; helps hold head in glenoid cavity |
| Teres major | Posterior surface of inferior angle of scapula | Medial lip of intertubercular groove of humerus | Lower subscapular nerve (C5-C6) | Adducts arm and medially rotates shoulder |
| Subscapularis (rotator cuff muscle) | Subscapular fossa of scapula | Lesser tubercle of humerus | Upper and lower subscapular nerves (C5-C6) | Medially rotates arm at shoulder and adducts it; helps hold humeral head in glenoid cavity |

**COLOR** the following muscles, using a different color for each muscle:

- ☐ 1. **Trapezius**
- ☐ 2. **Levator scapulae**
- ☐ 3. **Supraspinatus**
- ☐ 4. **Infraspinatus**
- ☐ 5. **Teres minor (may blend with the infraspinatus muscle)**
- ☐ 6. **Teres major**
- ☐ 7. **Subscapularis (on the anterior surface of the scapula)**

***Clinical Note:***

The musculotendinous **rotator cuff** strengthens the shoulder joint on its superior, posterior, and anterior aspects; hence about 95% of shoulder dislocations occur in an anteroinferior direction. Repetitive abduction, extension, lateral (external) rotation, and flexion of the arm at the shoulder, the motion used in throwing a ball, places stress on the elements of the rotator cuff, especially the tendon of the supraspinatus muscle as it rubs on the acromion and coracoacromial ligament. Tears or ruptures of this tendon are relatively common athletic injuries.

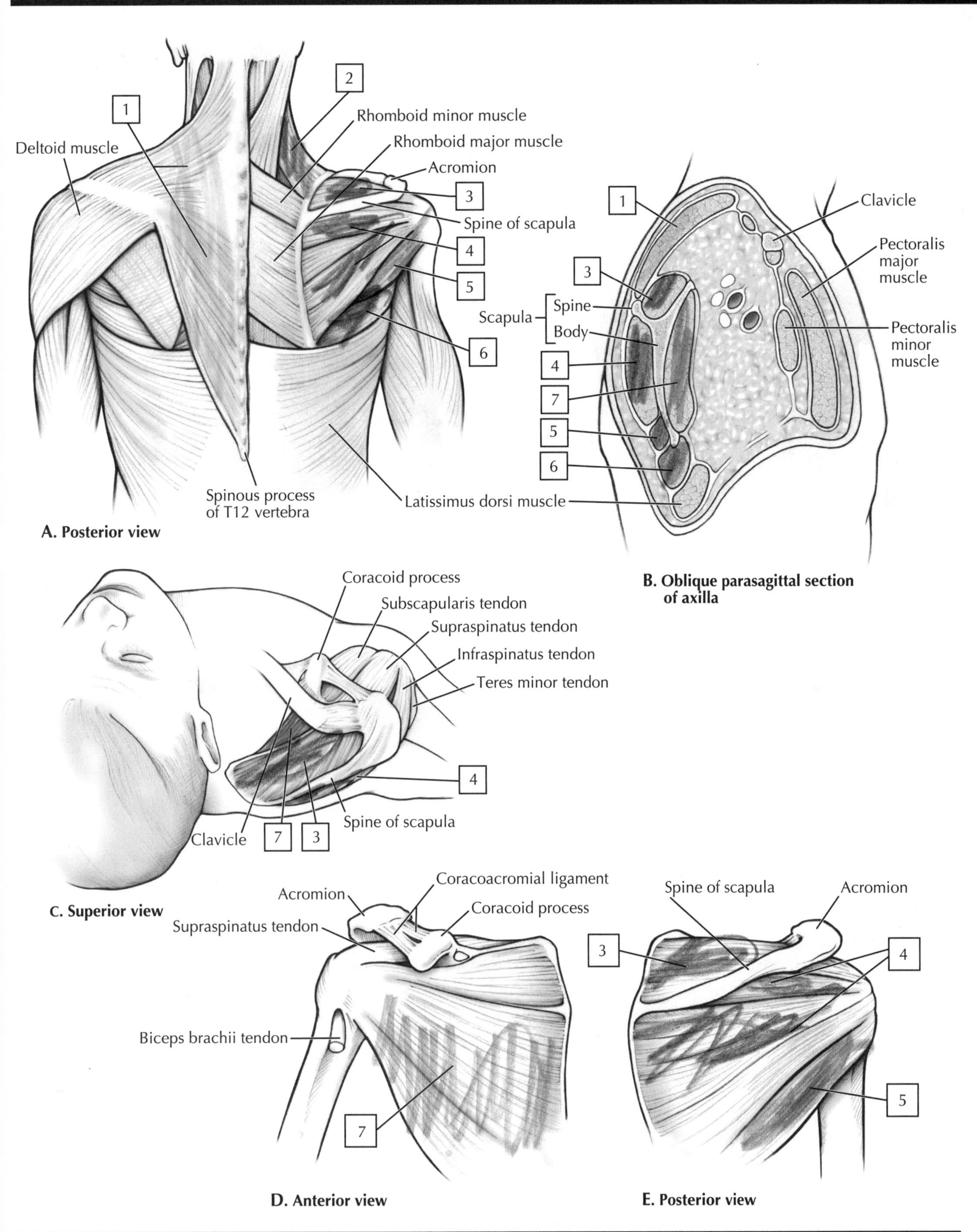


A. Posterior view

B. Oblique parasagittal section of axilla

C. Superior view

D. Anterior view

E. Posterior view

# 3 Anterior Shoulder Muscles

Muscles of the anterior shoulder have attachments to the pectoral girdle (scapula and clavicle) or the humerus, and assist in movements of the pectoral girdle and shoulder. These muscles "cap" the shoulder (deltoid muscle) or arise from the anterior or lateral thoracic wall, and are summarized in the table below.

**COLOR** the following muscles, using a different color for each muscle:

- ☐ 1. **Deltoid**

- ☐ 2. **Pectoralis major**

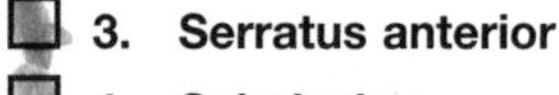

- ☐ 3. **Serratus anterior**

- ☐ 4. **Subclavius**
- ☐ 5. **Pectoralis minor**

The anterior and posterior muscles define the "axilla" (armpit) region, a pyramid-shaped area containing important neurovascular structures that pass through the shoulder region. The six boundaries of the axilla include the:

- **Base**: axillary fascia and skin of the armpit
- **Apex**: bounded by the 1st rib, clavicle, and superior part of the scapula; a passageway for structures entering or leaving the shoulder and arm
- **Anterior wall**: pectoralis major and minor muscles
- **Posterior wall**: subscapularis, teres major, and latissimus dorsi muscles
- **Medial wall**: upper rib cage, intercostal and serratus anterior muscles
- **Lateral wall**: proximal humerus (intertubercular groove)

| MUSCLE | PROXIMAL ATTACHMENT | DISTAL ATTACHMENT | INNERVATION | MAIN ACTIONS |
|---|---|---|---|---|
| Pectoralis major | Medial half of clavicle; sternum; superior six costal cartilages; aponeurosis of external abdominal oblique | Intertubercular groove of humerus | Lateral (C5-C7) and medial pectoral nerves (C8-T1) | Flexes, adducts, and medially rotates arm at shoulder |
| Pectoralis minor | 3rd to 5th ribs | Coracoid process of scapula | Medial pectoral nerve (C8-T1) | Lowers lateral angle of scapula and protracts it |
| Serratus anterior | Lateral surface of upper eight ribs | Medial border of scapula | Long thoracic nerve (C5-C7) | Protracts and rotates scapula and pulls it anteriorly toward thoracic wall |
| Subclavius | Junction of 1st rib and costal cartilage | Inferior surface of clavicle | Nerve to subclavius (C5-C6) | Depresses clavicle and anchors clavicle |
| Deltoid | Lateral third of clavicle, acromion, and spine of scapula | Deltoid tuberosity of humerus | Axillary nerve (C5-C6) | *Anterior part*: flexes and medially rotates arm at shoulder<br>*Middle part*: abducts arm at shoulder<br>*Posterior part*: extends and laterally rotates arm at shoulder |

***Clinical Note:***

**Impingement syndrome** is a term that refers to a variety of soft tissue injuries in the subacromial space, which is that narrow space between the head of the humerus and the underside of the acromion of the scapula. The supraspinatus muscle tendon and the subacromial bursa are in this narrow space, and repetitive overhead motions, weak rotator cuff muscles, and anatomical anomalies or degeneration of structures can lead to microtrauma. Repetitive motion can lead to bursitis, tendonitis, and rotator cuff tears. Athletes (repetitive movements) and the elderly (underuse of abduction) may be prone to this condition.

If the **axillary nerve** is damaged, for example by a fracture of the upper portion of the humerus, specifically its surgical neck, the deltoid muscle will **atrophy**. This will become obvious when the affected shoulder is compared with the unaffected side, as the deltoid muscle will atrophy and the shoulder will appear flattened or asymmetrical when compared to the unaffected shoulder contour. There will also be a loss of sensation over the skin of the lateral aspect of the proximal arm and the patient will be unable to raise the affected limb above about 15 degrees of abduction against resistance.

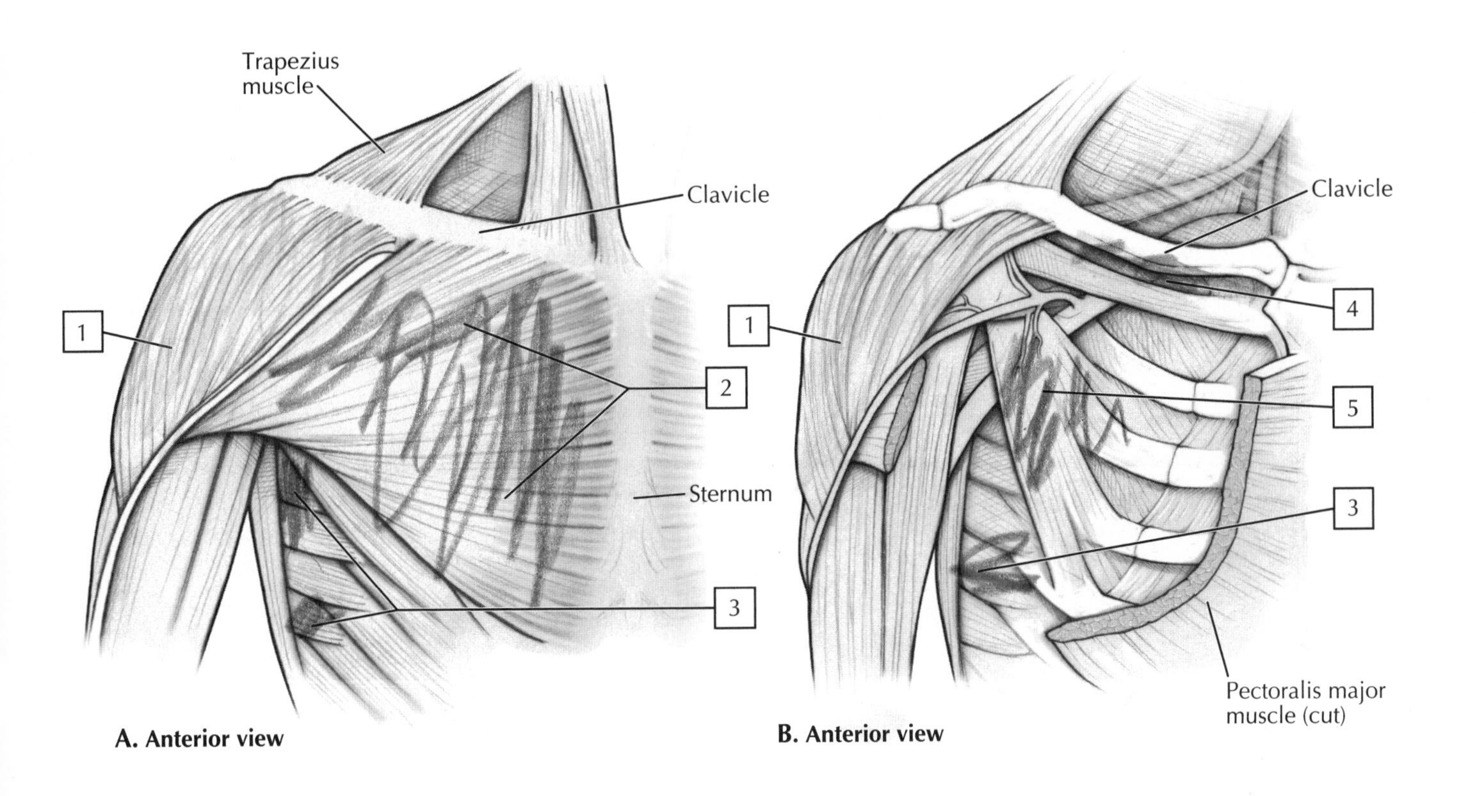


A. Anterior view

B. Anterior view

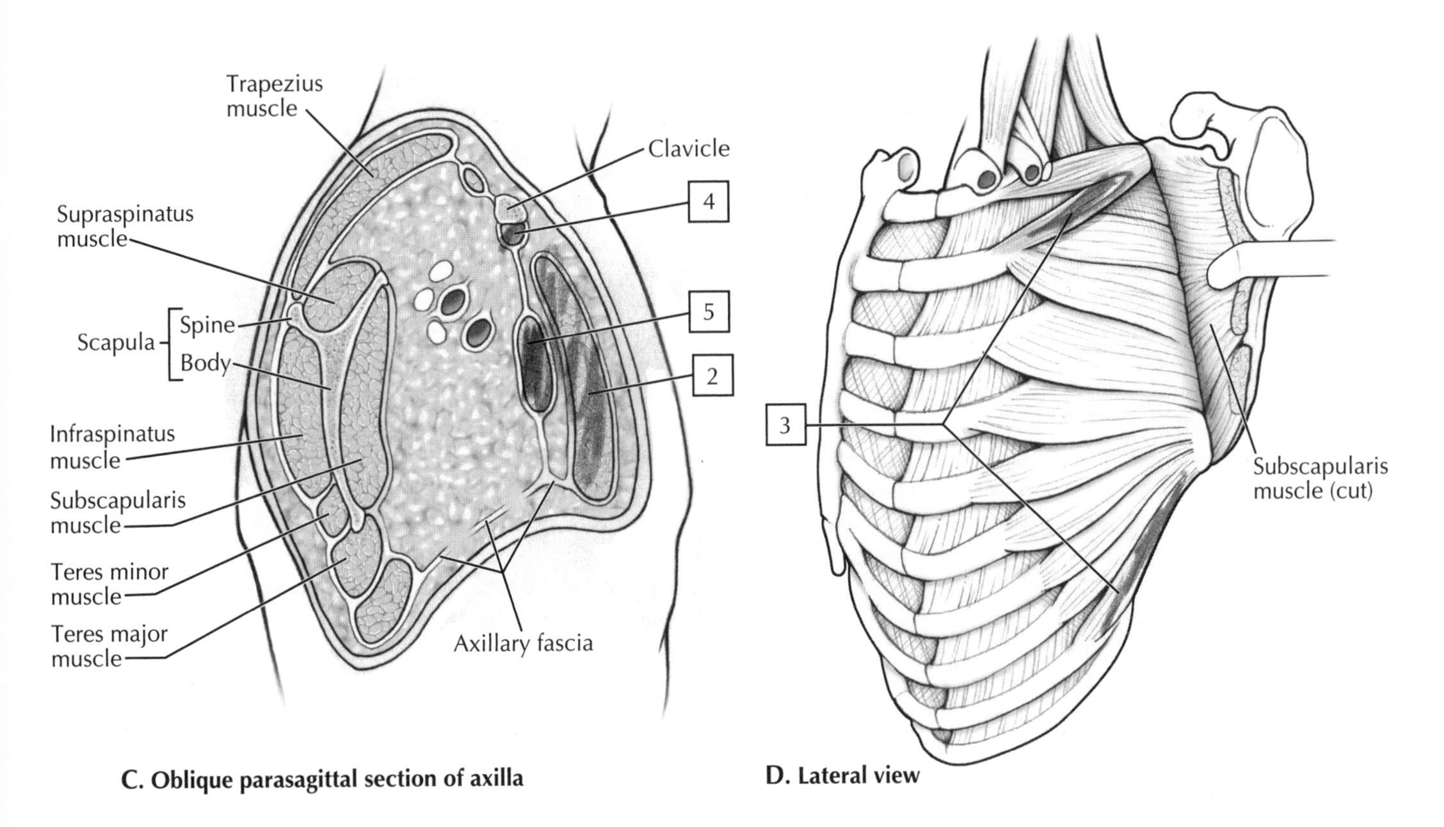


C. Oblique parasagittal section of axilla

D. Lateral view

# 3 Arm Muscles

The arm (region between the shoulder and elbow) is divided by a connective tissue intermuscular septum into two compartments:

- **Anterior**: contains muscles that primarily flex the elbow and/or the shoulder
- **Posterior**: contains muscles that primarily extend the elbow

Additionally, the biceps is a powerful supinator of the flexed forearm, used when turning a screw into wood, if right-handed, or for removing a screw, if left-handed. Of the arm flexors, the brachialis is the most powerful flexor of the forearm at the elbow, not the biceps, although it is the biceps that most weight lifters focus on, because it is the more visible of the two muscles. The muscles of the anterior and posterior compartments are summarized in the table below.

**COLOR** the following muscles, using a different color for each muscle:

- ☐ 1. **Biceps brachii (has a long and a short head)**
- ☐ 2. **Coracobrachialis**
- ☐ 3. **Brachialis**
- ☐ 4. **Triceps: has three components; its medial head lies deep to the overlying long and lateral heads**
- ☐ 5. **Anconeus: sometimes grouped with the forearm extensor muscles**

| MUSCLE | PROXIMAL ATTACHMENT | DISTAL ATTACHMENT | INNERVATION | MAIN ACTIONS |
|---|---|---|---|---|
| Biceps brachii | *Short head:* apex of coracoid process of scapula<br>*Long head:* supraglenoid tubercle of scapula | Tuberosity of radius and fascia of forearm via bicipital aponeurosis | Musculocutaneous nerve (C5-C6) | Supinates flexed forearm; flexes forearm at elbow |
| Brachialis | Distal half of anterior humerus | Coronoid process and tuberosity of ulna | Musculocutaneous nerve (C5-C6) | Flexes forearm at elbow in all positions |
| Coracobrachialis | Tip of coracoid process of scapula | Middle third of medial surface of humerus | Musculocutaneous nerve (C5-C7) | Helps flex and adduct arm at shoulder |
| Triceps brachii | *Long head:* infraglenoid tubercle of scapula<br>*Lateral head:* posterior humerus<br>*Medial head:* posterior surface of humerus, inferior to radial groove | Posterior surface of olecranon of ulna and fascia of forearm | Radial nerve (C6-C8) | Extends forearm at elbow; is chief extensor of elbow; steadies head of abducted humerus (long head) |
| Anconeus | Posterior surface of epicondyle of humerus | Lateral surface of olecranon and superior part of posterior surface of ulna | Radial nerve (C5, C6-C7) | Assists triceps in extending elbow; abducts ulna during pronation |

***Clinical Note:***

**Rupture** of the biceps brachii may occur at the proximal tendon or, rarely, the muscle belly. The biceps tendon has the highest rate of spontaneous rupture of any tendon in the body. Its rupture is seen most commonly in people older than 40 years, in association with rotator cuff injuries or repetitive lifting (in weight lifters). Rupture of the tendon of the long head of the biceps is most common and may occur at the shoulder joint, along the intertubercular (bicipital) sulcus of the humerus or at the musculotendinous junction.

**Biceps tendinitis** results from inflammation of the biceps, usually from overuse and repetitive movements. Athletes who use a throwing motion are most at risk for this inflammation, e.g., baseball pitchers, football quarterbacks.

The **biceps reflex** is a deep tendon reflex that tests the integrity of the biceps muscle. With the limb relaxed and supinated, the physician can use a reflex hammer to lightly tap the biceps tendon, or place their thumb on the tendon and tap their thumb at the nail bed; this will elicit a reflex contraction of the biceps (a sudden jerk) and test not only the integrity of the muscle but also the musculocutaneous nerve (C5-6).

**Stretch reflexes** tend to be either hypoactive or absent if there is peripheral nerve damage or a spinal cord injury (the ventral horn where the efferent motor axons arise from the spinal cord).

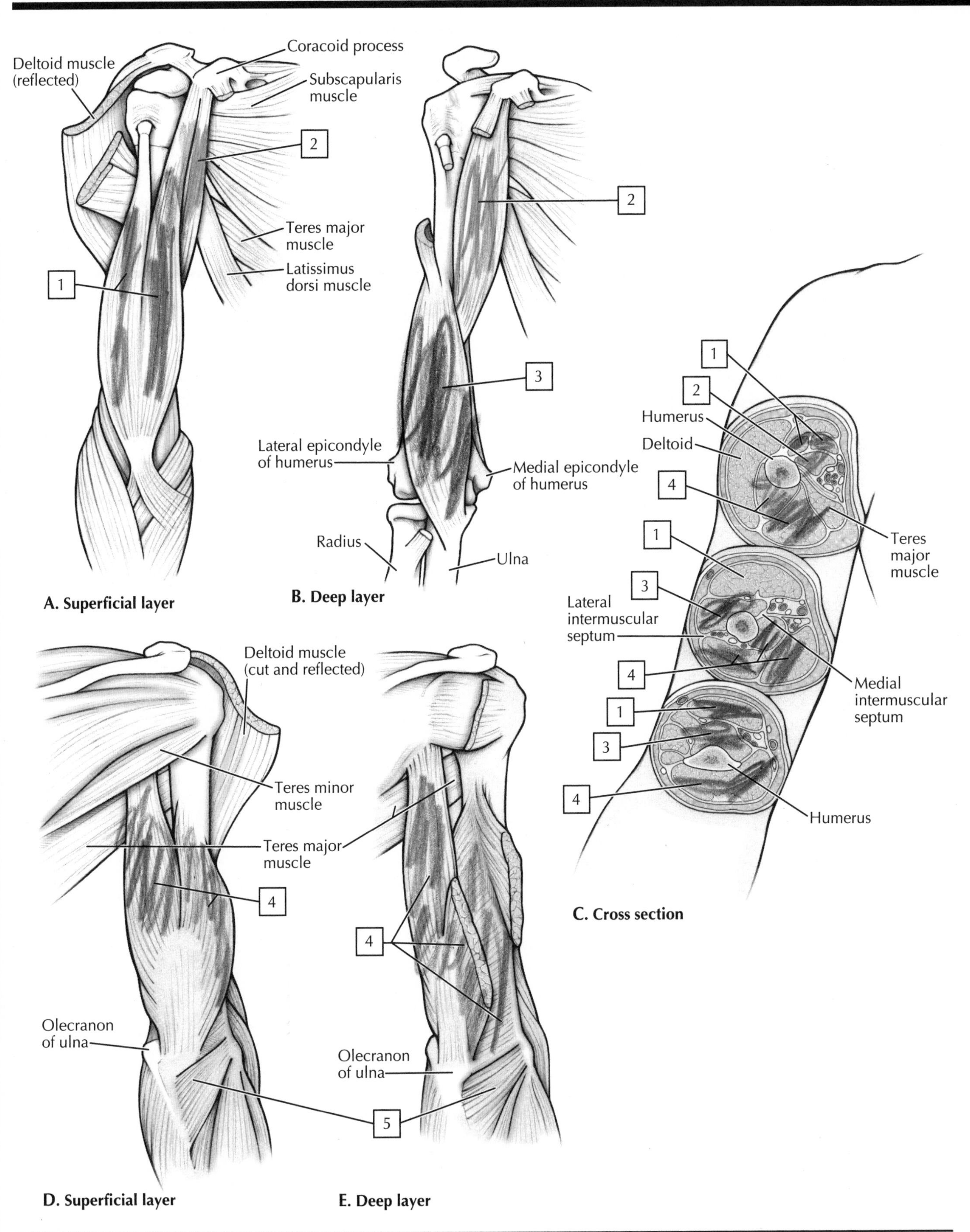
Deltoid muscle
(reflected)
Coracoid process
Subscapularis
muscle
2
Teres major
muscle
Latissimus
dorsi muscle
1
A. Superficial layer
2
3
Lateral epicondyle
of humerus
Medial epicondyle
of humerus
Radius
Ulna
B. Deep layer
1
2
Humerus
Deltoid
4
1
3
Lateral
intermuscular
septum
4
1
3
4
Teres
major
muscle
Medial
intermuscular
septum
Humerus
C. Cross section
Deltoid muscle
(cut and reflected)
Teres minor
muscle
Teres major
muscle
4
Olecranon
of ulna
4
Olecranon
of ulna
5
D. Superficial layer
E. Deep layer

# 3 Pronation and Supination of the Radioulnar Joints

Two muscles pronate and two muscles supinate the radioulnar joints. The forearm in anatomical position, with the palm facing forward, is **supinated,** and the radius and ulna lie side by side in the forearm. Rotation of the palm medially so it faces backward, or toward the ground if the elbow is flexed 90 degrees, is **pronation**.

The pronator muscles lie in the forearm; one is more superficial and lies near the elbow (pronator teres) and the other lies deep beneath other forearm muscles distally near the wrist (pronator quadratus). The word *teres* refers to "round earth" (in pronation of the flexed forearm at 90 degrees, the hand faces the ground or earth), whereas the word *quadratus* refers to the quadrangular shape of the wrist pronator. When the pronators contract, they wrap or pull the radius across the stable ulna, proximally by the pronator teres and distally by the pronator quadratus. The ulna is stabilized by its articulation at the elbow with the distal end of the humerus and moves very little.

The supinator muscles include the biceps brachii of the arm, which is a powerful supinator with the elbow flexed, but with the forearm straight, the supinator, a muscle of the extensor compartment of the forearm, executes supination. From the illustrations on the facing page, note that when the supinator contracts, it unwraps the crossed radius and brings it back into alignment with the medially placed ulna.

**COLOR** the following muscles, using a different color for each muscle:

- ☐ 1. **Supinator**
- ☐ 2. **Pronator teres**
- ☐ 3. **Pronator quadratus**
- ☐ 4. **Biceps brachii**

***Clinical Note:***

When the **radius is fractured**, the muscles attaching to the bone deform the normal alignment of the radius and ulna. If the fracture of the radius is above the insertion of the pronator teres, the proximal fragment will be flexed and supinated by the action and pull of the biceps brachii and supinator muscles. The distal fragment will be pronated by the pronator teres and quadratus muscles (see part ***D***).

In fractures of the middle or distal radius that are distal to the insertion of the pronator teres, the supinator and pronator teres will keep the proximal bone fragment of the radius in the neutral position. The distal fragment, however, will be pronated by the pronator quadratus muscle, because it is unopposed by either supinator muscle (see part ***E***).

# Pronation and Supination of the Radioulnar Joints

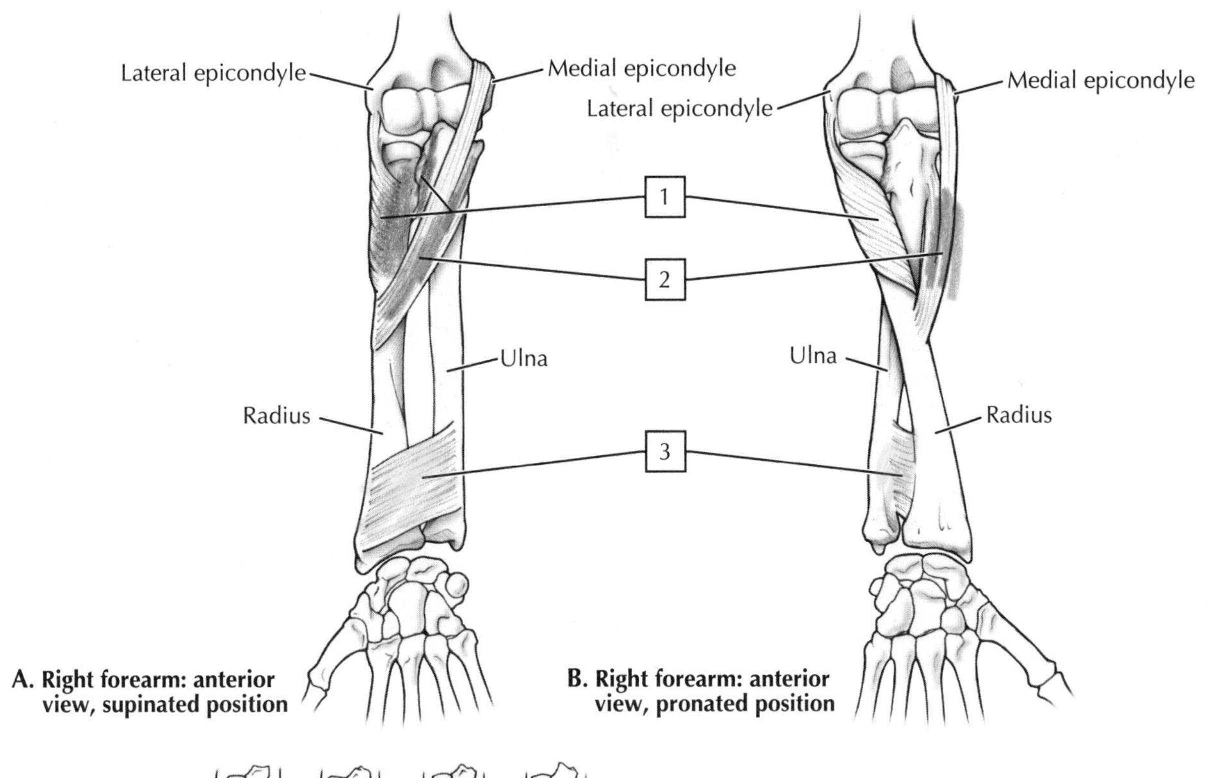


**A. Right forearm: anterior view, supinated position**

**B. Right forearm: anterior view, pronated position**

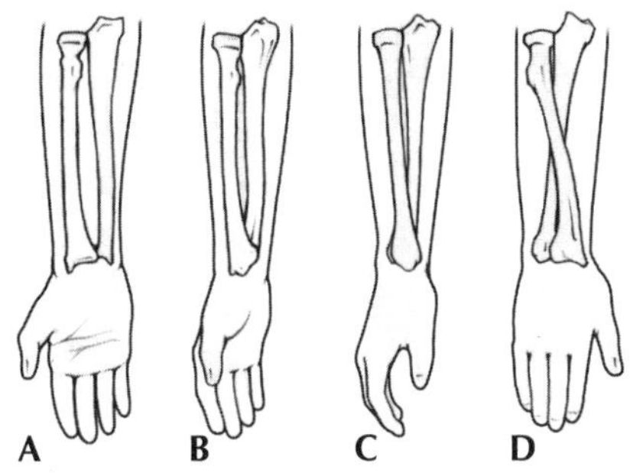


**C. Biomechanics of forearm**

Tuberosity of radius useful indicator of degree of pronation or supination of radius

A. In full supination, tuberosity directed toward ulna
B. In about 40 degrees of supination, tuberosity primarily posterior
C. In neutral position, tuberosity directly posterior
D. In full pronation, tuberosity directed laterally

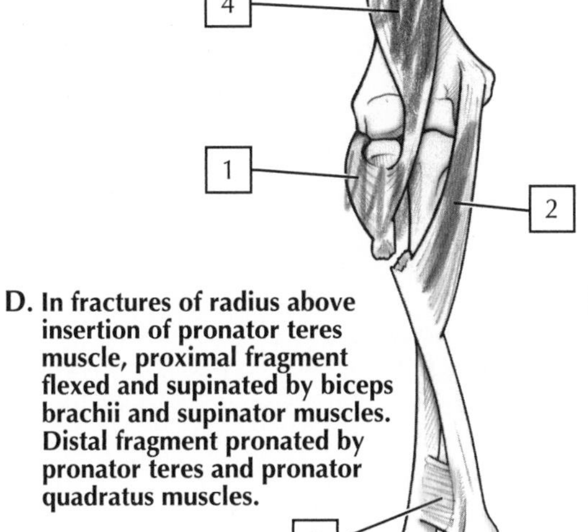


**D. In fractures of radius above insertion of pronator teres muscle, proximal fragment flexed and supinated by biceps brachii and supinator muscles. Distal fragment pronated by pronator teres and pronator quadratus muscles.**

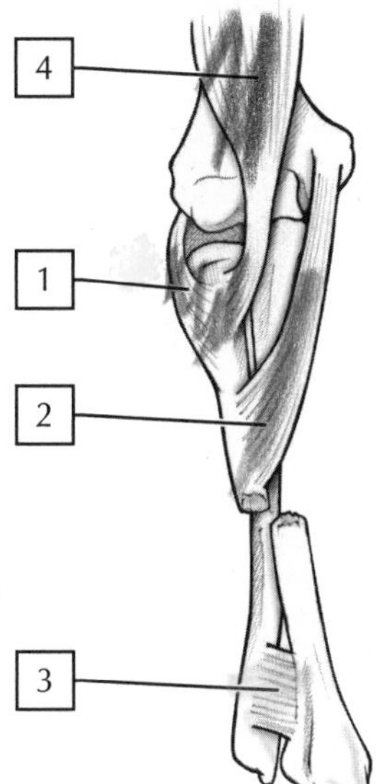


**E. In fractures of middle or distal radius that are distal to insertion of pronator teres muscle, supinator and pronator teres muscles keep proximal fragment in neutral position. Distal fragment pronated by pronator quadratus muscle.**

# Anterior Forearm Muscles

The forearm is divided into two muscle compartments by a connective tissue intermuscular septum. The anterior compartment contains muscles that primarily flex the wrist and fingers. In the anterior compartment, a superficial layer of muscles arises from the medial epicondyle of the humerus, whereas a deep layer of muscles arises from the bones (radius and ulna) of the forearm or the interosseous membrane connecting these bones. If you squeeze your hand very tightly to make a fist and flex your wrist, you will note the contraction of these muscles in your own anterior forearm. These muscles are summarized in the table below.

**COLOR** each of the following muscles, using a different color for each muscle:

- ☐ **1. Pronator teres**
- ☐ **2. Flexor carpi radialis (also abducts the wrist)**
- ☐ **3. Palmaris longus: absent in about 10% of humans, this muscle is of little importance in us but is the muscle in cats that allows them to retract their claws**
- ☐ **4. Flexor carpi ulnaris (also adducts the wrist)**
- ☐ **5. Flexor digitorum superficialis**
- ☐ **6. Flexor digitorum profundus: "profundus" means deep, as in a profound comment**
- ☐ **7. Flexor pollicis longus: "pollicis" refers to the thumb**

| MUSCLE | PROXIMAL ATTACHMENT | DISTAL ATTACHMENT | INNERVATION | MAIN ACTIONS |
|---|---|---|---|---|
| Pronator teres | Medial epicondyle of humerus and coronoid process of ulna | Middle of lateral surface of radius | Median nerve (C6-C7) | Pronates forearm and flexes elbow |
| Flexor carpi radialis | Medial epicondyle of humerus | Base of 2nd metacarpal bone | Median nerve (C6-C7) | Flexes hand at wrist and abducts it |
| Palmaris longus | Medial epicondyle of humerus | Distal half of flexor retinaculum and palmar aponeurosis | Median nerve (C7-C8) | Flexes hand at wrist and tightens palmar aponeurosis |
| Flexor carpi ulnaris | *Humeral head:* medial epicondyle of humerus *Ulnar head:* olecranon and posterior border of ulna | Pisiform bone, hook of hamate bone, and 5th metacarpal bone | Ulnar nerve (C7-T1) | Flexes hand at wrist and adducts it |
| Flexor digitorum superficialis | *Humeroulnar head*: medial epicondyle of humerus, ulnar collateral ligament, and coronoid process of ulna *Radial head*: superior half of anterior radius | Bodies of middle phalanges of medial four digits on the palmar aspect | Median nerve (C8-T1) | Flexes middle phalanges of medial four digits; also weakly flexes proximal phalanges, forearm, and wrist |
| Flexor digitorum profundus | Proximal three-fourths of medial and anterior surfaces of ulna and interosseous membrane | Palmar bases of distal phalanges of medial four digits | *Medial part*: ulnar nerve (C8-T1) *Lateral part*: median nerve (C8-T1) | Flexes distal phalanges of medial four digits; assists with flexion of wrist |
| Flexor pollicis longus | Anterior surface of radius and adjacent interosseous membrane | Base of distal phalanx of thumb on the palmar aspect | Median nerve (anterior interosseous) (C7-C8) | Flexes phalanges of 1st digit (thumb) |
| Pronator quadratus | Distal fourth of anterior surface of ulna | Distal fourth of anterior surface of radius | Median nerve (anterior interosseous) (C8-T1) | Pronates forearm |

***Clinical Note:***

The flexor digitorum superficialis (FDS) muscle may be tested by holding a patient's index, ring, and little finger in an extended position, thus inactivating the flexor digitorum profundus (FDP) muscle, and then asking the patient to flex their middle finger at the proximal interphalangeal joint against resistance. To test of the FDP, the proximal interphalangeal joint of the middle finger is held in the extended position while asking the patient to flex their distal interphalangeal joint, thus flexing the tip of their middle finger.

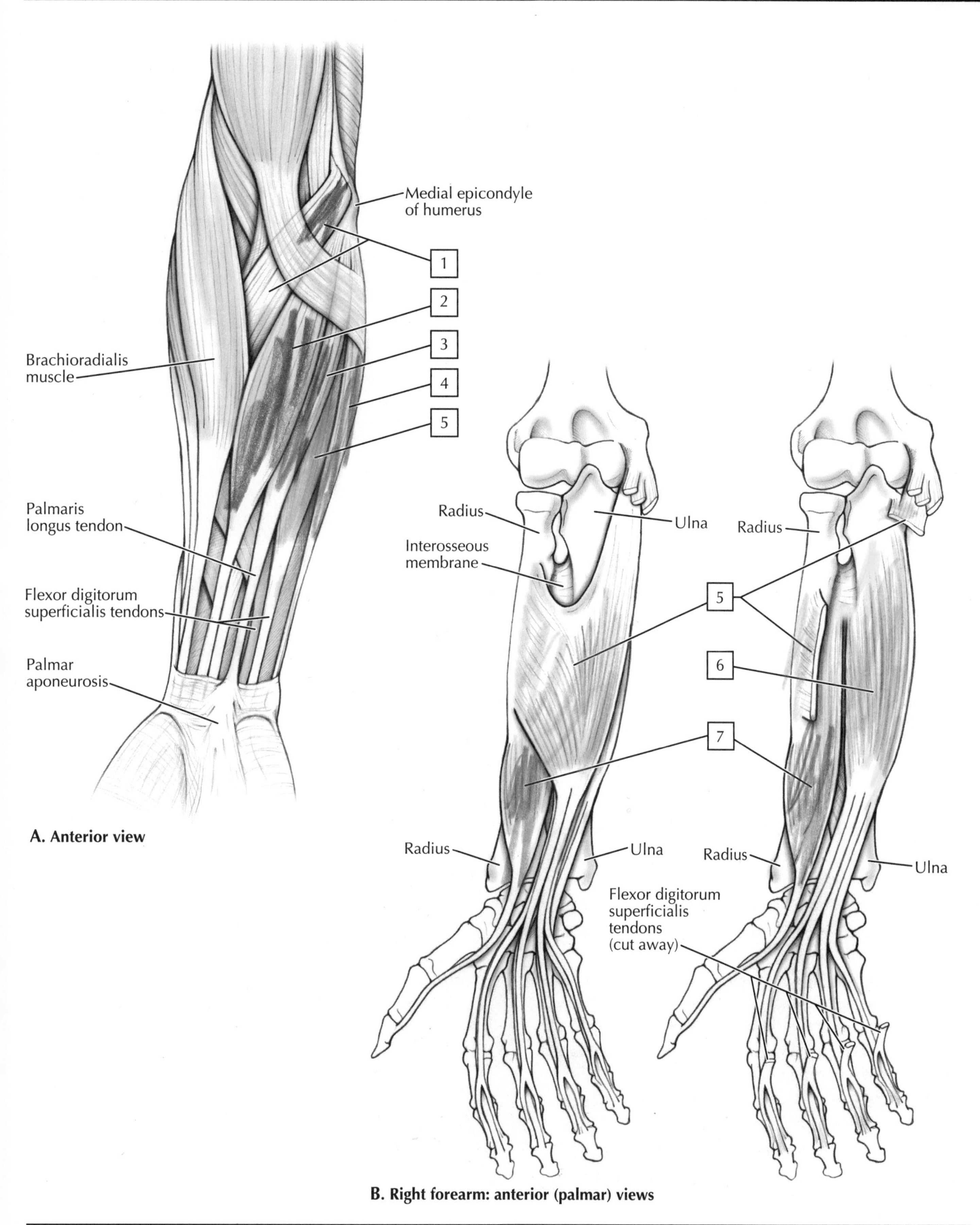


A. Anterior view

B. Right forearm: anterior (palmar) views

# Posterior Forearm Muscles

The forearm is divided into two muscle compartments by a connective tissue intermuscular septum. The posterior compartment contains muscles that primarily extend the wrist and fingers. In the posterior compartment, a superficial layer of muscles arises largely from the lateral epicondyle of the humerus, whereas a deep layer of muscles arises from the bones of the forearm (radius and ulna) or the interosseous membrane connecting these bones. If you hyperextend your fingers and wrist and pronate your forearm, you will note the contraction of these muscles in your own posterior forearm. Extending the wrist when gripping an object adds extra strength to our grip (the power grip). These muscles are summarized in the table below.

| MUSCLE | PROXIMAL ATTACHMENT | DISTAL ATTACHMENT | INNERVATION | MAIN ACTIONS |
|---|---|---|---|---|
| Brachioradialis | Proximal two-thirds of lateral supracondylar ridge of humerus | Lateral surface of distal end of radius | Radial nerve (C5-C6) | Flexes forearm at elbow, especially in midpronation |
| Extensor carpi radialis longus | Lateral supracondylar ridge of humerus | Base of 2nd metacarpal bone | Radial nerve (C6-C7) | Extends and abducts hand at wrist |
| Extensor carpi radialis brevis | Lateral epicondyle of humerus | Base of 3rd metacarpal bone | Radial nerve (deep branch) (C7-C8) | Extends and abducts hand at wrist |
| Extensor digitorum | Lateral epicondyle of humerus | Extensor expansions of medial four digits | Radial nerve (posterior interosseous) (C7-C8) | Extends medial four digits at metacarpophalangeal joints; extends hand at wrist joint |
| Extensor digiti minimi | Lateral epicondyle of humerus | Extensor expansion of 5th digit | Radial nerve (posterior interosseous) (C7-C8) | Extends 5th digit at metacarpophalangeal and interphalangeal joints |
| Extensor carpi ulnaris | Lateral epicondyle of humerus and posterior border of ulna | Base of 5th metacarpal bone | Radial nerve (posterior interosseous) (C7-C8) | Extends and adducts hand at wrist |
| Supinator | Lateral epicondyle of humerus; radial collateral and anular ligaments; supinator fossa; and crest of ulna | Lateral, posterior, and anterior surfaces of proximal third of radius | Radial nerve (deep branch) (C7-C8) | Supinates forearm |
| Abductor pollicis longus | Posterior surfaces of ulna, radius, and interosseous membrane | Base of 1st metacarpal bone on the lateral aspect | Radial nerve (posterior interosseous) (C7-C8) | Abducts thumb and extends it at carpometacarpal joint |
| Extensor pollicis brevis | Posterior surfaces of radius and interosseous membrane | Base of proximal phalanx of thumb on the dorsal aspect | Radial nerve (posterior interosseous) (C7-C8) | Extends proximal phalanx of thumb at carpometacarpal joint |
| Extensor pollicis longus | Posterior surface of middle third of ulna and interosseous membrane | Base of distal phalanx of thumb on the dorsal aspect | Radial nerve (posterior interosseous) (C7-C8) | Extends distal phalanx of thumb at metacarpophalangeal and interphalangeal joints |
| Extensor indicis | Posterior surface of ulna and interosseous membrane | Extensor expansion of 2nd digit | Radial nerve (posterior interosseous) (C7-C8) | Extends 2nd digit and helps extend hand at wrist |

**COLOR** each of the following muscles, using a different color for each muscle:

☐ 1. **Extensor carpi ulnaris (also adducts the wrist)**

☐ 2. **Extensor digiti minimi ("minimi" refers to the little finger)**

☐ 3. **Brachioradialis: lumped with the posterior forearm muscles because of its innervation, it actually flexes the forearm at the elbow**

☐ 4. **Extensor carpi radialis longus (also abducts the wrist; important in power grip)**

☐ 5. **Extensor carpi radialis brevis (also abducts the wrist; important in power grip)**

☐ 6. **Extensor digitorum**

☐ 7. **Abductor pollicis longus ("pollicis" refers to the thumb)**

☐ 8. **Extensor pollicis brevis**

☐ 9. **Extensor pollicis longus**

☐ 10. **Extensor indicis ("indicis" refers to the index finger)**

***Clinical Note:***

**Tennis elbow** is a condition that clinicians call lateral epicondylitis; the term is somewhat misleading because the problem really involves a tendinosis of the extensor carpi radialis brevis (probably the most important wrist extensor), which arises just proximal to this epicondyle. Moreover, most sufferers are not tennis players! The elbow pain experienced in tennis elbow occurs just distal and posterior to the lateral epicondyle and is exacerbated during wrist extension, especially against resistance. The pain may be due to the muscle, its innervating nerve, and/or something within the elbow joint itself.

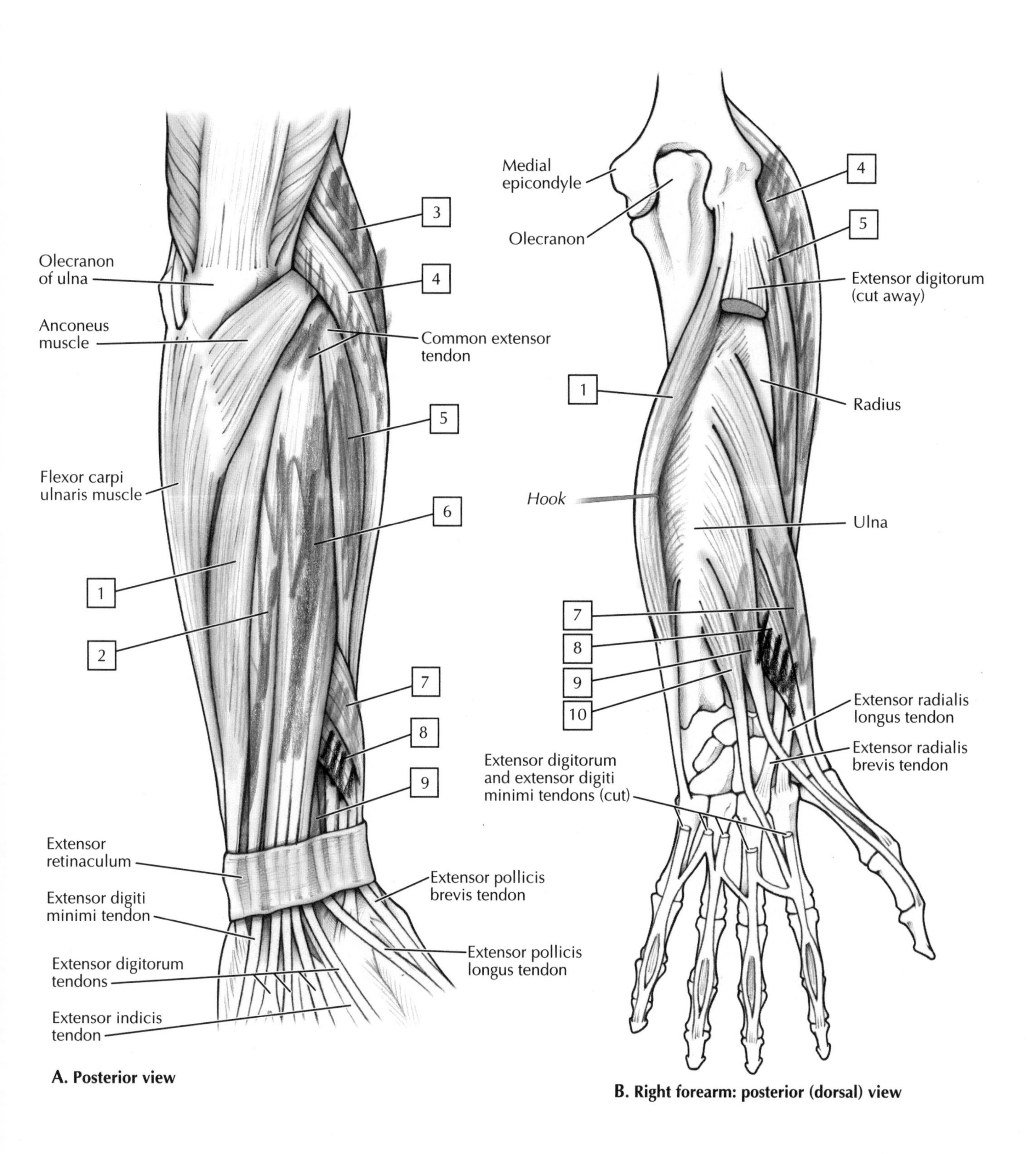


**A. Posterior view**

**B. Right forearm: posterior (dorsal) view**

# 3 Intrinsic Hand Muscles

The intrinsic hand muscles move the fingers, complementing the long flexor and extensor forearm muscles that also move the fingers. Two groups of muscles lie most superficially:

- **Thenar eminence**: a cone of three thenar muscles at the base of the thumb
- **Hypothenar eminence**: a cone of three hypothenar muscles at the base of the little finger

Deeper intrinsic muscles include the:

- **Adductor pollicis**: deep in the palm, it adducts the thumb
- **Lumbricals**: four small muscles attached to the flexor digitorum profundus tendons
- **Interossei**: three palmar and four dorsal interosseous muscles between the metacarpals; palmar interossei adduct the digits (**PAD**) and dorsal interossei abduct the digits (**DAB**)

**COLOR** each of the following muscles, using a different color for each muscle:

- ☐ **1. Opponens pollicis (thenar muscle)**
- ☐ **2. Abductor pollicis brevis (thenar muscle)**
- ☐ **3. Flexor pollicis brevis (thenar muscle)**
- ☐ **4. Adductor pollicis**
- ☐ **5. Abductor digiti minimi (hypothenar muscle)**
- ☐ **6. Flexor digiti minimi (hypothenar muscle)**
- ☐ **7. Opponens digiti minimi (hypothenar muscle)**
- ☐ **8. Dorsal interossei**
- ☐ **9. Palmar interossei**

These intrinsic muscles are summarized in the table below.

| MUSCLE | PROXIMAL ATTACHMENT | DISTAL ATTACHMENT | INNERVATION | MAIN ACTIONS |
|---|---|---|---|---|
| Abductor pollicis brevis | Flexor retinaculum and tubercles of scaphoid and trapezium | Base of proximal phalanx of thumb | Median nerve (recurrent branch) (C8-T1) | Abducts thumb at metacarpophalangeal joint |
| Flexor pollicis brevis | Flexor retinaculum and tubercle of trapezium | Lateral side of base of proximal phalanx of thumb | Median nerve (recurrent branch) (C8-T1) | Flexes proximal phalanx of thumb |
| Opponens pollicis | Flexor retinaculum and tubercle of trapezium | Lateral side of 1st metacarpal bone | Median nerve (recurrent branch) (C8-T1) | Opposes thumb toward center of palm and rotates it medially |
| Adductor pollicis | *Oblique head:* bases of 2nd and 3rd metacarpals and capitate<br>*Transverse head:* anterior surface of body of 3rd metacarpal bone | Medial side of base of proximal phalanx of thumb | Ulnar nerve (deep branch) (C8-T1) | Adducts thumb toward middle digit |
| Abductor digiti minimi | Pisiform and tendon of flexor carpi ulnaris | Medial side of base of proximal phalanx of 5th digit | Ulnar nerve (deep branch) (C8-T1) | Abducts 5th digit |
| Flexor digiti minimi brevis | Hook of hamate and flexor retinaculum | Medial side of base of proximal phalanx of 5th digit | Ulnar nerve (deep branch) (C8-T1) | Flexes proximal phalanx of 5th digit |
| Opponens digiti minimi | Hook of hamate and flexor retinaculum | Palmar surface of 5th metacarpal bone | Ulnar nerve (deep branch) (C8-T1) | Draws 5th metacarpal bone anteriorly and rotates it, bringing it into opposition with thumb |
| Lumbricals 1 and 2 | Lateral two tendons of flexor digitorum profundus | Lateral sides of extensor expansions of 2nd and 3rd digits | Median nerve (C8-T1) | Flex digits at metacarpophalangeal joints and extend interphalangeal joints |
| Lumbricals 3 and 4 | Medial three tendons of flexor digitorum profundus | Lateral sides of extensor expansions of 4th and 5th digits | Ulnar nerve (deep branch) (C8-T1) | Flex digits at metacarpophalangeal joints and extend interphalangeal joints |
| Dorsal interossei | Sides of two metacarpal bones | Extensor expansions and bases of proximal phalanges of 2nd to 4th digits | Ulnar nerve (deep branch) (C8-T1) | Dorsal interossei abduct digits; flex digits at metacarpophalangeal joints and extend interphalangeal joints |
| Palmar interossei | Palmar surfaces of 2nd, 4th, and 5th metacarpal bones | Extensor expansions of digits and bases of proximal phalanges of 2nd, 4th, and 5th digits | Ulnar nerve (deep branch) (C8-T1) | Palmar interossei adduct digits; flex digits at metacarpophalangeal joints and extend interphalangeal joints |

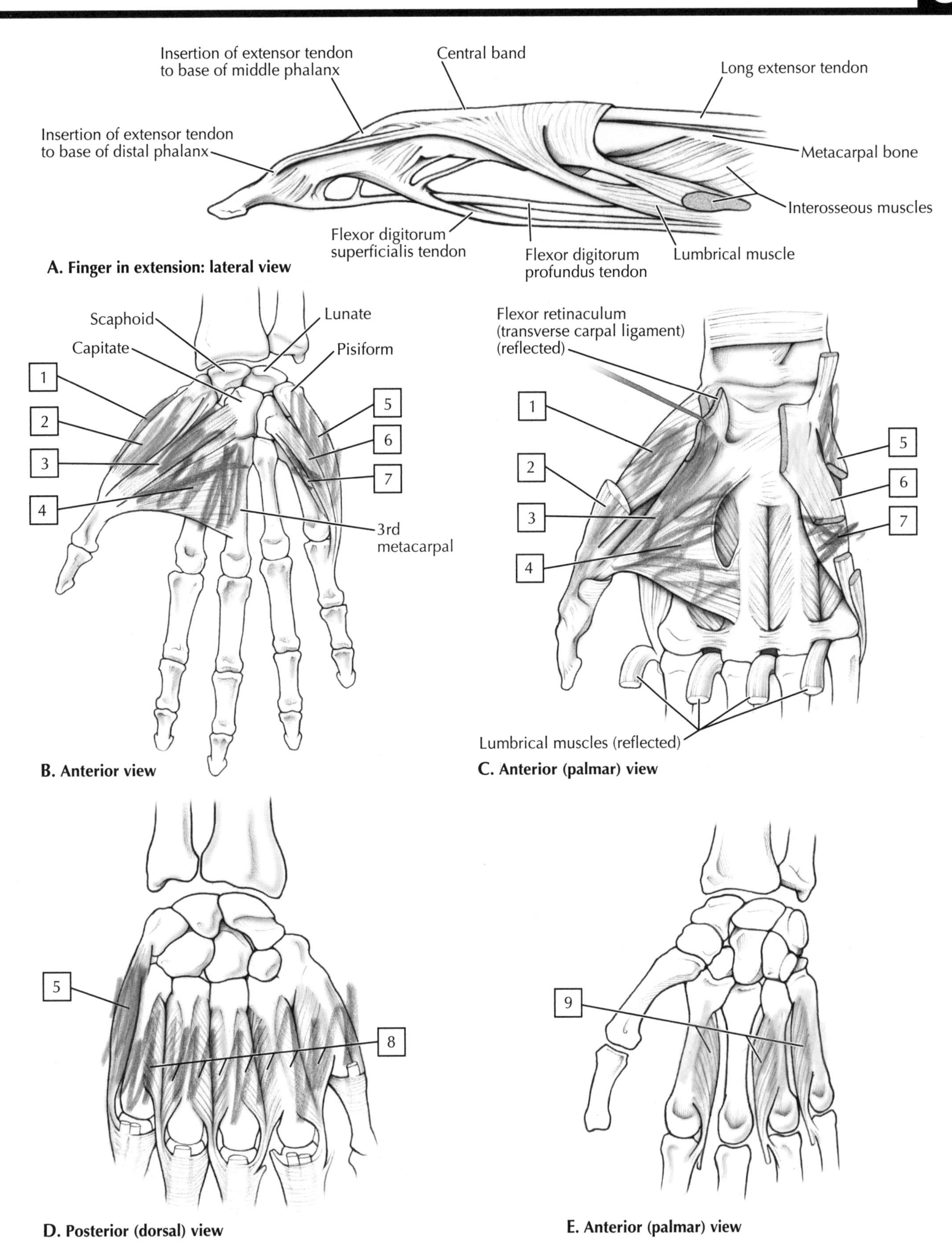


**A. Finger in extension: lateral view**

**B. Anterior view**

**C. Anterior (palmar) view**

**D. Posterior (dorsal) view**

**E. Anterior (palmar) view**

# 3 Summary of Upper Limb Muscles

It is best to learn the actions of the muscles by knowing which **compartment** (anterior or posterior) they reside in and then knowing the primary action of the muscles in that compartment. Few muscles act in isolation; more often, they act as a group. In general, muscles of the upper back and anterior chest wall primarily act on the shoulder, muscles of the arm primarily act on the elbow (with some shoulder movement), and muscles of the forearm act primarily on the wrist and fingers. The table below summarizes some of the major muscles acting on the joints of the upper limb (this table is not comprehensive, but highlights the major muscles*).

| SCAPULA | SHOULDER |
|---|---|
| **Elevate**: levator scapulae, trapezius<br>**Depress**: pectoralis minor<br>**Protrude**: serratus anterior<br>**Depress glenoid**: rhomboids<br>**Elevate glenoid**: serratus anterior, trapezius<br>**Retract**: rhomboids, trapezius | **Flex**: pectoralis major, coracobrachialis<br>**Extend**: latissimus dorsi, teres major<br>**Abduct**: deltoid, supraspinatus<br>**Adduct**: pectoralis major, latissimus dorsi<br>**Rotate medially**: subscapularis, teres major, pectoralis major, latissimus dorsi<br>**Rotate laterally**: infraspinatus, teres minor |
| **ELBOW** | **RADIOULNAR** |
| **Flex**: brachialis, biceps<br>**Extend**: triceps, anconeus | **Pronate**: pronators (teres and quadratus)<br>**Supinate**: supinator, biceps brachii |
| **WRIST** | **METACARPOPHALANGEAL** |
| **Flex**: flexor carpi radialis, ulnaris<br>**Extend**: all extensor carpi muscles<br>**Abduct**: flexor/extensor carpi radialis muscles<br>**Adduct**: flexor and extensor carpi ulnaris<br>**Circumduct**: combination of all movements | **Flex**: interossei and lumbricals<br>**Extend**: extensor digitorum<br>**Abduct**: dorsal interossei<br>**Adduct**: palmar interossei<br>**Circumduct**: combination of all movements |
| **INTERPHALANGEAL-PROXIMAL** | **INTERPHALANGEAL-DISTAL** |
| **Flex**: flexor digitorum superficialis<br>**Extend**: interossei and lumbricals | **Flex**: flexor digitorum profundus<br>**Extend**: interossei and lumbricals |

*Accessory actions of muscles are detailed in the muscle tables.

**COLOR** the following muscles, using a different color for each muscle:

- ☐ 1. **Biceps brachii**
- ☐ 2. **Brachialis**
- ☐ 3. **Triceps**
- ☐ 4. **Brachioradialis**
- ☐ 5. **Extensor carpi radialis longus**
- ☐ 6. **Extensor digitorum**
- ☐ 7. **Extensor digiti minimi**
- ☐ 8. **Flexor carpi radialis**
- ☐ 9. **Flexor carpi ulnaris**
- ☐ 10. **Flexor digitorum superficialis**

***Clinical Note:***

**Upper Limb Nerve Injury Summary**

**Suprascapular nerve injury** may result in shoulder pain, radiating to the arm and neck, with weakness in shoulder rotation.

**Musculocutaneous nerve injury** may result in weakened elbow flexion, with hypesthesia of the lateral forearm and weakened supination with the elbow flexed.

**Long thoracic nerve injury** may lead to a "winged scapula."

**Axillary nerve injury** (less common) can produce weakness of the deltoid muscle and abduction.

**Radial nerve injury** can lead to wrist drop, pain and tenderness over the lateral elbow and radiating distally along the lateral aspect of the forearm. Weakened elbow, wrist, and finger extension may occur, along with weakened supination.

**Median nerve injury** in the forearm may lead to loss of cutaneous sensation over the lateral two-thirds of the palm, and the palmar aspect of the thumb, index, middle, and lateral aspect of the ring finger. It is especially vulnerable in the carpal tunnel at the wrist, resulting in thenar atrophy and weakness of the thumb. Wrist flexion, finger flexion and forearm pronation may be affected.

**Ulnar nerve injury** may lead to loss of volar sensation of the medial half of the ring finger and all of the little finger. "Claw hand" deformity may result from motor loss to the ring and little fingers. The ulnar nerve is especially vulnerable as it passes posterior to the medial epicondyle of the humerus (hit one's "funny bone"), through the two heads of the flexor carpi ulnaris muscle, and as it passes through the cubital tunnel at the wrist.

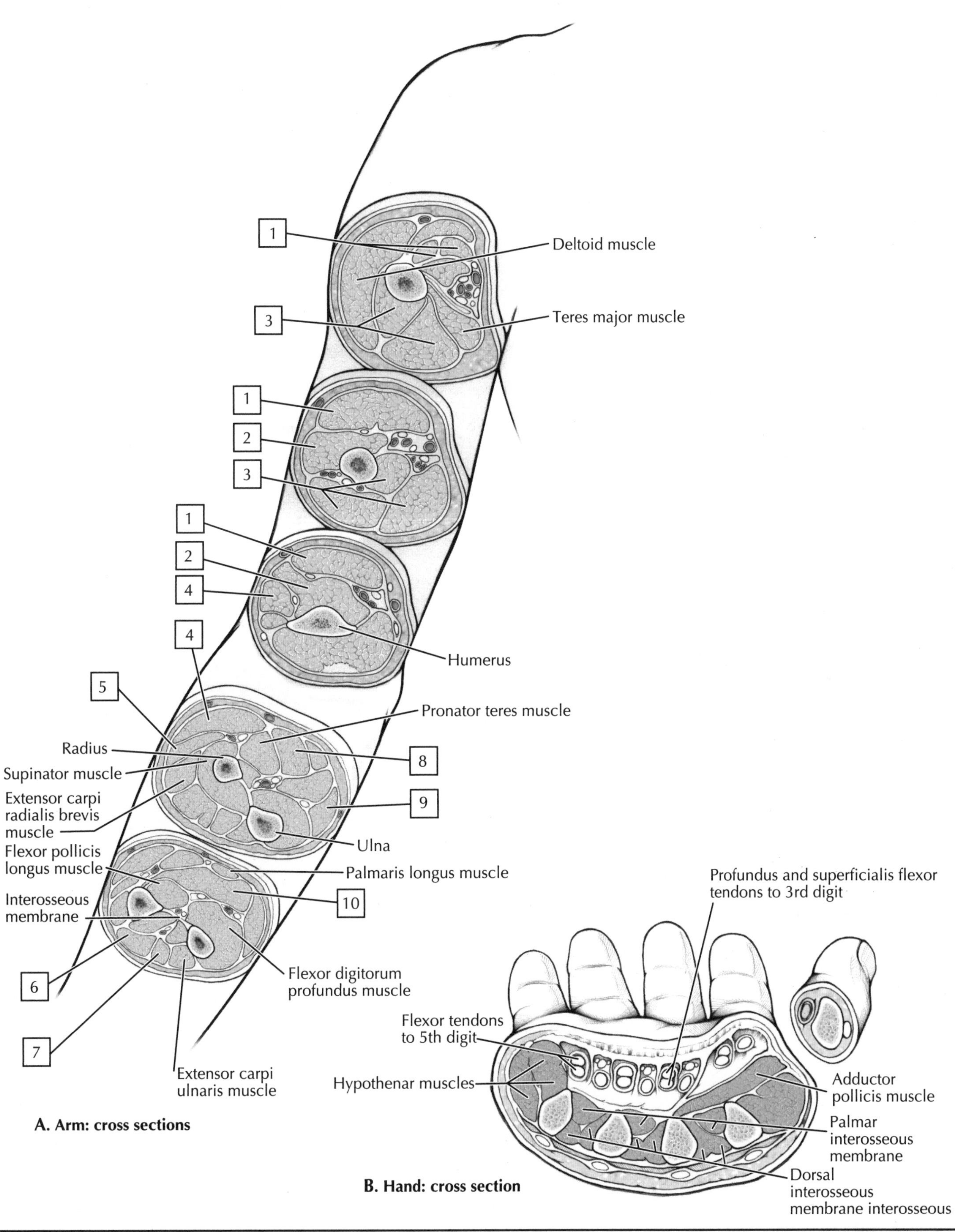


**A. Arm: cross sections**

**B. Hand: cross section**

3

# Gluteal Muscles

The gluteal muscles (muscles of the buttock) extend, abduct, and laterally rotate the femur (thigh bone) at the hip joint. The gluteus maximus is the strongest muscle, in total strength, in the body and is especially important in extension, where it is used to rise from a sitting position or to climb stairs (exercising this muscle on a Stairmaster exercise unit will give you "buns of steel!"). A number of other gluteal muscles lie deep to the maximus and are summarized in the table below.

**COLOR** the following muscles, using a different color for each muscle:

- ☐ 1. **Gluteus medius**
- ☐ 2. **Gluteus maximus**
- ☐ 3. **Gluteus minimus**
- ☐ 4. **Piriformis: arises from inside the pelvic wall off of the anterior sacrum and sacrotuberous ligament**
- ☐ 5. **Obturator internus: also arises from inside the pelvic cavity**
- ☐ 6. **Gemelli: superior and inferior heads; "Gemini" refers to these twin muscles; separated by the tendon of the obturator internus**
- ☐ 7. **Quadratus femoris**

| MUSCLE | PROXIMAL ATTACHMENT | DISTAL ATTACHMENT | INNERVATION | MAIN ACTIONS |
|---|---|---|---|---|
| Gluteus maximus | Ilium posterior to posterior gluteal line, dorsal surface of sacrum and coccyx, and sacrotuberous ligament | Most fibers end in iliotibial tract that inserts into lateral condyle of tibia; some fibers insert on gluteal tuberosity of femur | Inferior gluteal nerve (L5-S2) | Extends thigh at the hip and assists in its lateral rotation; steadies thigh and assists in raising trunk from flexed position |
| Gluteus medius | External surface of ilium | Lateral surface of greater trochanter of femur | Superior gluteal nerve (L4-S1) | Abducts and medially rotates thigh at hip; steadies pelvis on limb when opposite limb is raised |
| Gluteus minimus | Lateral surface of ilium | Anterior surface of greater trochanter | Superior gluteal nerve (L4-S1) | Abducts and medially rotates thigh at hip; steadies pelvis on limb when opposite limb is raised |
| Piriformis | Anterior surface of sacrum and sacrotuberous ligament | Superior border of greater trochanter | Branches of anterior rami L5-S2 | Laterally rotates extended thigh at hip and abducts flexed thigh at hip; steadies femoral head in acetabulum |
| Obturator internus | Pelvic surface of obturator membrane and surrounding bones | Medial aspect of greater trochanter of femur | Nerve to obturator internus (L5-S2) | Laterally rotates extended thigh at hip and abducts flexed thigh at hip; steadies femoral head in acetabulum |
| Gemelli, superior and inferior | *Superior:* ischial spine *Inferior:* ischial tuberosity | Medial aspect of greater trochanter of femur | Superior gemellus: same nerve supply as obturator internus; inferior gemellus: same nerve supply as quadratus femoris | Laterally rotate extended thigh at the hip and abduct flexed thigh at the hip; steady femoral head in acetabulum |
| Quadratus femoris | Lateral border of ischial tuberosity | Quadrate tubercle on intertrochanteric crest of femur | Nerve to quadratus femoris (L4-S1) | Laterally rotates thigh at hip |

***Clinical Note:***

Weakness or paralysis of the gluteus medius and minimus muscles can lead to an unstable pelvis, because these muscles stabilize the pelvis while walking by abducting and keeping the pelvis level when the opposite foot is off the ground and in its swing phase. If weakened, the pelvis becomes unstable during walking and tilts to the unaffected side.

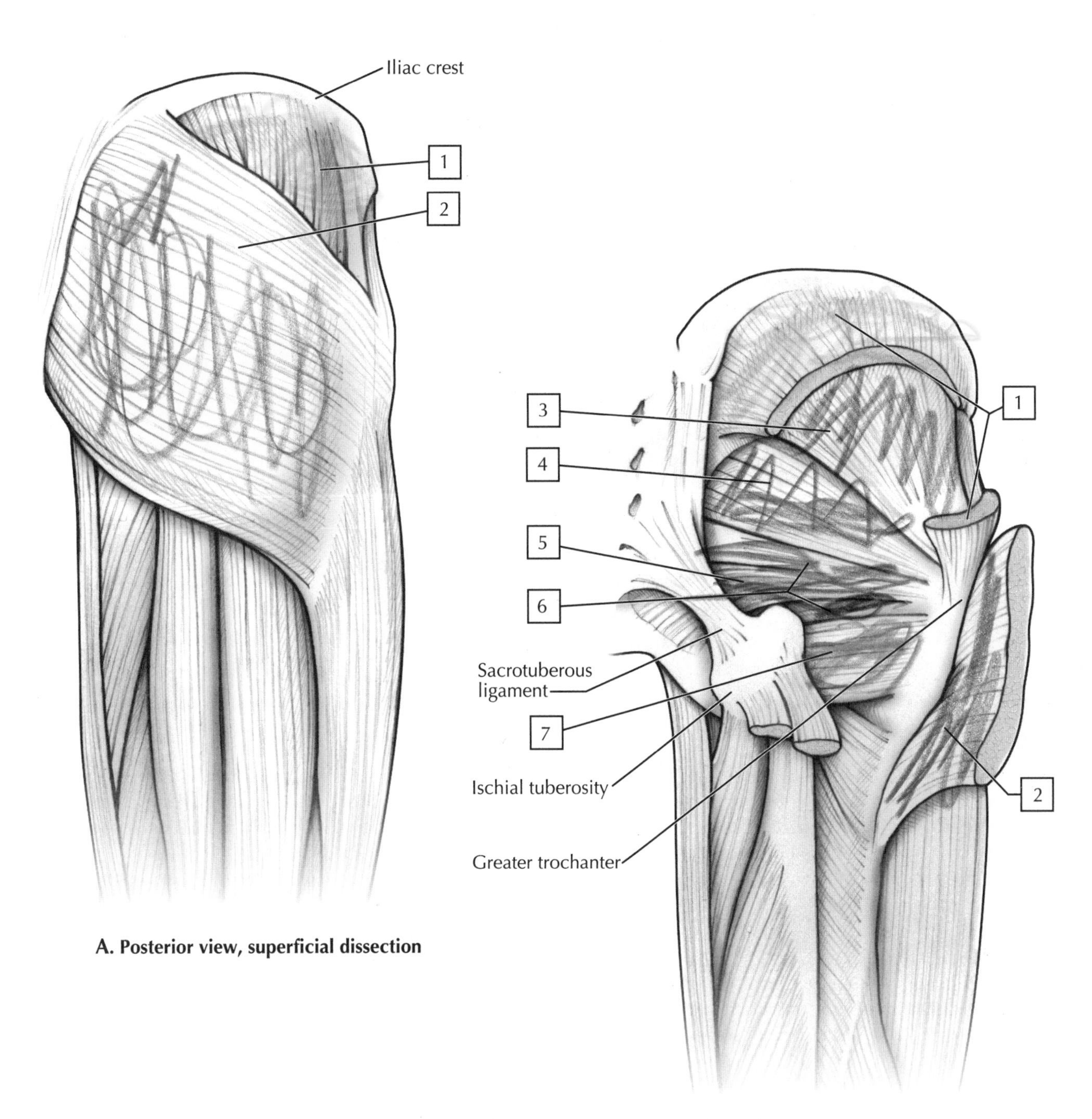


A. Posterior view, superficial dissection

B. Posterior view, deeper dissection

# Posterior Thigh Muscles

The thigh is divided into three muscle compartments by connective tissue intermuscular septae. The muscles of the posterior compartment primarily extend the hip and flex the knee. Three of the four muscles in this compartment are the hamstrings:

These muscles all arise from the ischial tuberosity, and they extend the hip and flex the knee. The short head of the biceps femoris is not a hamstring muscle and primarily flexes the knee. These muscles are summarized in the table below.

**COLOR** the following muscles, using a different color for each muscle:

- ☐ **1. Semitendinosus**
- ☐ **2. Semimembranosus**

- ☐ **3. Biceps femoris, short head (not a hamstring muscle)**

- ☐ **4. Biceps femoris, long head**

| MUSCLE | PROXIMAL ATTACHMENT | DISTAL ATTACHMENT | INNERVATION | MAIN ACTIONS |
|---|---|---|---|---|
| Semitendinosus | Ischial tuberosity | Medial surface of superior part of tibia | Tibial division of sciatic nerve (L2-S2) | Extends thigh at hip; flexes leg at knee and rotates it medially; with flexed hip and knee, extends trunk |
| Semimembranosus | Ischial tuberosity | Posterior part of medial condyle of tibia | Tibial division of sciatic nerve (L5-S2) | Extends thigh at hip; flexes leg at knee and rotates it medially; with flexed hip and knee, extends trunk |
| Biceps femoris | *Long head:* ischial tuberosity<br>*Short head:* linea aspera and lateral supracondylar line of femur | Lateral side of head of fibula; tendon at this site split by fibular collateral ligament of knee | *Long head:* tibial division of sciatic nerve (L5-S2)<br>*Short head:* common fibular (peroneal) division of sciatic nerve (L5-S2) | Flexes leg at knee and rotates it laterally; extends thigh at hip (e.g., when one is starting to walk [long head only]) |

***Clinical Note:***

The **hamstrings** cross two joints, extending at the hip and flexing at the knee. Hence, it is important to warm these muscles up before rigorous exercise by stretching the muscles, getting adequate blood flow into the muscle tissue, and activating the muscle fiber units.

Athletically active individuals may report **hip pain** when the injury may actually be related to the lumbar spine, e.g., herniated intervertebral disc. Pain in the buttocks may originate from bursitis or a hamstring muscle injury. **Pelvic pain** may suggest an intrapelvic disorder. Careful follow-up should examine all potential causes of the pain to determine whether it is referred pain and thus originates from another source.

**Intragluteal injections** (an intramuscular injection) are a common site for injections as the large muscle mass allows for venous absorption of the drug. Such injections must be given in the upper-outer quadrant (superolaterally) to avoid injury to the large sciatic nerve or a hematoma related to the rich vascular supply to the massive gluteal muscles.

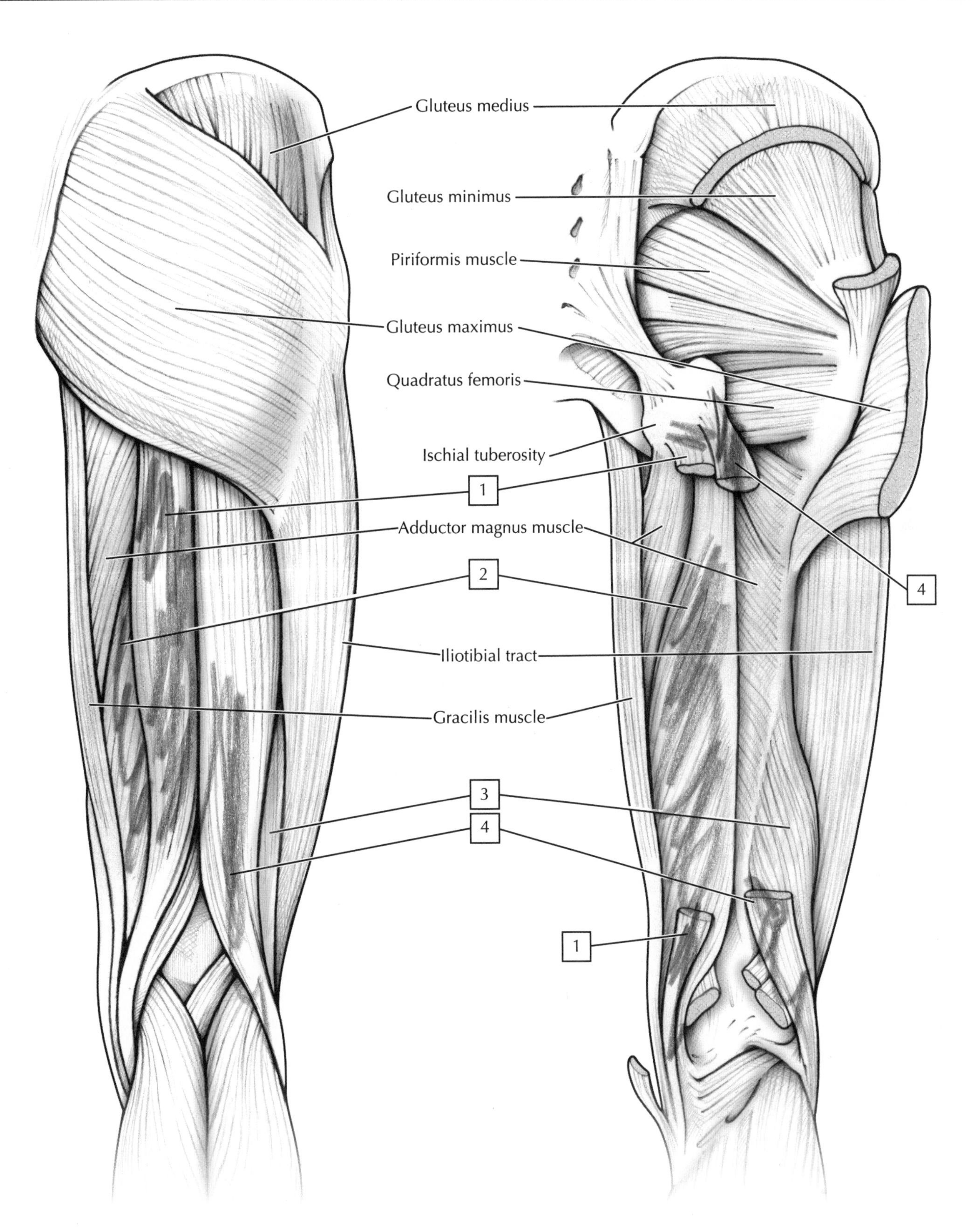


A. Posterior view, superficial dissection

B. Posterior view, deeper dissection

# 3 Anterior Thigh Muscles

The thigh is divided into three muscle compartments by connective tissue intermuscular septae. The muscles of the anterior compartment primarily extend the knee, although several muscles cross both the hip and knee and act on both joints. Additionally, two muscles of the posterior abdominal wall, the psoas and iliacus (iliopsoas) muscles, pass into the upper thigh and are the most powerful flexors of the hip joint (see Plate 3-14 for the origins, insertions, innervation, and actions of these two muscles). The anterior thigh muscles are summarized in the table below.

**COLOR** each of the following muscles, using a different color for each muscle:

- ☐ 1. **Psoas**
- ☐ 2. **Iliacus: psoas and iliacus fuse to form the iliopsoas muscle**
- ☐ 3. **Tensor fasciae latae**
- ☐ 4. **Sartorius: "sartorius" refers to a tailor, who sits cross-legged when sewing, so that the hip and knee are flexed; this is the action of the sartorius**
- ☐ 5. **Rectus femoris: muscles 5 to 8 in this list form the quadriceps femoris group; they all fuse to form the quadriceps femoris tendon, which is continuous with the patellar ligament**
- ☐ 6. **Vastus lateralis**
- ☐ 7. **Vastus medialis**
- ☐ 8. **Vastus intermedius**

| MUSCLE | PROXIMAL ATTACHMENT | DISTAL ATTACHMENT | INNERVATION | MAIN ACTIONS |
|---|---|---|---|---|
| Tensor fasciae latae | Anterior superior iliac spine and anterior iliac crest | Iliotibial tract that attaches to lateral condyle of tibia | Superior gluteal nerve (L4-S1) | Abducts, medially rotates, and flexes thigh at hip; helps keep knee extended |
| Sartorius | Anterior superior iliac spine and superior part of notch inferior to it | Superior part of medial surface of tibia | Femoral nerve (L2-L3) | Flexes, abducts, and laterally rotates thigh at hip joint; flexes knee joint |
| | | **Quadriceps Femoris** | | |
| Rectus femoris | Anterior inferior iliac spine and ilium superior to acetabulum | Base of patella and by patellar ligament to tibial tuberosity | Femoral nerve (L2-L4) | Extends leg at knee joint; rectus femoris also steadies hip joint and helps iliopsoas flex thigh at hip |
| Vastus lateralis | Greater trochanter and lateral lip of linea aspera of femur and gluteal tuberosity | Base of patella and by patellar ligament to tibial tuberosity | Femoral nerve (L2-L4) | Extends leg at knee joint |
| Vastus medialis | Intertrochanteric line and medial lip of linea aspera, and greater trochanter of femur | Base of patella and by patellar ligament to tibial tuberosity | Femoral nerve (L2-L4) | Extends leg at knee joint |
| Vastus intermedius | Anterior and lateral surfaces of femoral shaft | Base of patella and by patellar ligament to tibial tuberosity | Femoral nerve (L2-L4) | Extends leg at knee joint |

***Clinical Note:***

Tapping the **patellar ligament** with a reflex hammer elicits the patellar reflex, causing the flexed knee to jerk upward in extension. This maneuver tests the integrity of the muscle and its innervation by the femoral nerve.

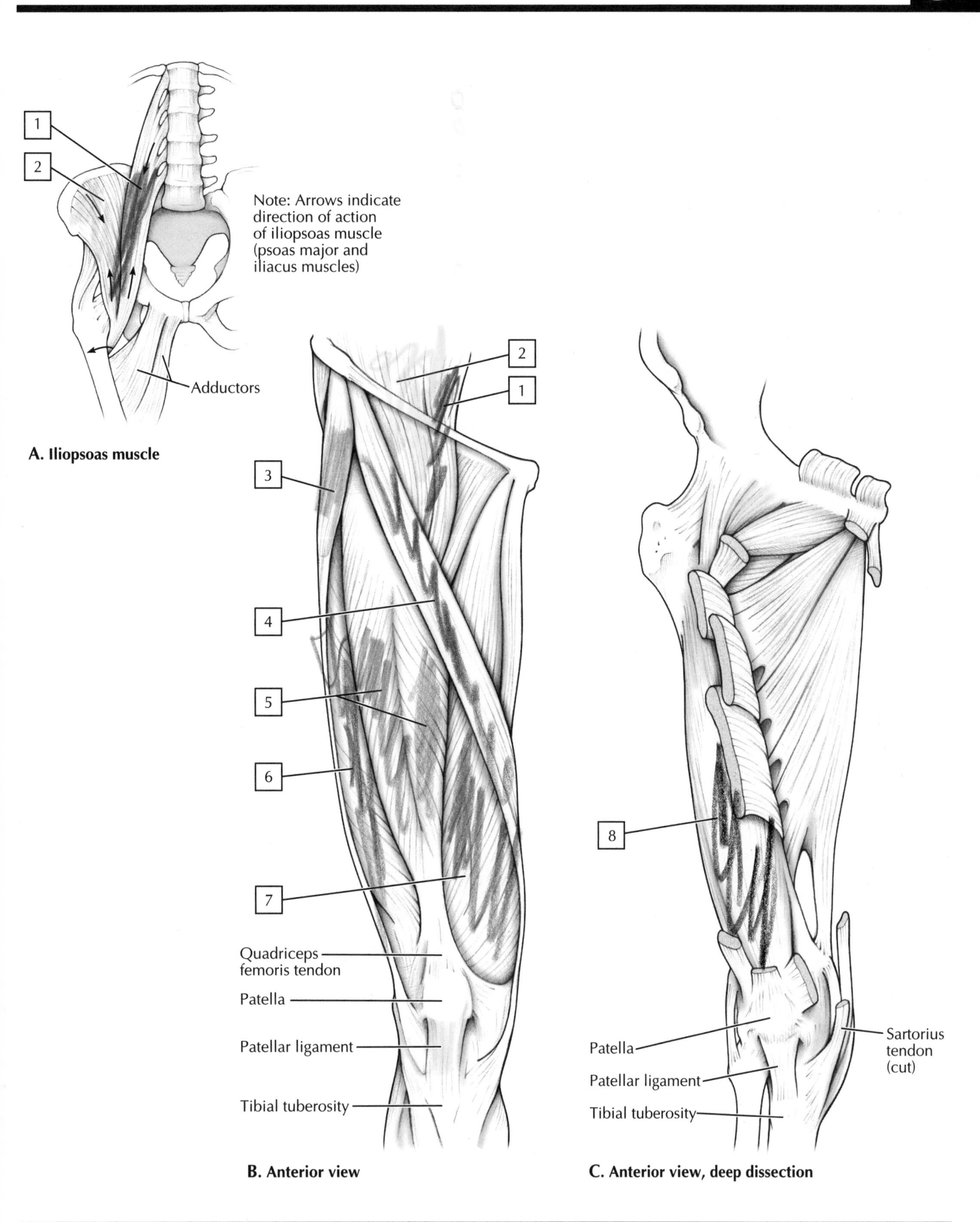


**A. Iliopsoas muscle**

**B. Anterior view**

**C. Anterior view, deep dissection**

# 3 Medial Thigh Muscles

The thigh is divided into three muscle compartments by connective tissue intermuscular septae. The muscles of the medial compartment primarily adduct the lower limb at the hip. Several muscles cross both the hip and knee joints and act on both joints. These muscles are summarized in the table below.

**COLOR** the following muscles, using a different color for each muscle:

- ☐ **1. Pectineus**
- ☐ **2. Adductor longus**
- ☐ **3. Gracilis**
- ☐ **4. Adductor brevis: lies deep to the adductor longus (cut in illustration)**
- ☐ **5. Obturator externus: lies very deep in the thigh**
- ☐ **6. Adductor magnus: the most powerful adductor of the hip**

| MUSCLE | PROXIMAL ATTACHMENT | DISTAL ATTACHMENT | INNERVATION | MAIN ACTIONS |
|---|---|---|---|---|
| Pectineus | Superior ramus of pubis | Pectineal line of femur, just inferior to lesser trochanter | Femoral nerve; may receive a branch from obturator nerve (L2-L4) | Adducts and flexes thigh at hip |
| Adductor longus | Body of pubis inferior to pubic crest | Middle third of linea aspera of femur | Obturator nerve (L2-L4) | Adducts thigh at hip |
| Adductor brevis | Body and inferior ramus of pubis | Pectineal line and proximal part of linea aspera of femur | Obturator nerve (L2-L4) | Adducts thigh at hip and to some extent flexes it |
| Adductor magnus | Inferior ramus of pubis, ramus of ischium, and ischial tuberosity | Gluteal tuberosity, linea aspera, medial supracondylar line (adductor part), and adductor tubercle of femur (hamstring part) | *Adductor part:* obturator nerve (L2-L4) *Hamstring part:* tibial part of sciatic nerve | Adducts thigh at hip; *Adductor part*: also flexes thigh at hip; *Hamstring part*: extends thigh |
| Gracilis | Body and inferior ramus of pubis | Superior part of medial surface of tibia | Obturator nerve (L2-L3) | Adducts thigh at hip, flexes leg at knee and helps rotate it medially |
| Obturator externus | Margins of obturator foramen and obturator membrane | Trochanteric fossa of femur | Obturator nerve (L3-L4) | Rotates thigh laterally at hip; steadies femoral head in acetabulum |

***Clinical Note:***

A **"groin pull"** is a common athletic injury and is a stretching or tearing of one or more of the adductor muscles in the medial compartment of the thigh. The adductor longus and magnus are especially vulnerable. Because the hamstring muscles cross two joints and are actively used in walking and running, they can become pulled or torn if not adequately stretched and loosened before vigorous use.

Likewise, a "**charley horse**" is a muscle pain or stiffness often felt in the quadriceps muscles of the anterior compartment or in the hamstring muscles.

**Muscle tears and tendon disruptions** are also injuries that may commonly occur in athletes.

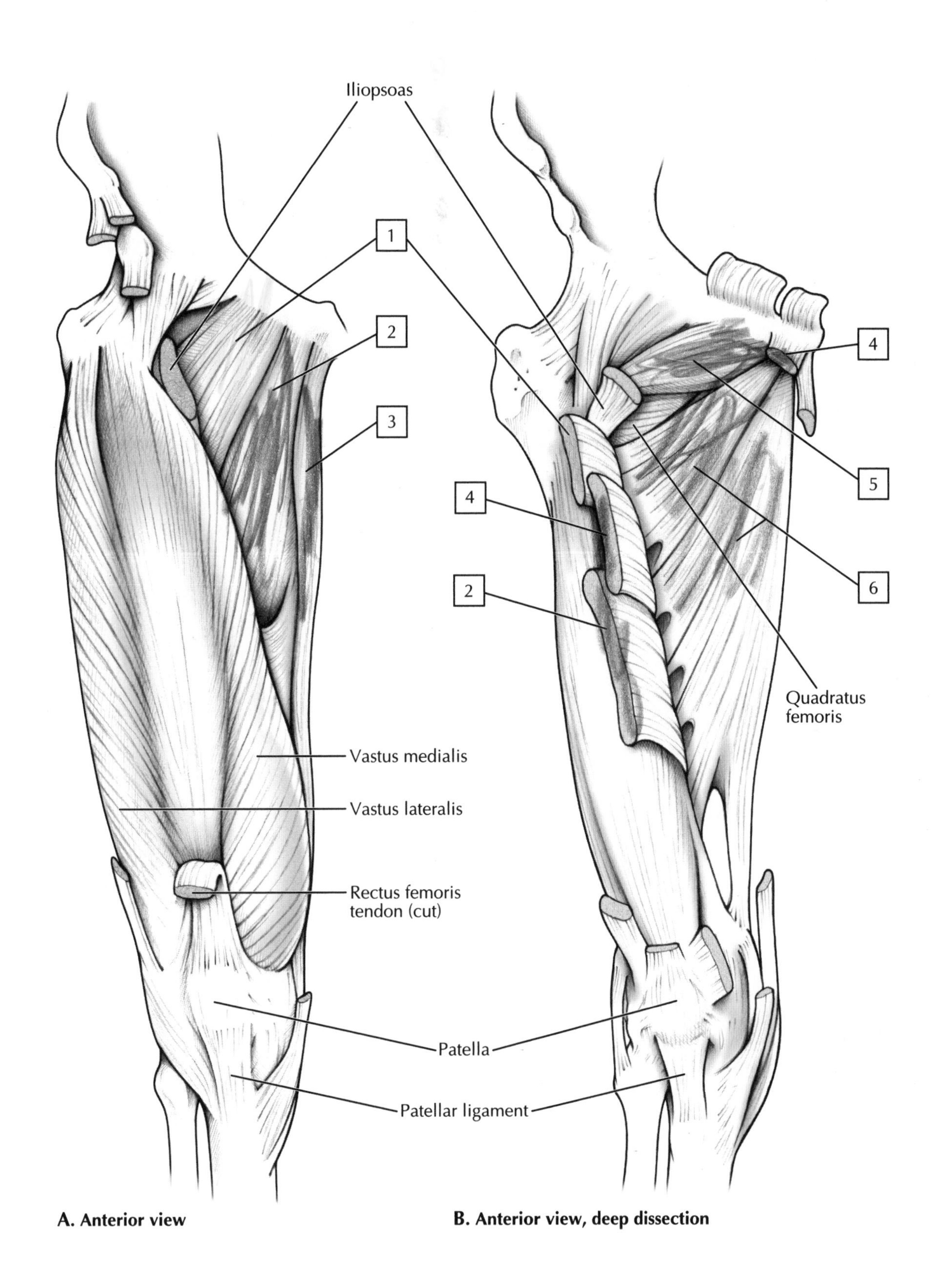


**A. Anterior view**

**B. Anterior view, deep dissection**

# Anterior and Lateral Leg Muscles

The leg is divided into three muscle compartments by connective tissue intermuscular septae. The muscles of the anterior compartment:

- Dorsiflex the foot at the ankle joint
- Extend the toes
- Invert the foot (turn the sole inward)

Realize that the muscles of the lower limb are just the reverse of the upper limb. Lower limb flexors are in the **posterior** compartments (anterior compartment in the upper limb) and extensors are in the **anterior** compartments (posterior compartment in the upper limb). This arrangement occurs because of the different ways the limbs rotate during embryonic development.

The muscles of the lateral compartment primarily evert the foot (turn the sole outward). The muscles of these two compartments are summarized in the table below.

**COLOR** the following muscles, using a different color for each muscle:

- ☐ 1. **Fibularis longus: the tendon crosses deep within the sole and inserts into the 1st metatarsal**
- ☐ 2. **Tibialis anterior**
- ☐ 3. **Fibularis brevis: the tendon inserts into the 5th metatarsal**
- ☐ 4. **Extensor digitorum longus**
- ☐ 5. **Extensor hallucis longus ("hallucis" refers to the big toe)**
- ☐ 6. **Fibularis tertius: tendon only; muscle deep to extensor digitorum longus**

| MUSCLE | PROXIMAL ATTACHMENT | DISTAL ATTACHMENT | INNERVATION | MAIN ACTIONS |
|---|---|---|---|---|
| Tibialis anterior | Lateral condyle and superior half of lateral surface of tibia and interosseous membrane | Medial plantar surfaces of medial cuneiform and base of 1st metatarsal | Deep fibular (peroneal) nerve (L4-L5) | Dorsiflexes foot at ankle and inverts foot |
| Extensor hallucis longus | Middle part of anterior surface of fibula and interosseous membrane | Dorsal aspect of base of distal phalanx of great toe | Deep fibular (peroneal) nerve (L5-S1) | Extends great toe and dorsiflexes foot at ankle |
| Extensor digitorum longus | Lateral condyle of tibia and superior three-fourths of anterior surface of interosseous membrane and fibula | Middle and distal phalanges of lateral four digits | Deep fibular (peroneal) nerve (L5-S1) | Extends lateral four digits and dorsiflexes foot at ankle |
| Fibularis (peroneus) tertius | Inferior third of anterior surface of fibula and interosseous membrane | Dorsum of base of 5th metatarsal | Deep fibular (peroneal) nerve (L5-S1) | Dorsiflexes foot at ankle and aids in eversion of foot |
| Fibularis (peroneus) longus | Head and superior two-thirds of lateral surface of fibula | Plantar base of 1st metatarsal and medial cuneiform | Superficial fibular (peroneal) nerve (L5-S2) | Everts foot and weakly plantarflexes foot at ankle |
| Fibularis (peroneus) brevis | Inferior two-thirds of lateral surface of fibula | Dorsal aspect of tuberosity on lateral side of 5th metatarsal | Superficial fibular (peroneal) nerve (L5-S2) | Everts foot and weakly plantarflexes foot at ankle |

***Clinical Note:***

**Anterior compartment syndrome** (sometimes called **anterior shin splints**) occurs from excessive contraction of anterior compartment muscles. The pain over these muscles radiates down the ankle and onto the dorsum of the foot overlying the extensor tendons. This condition is usually chronic, and swelling of the muscle in the tightly ensheathed muscular compartment may lead to nerve and vascular compression.

In the acute syndrome (rapid, unrelenting swelling), the compartment may have to be opened surgically (fasciotomy) to relieve the pressure.

**Lateral compartment syndrome** (exertional syndrome) from overuse may involve the fibularis longus and brevis muscles, and the pain is often felt on the inferior third of the lateral leg just above the ankle.

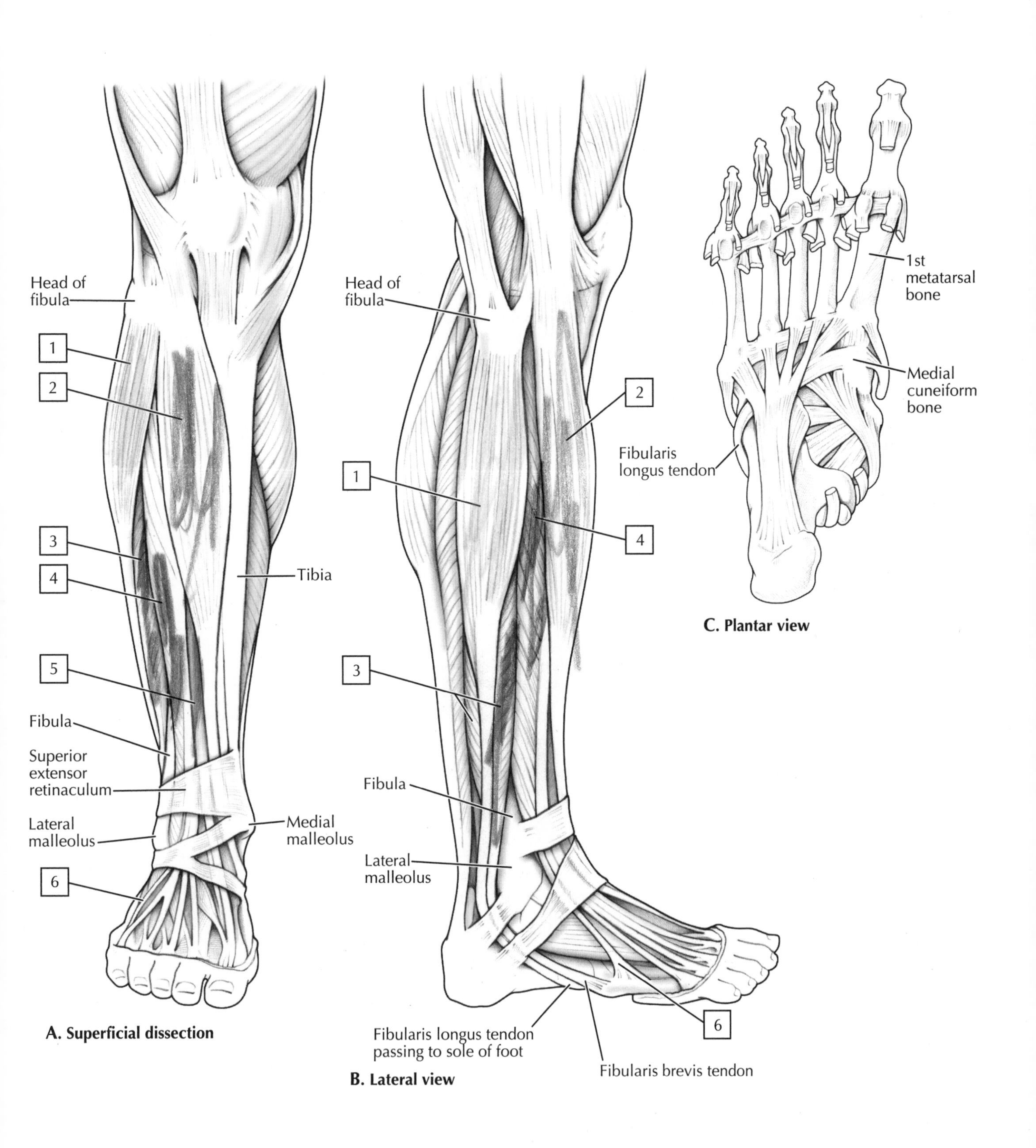


A. Superficial dissection

B. Lateral view

C. Plantar view

# 3 Posterior Leg Muscles

The leg is divided into three muscle compartments by connective tissue intermuscular septae. The muscles of the posterior compartment:

- Plantarflex the foot at the ankle joint
- Flex the toes
- Invert the foot (turn the sole inward)

The muscles of the posterior compartment are arranged into a superficial and a deep group. The superficial group of muscles (gastrocnemius, plantaris, soleus muscles) all merge their tendons of insertion into a strong calcaneal (Achilles) tendon that attaches to the heel (calcaneal tuberosity). The remaining muscles of the posterior compartment form the deep group. These muscles are summarized in the table below.

**COLOR** the following muscles, using a different color for each muscle:

☐ 1. **Plantaris (muscles 1 to 3 of this list form the superficial group)**

☐ 2. **Gastrocnemius: lateral and medial heads, the "calf" muscle**

☐ 3. **Soleus**

☐ 4. **Popliteus**

☐ 5. **Flexor digitorum longus**

☐ 6. **Tibialis posterior**

☐ 7. **Flexor hallucis longus ("hallucis" refers to the big toe)**

| MUSCLE | PROXIMAL ATTACHMENT | DISTAL ATTACHMENT | INNERVATION | MAIN ACTIONS |
|---|---|---|---|---|
| Gastrocnemius | *Lateral head:* lateral aspect of lateral condyle of femur<br>*Medial head:* popliteal surface of femur, superior to medial condyle | Posterior aspect of calcaneus via calcaneal tendon | Tibial nerve (S1-S2) | Plantarflexes foot at ankle; flexes leg at knee joint |
| Soleus | Posterior aspect of head of fibula, superior fourth of posterior surface of fibula, soleal line, and medial border of tibia | Posterior aspect of calcaneus via calcaneal tendon | Tibial nerve (S1-S2) | Plantarflexes foot at ankle; steadies leg on foot |
| Plantaris | Inferior end of lateral supracondylar line of femur and oblique popliteal ligament | Posterior aspect of calcaneus via calcaneal tendon | Tibial nerve (S1-S2) | Weakly assists gastrocnemius in plantarflexing foot at ankle and flexing knee |
| Popliteus | Lateral condyle of femur and lateral meniscus | Posterior surface of tibia, superior to soleal line | Tibial nerve (L4-S1) | Weakly flexes leg at knee and unlocks it and fixed tibia |
| Flexor hallucis longus | Inferior two-thirds of posterior surface of fibula and inferior interosseous membrane | Base of distal phalanx of great toe (big toe) | Tibial nerve (S2-S3) | Flexes great toe at all joints and weakly plantarflexes foot at ankle |
| Flexor digitorum longus | Medial part of posterior surface of tibia inferior to soleal line | Plantar bases of distal phalanges of lateral four digits | Tibial nerve (S2-S3) | Flexes lateral four digits and plantarflexes foot at ankle; supports longitudinal arch of foot |
| Tibialis posterior | Interosseous membrane, posterior surface of tibia inferior to soleal line, and posterior surface of fibula | Tuberosity of navicular, cuneiform, and cuboid and bases of metatarsals 2, 3, and 4 | Tibial nerve (L4-L5) | Plantarflexes foot at ankle and inverts foot |

***Clinical Note:***

"**Shin splints**" refers to pain along the inner distal two-thirds of the tibial shaft and is a common syndrome in athletes. The primary cause is repetitive pulling of the tibialis posterior tendon as one pushes off the foot during running.

**Tendinitis** of the calcaneal (Achilles) tendon is a painful inflammation that often occurs in runners who run on hills or uneven surfaces. Repetitive stress on the tendon occurs as the heel strikes the ground and when plantarflexion lifts the foot and toes. This is the strongest muscle tendon in the body. Rupture of the tendon is a serious injury, because the avascular tendon heals slowly. In general, most tendon injuries heal more slowly because of their avascular nature.

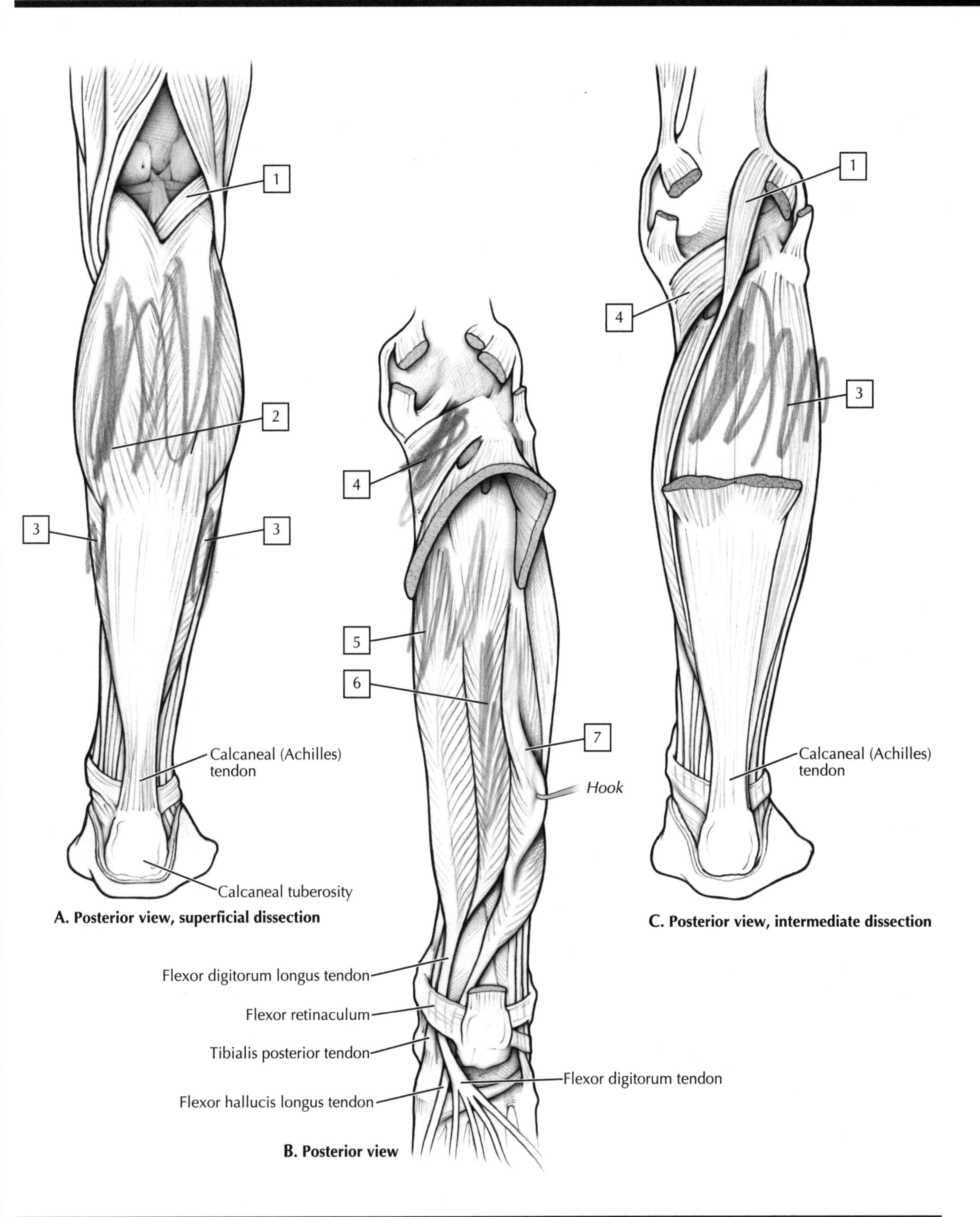


**A. Posterior view, superficial dissection**

**B. Posterior view**

**C. Posterior view, intermediate dissection**

# 3 Intrinsic Foot Muscles

The intrinsic muscles are arranged in four layers on the sole of the foot and complement the actions of the leg's long flexor tendons as they pass into the foot. These muscles are summarized in the table below.

| MUSCLE | PROXIMAL ATTACHMENT | DISTAL ATTACHMENT | INNERVATION | MAIN ACTIONS |
|---|---|---|---|---|
| Abductor hallucis | Medial tubercle of tuberosity of calcaneus, flexor retinaculum, and plantar aponeurosis | Medial aspect of base of proximal phalanx of 1st digit | Medial plantar nerve (S1-S2) | Abducts and flexes great toe |
| Flexor digitorum brevis | Medial tubercle of tuberosity of calcaneus, plantar aponeurosis, and intermuscular septa | Both sides of middle phalanges of lateral four digits | Medial plantar nerve (S1-S2) | Flexes lateral four digits at interphalangeal joints |
| Abductor digiti minimi | Medial and lateral tubercles of tuberosity of calcaneus, plantar aponeurosis, and intermuscular septa | Lateral aspect of base of proximal phalanx of 5th digit | Lateral plantar nerve (S1-S3) | Abducts and flexes little toe |
| Quadratus plantae | Medial aspect and lateral margin of plantar surface of calcaneus | Posterolateral margin of tendon of flexor digitorum longus | Lateral plantar nerve (S1-S3) | Assists flexor digitorum longus in flexing lateral four digits |
| Lumbricals | Tendons of flexor digitorum longus | Medial aspect of expansion over lateral four digits | *Medial one:* medial plantar nerve<br>*Lateral three:* lateral plantar nerve (S2-S3) | Flex proximal phalanges and extend middle and distal phalanges of lateral four digits |
| Flexor hallucis brevis | Plantar surfaces of cuboid and lateral cuneiform | Both sides of base of proximal phalanx of 1st digit | Medial plantar nerve (S1-S2) | Flexes proximal phalanx of great toe |
| Adductor hallucis | *Oblique head:* bases of metatarsals 2-4<br>*Transverse head:* plantar ligaments of metatarsophalangeal joints of digits 3-5 | Tendons of both heads attach to lateral side of base of proximal phalanx of 1st digit | Deep branch of lateral plantar nerve (S2-S3) | Adducts great toe; assists in maintaining transverse arch of foot |
| Flexor digiti minimi brevis | Base of 5th metatarsal | Lateral base of proximal phalanx of 5th digit | Superficial branch of lateral plantar nerve (S2-S3) | Flexes proximal phalanx of little toe, thereby assisting with its flexion |
| Plantar interossei (three muscles) | Bases and medial sides of metatarsals 3-5 | Medial sides of bases of proximal phalanges of digits 3-5 | Lateral plantar nerve (S2-S3) | Adduct digits (3-5) and flex metatarsophalangeal joints and extend phalangeal bones |
| Dorsal interossei (four muscles) | Adjacent sides of metatarsals 1-5 | *First*: medial side of proximal phalanx of 2nd digit<br>*Second to fourth*: lateral sides of digits 2-4 | Lateral plantar nerve (S2-S3) | Abduct digits (2-4) and flex metatarsophalangeal joints and extend phalanges |

**COLOR** the following muscles, using a different color for each muscle (the muscles of the sole are organized into several layers beneath a tough plantar aponeurosis, as seen in the illustrations):

- ☐ 1. **Flexor digiti minimi brevis**
- ☐ 2. **Abductor digiti minimi**
- ☐ 3. **Lumbricals: four small muscles that attach to the long flexor tendons**
- ☐ 4. **Flexor hallucis brevis: has two heads whose tendons contain two small sesamoid bones**
- ☐ 5. **Abductor hallucis**
- ☐ 6. **Flexor digitorum brevis**
- ☐ 7. **Quadratus plantae**
- ☐ 8. **Plantar interossei: three muscles that adduct the toes (plantar adduct, PAD)**
-  9. **Adductor hallucis: has two heads (transverse and oblique)**
- ☐ 10. **Dorsal interossei: four muscles that abduct the toes (dorsal abduct, DAB)**

***Clinical Note:***
Just beneath the skin of the sole of the foot and overlying the superficial layer of intrinsic muscles lies the plantar aponeurosis, a broad, flat tendon that stretches from the heel to the toes. **Plantar fasciitis** is a common cause of heel pain, especially in joggers, and results from inflammation of the plantar aponeurosis at its point of attachment for the calcaneus, with pain often radiating distally toward the toes.

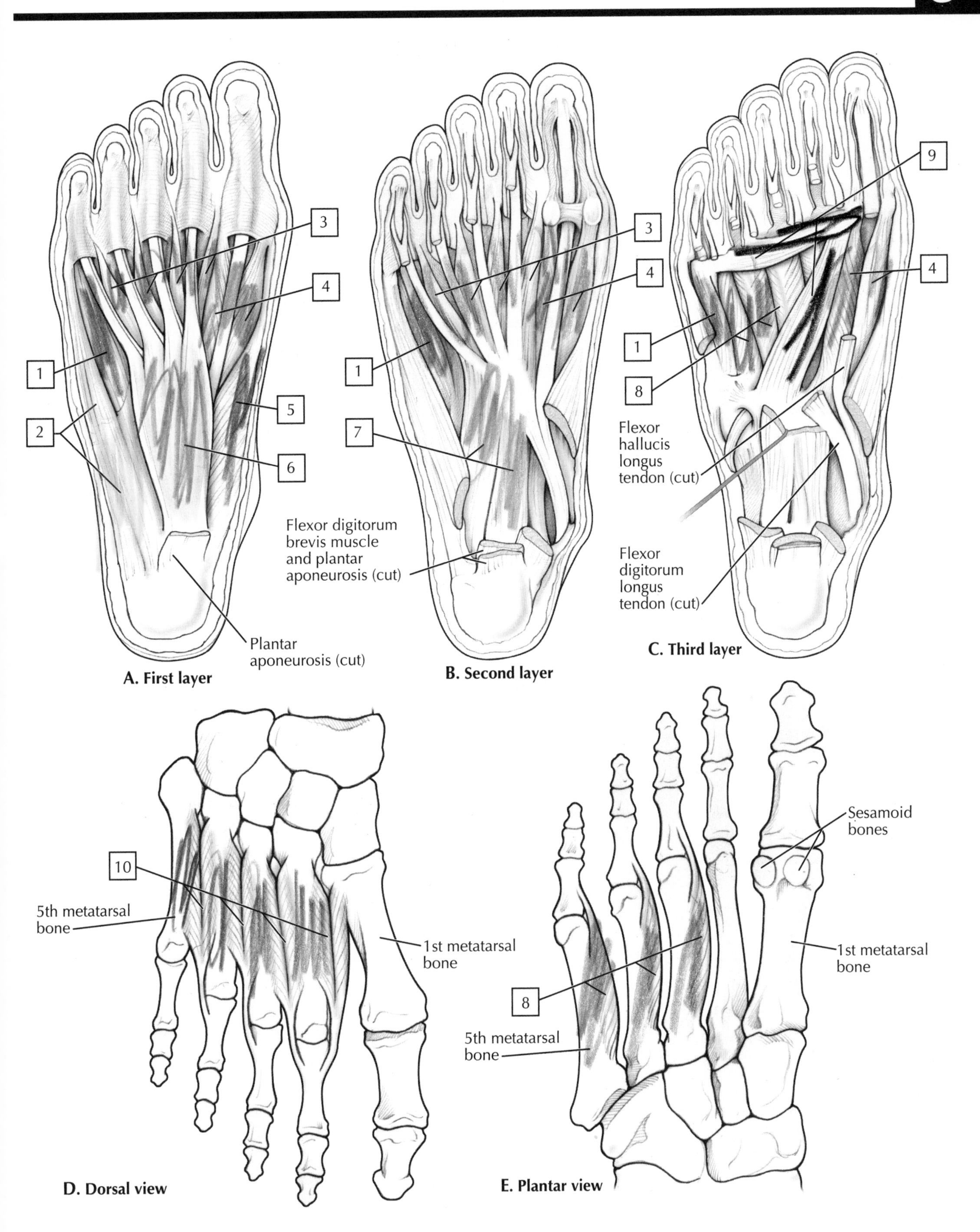
3
4
1
2
5
6
Plantar aponeurosis (cut)
A. First layer
3
4
1
7
Flexor digitorum brevis muscle and plantar aponeurosis (cut)
B. Second layer
9
4
1
8
Flexor hallucis longus tendon (cut)
Flexor digitorum longus tendon (cut)
C. Third layer
10
5th metatarsal bone
1st metatarsal bone
D. Dorsal view
Sesamoid bones
1st metatarsal bone
8
5th metatarsal bone
E. Plantar view

# 3 Summary of Lower Limb Muscles

It is best to learn the action of the muscles by knowing which **compartment** they reside in and then knowing the primary action of the muscles in that compartment. Few muscles act in isolation; more often, they act as a group. Generally, muscles of the gluteal region extend the hip, abduct the limb, and rotate it. Muscles of the anterior thigh act on the knee to extend it, whereas the muscles of the medial thigh adduct the limb at the hip. Muscles of the posterior thigh extend the hip and flex the knee. Muscles of the lateral leg evert the foot, muscles of the anterior leg dorsiflex the ankle and extend the toes, and muscles of the posterior leg plantarflex the ankle and flex the toes.

| HIP | KNEE |
|---|---|
| **Flex**: iliopsoas, rectus femoris, sartorius<br>**Extend**: hamstrings, gluteus maximus<br>**Abduct**: gluteus medius and minimus, tensor fasciae latae<br>**Rotate medially**: gluteus medius and minimus<br>**Rotate laterally**: obturator internus, gemelli, piriformis<br>**Adduct**: adductor group of muscles | **Flex**: hamstrings, gracilis, sartorius, gastrocnemius<br>**Extend**: quadriceps femoris<br>**Rotate medially**: semitendinosus, semimembranosus<br>**Rotate laterally**: biceps femoris |
| **ANKLE** | **METATARSOPHALANGEAL** |
| **Plantarflex**: gastrocnemius, soleus, tibialis posterior, flexor digitorum longus, flexor hallucis longus<br>**Dorsiflex**: tibialis anterior, extensor digitorum longus, extensor hallucis longus, fibularis tertius | **Flex**: interossei and lumbricals<br>**Extend**: extensor digitorum longus, brevis<br>**Abduct**: dorsal interossei<br>**Adduct**: plantar interossei |
| **INTERPHALANGEAL** | **INTERTARSAL** |
| **Flex**: flexor digitorum longus, brevis<br>**Extend**: extensor digitorum longus, brevis | **Evert**: fibularis longus, brevis, tertius<br>**Invert**: tibialis anterior and posterior |

**COLOR** the following muscles, using a different color for each muscle:

- ☐ 1. **Rectus femoris**
- ☐ 2. **Sartorius**
- ☐ 3. **Gracilis**
- ☐ 4. **Adductor magnus**
- ☐ 5. **Tibialis anterior**
- ☐ 6. **Soleus**
- ☐ 7. **Tibialis posterior**
- ☐ 8. **Fibularis longus**
- ☐ 9. **Adductor hallucis**
- ☐ 10. **Abductor digiti minimi**

***Clinical Note:***

**Knee extension** is an inability to dorsiflex the foot at the ankle, resulting in a foot that cannot be raised when walking. A person with knee extension must raise the knee during the swing phase of walking to avoid dragging the affected foot on the ground or to avoid tripping (steppage gait). Typically, knee extension results from injury to the common fibular nerve where it lies superficially beneath the skin and passes around the fibular neck (coffee table or car bumper height) or deep fibular nerve. The nerve also may be affected by a herniated disc that compresses the L5 nerve root (an L4-L5 herniated disc).

**Lower Limb Nerve Clinical Summary:**

The **femoral nerve integrity** may be checked with the **patellar tendon reflex** (L3-L4) (knee extension).

The **obturator nerve damage** (herniated disc) results in a weakened ability to adduct the thigh.

The **sciatic nerve** (largest nerve in the body) is composed of the tibial and common fibular nerves. The **tibial nerve injury** may result in the loss of plantarflexion and weakened inversion of the foot, resulting in a shuffling gait. The **common fibular nerve injury** may result in knee extension and steppage gait (high stepping).

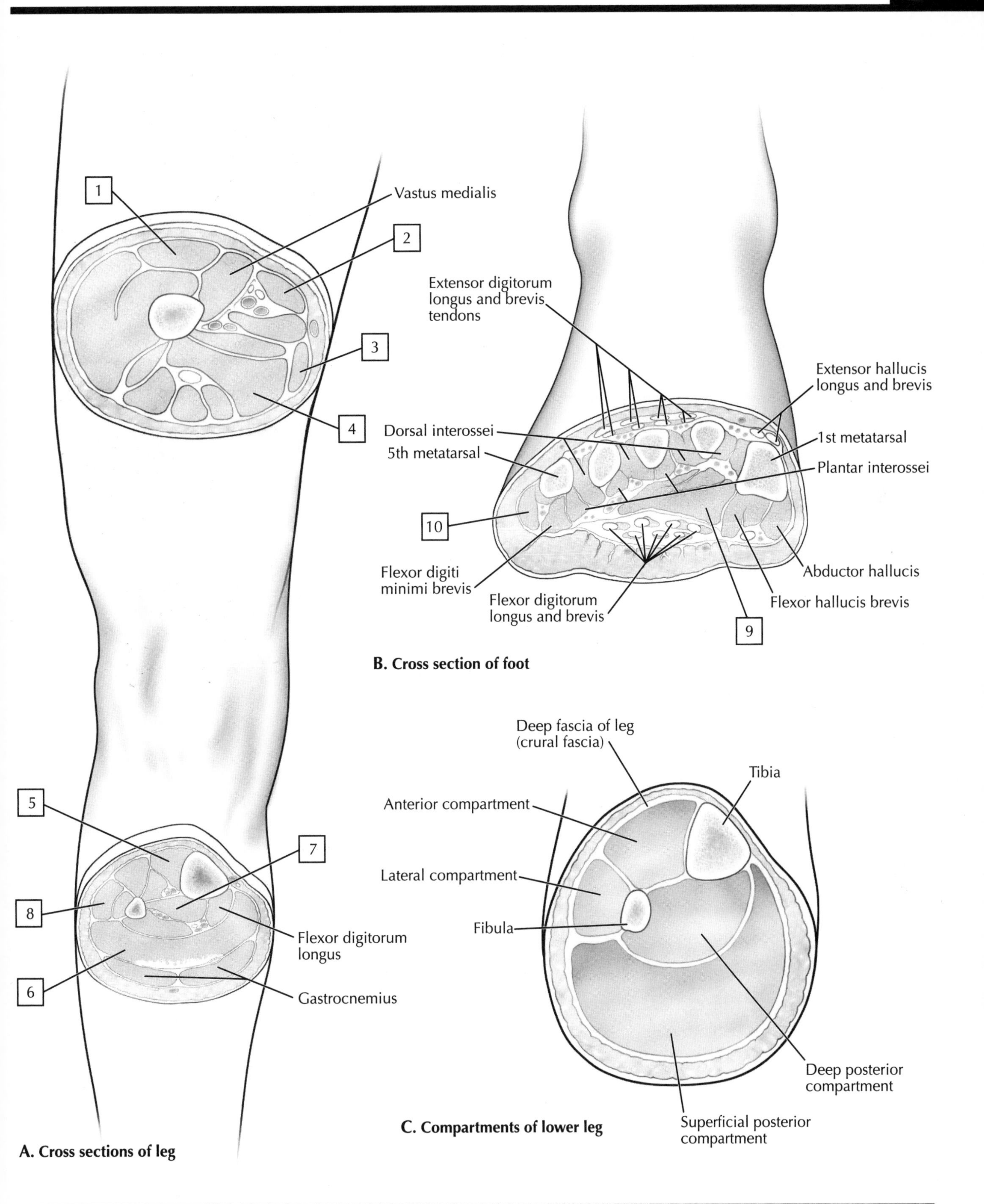


**A. Cross sections of leg**

**B. Cross section of foot**

**C. Compartments of lower leg**

## REVIEW QUESTIONS

1. Why might a patient with Bell's palsy (unilateral facial nerve inflammation) be unable to close his or her ipsilateral eye?________
________________________________________________________________________________

2. Which muscle might be paralyzed if, during an eye exam (clinical testing), an inability to adduct and depress the eyeball was demonstrated? ________________________________________________________________

3. What are the three muscles that line the posterior wall of the pharynx and assist in swallowing?________________________
________________________________________________________________________________

4. The deep intrinsic muscles of the back innervated by a posterior ramus of the spinal nerve include which of the following muscle groups?
   A. Erector spinae
   B. Latissimus dorsi
   C. Levator scapulae
   D. Rhomboid major
   E. Serratus posterior inferior

5. A hernia occurs in the groin region, and a portion of bowel and mesentery descends into the scrotum. This patient most likely has which of the following types of hernias?
   A. Femoral
   B. Direct inguinal
   C. Hiatal
   D. Indirect inguinal
   E. Umbilical

6. An athlete suffers a rotator cuff injury. Which of the following muscles is most likely torn?
   A. Infraspinatus
   B. Subscapularis
   C. Supraspinatus
   D. Teres major
   E. Teres minor

7. A groin pull injury would most likely involve which of the following muscles?
   A. Adductor longus
   B. Rectus femoris
   C. Sartorius
   D. Semitendinosus
   E. Vastus medialis

**Color each muscle described below:**

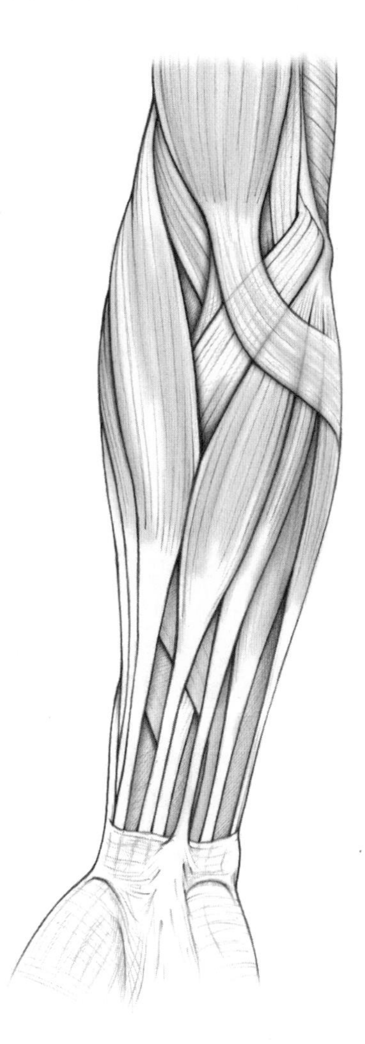

8. This muscle is absent in a small percentage of the population (color it red).
9. This muscle is innervated by the radial nerve (color it blue).
10. This muscle flexes the wrist and is innervated by the ulnar nerve (color it green).

## ANSWER KEY

1. Paralysis of the orbicularis oculi muscle of facial expression
2. Superior oblique muscle
3. Superior, middle, and inferior pharyngeal constrictor muscles
4. A
5. D
6. C

7. A
8. Palmaris longus muscle
9. Brachioradialis muscle
10. Flexor carpi ulnaris muscle

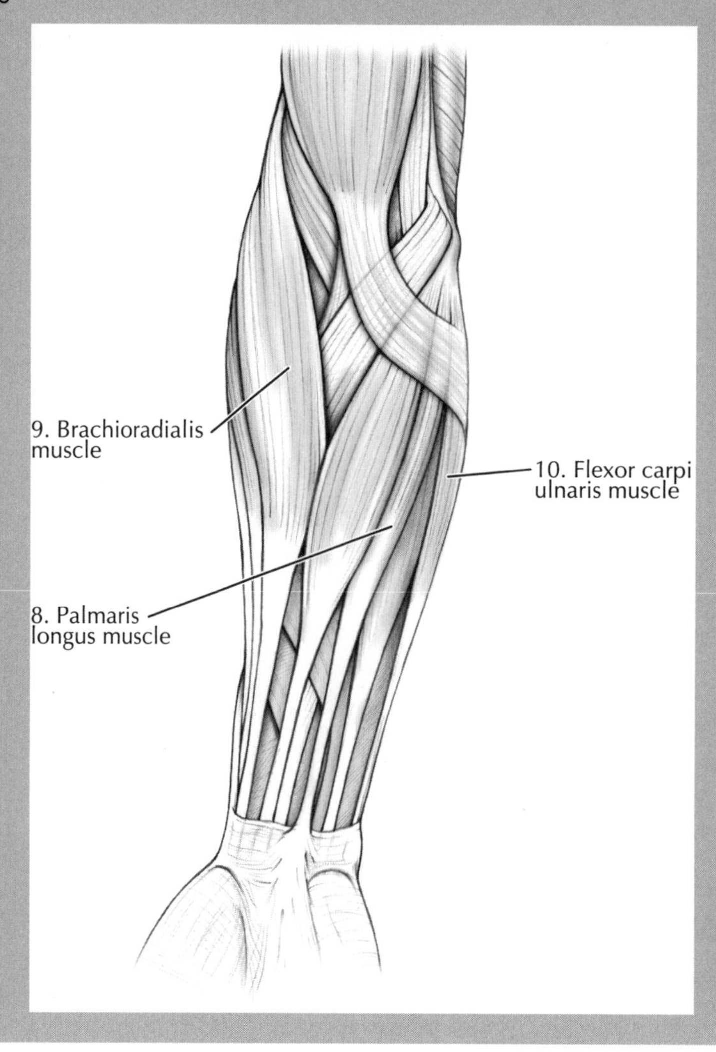

# Chapter 4 Nervous System and Sense Organs

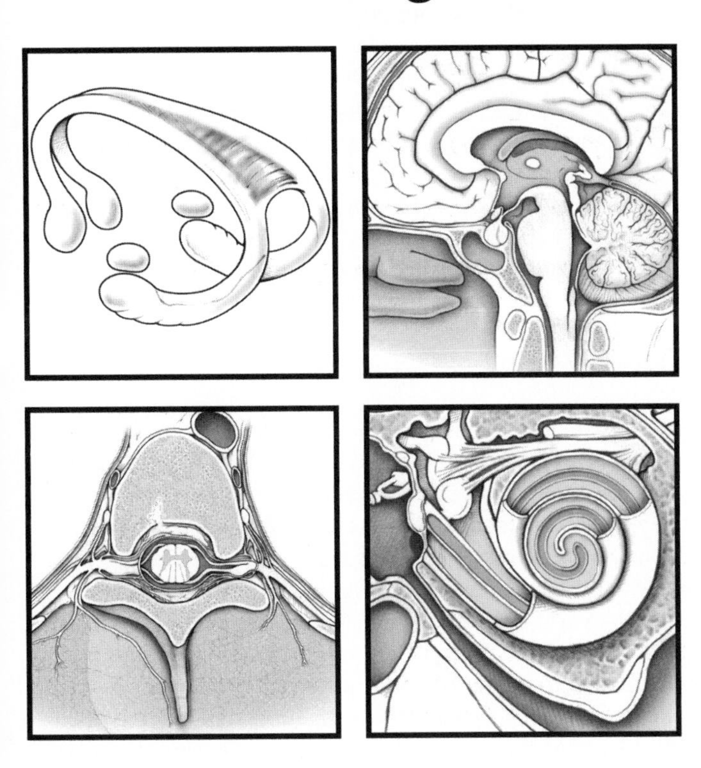

# 4 Neuronal Structure

Nerve cells are called neurons. Each neuron has a structure that reflects its individual functional characteristics. Information comes to the neuron largely via processes called **axons**, which terminate on the neuron at specialized junctions called **synapses**. Synapses can occur on neuronal processes called **dendrites** or on the neuronal cell body, called a **soma** or **perikaryon**.

Typically, a neuron has multiple dendrites and one axon (or sometimes none). Like most cells, the neurons contain a nucleus (and its nucleolus) containing its neuronal genome, cytoplasm containing many mitochondria, an extensive rough endoplasmic reticulum, a Golgi apparatus, and other cellular inclusions (see Plate 1-4). The neuronal dendrites provide an extensive network of branched processes that receive incoming axons that terminate on them and provide neurotransmitter influences on the receiving neuron (sometimes terminating on **dendritic spines**).

Single axons may themselves branch extensively and contact a large number of neuronal cell bodies (**axosomatic synapses**) and/or dendritic synapses (**axodendritic synapses**).

Some neurons, such as the primary sensory neurons in a dorsal root ganglion and some of the cranial nerve ganglia have neuronal cell bodies with no dendrites, a few neurons lack an axon (amacrine cells in the retina of the eye), and some neuronal systems have extensive networks of **dendrodendritic** synaptic interactions.

**COLOR** each of the following features of a neuron, using a different color for each feature:

- ☐ **1. Dendrites**
- ☐ **2. Axon**
- ☐ **3. Soma, or cell body of the neuron**

Neurons convey efferent information via action potentials that course along a single axon arising from a soma, which then synapses on a selective target, usually another neuron or target cell, for example, muscle cells. There are many different types of neurons; some of the more common types include:

- **Unipolar (often called pseudounipolar)**: possesses one axon that divides T-like into two long processes, one distal and one central; usually sensory neurons such as a primary sensory neuron in a cranial nerve ganglion or in a dorsal root ganglion
- **Bipolar**: possesses one axon and one dendrite; rare, but found in the retina and olfactory epithelium
- **Multipolar**: possesses one axon and two, or more often, many more dendrites (a dendritic tree); most common and probably accounting for about 99% of all neurons

**COLOR** each different type of neuron, using a different color for each type:

- ☐ **4. Unipolar (pseudounipolar)**
- ☐ **5. Bipolar**
- ☐ **6. Multipolar**

Although the human nervous system contains billions of neurons, they can be classified largely into one of three functional types:

- **Motor neurons**: convey efferent impulses from the central nervous system (CNS) or ganglia (collections of neurons outside the CNS) to target (effector) cells; somatic efferent axons target skeletal muscle, and visceral efferent axons target smooth muscle, cardiac muscle, and glands
- **Sensory neurons**: convey afferent impulses from receptors to the CNS; somatic afferent axons convey pain, temperature, touch, pressure, and proprioception (nonconscious) sensations, and visceral afferent axons convey pain and other sensations (e.g., nausea) from organs, glands, and smooth muscle to the CNS
- **Interneurons**: convey impulses between sensory and motor neurons, thus forming integrating networks between cells; more than 99% of all neurons in the body are probably interneurons

Neurons can vary considerably in size, ranging from several micrometers to over 100 μm in diameter. They may possess numerous branching dendrites, studded with **dendritic spines** that increase the receptive area of the neuron many times over. The neuron's axon may be quite short or over a meter long, and the axonal diameter may vary, with axons that measure larger than 1 to 2 μm in diameter being insulated by **myelin sheaths**. In the CNS, axons are myelinated by a special glial cell called an **oligodendrocyte**, whereas all the axons in the peripheral nervous system (PNS) are surrounded by a type of glial cell called **Schwann cells**. Schwann cells also myelinate many of the PNS axons they surround.

---

***Clinical Note:***
Neurons are metabolically very active aerobic cells requiring high levels of glucose and oxygen. Neurons have very little energy reserves, so they may easily become **ischemic** if their arterial blood supply is compromised. Neurons also are highly active genomically.

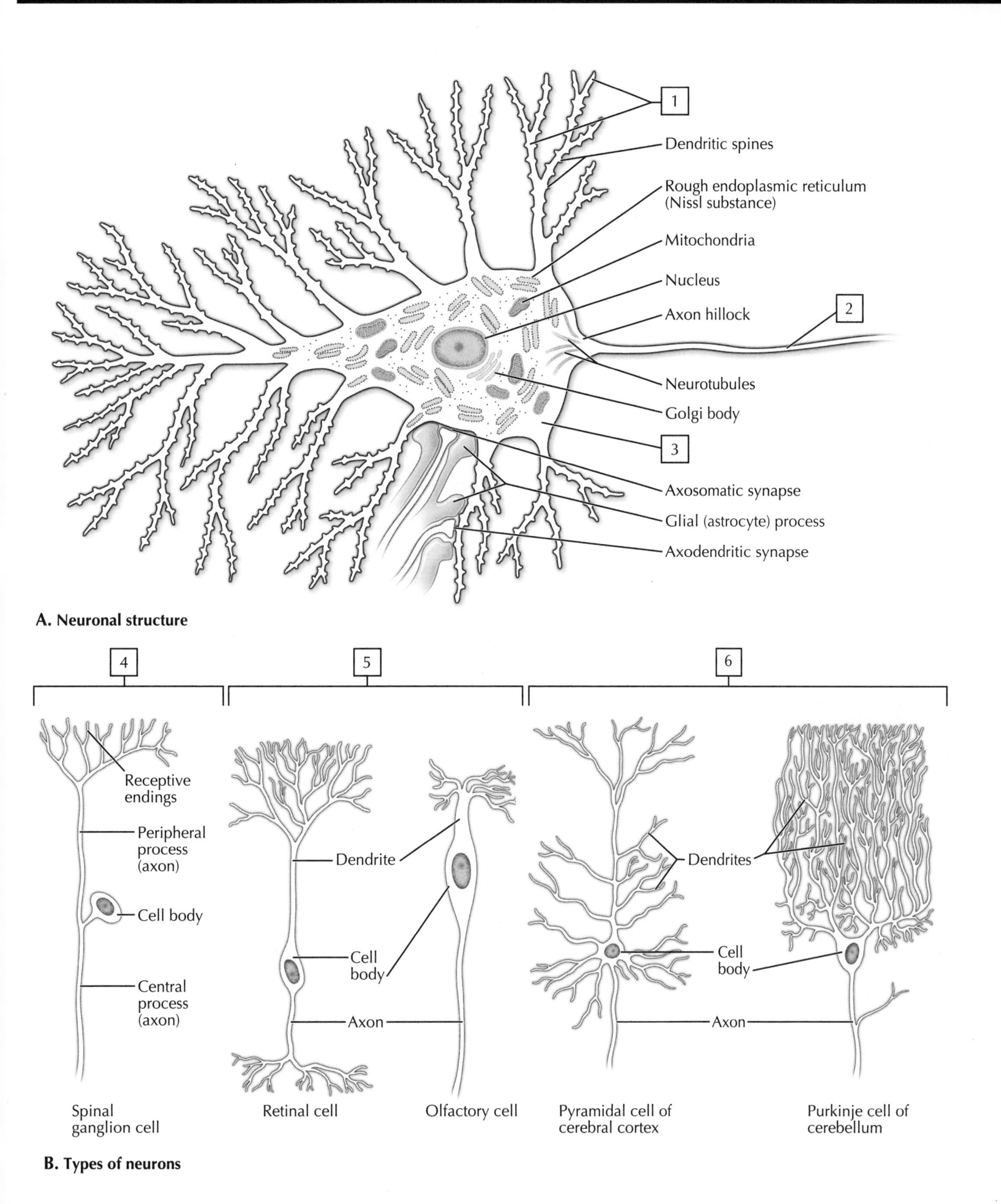


**A. Neuronal structure**

**B. Types of neurons**

# 4 Glial Cells

Glia are the cells that support neurons, by about 10 to 1, in the central nervous system. They myelinate new axons and contribute to most of the postnatal growth seen in the CNS. Functionally, glia:

- Provide structural isolation of neurons and their synapses
- Sequester ions in the extracellular compartment
- Provide trophic support to the neurons and their processes
- Support growth and secrete growth factors
- Support some of the signaling functions of neurons
- Myelinate axons
- Phagocytize debris and participate in inflammatory responses
- Participate in the formation of the blood-brain barrier

The different types of glial cells include the:

- **Astrocytes**: the most numerous of the glial cells, they provide physical and metabolic support for CNS neurons, and contribute to the formation of the blood-brain barrier via end-foot processes that support capillary endothelial cell tight junctions, and contact pial cells (pia mater) that form a pial-glial membrane to protect the outer surface of the brain
- **Oligodendrocytes**: smaller glial cells that are responsible for the formation and maintenance of myelin in the CNS; a single oligodendrocyte may myelinate a single axonal segment for each of over 30 or more axons
- **Microglia**: smallest and most rare of the CNS glia (still more numerous than neurons in the CNS), they are phagocytic cells and participate in inflammatory reactions; they may remove unneeded synapses, remodel synaptic sites, and can respond to cell injury by phagocytosis, release of interleukins and cytokines, and participate in an immune response
- **Ependymal cells**: line the ventricles of the brain and the central canal of the spinal cord that contain cerebrospinal fluid (CSF); specialized ependymal cells called **tanycytes** can sequester substances in the CSF and transport them to specific CNS sites where they may influence CNS neuronal function
- **Schwann cells**: glial cells of the PNS, they surround all axons, myelinating many of them, and provide trophic support, facilitate regrowth of PNS axons, and clean away cellular debris

While ependymal cells line the brain's ventricles, the surface of the brain and spinal cord is lined by the pia mater.

**COLOR** each of the different types of CNS glia, using a different color for each glial cell:

- ☐ **1. Astrocytes**
- ☐ **2. Oligodendrocyte (with myelinating processes)**
- ☐ **3. Microglial cell**
- ☐ **4. Ependymal cells**

***Clinical Note:***

Oligodendrocyte dysfunction occurs in **multiple sclerosis (MS)**, an autoimmune reaction that results in demyelination, a decrease in axonal conduction, and neurological dysfunction (motor and sensory activity, vision problems, eye movement disorders, and emotional changes).

Unfortunately, CNS glial cells are a major source of **brain tumors**. Astrocytomas, oligodendrogliomas, ependymomas, and glioblastomas are highly invasive tumors that damage neurons, can grow large, and are quite resistant to chemotherapy and radiation therapy.

**Microglia** may become "activated" if there is cell injury or the presence of pathogens. They may release a host of substances such as proinflammatory cytokines and other mediators that lead to inflammation and neuronal apoptosis. One example of the destruction of neurons is that which occurs in **Alzheimer's disease**. The inflammatory response occurs from both microglial and astrocytic reactivity.

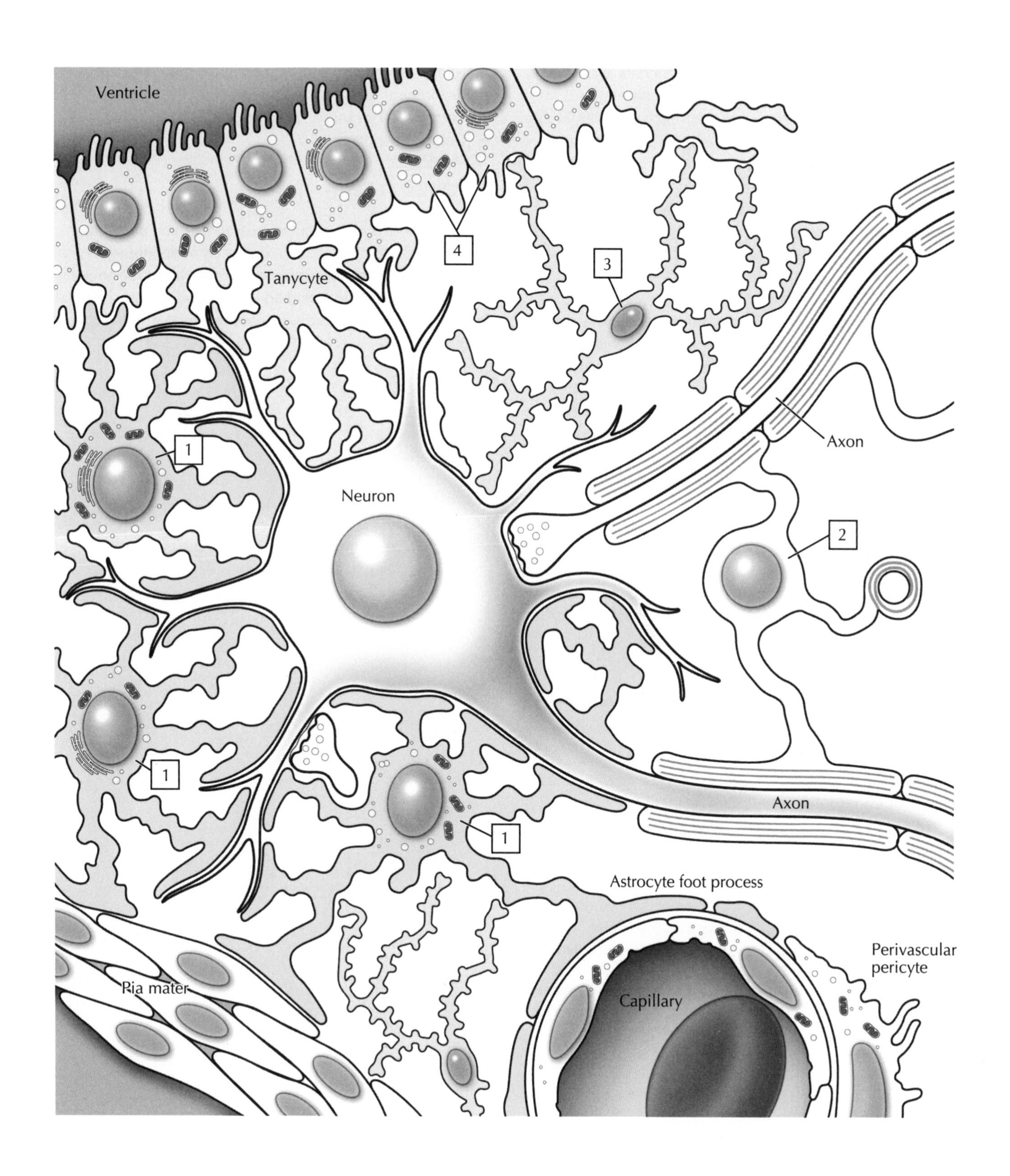
Ventricle
4
3
Tanycyte
1
Axon
Neuron
2
1
1
Axon
Astrocyte foot process
Perivascular
pericyte
Pia mater
Capillary

# 4 Types of Synapses

The major form of communication in the nervous system is by synapses, discrete sites where the axon, or its extensive branching of axonal terminals, sometimes numbering in the thousands, abuts another neuron or target cell. Typically, a neuron receives numerous synaptic contacts on its arborization of dendrites (its "dendritic tree") and dendritic spines or on the soma (neuron cell body). As the axon approaches its target site, it loses its myelin sheath, often undergoes extensive branching, and then terminates on the target as **synaptic boutons.** Communication is by electrochemical transmission, triggering the release of neurotransmitter(s) into the **synaptic cleft**. The transmitter(s) bind to receptors on the postsynaptic membrane and initiate a graded excitatory or inhibitory response, or neuromodulatory effect, on the target cell.

The most common types of synapses are the **axosomatic** (axon with cell body of a target neuron) and **axodendritic** (axon with dendrite(s) or dendritic spines of the target neuron) synapses. Other synapses may target another axon terminal (**axoaxonic synapses**); these synapses can prevent the target axon terminal from releasing its neurotransmitter, thereby causing presynaptic inhibition. Another type of synapse is the **reciprocal synapse** in which the neuronal element communicates with its counterpart. A **serial synapse** involves multiple sites in a single axon where clusters of synaptic vesicles are found at many adjacent sites, thus providing transmitter release along a long stretch of the target membrane. **Dendrodendritic synapses** in dendrite bundles provide a mechanism for neurons to provide coordinated activation of their targets.

**COLOR** the features of the typical synapse, using a different color for each feature:

- ☐ **1. Synaptic vesicles: contain the neurotransmitter and/ or neuromodulatory substance**
- ☐ **2. Vesicle exocytosis: fusion of the synaptic vesicle membrane with the presynaptic membrane, thus releasing the transmitter**
- ☐ **3. Postsynaptic membrane: thickened site where membrane postsynaptic receptors bind the neurotransmitter and initiate an appropriate graded response**

A variety of morphological synaptic types can be identified:

- Simple axodendritic or axosomatic (most common synapses)
- Dendritic spine
- Dendritic crest
- Simple synapse along with an axoaxonic synapse
- Combined axoaxonic and axodendritic
- Varicosities (*boutons en passant*)
- Dendrodendritic
- Reciprocal
- Serial

Synapses are dynamic structures and exhibit significant "plasticity." New synapses are formed continuously in many regions, and some are "pruned" or eliminated for any one of a variety of reasons, including lack of use, atrophy or loss of target cells, or degenerative processes due to normal aging or pathology.

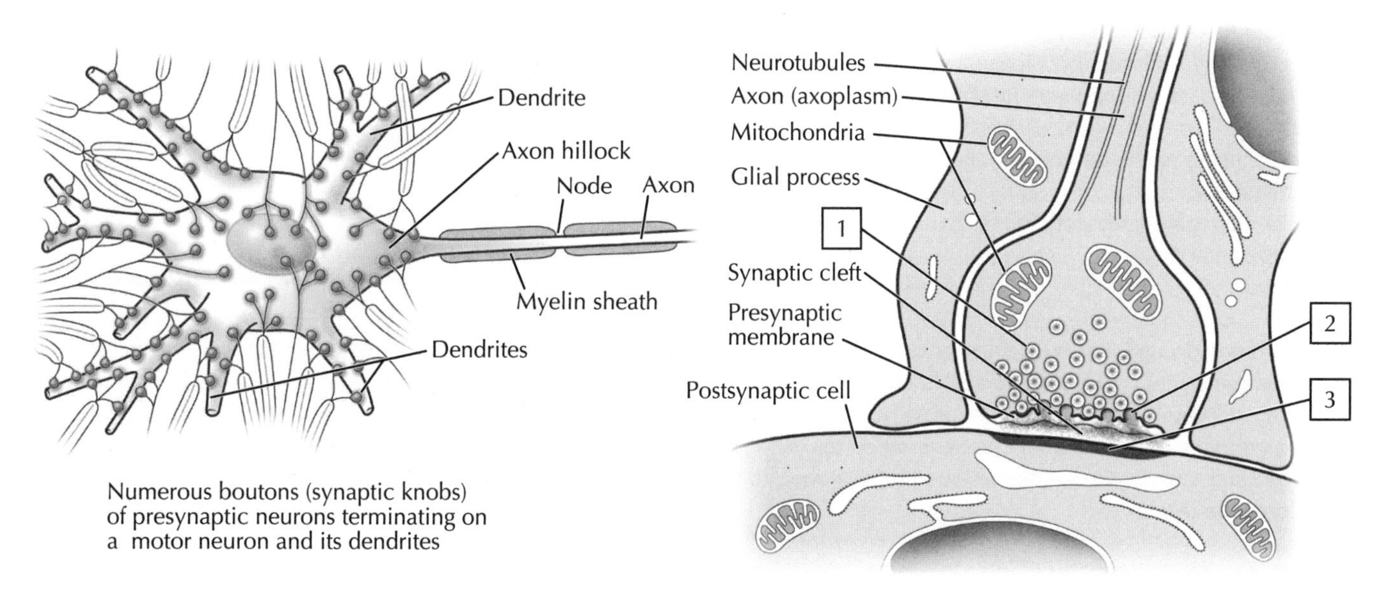


A. Schematic of synaptic endings

B. Enlarged section of bouton

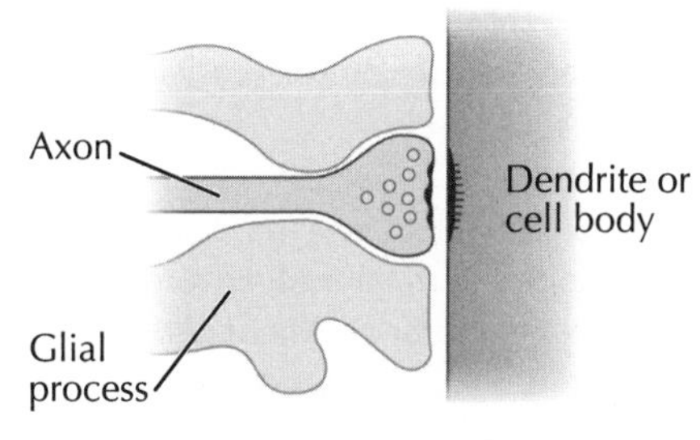


C. Simple axodendritic or axosomatic synapse

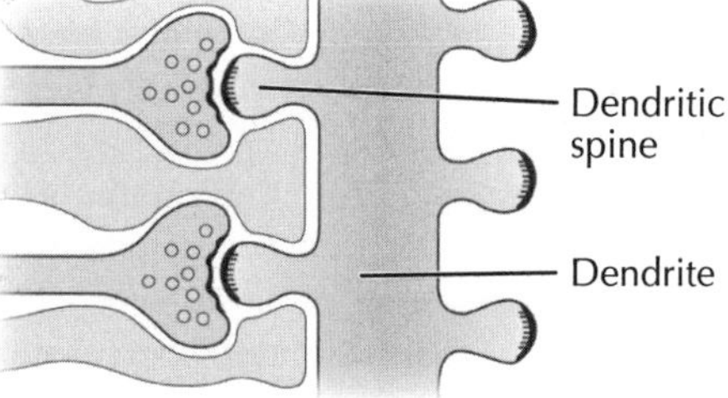


D. Dendritic spine synapse

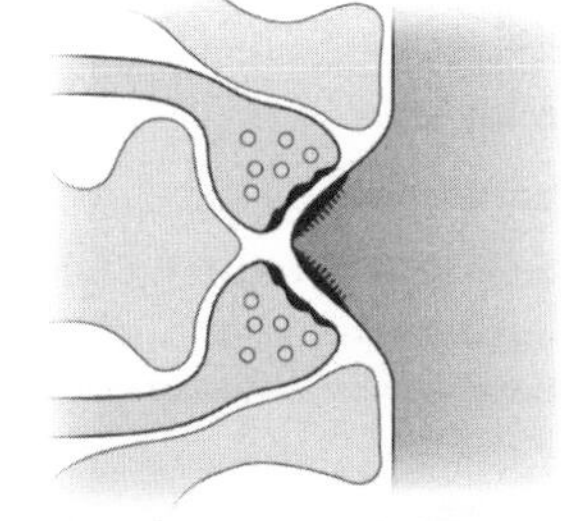

E. Dendritic crest synapse

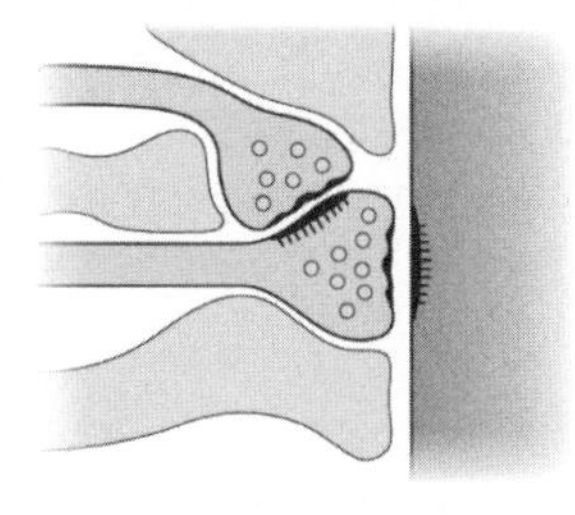

F. Simple synapse plus axoaxonic synapse

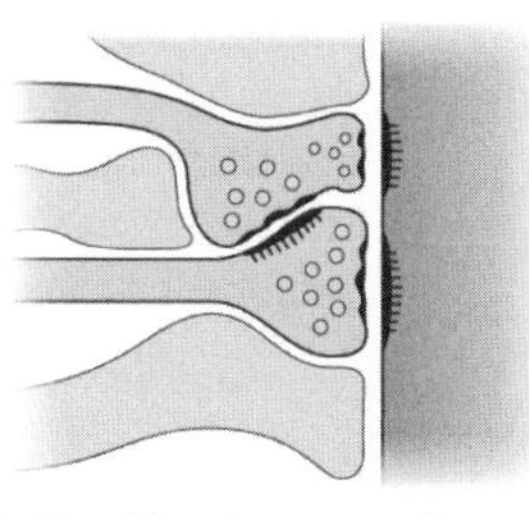

G. Combined axoaxonic and axodendritic synapse

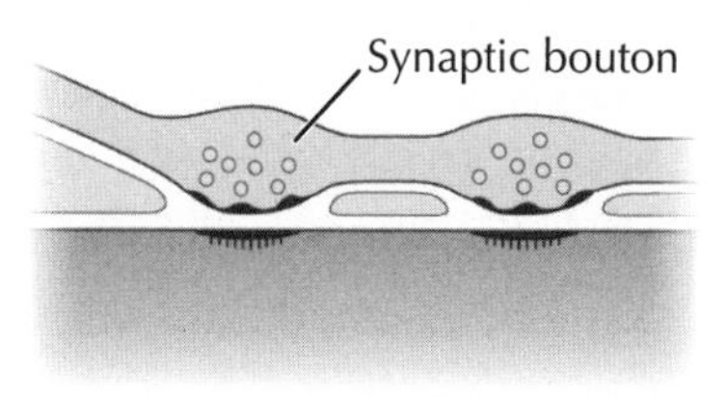


H. Varicosities (*boutons en passant*)

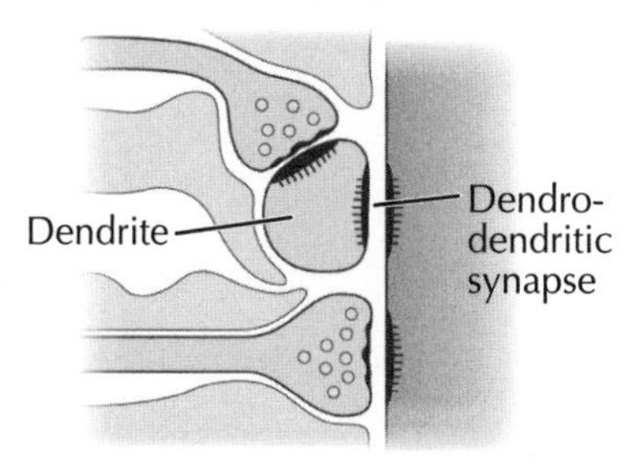


I. Dendrodendritic synapse

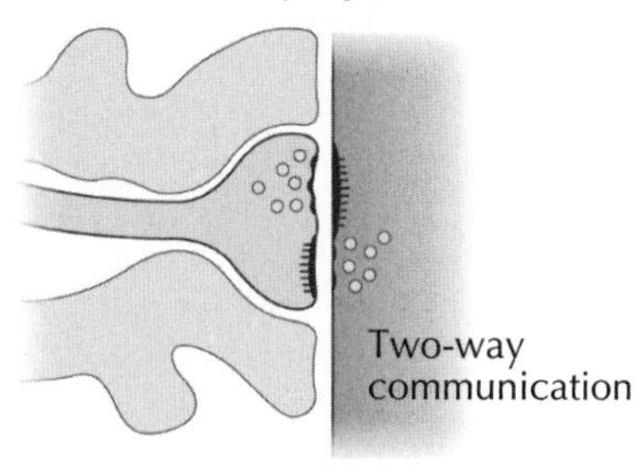


J. Reciprocal synapse

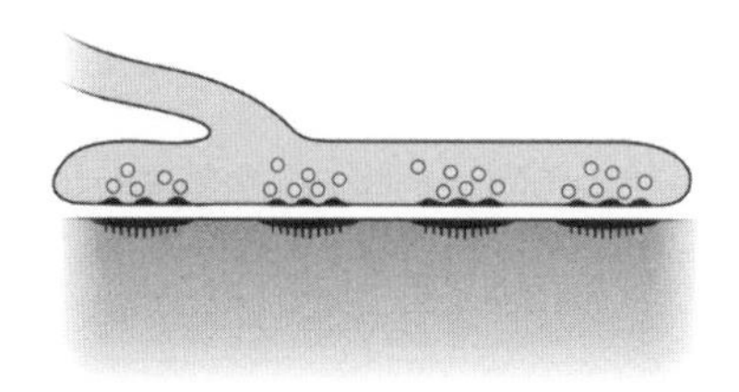

K. Serial synapse

# 4 Cerebrum

As reviewed in Chapter 1, the human brain consists of the following parts:

- Cerebrum (cerebral cortex)
- Diencephalon (thalamus, hypothalamus, and pineal gland)
- Midbrain (also called the mesencephalon, a part of the brainstem)
- Pons (connects to the cerebellum and medulla and is part of the brainstem)
- Medulla oblongata (connects to the spinal cord and is part of the brainstem)
- Cerebellum

The cerebrum is divided into two large hemispheres and is characterized by its convoluted cerebral cortex, which significantly increases the surface area for neurons by folding the tissue into a compact volume. The cerebral cortex is divided into four visible lobes and one lobe that lies deep to the outer cortex (the insula).

**COLOR** the five lobes of the cerebral cortex, using a different color for each lobe:

- ☐ **1. Frontal lobe**
- ☐ **2. Parietal lobe**
- ☐ **3. Occipital lobe**
- ☐ **4. Temporal lobe**
- ☐ **5. Insula: a fifth, deep lobe lying medial to the temporal lobe**

Regions of the cerebral cortex are associated with specific functional attributes. Many of these areas overlap, and some may be more or less developed in individuals with specific talents (e.g., musical or artistic talent, highly trained athletes, etc.), or with specific deficits, either from congenital anomalies (birth defects) or from pathology, such as a stroke.

**COLOR** the following functional regions of the cerebral hemisphere, using a different color for each region:

- ☐ **6. Primary motor cortex (just anterior to the central sulcus)**
- ☐ **7. Primary somatosensory cortex (just posterior to the central sulcus)**
- ☐ **8. Primary visual cortex**
- ☐ **9. Primary auditory cortex**

The fold of cortical tissue just anterior to the central sulcus is the **precentral gyrus** of the frontal lobes. The primary motor cortex is located in this gyrus, and the human body is represented topographically over this cortical area. That is, the cortical neurons concerned with certain motor functions associated with a region of the human body, such as the thumb, can be identified in a particular region of the precentral gyrus. To represent this topographical relationship, a motor homunculus ("little man") is drawn over the motor cortex (see part ***E***), and the size of each body part is representative of the portion of the cortex devoted to innervating this body part. Note that the motor cortex is disproportionately large for the face, oral cavity, and hand. The sensory cortex (see part ***D***) is especially large over the face and hand.

The **postcentral gyrus** of the parietal lobe is the primary sensory cortex and represents the cortical area devoted to sensory function. Similar to the motor cortex, a sensory homunculus can be represented over this cortical region (see part ***E***).

---

***Clinical Note:***
Because the motor and sensory functions are "mapped" to specific regions of the cerebral cortex, damage by trauma, CNS lesions, or vascular problems may result in specific motor and/or sensory loss of function.

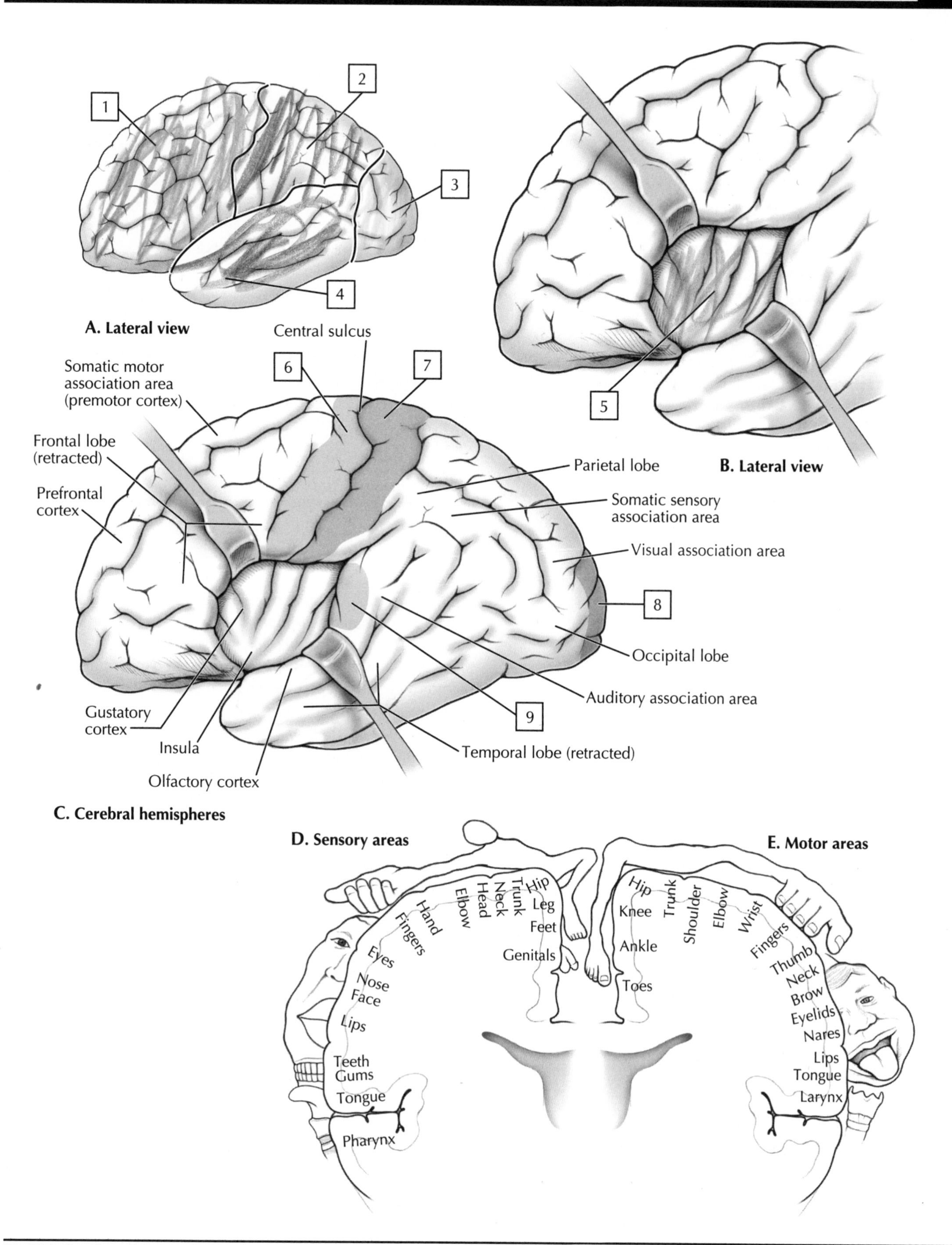
1
2
3
4
A. Lateral view
5
B. Lateral view
Central sulcus
Somatic motor association area (premotor cortex)
6
7
Frontal lobe (retracted)
Prefrontal cortex
Parietal lobe
Somatic sensory association area
Visual association area
8
Occipital lobe
Auditory association area
9
Temporal lobe (retracted)
Gustatory cortex
Insula
Olfactory cortex
C. Cerebral hemispheres
D. Sensory areas
Hip
Trunk
Neck
Head
Elbow
Hand
Fingers
Leg
Feet
Genitals
Eyes
Nose
Face
Lips
Teeth
Gums
Tongue
Pharynx
E. Motor areas
Hip
Trunk
Knee
Shoulder
Elbow
Wrist
Ankle
Toes
Fingers
Thumb
Neck
Brow
Eyelids
Nares
Lips
Tongue
Larynx

# 4 Cortical Connections

The convoluted surface of the cerebral hemispheres containing the cortical neurons is the **gray matter**. It lies above the more deeply situated **white matter**, which contains fiber connections that course from deeper brain regions and also interconnections that permit communication between the two hemispheres. These fiber pathways are called white matter because the myelin that insulates most of the fiber connections causes them to appear to be more of a whitish color than the gray matter. The major white matter tracts forming these connections include the:

- **Corpus callosum**: commissural fibers that interconnect the two hemispheres (see parts ***A*** and ***C***)
- **Association tracts**: connect cortical regions within the same hemisphere (see part ***B***)
- **Corona radiata**: two-way connections between the cortex and subcortical nuclei and the spinal cord; narrows into the **internal capsule** as it passes by the thalamus and basal ganglia (see parts ***C*** and ***D***)

The major fiber pathway that interconnects the two hemispheres is called the **corpus callosum**. These commissural fibers provide important coordination of functional activity between the two separate hemispheres. The fibers interconnecting the frontal and occipital lobes in particular curve rostrally and caudally after they cross the midline. In essence, the corpus callosum forms a roof over the subcortical nuclei (*nucleus* in the CNS is a term used to describe collections of neurons that subserve similar functions).

Additionally, **association tracts** of fibers connect the anterior and posterior aspects of the cerebral cortex; these tracts can exist as very long tracts connecting frontal lobe regions with the occipital lobes or as shorter association tracts.

Finally, a fan-shaped white matter tract of fibers called the **corona radiata** provides a projection system that "radiates" inferiorly and caudally from the cortex. It descends between the caudate nucleus and thalamus medially and the putamen, which lies lateral to this projection (at this point the radiation is called the **internal capsule**). The axons in this projection tract both ascend and descend from lower brainstem and spinal cord areas, providing connections to and from these regions to the cerebral cortex.

**COLOR** the following white matter fiber pathways, using a different color for each pathway:

- ☐ **1. Corpus callosum**
- ☐ **2. Corona radiata**
- ☐ **3. Internal capsule**

***Clinical Note:***
Regions of the cerebral cortex are well-organized and may be referred to as the **primary cortex**. These regions include the somatosensory cortex, trigeminal sensory cortex (head, face, and jaw sensation), visual cortex, auditory cortex, and motor cortex. **Vascular lesions** (hemorrhages or infarcts leading to ischemia or anoxia) or other **mass lesions** (trauma, degenerative processes) can result in global dysfunction and cognitive deficits, even resulting in **coma**. For example, cortical damage in the dominant hemisphere (usually the left hemisphere in right-handed people and most left-handed people) can result in **expressive aphasia** (Broca area: responsible for initiating expressive language), **receptive aphasia** (Wernicke area: understanding of spoken language), or **global aphasia** (all aspects of speech and communication are severely impaired).

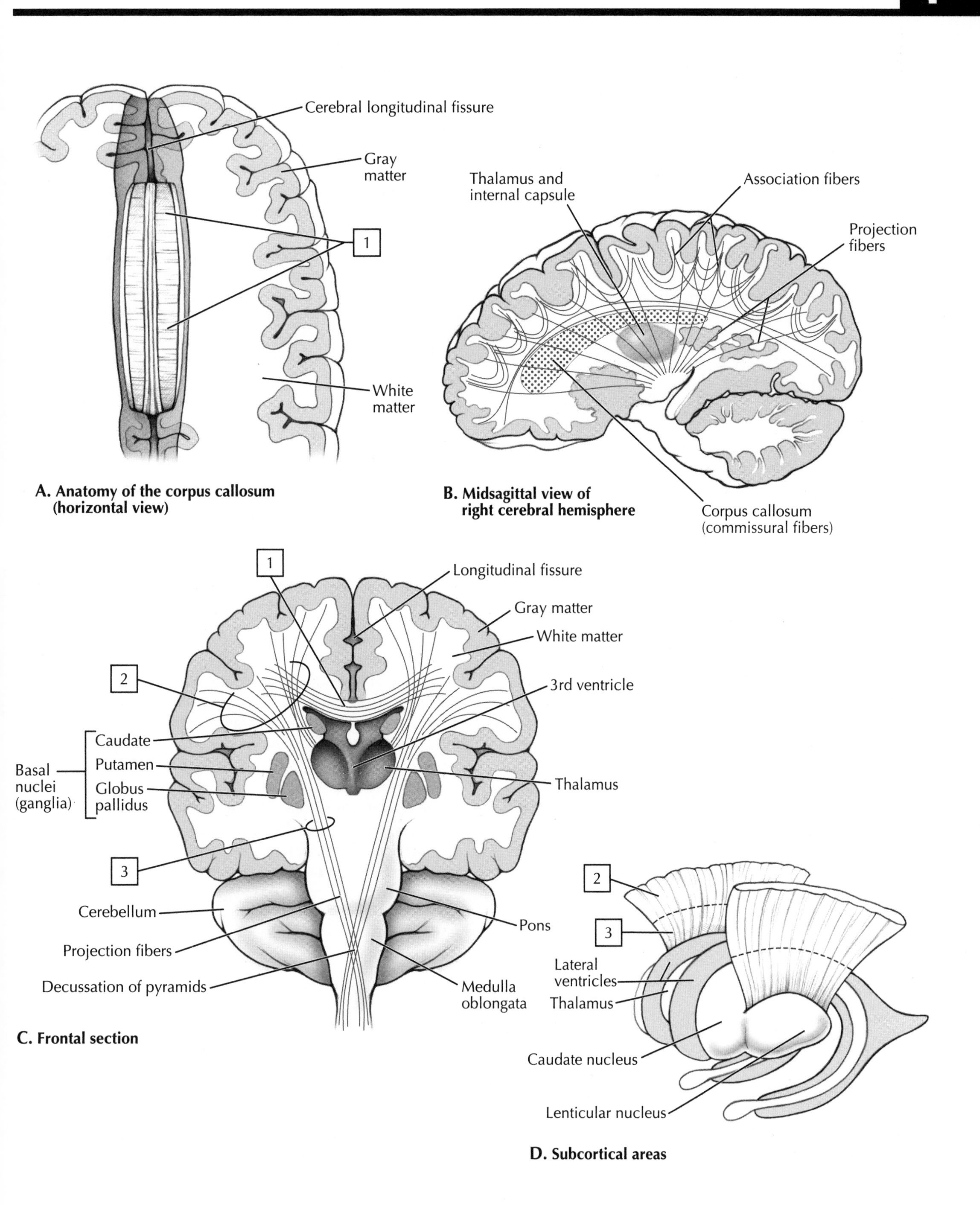


**A.** **Anatomy of the corpus callosum (horizontal view)**

**B.** **Midsagittal view of right cerebral hemisphere**

**C.** **Frontal section**

**D.** **Subcortical areas**

# 4 Midsagittal and Basal Brain Anatomy

Many of the deeper midline structures of the brain are visible if the brain is viewed in a midsagittal section between the cerebral hemispheres and through the **diencephalon** (its major components are the thalamus, hypothalamus, and pineal gland), and the **brainstem**, which includes the midbrain, pons, cerebellum, and medulla oblongata, and then the medulla's connection with the upper spinal cord. Likewise, basal views of the brain and isolated views of the brainstem help one to delineate the individual regions of the brain located below the level of the cerebrum.

First, note the prominent corpus callosum, the commissural connection between the two cerebral hemispheres. Its major parts include the:

- **Genu**: anterior portion
- **Body**: larger midsection
- **Splenium**: posterior portion

Just beneath the corpus callosum lie the **diencephalic structures**, including the:

- **Thalamus**: the "executive secretary" of the cortex, because it is reciprocally connected to the cortex and conveys motor, sensory, and autonomic information from the brainstem and spinal cord
- **Hypothalamus**: lies beneath the thalamus, and its connections with the pituitary gland reflect its important role in neuroendocrine function
- **Pineal gland**: an endocrine organ that secrets melatonin and is important in regulating circadian (day-night) rhythms

The **midbrain** contains fiber tracts that ascend and descend through the thalamus; it also includes the:

- **Colliculi (*colliculus*, "small hill")**: superior and inferior colliculi are sensory nuclei associated with visual reflexes and auditory reflexes, respectively
- **Cerebral peduncles (*pedunculus*, "little feet")**: convey descending motor fibers to the spinal cord and connections to the cerebellum

The **pons**, meaning "bridge," literally connects the cerebellum with the other portions of the brain (midbrain) and spinal cord. Some deep fiber tracts connect higher brain centers with the spinal cord, whereas more superficial tracts relay information between the cortex and cerebellum via three cerebellar peduncles.

The **medulla** links the brainstem with the spinal cord, and all of the ascending and descending fiber pathways pass through the medulla and/or synapse on sensory and motor nuclei within this region. Important regulatory cardiopulmonary centers also are located in the medulla oblongata.

**COLOR** each of the following features of the diencephalon, midbrain, pons, and medulla, using a different color for each feature:

- ☐ 1. **Corpus callosum**
- ☐ 2. **Pineal gland**
- ☐ 3. **Colliculi of the midbrain (superior and inferior)**
- ☐ 4. **Mammillary bodies of the hypothalamus**
- ☐ 5. **Thalamus**
- ☐ 6. **Cerebellar peduncles (superior, middle, and inferior)**
- ☐ 7. **Medulla oblongata (often just called the medulla)**

***Clinical Note:***

**Brainstem lesions** often involve specific dysfunction of cranial nerves III to XII (CN III-XII) and may result in loss of somatosensory and somatic motor functions, as well as ipsilateral cerebellar dysfunction.

**Forebrain lesions** usually involve motor and sensory deficits on the opposite side (contralateral) of the body. Large lesions to the cerebral hemispheres may result in cognitive dysfunction and damage to specific regions like the basal ganglia can result in movement disorders (see Plate 4-7). **Hippocampal damage** bilaterally may lead to the loss of short-term memory, confusion, and disorientation. **Limbic lesions** may result in fear, anxiety, compulsive disorders, and emotional changes.

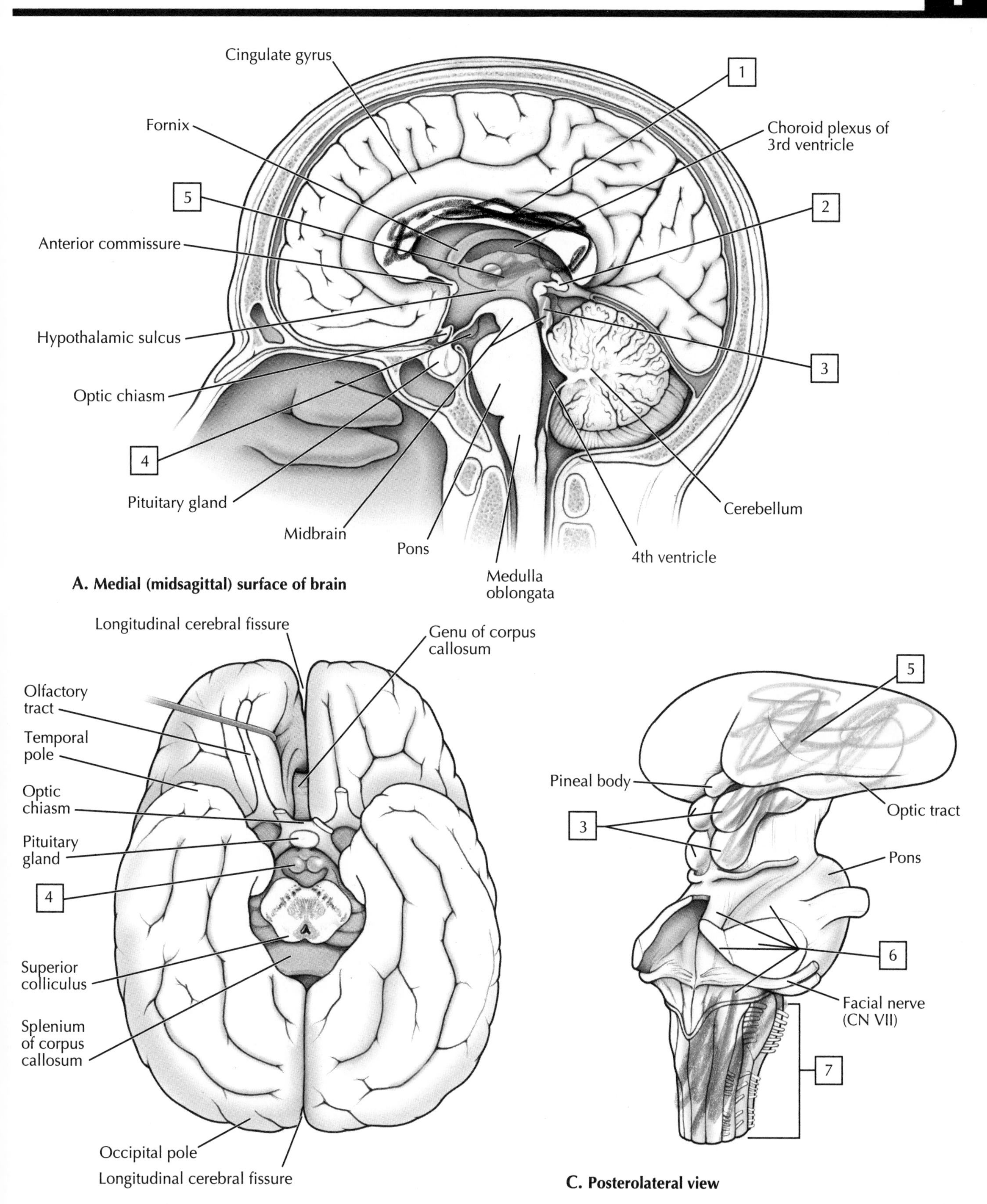


**A. Medial (midsagittal) surface of brain**

**B. Basal surface of the brain (brainstem and cerebellum removed)**

**C. Posterolateral view**

# Basal Ganglia

The basal ganglia (basal nuclei) provide subconscious control of skeletal muscle tone and coordination of learned movements. Once a voluntary movement is initiated cortically, the natural rhythm and patterns that we take for granted in walking or reaching for objects are controlled subconsciously by the basal ganglia. Additionally, they inhibit unnecessary movements. The interconnections of the basal ganglia are complex, involve both excitatory and inhibitory pathways, and use multiple transmitters (dopamine [DA], glutamate [GLUT], gamma aminobutyric acid [GABA], acetylcholine [ACh], 5-hydroxytryptamine [5HT], and substance P [SUB P] summarized in the complex flow diagram below). While it is probably not important to memorize this schematic diagram, it does illustrate the complexity of interconnections in this network.

The basal ganglia (nuclei) include the:

- **Caudate nucleus**: descriptively, it has a large head and a slender tail, which arches over the diencephalon
- **Putamen**: the putamen and globus pallidus together form the **lentiform nucleus**
- **Globus pallidus**: the putamen and globus pallidus together form the lentiform nucleus

**COLOR** the nuclei associated with the basal ganglia, using a different color for each nucleus:

- ☐ **1. Caudate (head and tail)**
- ☐ **2. Putamen**

- ☐ **3. Globus pallidus**
- ☐ **4. Lentiform nucleus**

***Clinical Note:***

Disorders affecting the basal ganglia involve defects that either result in too much movement or not enough movement. **Huntington's disease** (Huntington's chorea) results in the degeneration of neurons in the caudate nucleus (and several other structures) that leads to a hyperactive state of involuntary movements, emotional dysfunction, and cognitive decline. The jerky movements of this disease almost resemble a dancer out of control (choreiform movements), and the term **chorea** ("dance") aptly characterizes this fatal condition.

A contrasting disease to Huntington's chorea is **Parkinson's disease.** Resulting from the degeneration of dopamine-secreting neurons of the substantia nigra, pars compacta, and the loss of dopaminergic input to the caudate nucleus and the putamen. This progressive disease results in bradykinesia (slow movements), resting rhythmic muscular tremor, muscular rigidity, stooped posture, a masked or expressionless face, and a shuffling gait.

Other basal ganglia disorders may result in **athetosis** (writhing movements), **spasmodic torticollis** (involuntary neck rotational movements), **dystonia**, and **hemiballismus** (flinging movements). Treatment options include dopamine replacement, deep brain stimulation, and even some surgical ablative procedures.

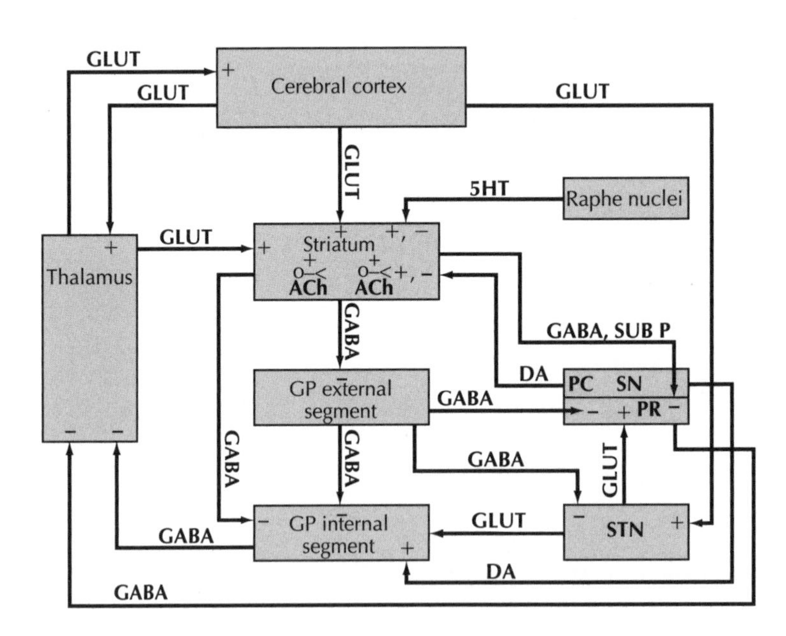


**ACh:** Acetylcholine
**DA:** Dopamine
**GABA:** Gamma aminobutyric acid
**GLUT:** Glutamate
**GP:** Globus pallidus
**5HT:** 5-Hydroxytryptamine (serotonin)
**PC:** Pars compacta
**PR:** Pars reticularis
**SN:** Substantia nigra
**STN:** Subthalamic nucleus
**SUB P:** Substance P

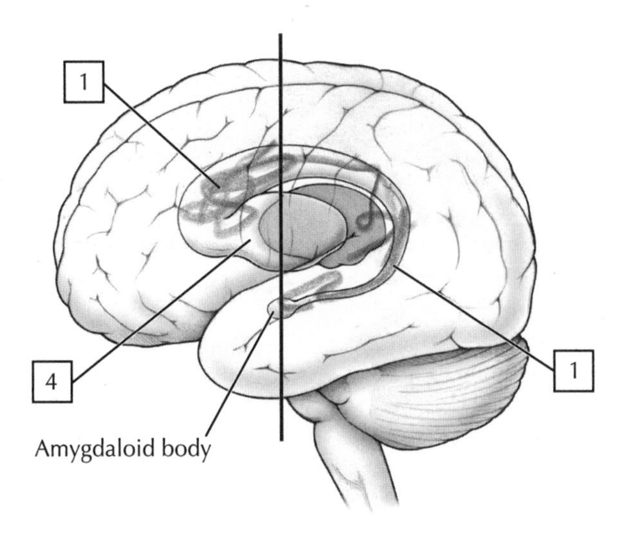


A. Level of section for C below

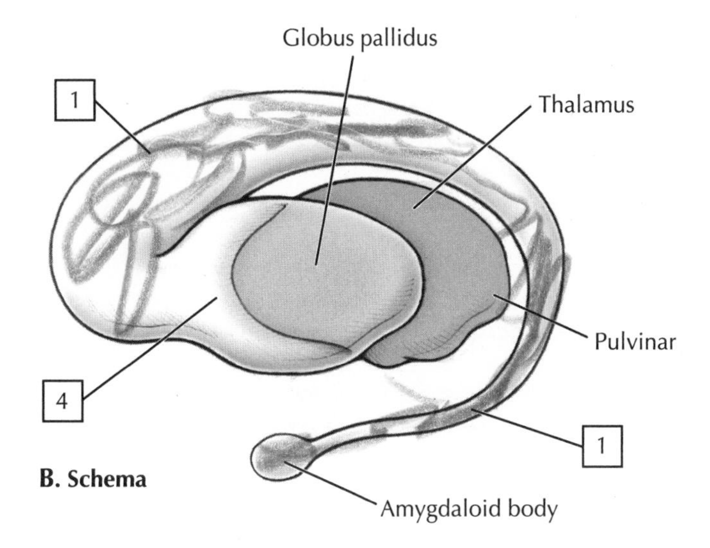


B. Schema

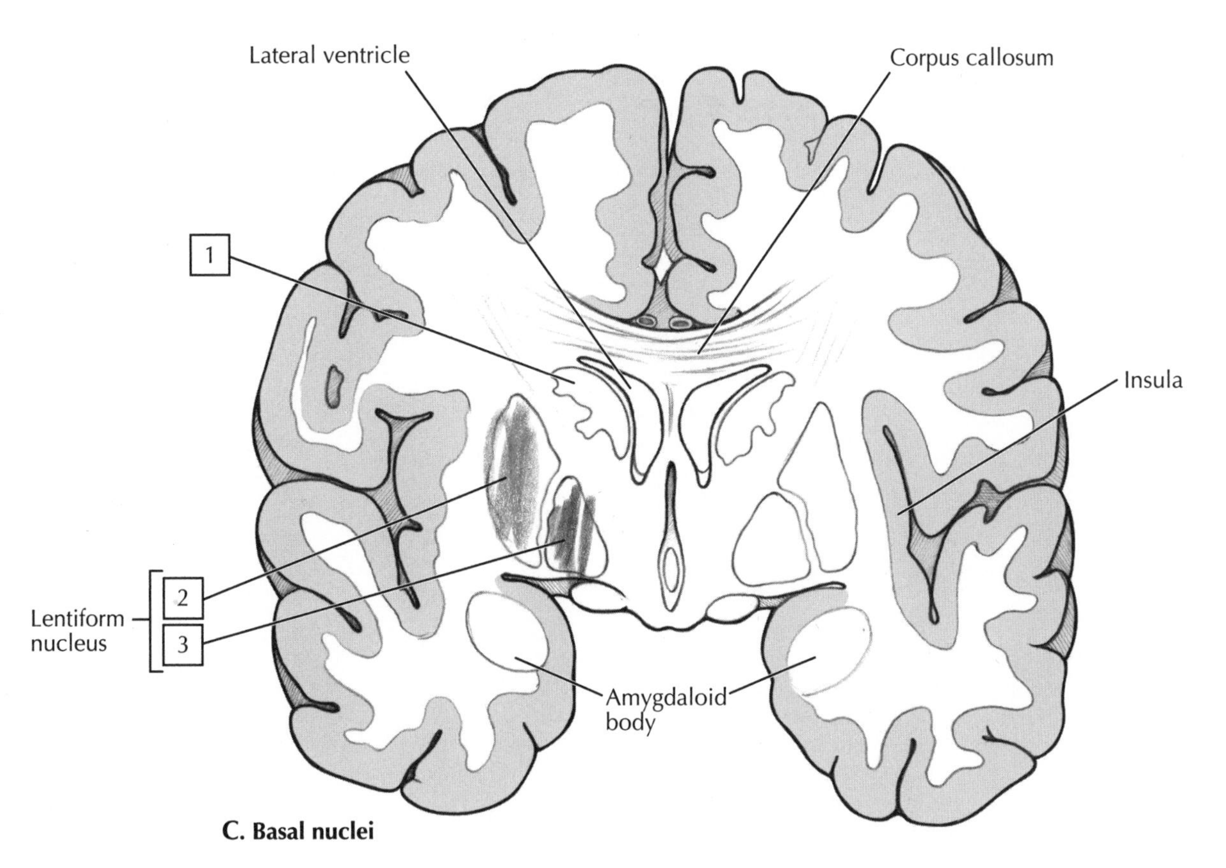


C. Basal nuclei

# 4 Limbic System

The limbic system is a functional group of structures that form a ring ("limbus") around the diencephalon. The limbic system participates in emotional behaviors (fear, rage, pleasure, and sexual arousal) and the interpretation of internal and external stimuli (linking conscious functions with autonomic functions, and aspects of memory and retrieval). There are varying views about which structures are actually part of the limbic system and which structures merely communicate with the system. One can say, however, that typical structures of the system include the:

- Cingulate gyrus
- Parahippocampal gyrus
- Hippocampus (functions in memory)
- Amygdala (and its axonal projection called the stria terminalis, which projects to the hypothalamus and basal forebrain structures)
- Septal nuclei: lie just rostral to the hippocampus; regulate emotions
- Hypothalamus (has autonomic and neuroendocrine functions)
- Olfactory area (involved in the sense of smell)

The limbic forebrain expresses its responses via the hypothalamus and its neuroendocrine and visceral autonomic connectivity. This involves an extensive circuitry through the brainstem and the control of sympathetic and parasympathetic efferent systems. Thus the limbic system forms extensive connections with cortical regions and the brainstem, allowing for extensive integration of stimuli, emotional states, and conscious behaviors linked to these stimuli and emotions.

**COLOR** the following structures associated with the limbic system, using a different color for each structure:

- ☐ **1. Cingulate gyrus**
- ☐ **2. Hippocampus**
- ☐ **3. Amygdala and stria terminalis**
- ☐ **4. Septal nuclei**
- ☐ **5. Olfactory tract**

***Clinical Note:***

The **hypothalamus**, as a center for neuroendocrine and autonomic functioning and as a processing center for smell and emotions, along with other limbic structures, plays a key role in **psychosomatic illnesses.** Stress and its accompanying emotions can trigger autonomic visceral reactions that are the hallmark of psychosomatic, or emotion-driven, illnesses.

**Forebrain trauma** may result in bilateral damage to the hippocampal formation, the amygdala, and surrounding structures. Deficits include visual agnosias with an inability to recognize objects or faces, hyperorality (exploring objects by mouth), a decrease in emotional responsiveness, compulsive food consumption, short-term memory loss, and hypersexual responsiveness. These neurobehavioral changes are attributed to bilateral damage to the anteromedial temporal lobes, the amygdala, and its cortical and hippocampal formation for memory-related symptoms.

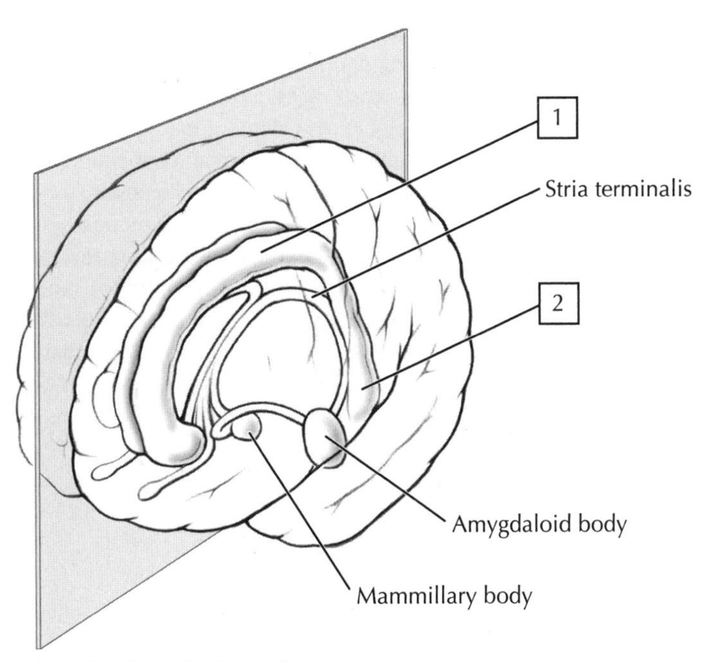


**A. Anterolateral schematic**

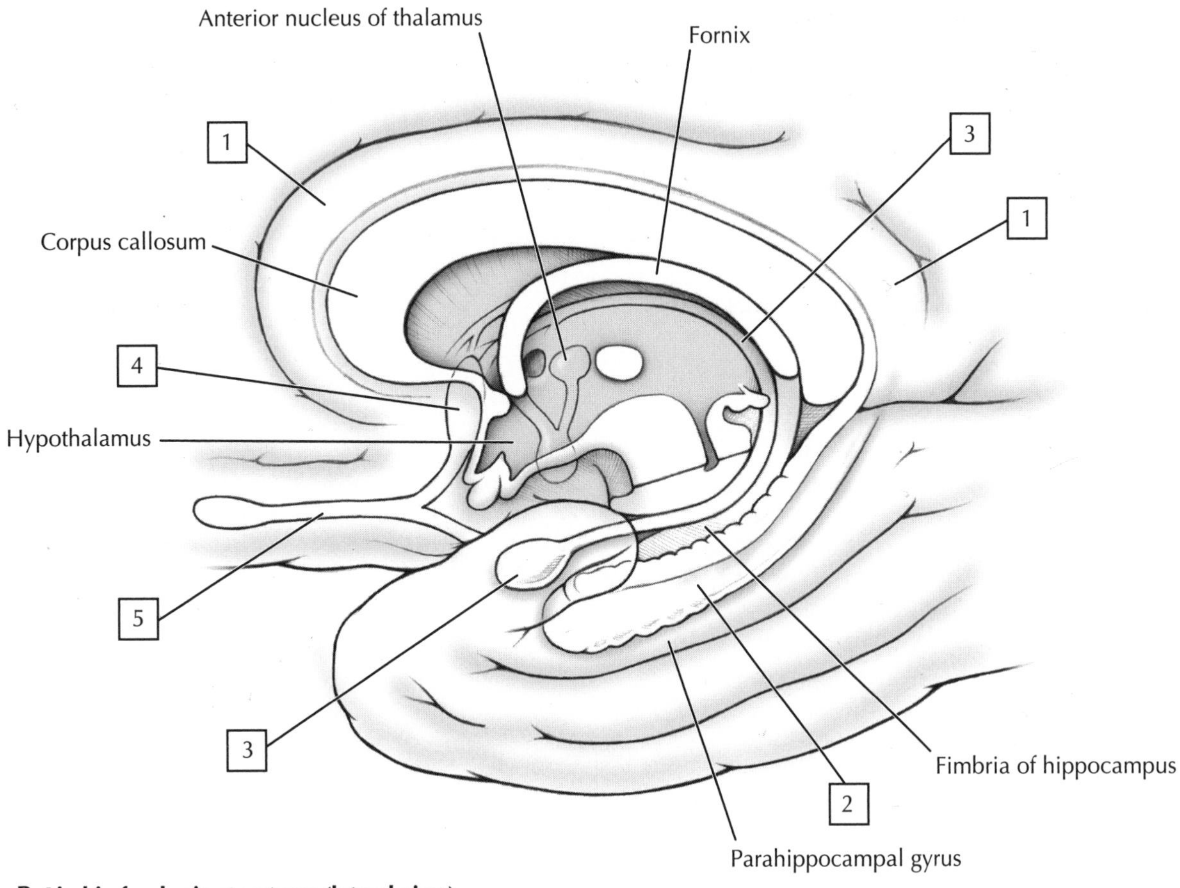


**B. Limbic forebrain structures (lateral view)**

# 4 Hippocampus

The hippocampus extends from the amygdala and arches up and forward into the diencephalon in close association with the dentate gyrus. It consists of three structures: the dentate gyrus, the hippocampus proper, and the subiculum. Its appearance resembles a seahorse (in coronal sections), which is what the term *hippocampus* actually means. It occupies a portion of the medial temporal lobes, lying just medial to the temporal pole of the lateral ventricles. The efferent fiber tract of the hippocampus is the **fornix**, which arches forward under the corpus callosum and toward the mammillary bodies of the hypothalamus, where many of its fibers terminate. The hippocampal formation (dentate gyrus, hippocampus proper, and subiculum) has many interconnections with the limbic system, hypothalamus, and cortical association areas.

Functionally, the hippocampus and amygdala are important in memory consolidation and access. Moreover, the hippocampus plays a role in spatial relationships, whereas the amygdala associates a variety of sensory memories and links them to our emotional responses, especially fear and aversion.

The CA1-CA4 subregions of the hippocampus reflect the unique microarchitecture that distinguish these four regions, which communicate with one another.

**COLOR** the following structures associated with the hippocampal formation, using a different color for each structure:

- ☐ **1. Body of the fornix**
- ☐ **2. Crura ("legs") of the fornix**
- ☐ **3. Dentate gyrus**
- ☐ **4. Hippocampus**

***Clinical Note:***

**Alzheimer's disease** is a common cause of dementia in the elderly and is characterized by the progressive degeneration of neurons, especially evident in the frontal, temporal, and parietal lobes. Many of the cortical regions of the hippocampal formation and its cortical connections are susceptible to neuronal degeneration in Alzheimer's disease. The presence of neurofibrillary tangles (filamentous aggregates in the cytoplasm of neurons) is common in the cortex, hippocampus, basal forebrain, and some regions of the brainstem. Disruption of the hippocampal formation circuitry results in an inability to consolidate short-term and intermediate-term memory into long-term memory. Memory loss and cognitive impairments lead to progressive loss of orientation, language, and other higher cortical functions.

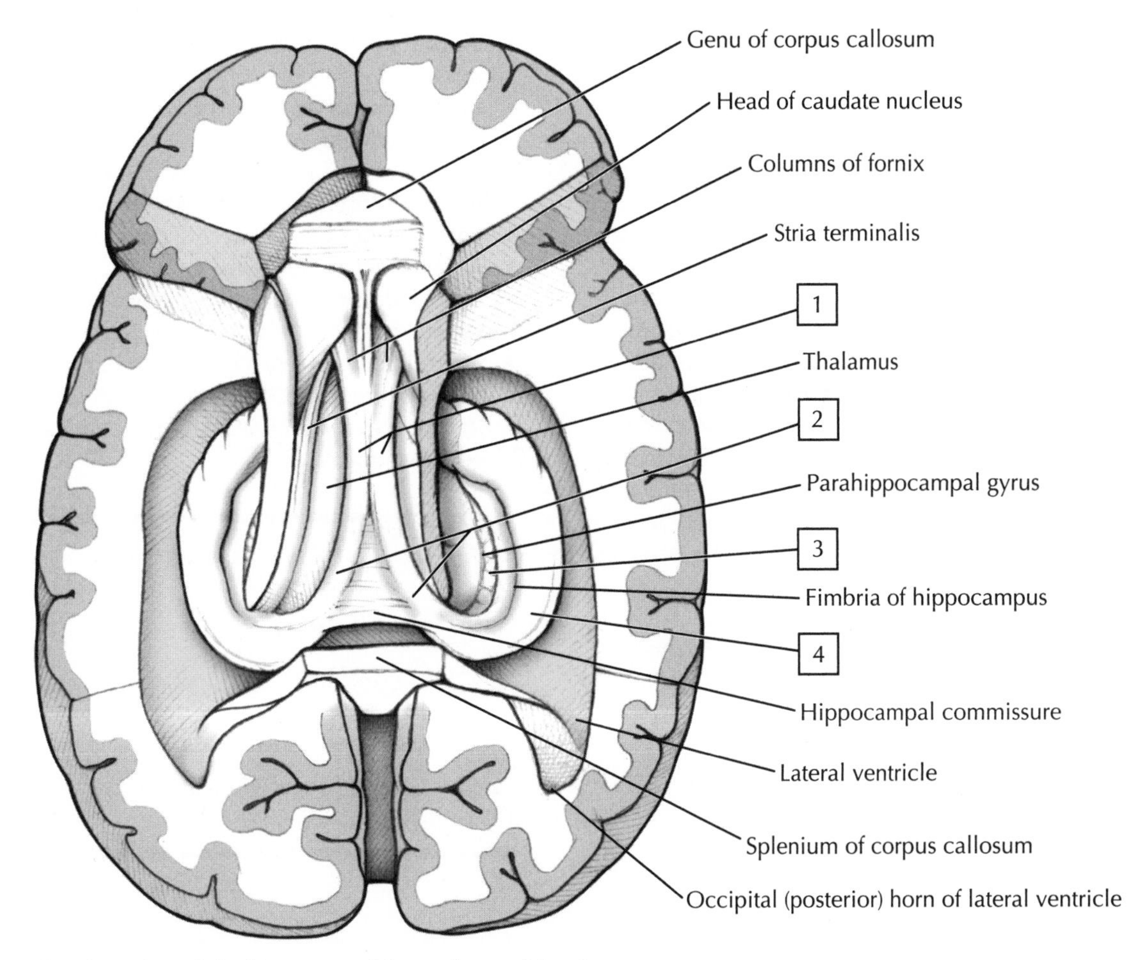


**A. Dissection of the hippocampal formation and fornix**

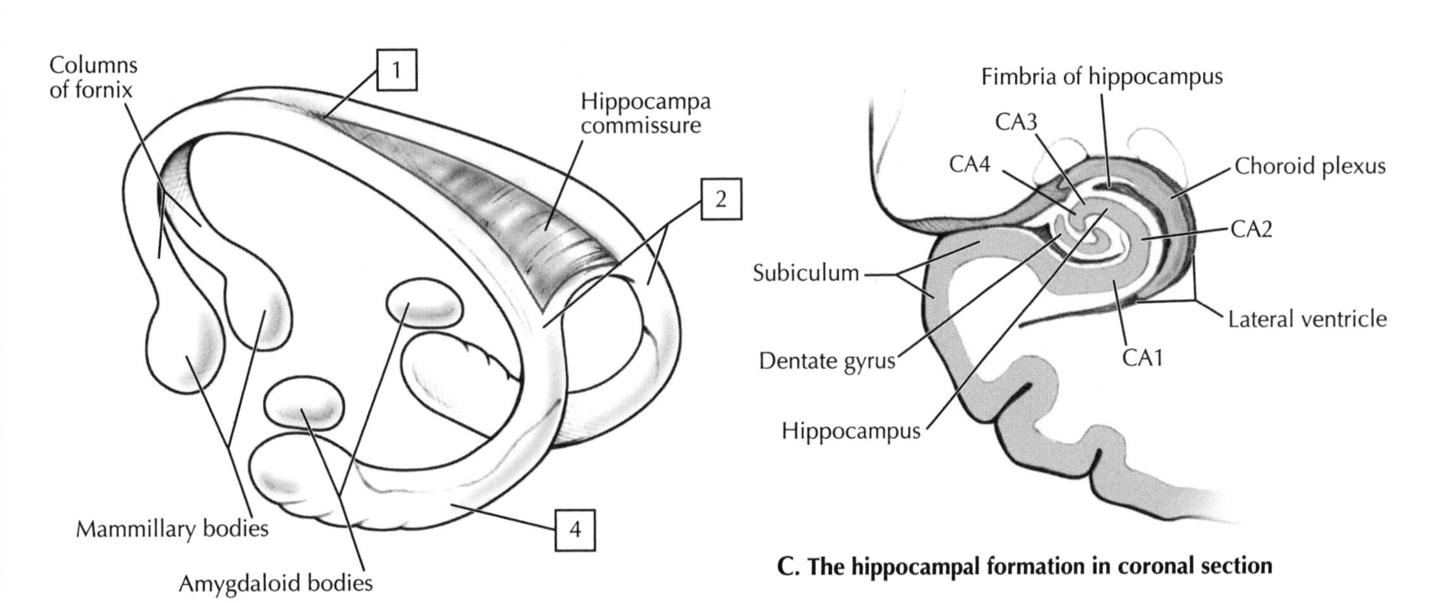


**C. The hippocampal formation in coronal section**

**B. 3-D reconstruction of the fornix**

# 4 Thalamus

The thalamus ("inner room") has right and left halves, which are separated by the 3rd ventricle; they form the major portion of the diencephalon (about 80%). The thalamic nuclei are consolidated into an ovoid mass and divided into three major groups:

- Anterior
- Medial
- Lateral

The central location of the thalamus is representative of its importance; essentially, no sensory information, except olfactory information, passes to the higher cortical regions without synapsing in the thalamus. Thus the thalamus has been characterized as the "**executive secretary**" of the brain because it sorts and edits information. Sensory, motor, and autonomic information from the spinal cord and brainstem is conveyed to the cortex via the thalamus. Likewise, the thalamic nuclei are reciprocally interconnected with the cortex. A white matter tract, the medullary lamina, runs through the thalamus and relays information to the cortex.

Inputs passing through the thalamus on their way to the cerebral cortex include those that:

- Regulate emotion and visceral functions from the hypothalamus
- Direct motor activity from the cerebellum and basal ganglia
- Integrate sensory function
- Relay visual and auditory information
- Participate in autonomic and limbic-related functions

In general, the thalamic nuclei project to the following cortical areas (many of these connections are reciprocal):

- **Ventral posterolateral**: primary sensory cortex (postcentral gyrus)
- **Ventral posteromedial**: primary sensory cortex and primary somesthetic cortex
- **Ventral lateral**: primary motor cortex (precentral gyrus)
- **Ventral intermedial**: primary motor cortex (precentral gyrus)
- **Ventral anterior**: premotor and supplementary motor cortex
- **Anterior**: cingulate gyrus
- **Lateral dorsal**: cingulate gyrus and precuneus, anterior parietal cortex
- **Lateral posterior**: precuneus and superior parietal lobe
- **Medial dorsal**: prefrontal cortex and frontal lobe
- **Pulvinar**: association areas of the parietal, temporal, and occipital lobes

**COLOR** the following thalamic nuclei, using a different color for each nucleus:

☐ **1. Medial dorsal**
☐ **2. Pulvinar**
☐ **3. Lateral posterior**
☐ **4. Ventral posterolateral**
☐ **5. Ventral posteromedial**
☐ **6. Anterior cell mass**

***Clinical Note:***
Thalamus receives its blood supply by several small arteries and, fortunately, these arteries are rarely damaged selectively by infarcts. If their blood supply is compromised, the resulting symptoms may include changes in consciousness, altertness, affective disorders, memory dysfunction, motor disorders, altered somatic sensation, visual dysfunction, and hallucinations.

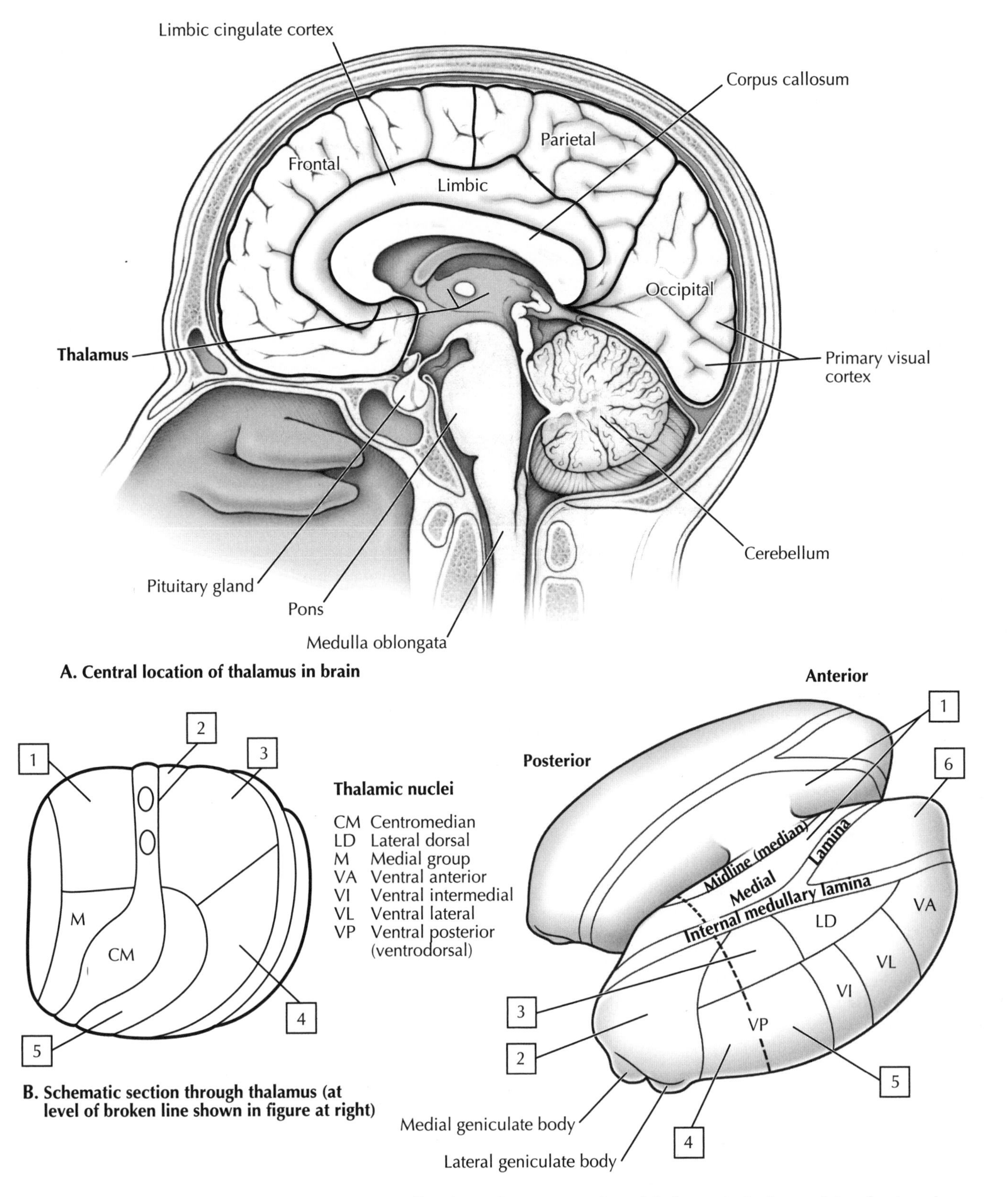


A. Central location of thalamus in brain

B. Schematic section through thalamus (at level of broken line shown in figure at right)

C. Schematic representation of thalamus (reticular nuclei and external medullary lamina removed)

# 4 Hypothalamus

The hypothalamus lies below the thalamus and the 3rd ventricle, forming most of the remainder of the diencephalon besides the thalamus and the small epithalamus (pineal gland). The hypothalamus can be divided anteroposteriorly into chiasmatic (supraoptic), tuberal (infundibulotuberal) and posterior (mammillary) regions. It also can be divided mediolaterally into periventricular, intermediate (medial), and lateral zones. Between the intermediate and lateral zones is a paramedian plane which contains the myelinated fibers of the column of the fornix. The main nuclei and their functions are shown in the image.

**COLOR** each of the principal hypothalamic nuclei, using a different color for each nucleus:

- ☐ **1. Paraventricular nucleus**
- ☐ **2. Lateral hypothalamic nucleus (area)**
- ☐ **3. Dorsomedial nucleus**
- ☐ **4. Lateral preoptic nucleus (area)**
- ☐ **5. Anterior hypothalamic area**
- ☐ **6. Medial preoptic nucleus (area)**
- ☐ **7. Supraoptic nucleus**
- ☐ **8. Suprachiasmatic nucleus**
- ☐ **9. Ventromedial nucleus**
- ☐ **10. Mammillary nuclei (complex)**
- ☐ **11. Periventricular nucleus**
- ☐ **12. Posterior hypothalamic area**

Functionally, the hypothalamus is very important in visceral control and homeostasis and possesses extensive connections with other brain regions (septal nuclei, hippocampus, amygdala, brainstem, and spinal cord). Specifically, its main functions include:

- Regulation of the autonomic nervous system (heart rate, blood pressure, respiration, and digestion)
- Expression and regulation of emotional responses
- Water balance and thirst
- Sleep and wakefulness related to our daily biological cycles
- Temperature regulation
- Food intake and appetite regulation
- Reproductive and sexual behaviors
- Endocrine control

***Clinical Note:***
Because the hypothalamus has such far-reaching regulatory effects on so many functions, impairments to this brain region can have significant consequences. **Disorders** can include emotional imbalance, sexual dysfunction, obesity, sleep disturbances, body wasting, dehydration, and temperature disturbances, to name a few.

The hypothalamus **regulates many functions** (neuroendocrine output and visceral functions). Some of these regulations are coordinated with motor and sensory processing, for example, feeding behavior, drinking behavior, and even reproductive behavior. Some of these regulatory functions are via direct connections and some via polysynaptic connections. For example, if one perceives an outside threat in their environment, cortical and limbic structures interpret that threat and immediately recruit appropriate hypothalamic and motor circuits to achieve an appropriate behavioral response (fleeing, freezing, or acting calmly). Likewise, such situations may recall similar past events of joy, fear, or rage. These emotions may elevate one's blood pressure, produce anxiety and activate "fight or flight" autonomic functions, and stimulate neuroendocrine responses. Thus, the hypothalamus is at the crossroads of limbic-hypothalamic-autonomic and neuroendocrine regulation.

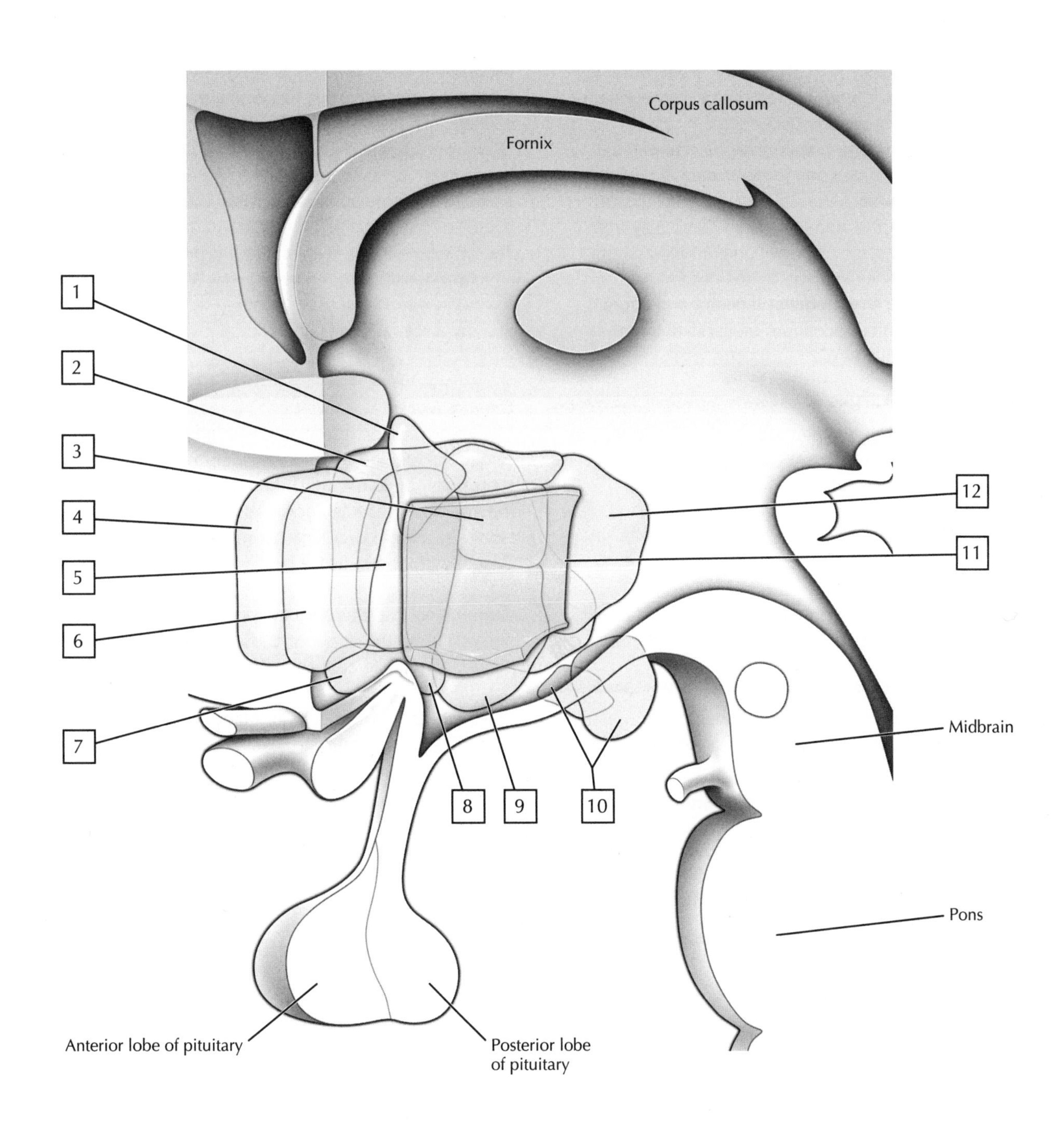
Corpus callosum
Fornix
1
2
3
4
5
6
7
8
9
10
11
12
Midbrain
Pons
Anterior lobe of pituitary
Posterior lobe of pituitary

# 4 Cerebellum

The cerebellum consists of two hemispheres, connected in the middle by the **vermis**, with gray matter (neurons) on the surface, like the cerebral cortex. Deep nuclei also are embedded in the white matter, which forms an arbor-like pattern when viewed grossly in section. The cerebellum overlies the pons and medulla and is connected to the diencephalon and brainstem by three cerebellar peduncles:

- **Superior (connects to the diencephalon)**: afferent and efferent fibers, with connections to the thalamus and then the cerebral motor cortex
- **Middle (connects to the pons)**: afferent fibers only from the pons to the cerebellum, conveying information about voluntary motor activities initiated by the cortex
- **Inferior (connects to the medulla)**: afferent and efferent fibers, with sensory information (proprioceptive) from the body and vestibular system

**COLOR** each of the three cerebellar anatomical lobes (on the right side only), using a different color for each lobe:

- ☐ 1. **Anterior lobe**
- ☐ 2. **Posterior lobe**
- ☐ 3. **Flocculonodular lobe**

Functionally, the cerebellum is organized in a vertical fashion, so that each hemisphere contains three functional zones.

**COLOR** each of the three functional zones of the cerebellum (on the left side only), using a different color for each zone:

- ☐ 4. **Lateral hemisphere: planning movements**
- ☐ 5. **Paravermis (intermediate) zone: adjusts limb movements**
- ☐ 6. **Vermis (in the midline): postural adjustments and eye movements**

Each of these functional divisions is associated with specific deep nuclei.

Functionally, the deep cerebellar nuclei provide the course adjustment upon which is layered the finer adjustment provided by the cerebellar cortex. Generally, the cerebellum functions to:

- Regulate the postural muscles of the body to maintain balance and stereotyped movements associated with walking
- Adjust limb movements initiated by the cerebral motor cortex
- Participate in the planning and programming of voluntary, learned, skilled movements
- Play a role in eye movements
- Play a role in cognition

***Clinical Note:***
Malnutrition, often associated with chronic alcoholism, can lead to degeneration of the cerebellar cortex, often starting anteriorly (**anterior lobe syndrome**). An uncoordinated or staggering gait may result and this is known as **ataxia**. Damage to this lobe also may affect lower limb coordination resulting in a wide-based, stiff-legged gait.

**Damage to the lateral hemispheres** of the middle lobe causes ataxia in both the upper and lower extremities, mild hypotonia, dysarthria (affects speech), and oculomotor dysfunction.

Selective **damage to the flocculonodular lobe** may result in impaired visual tracking, nystagmus (involuntary rhythmic oscillation of the eyeballs), vertigo, and loss of balance. For example, the nodule of the flocculonodular lobe overlies the 4th ventricle, where tumors called **medulloblastomas** arising from the roof of the ventricle can impinge on the nodule and affect balance and eye movements.

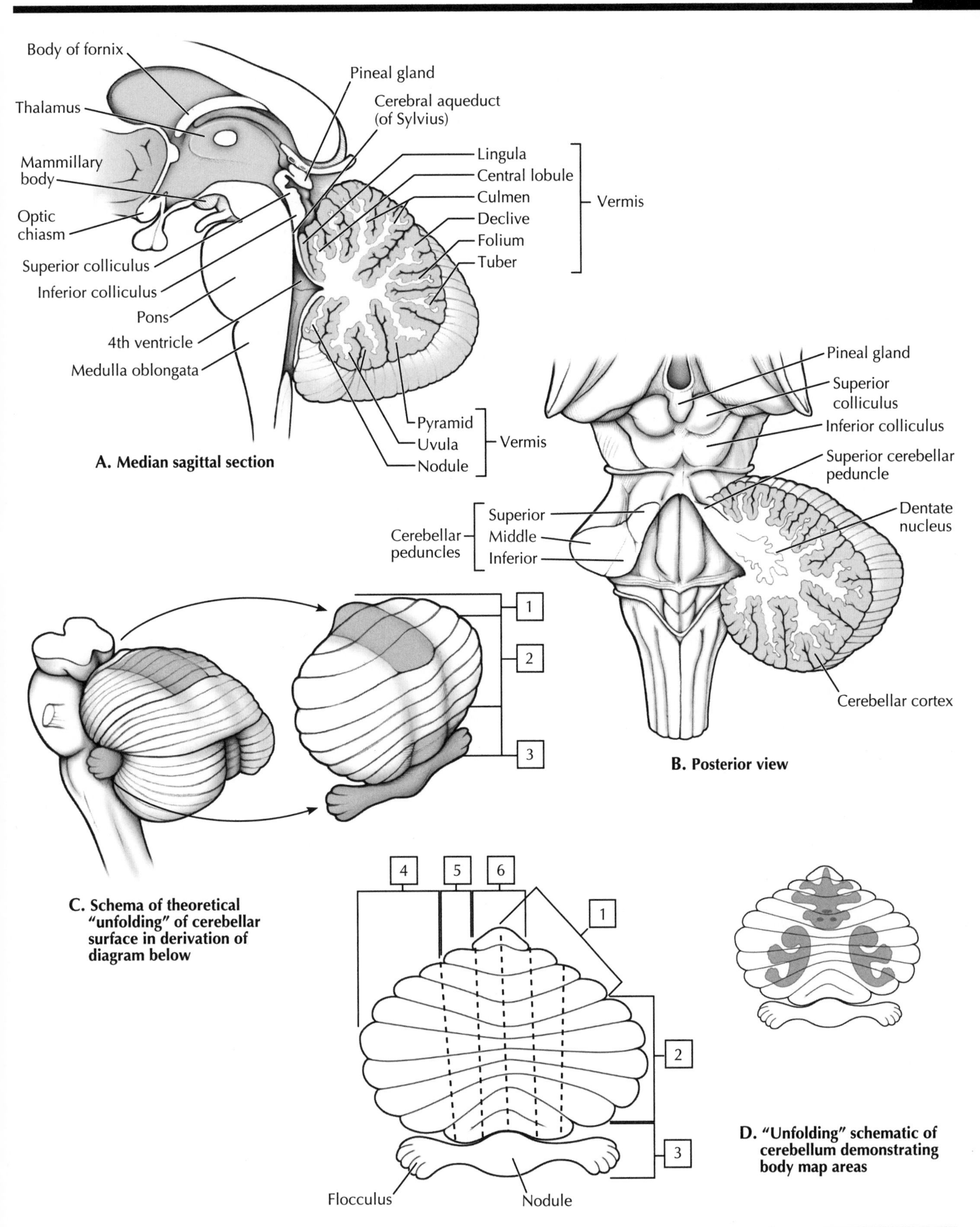


A. Median sagittal section

B. Posterior view

C. Schema of theoretical "unfolding" of cerebellar surface in derivation of diagram below

D. "Unfolding" schematic of cerebellum demonstrating body map areas

# 4 Spinal Cord I

The spinal cord is a direct continuation of the medulla oblongata, extending below the foramen magnum at the base of the skull and passing through the vertebral (spinal) canal formed by the articulated vertebrae.

The spinal cord has a slightly larger diameter in the cervical and lumbar regions, owing in large measure to the increased presence of neurons and axons in these regions related to the innervation of the large number of muscles in the upper and lower limbs. The spinal cord ends at a tapered region called the **conus medullaris**, which is situated at about the level of the L1-L2 vertebrae. From this point inferiorly, the nerve rootlets course to their respective levels and form a bundle called the **cauda equina**, so named because it resembles a horse's tail. The spinal cord is anchored inferiorly by the **terminal filum**, which is attached to the coccyx. Features of the spinal cord include:

- 31 pairs of spinal nerves (8 cervical pairs, 12 thoracic pairs, 5 lumbar pairs, 5 sacral pairs, and 1 coccygeal pair)
- Each spinal nerve is formed by posterior and anterior roots
- Motor neurons reside in the spinal cord gray matter (anterior horn)
- Sensory neurons reside in the spinal ganglia (also called dorsal root ganglia)
- Anterior rami of spinal nerves often converge to form plexuses (mixed networks of nerve axons)

The typical scheme for a somatic peripheral nerve (innervates skin and skeletal muscle) shows a motor neuron in the spinal cord anterior horn (gray matter) sending a myelinated axon through an anterior root and into a peripheral nerve that ends at a neuromuscular junction on a skeletal muscle. Likewise, a nerve ending in the skin sends a sensory axon toward the spinal cord in a peripheral nerve. Thus each peripheral nerve contains hundreds or thousands of somatic motor and sensory axons. The sensory neuron is a pseudounipolar neuron with a cell body that resides in a spinal ganglion (a ganglion in the periphery is a collection of neurons, just as a nucleus is in the brain) and sends its central axon into the posterior horn (gray matter) of the spinal cord. At each level of the spinal cord, the gray matter is visible as a butterfly-shaped central collection of neurons, exhibiting a posterior horn and an anterior horn.

**COLOR** the following features of the spinal cord, using a different color for each feature:

☐ 1. **Spinal cord**
☐ 2. **Cauda equina: collection of nerve roots inferior to the spinal cord**
☐ 3. **White matter of spinal cord as seen in cross section: ascending and descending fiber tracts**
☐ 4. **Sensory axon and its pseudounipolar neuron (in the spinal ganglion)**
☐ 5. **Central gray matter of the spinal cord (as seen in cross section)**
☐ 6. **Motor neuron and its axon to a skeletal muscle**

***Clinical Note:***

The spinal nerve is vulnerable if one experiences a **herniated disc** (also see Plate 2-7). This may result in radiating pain and possible motor dysfunction.

Additionally, the nerve roots and peripheral nerves may be subject to an acute inflammatory autoimmune demyelinating condition (**Guillain-Barré syndrome**), which can also affect the cranial nerves. It commonly presents as a rapidly progressive, areflexive, relatively symmetric ascending weakness of the limb, truncal, respiratory, pharyngeal, and facial musculature, with variable sensory and autonomic dysfunction. Often, an autoimmune response to *Campylobacter jejuni* from an enteric (gut) infection triggers the acute response.

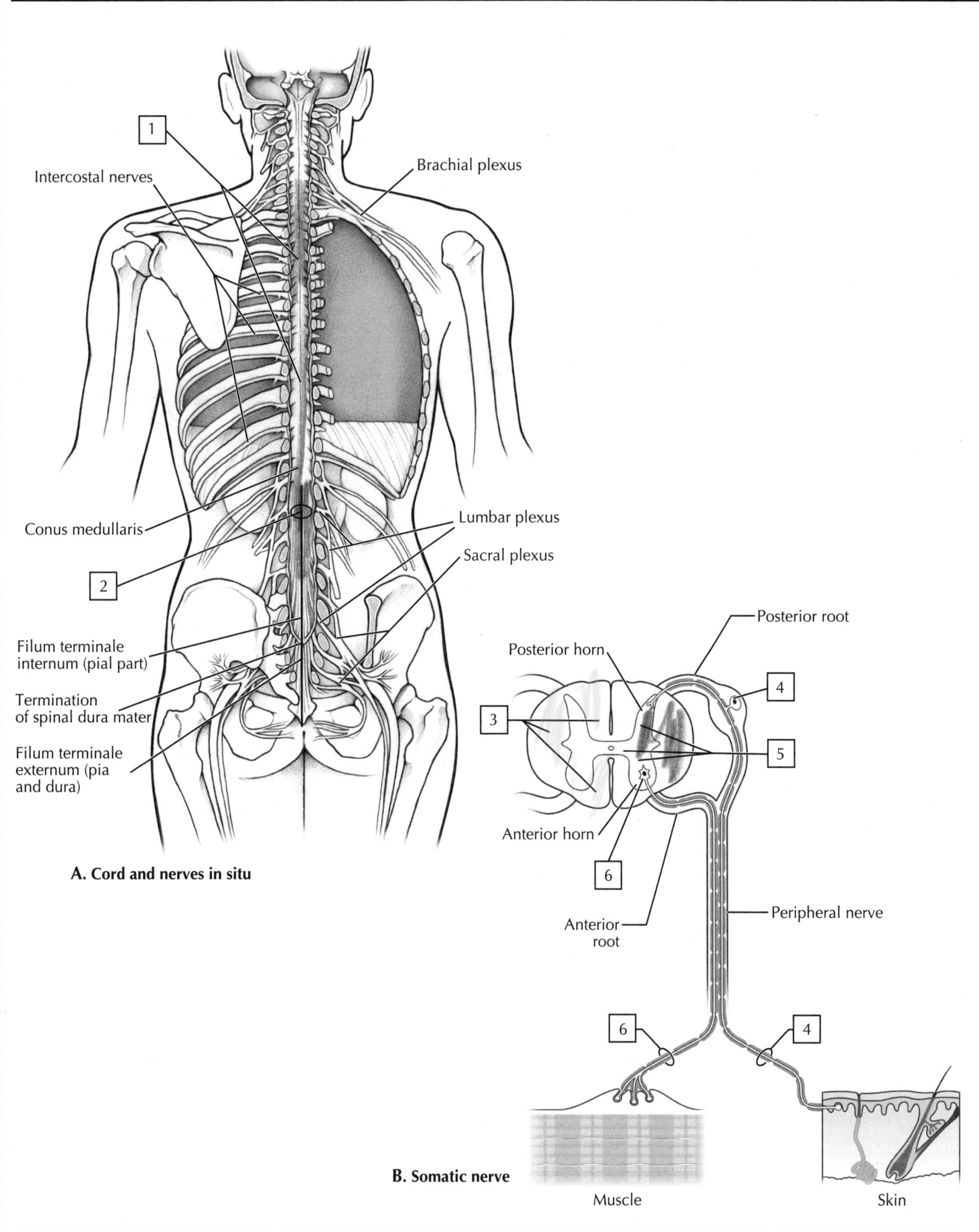


A. Cord and nerves in situ

B. Somatic nerve

# 4 Spinal Cord II

The gray matter of the cerebral cortex lies on the surface of the brain, whereas in the spinal cord the gray matter and its associated neurons lie in the center of the cord, where they form a butterfly- or H-shaped region that can be discerned from the surrounding white matter. Spinal cord levels associated with the innervation of the limbs possess a larger amount of gray matter (C5-T1 and L1-S4 levels, corresponding to the brachial and lumbosacral plexuses, respectively). The gray matter is divided into a **posterior horn**, which receives sensory axons from spinal neurons from the periphery, and an **anterior horn**, where efferent axons exit the cord to enter a spinal nerve. Between spinal cord levels T1 and L2, a lateral horn or cell column is present and is where the sympathetic preganglionic neurons of the autonomic nervous system (ANS) reside.

The white matter of the cord decreases as one continues inferiorly from rostral to caudal. The white matter is divided into posterior, lateral, and anterior funiculi (bundles; singular is funiculus) that contain multiple fiber tracts. In general, these tracts include:

- **Posterior (dorsal) funiculi**: ascending pathways that, generally speaking, convey proprioception (muscle and joint position), touch, and tactile discrimination (size and shape discrimination) from the leg (fasciculus gracilis) and arm (fasciculus cuneatus)
- **Lateral funiculi**: ascending pathways that convey proprioception, pain, temperature, and touch sensations to higher centers, and convey descending pathways concerned with skilled movements and autonomic information to preganglionic neurons
- **Anterior funiculi**: ascending and descending pathways; ascending pathways convey pain, temperature, and touch, and descending pathways convey information that facilitates or inhibits flexor and extensor muscles; reflex movements that control tone, posture, and head movements; and some skilled movements

**COLOR** each of the following white matter tracts, using a different color for each tract:

☐ 1. **Posterior funiculus (fasciculus cuneatus and fasciculus gracilis): ascending fibers conveying proprioception, touch, and tactile discrimination from limbs**

☐ 2. **Lateral corticospinal (pyramidal) tract: descending fibers conveying skilled movements**

☐ 3. **Rubrospinal tract: descending fibers that control movements of flexor muscle neurons**

☐ 4. **Lateral (medullary) reticulospinal tract: descending fibers that regulate autonomic preganglionic neurons**

☐ 5. **Anterior or medial (pontine) reticulospinal tract: descending fibers that control extensor muscle neurons**

☐ 6. **Anterior funiculus (vestibulospinal, tectospinal, and corticospinal) tracts: descending fibers conveying reflex movements that control tone, posture, and head movements, and some skilled movements**

☐ 7. **Anterior spinocerebellar tract: ascending fibers conveying proprioception**

☐ 8. **Spinothalamic and spinoreticular tracts: ascending fibers conveying pain, temperature, and touch**

☐ 9. **Posterior spinocerebellar tract: ascending fibers conveying proprioception**

***Clinical Note:***

**Lower motor neurons** are the neurons of the anterior horn that innervate skeletal muscle. Lesions of these neurons or their axons in the peripheral nerve result in the loss of voluntary and reflex responses of the muscles and cause muscle atrophy. The denervated muscles exhibit **fibrillations** (fine twitching) and **fasciculations** (brief contractions of muscle motor units).

**Upper motor neurons** are neurons at higher levels in the CNS that send axons either to the brainstem or spinal cord. In general, lesions of these neurons or their axons result in spastic paralysis, hyperactive muscle stretch reflexes, clonus (a series of rhythmic jerks), a "clasp-knife" response (muscle hypertonia) to passive movements, and lack of muscle atrophy (except by disuse).

**Amyotrophic lateral sclerosis** (ALS) is a progressive and fatal disease that results in the degeneration of motor neurons in cranial nerves and in the anterior horns of the spinal cord. Muscle weakness and atrophy occur in some muscles, whereas spasticity and hyperreflexia are present in other muscles.

Damage to the dorsal funiculus may occur in **pernicious anemia**, resulting in paresthesias in the feet and legs (and sometimes in the arms and hands), sensory ataxia and loss of fine discriminative touch, vibratory sensation, and joint position sense.

Damage to the lateral funiculus may result in ipsilateral (same side) **spastic paraparesis** with increased tone, muscle stretch reflexes (spasticity), and plantar extensor responses.

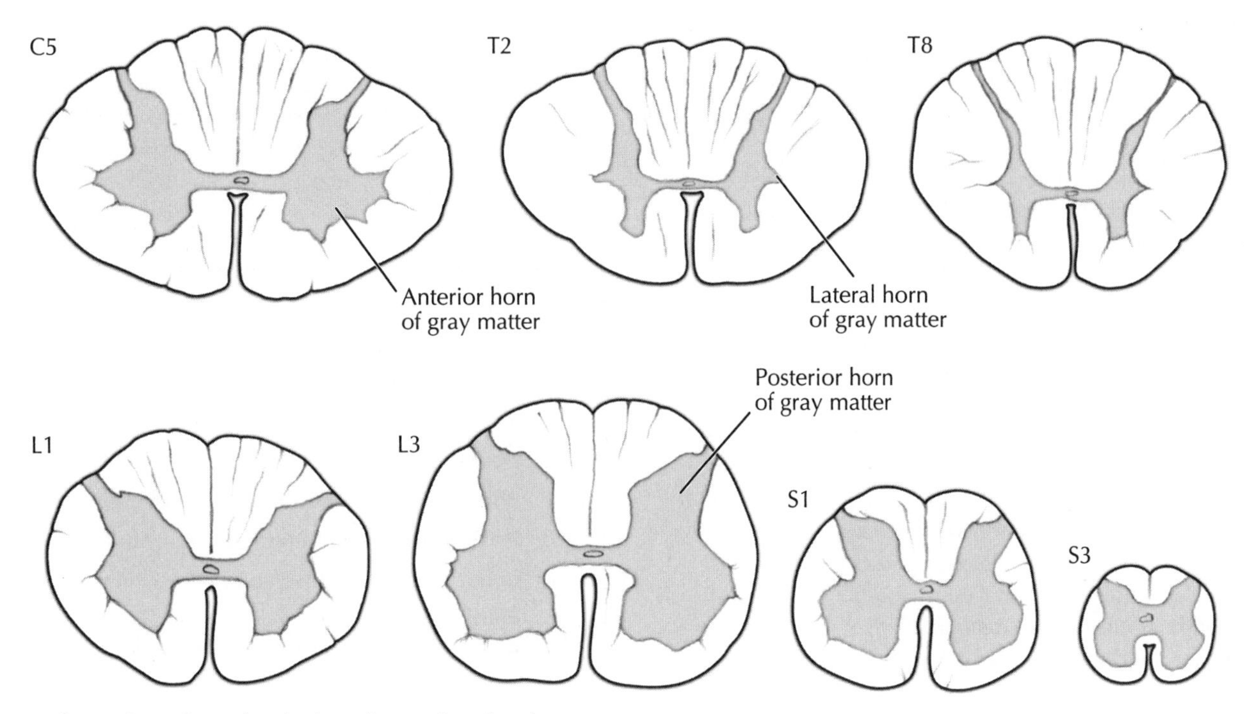


A. Sections through spinal cord at various levels

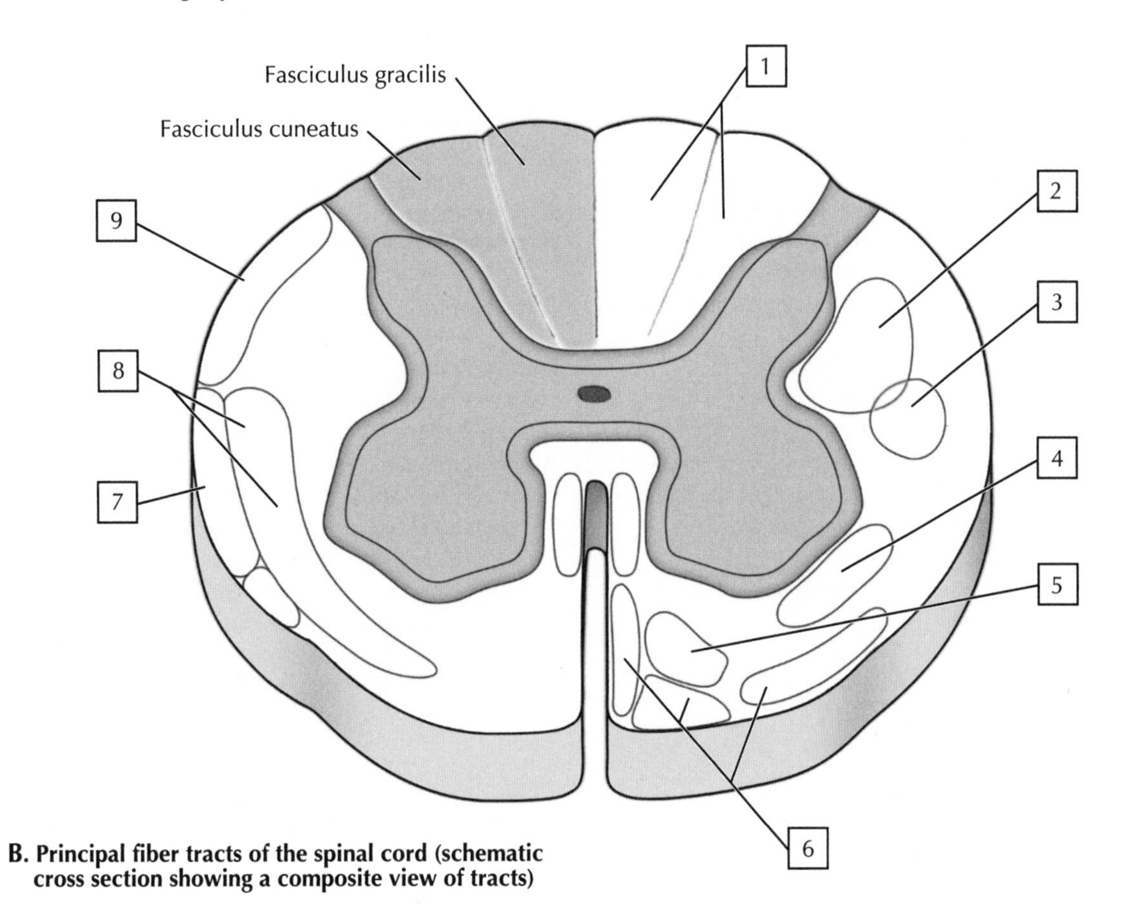


B. Principal fiber tracts of the spinal cord (schematic cross section showing a composite view of tracts)

# Spinal and Peripheral Nerves

The spinal cord gives rise to 31 pairs of spinal nerves, which then form two major branches (rami):

- **Posterior (dorsal) ramus:** a small ramus that courses posteriorly to the back and conveys motor and sensory information to and from the skin and intrinsic back skeletal muscles (erector spinae and transversospinalis muscles) (see Plate 3-10)
- **Anterior (ventral) ramus:** a much larger ramus that courses laterally and anteriorly and innervates all the remaining skin and skeletal muscles of the neck, limbs, and trunk

Once nerve fibers (sensory or motor) are beyond, or peripheral to, the spinal cord proper, the fibers then reside in nerves of the peripheral nervous system (PNS). Components of the PNS include the:

- **Somatic nervous system**: sensory and motor fibers to skin, skeletal muscle, and joints (illustrated in part ***B***, Somatic components)
- **Autonomic nervous system (ANS)**: sensory and motor fibers to all smooth muscle (including viscera and vasculature), cardiac muscle (heart), and glands (illustrated in part ***B***, efferent autonomic components)
- **Enteric nervous system**: plexuses and ganglia of the gastrointestinal tract that regulate bowel secretion, absorption, and motility (originally considered part of the ANS); linked to the ANS for optimal regulation (see Plate 4-21).

Features of the somatic nervous system include:

- It is a one-neuron motor system
- The motor (efferent) neuron is in the CNS, and an axon projects to a peripheral target, such as a skeletal muscle
- The sensory (afferent) neuron (pseudounipolar neuron) resides in a peripheral ganglion called the spinal ganglion and conveys sensory information from the skin, muscle, or joints to the CNS (spinal cord)

Features of the ANS division of the PNS include:

- It is a two-neuron motor system; the first neuron resides in the CNS and the second neuron in a peripheral autonomic ganglion
- The axon of the first neuron is termed *preganglionic* and the axon of the second neuron is termed *postganglionic*
- The ANS has two divisions: sympathetic and parasympathetic
- The sensory neuron (pseudounipolar neuron) resides in a spinal ganglion, just like the somatic system (not shown in part ***B***), and conveys sensory information from viscera to the CNS; most "visceral reflexes" (unconscious sensations) and some pain sensations travel in a retrograde fashion with the parasympathetic fibers, whereas "visceral pain" fibers from the heart and most organs in the peritoneal cavity travel centrally in visceral afferent fibers coursing with the sympathetic fibers

**COLOR** the following features of the PNS, using a different color for each feature:

- ☐ 1. **Anterior (ventral) root (contains efferent fibers)**
- ☐ 2. **Anterior ramus**
- ☐ 3. **Posterior (dorsal) ramus (to intrinsic back muscles)**
- ☐ 4. **Spinal ganglion (contains sensory neurons)**
- ☐ 5. **Posterior root (contains afferent fibers)**
- ☐ 6. **Sensory axon and nerve cell body in a spinal root ganglion (in part *B*)**
- ☐ 7. **Somatic motor axon and nerve cell body (in part *B*, somatic) in anterior horn**
- ☐ 8. **Autonomic preganglionic fiber in anterior root passing to a sympathetic chain ganglion (ANS ganglion) (in part *B*, efferent autonomic components)**
- ☐ 9. **Autonomic postganglionic fiber in posterior ramus passing from a sympathetic chain ganglion to the skin (in part *B*, efferent autonomic components) on back**

***Clinical Note:***

The sensory, motor, and autonomic neurons of the peripheral nervous system have some components in the central nervous system (brain and spinal cord). These central components are vulnerable to CNS demyelination, e.g., **multiple sclerosis**. The peripheral axons are protected by Schwann cells and are vulnerable to peripheral neuropathies and peripheral demyelinating disorders, e.g., **Guillain-Barré syndrome**. Damage to peripheral sensory axons can lead to the loss of sensation. Damage to peripheral motor axons can result in the loss of movement, strength, tone, and reflexes.

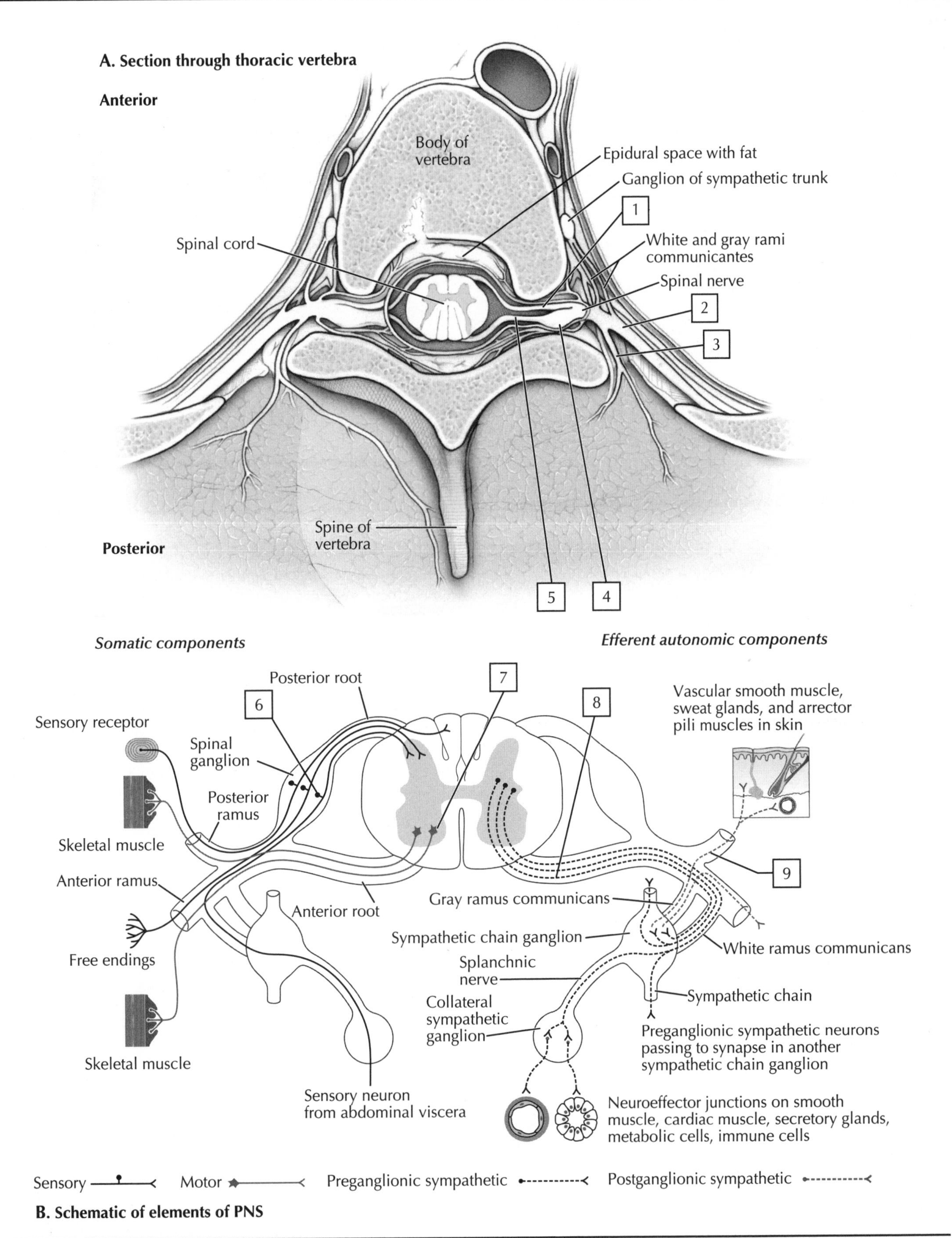
A. Section through thoracic vertebra
Anterior
Body of vertebra
Epidural space with fat
Ganglion of sympathetic trunk
1
Spinal cord
White and gray rami communicantes
Spinal nerve
2
3
Posterior
Spine of vertebra
5
4
Somatic components
Efferent autonomic components
Posterior root
7
6
8
Vascular smooth muscle, sweat glands, and arrector pili muscles in skin
Sensory receptor
Spinal ganglion
Posterior ramus
Skeletal muscle
9
Anterior ramus
Gray ramus communicans
Anterior root
Sympathetic chain ganglion
White ramus communicans
Free endings
Splanchnic nerve
Collateral sympathetic ganglion
Sympathetic chain
Preganglionic sympathetic neurons passing to synapse in another sympathetic chain ganglion
Skeletal muscle
Sensory neuron from abdominal viscera
Neuroeffector junctions on smooth muscle, cardiac muscle, secretory glands, metabolic cells, immune cells
Sensory
Motor
Preganglionic sympathetic
Postganglionic sympathetic
B. Schematic of elements of PNS

# 4 Dermatomes

The region of skin innervated by the somatic sensory nerve fibers associated with a single spinal cord level is called a **dermatome**. (Likewise, over the anterolateral head, the skin is innervated by one of the three divisions of the trigeminal cranial nerve, which will be discussed later.) The neurons that give rise to these sensory fibers are pseudounipolar neurons that reside in the single spinal ganglion associated with the specific spinal cord level (realize that for each level we are speaking of a pair of nerves, roots, and ganglia, because there are 31 pairs of spinal nerves, one pair for each spinal cord level). C1, the first cervical spinal cord level, does possess sensory fibers, but they provide little if any contribution to the skin, so at the top of the head the dermatome pattern begins with the C2 dermatome.

The dermatomes encircle the body segmentally, corresponding to the spinal cord level that receives sensory input from that segment of skin. The sensation conveyed by touching the skin is largely that of pressure and pain. Knowledge of the dermatome pattern is useful in localizing specific spinal cord segments and in assessing the integrity of the spinal cord at that level (an intact or lesioned spinal cord).

**COLOR** the dermatomes associated with the spinal cord segments of each region, using the color indicated for each region (the single coccygeal pair are not illustrated but encircle the anus):

- ☐ 1. **Cervical dermatomes: C2-C8 (green)**
- ☐ 2. **Thoracic dermatomes: T1-T12 (blue)**
- ☐ 3. **Lumbar dermatomes: L1-L5 (purple)**
- ☐ 4. **Sacral dermatomes: S1-S5 (red)**

The sensory nerve fibers that innervate a segment of skin and constitute the dermatome do exhibit some overlap of nerve fibers. Consequently, a segment of skin is innervated primarily by fibers from one level above and one level below the primary cord level. For example, dermatome T5 will have some overlap with sensory fibers associated with the T4 and T6 levels. Thus dermatomes give pretty good approximations of cord levels, but variation is common and overlap does exist.

Key dermatomes that are related to the body surface include the following:

| | | | |
|---|---|---|---|
| C4 | Level of clavicles | T10 | Level of umbilicus |
| C5, C6, C7 | Lateral sides of upper limbs | L1 | Inguinal region and proximal anterior thigh |
| C8, T1 | Medial sides of upper limbs | L1, L2, L3, L4 | Anterior and medial sides of lower limbs, gluteal region |
| C6 | Lateral digits | L4, L5, S1 | Foot |
| C6, C7, C8 | Hand | L4 | Medial leg |
| C8 | Medial digits | L5, S1 | Posterolateral lower limb and dorsal foot |
| T4 | Level of nipples | | |
| | | S1 | Lateral foot |

***Clinical Note:***
Should a **single dorsal root** be cut or damaged, the resulting sensation over the skin supplied by that dermatome will be hypoesthetic (displays diminished sensation) but will not exhibit total anesthesia because of the overlap from the dermatome above and below the affected region. In the case of a **herniated intervertebral disc** (see Plate 2-7), the dorsal root may be compressed and the individual would experience radiating pain in the distribution of that dermatome. In the cervical region, the C5-C6 and the C6-C7 dermatomes are most commonly involved. In the lumbar region, the L4-L5 and the L5-S1 dermatomes are most commonly involved.

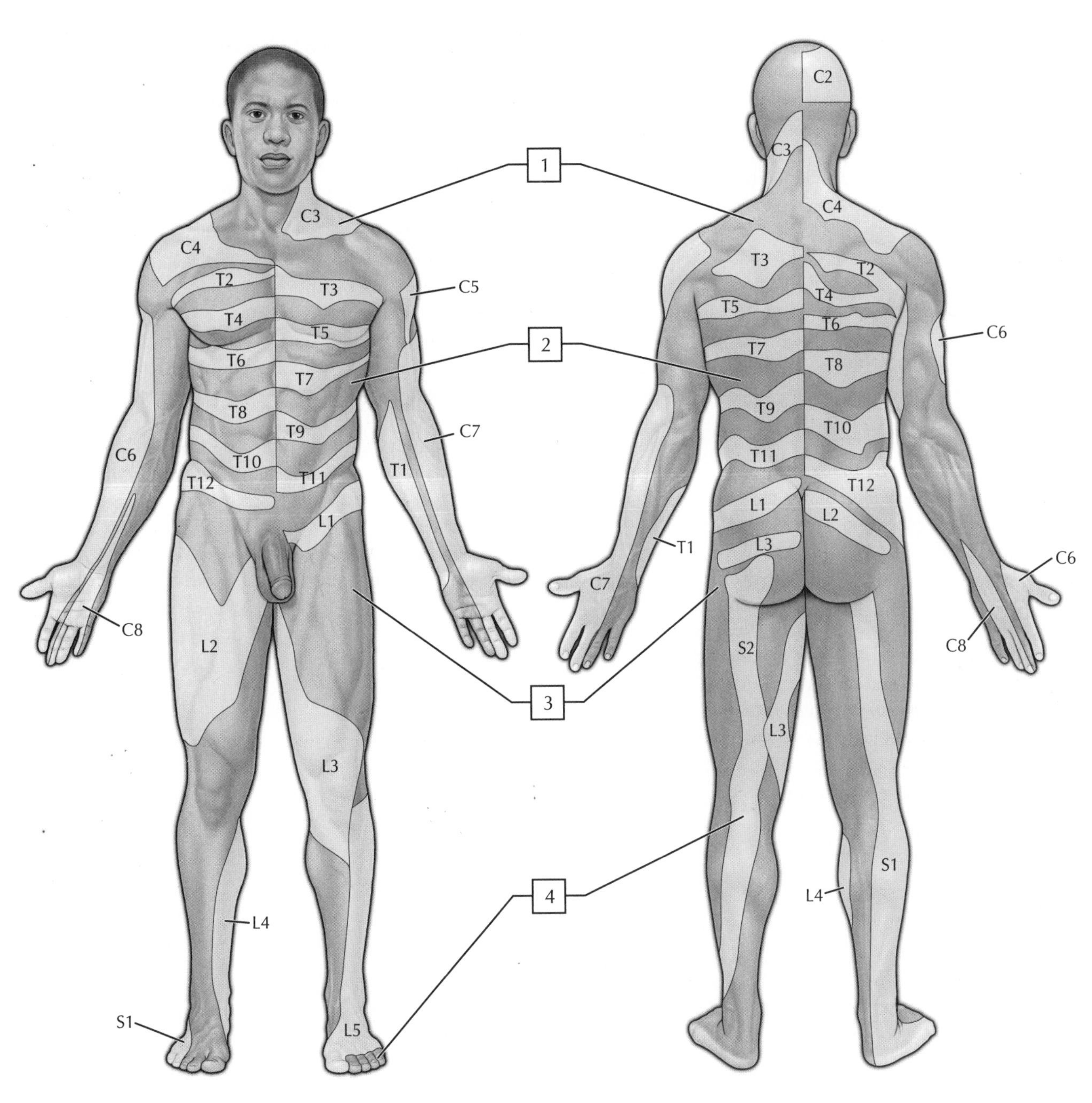


Spinal cord dermatomes

# 4 Brain Ventricles

The small central canal of the spinal cord contains cerebrospinal fluid (CSF) and continues rostrally to expand into four brain ventricles, which include the:

- **4th ventricle**: situated above the pons and rostral portion of the medulla oblongata
- **3rd ventricle**: situated in the midline diencephalon between the thalamic nuclei
- **Lateral ventricles**: two lateral ventricles in the cerebral hemispheres that are C-shaped and extend forward, upward, and back, and then downward and forward into the temporal lobes

CSF fills these ventricles and is produced by the **choroid plexus** (a capillary network and its secretory epithelium), which is found in the floor of each lateral ventricle, with smaller accumulations in the roof of the 3rd and 4th ventricles. About 500 ml of CSF is produced in a 24-hour period, and it functions to:

- Support and cushion the brain and spinal cord
- Fulfill some of the functions normally provided by the lymphatic system
- Fill the 150-ml volume of the subarachnoid space and ventricular cavities

CSF is reabsorbed largely by the arachnoid granulations that project into the superior sagittal dural venous sinus and by small pial veins of the brain and spinal cord.

The flow of CSF is from the choroid plexus of the lateral ventricles to the 3rd ventricle via the interventricular foramen (of Monro), then to the 4th ventricle via the narrow cerebral aqueduct (of Sylvius), and then into the spinal canal, or through openings (lateral and median apertures), to reach the subarachnoid space (between the pia mater and arachnoid mater) surrounding the brain and spinal cord. Secretion of CSF normally is matched by its absorption by the **arachnoid granulations** and small pial and spinal veins.

**COLOR** the following features of the ventricular system, using a different color for each feature:

- ☐ 1. **3rd ventricle**
- ☐ 2. **Lateral ventricles**
- ☐ 3. **4th ventricle**
- ☐ 4. **Choroid plexus of 3rd ventricle (in part *B*)**
- ☐ 5. **Spinal canal in the middle of the spinal cord**

***Clinical Note:***

The accumulation of excess CSF (overproduction or decreased absorption) within the brain's ventricular system is called **hydrocephalus** (see part ***C***). Clinically, three types of hydrocephalus are recognized:

- **Obstructive:** usually a congenital stenosis (narrowing) of the cerebral aqueduct, interventricular foramina, or lateral and median apertures; obstruction also may be caused by CNS tumors that block the normal flow of CSF through the ventricles
- **Communicating:** obstruction outside the ventricular system, perhaps because of pressure from hemorrhage (bleeding) in the subarachnoid space or around the arachnoid granulations
- **Normal pressure:** an adult syndrome that results in progressive dementia, gait disorders, and urinary incontinence

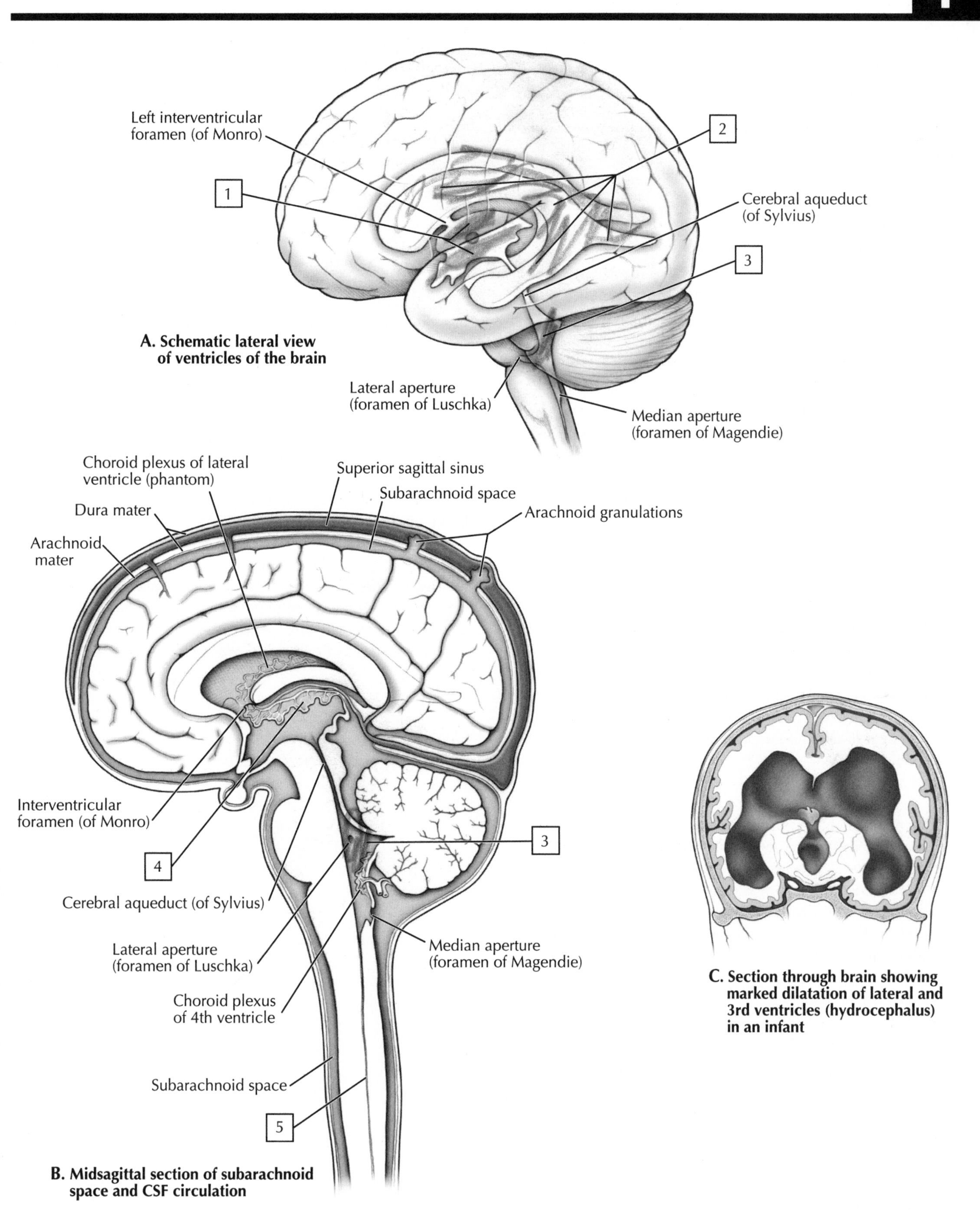


**A. Schematic lateral view of ventricles of the brain**

**B. Midsagittal section of subarachnoid space and CSF circulation**

**C. Section through brain showing marked dilatation of lateral and 3rd ventricles (hydrocephalus) in an infant**

# Subarachnoid Space

The brain and spinal cord are covered by three membranes called the **meninges** and are bathed in cerebrospinal fluid (CSF).

**COLOR** the dura mater of the brain and spinal cord, and then color all three meningeal layers of the spinal cord as seen in section, using a different color for each layer:

- ☐ 1. **Dura mater: a thick outer covering that is richly innervated by sensory nerve endings**
- ☐ 2. **Arachnoid mater: a fine, weblike membrane that is avascular and lies directly beneath the dura mater**
- ☐ 3. **Pia mater: a delicate, transparent inner layer that intimately covers the spinal cord**

The thick dura mater in the cranium is composed of two layers, a **periosteal layer** lining the inner aspect of the skull and a **meningeal layer** in close contact with the arachnoid mater. This layer also is continuous with the spinal dura. CSF fills a space, called the **subarachnoid space**, which lies between the arachnoid and pia meningeal layers. Thus CSF circulates through the brain ventricles and then gains access to the subarachnoid space via the lateral and median apertures, where it flows around and over the brain and spinal cord to the most caudal extent of the dural sac at the S2 vertebral level.

While CSF is secreted by the **choroid plexus**, it is absorbed largely by the **arachnoid granulations** associated with the superior sagittal dural venous sinus and, to a lesser degree, by small pial and spinal veins throughout the CNS. The arachnoid granulations are tufts of arachnoid mater that extend through a layer of split dura mater, which forms the dural venous sinus; they act as one-way valves that deliver CSF into the venous blood of the sinus.

Additionally, the brain possesses small **meningeal (dural) lymphatics**. The venous drainage of the brain assists in removing interstitial waste products from the brain and some of the clearance of these waste products are taken up into the small meningeal lymph vessels. Interstitial metabolites are cleared from the brain into the CSF, which then delivers them primarily to the dural venous system, but also to the small meningeal (dural) lymphatics. This process is especially important during resting periods (sleep) when the brain is less active. This meningeal lymphatic system is called the **glymphatic system** and is vitally important as a parallel system for clearing the brain of interstitial waste products.

**COLOR** the features of the arachnoid granulations, using the following color scheme:

- ☐ 4. **Pia mater covering the cerebrum (green)**
- ☐ 5. **Arachnoid mater and its arachnoid granulations (villi) (red)**
- ☐ 6. **Dura mater splitting to create the dural venous sinus (yellow)**
- ☐ 7. **Venous blood in the superior sagittal dural venous sinus: note the connections with small emissary veins that pass from the scalp through the bony skull to join the sinus (also color blue)**

***Clinical Note:***
CSF may be sampled and examined clinically by performing a **lumbar puncture** (spinal tap). A needle is inserted into the subarachnoid space of the lumbar cistern, in the midline between the L3-L4 or L4-L5 vertebral spinal processes to avoid sticking the spinal cord proper (the cord ends at about the L1-L2 vertebrae; see part ***D***).
Additionally, anesthetic agents may be administered into the epidural space (above the dura mater) to directly anesthetize the nerve fibers of the cauda equina. The **epidural anesthetic** infiltrates the dural sac to reach the nerve roots and is usually administered at the same vertebral levels as the lumbar puncture (see part ***E***).

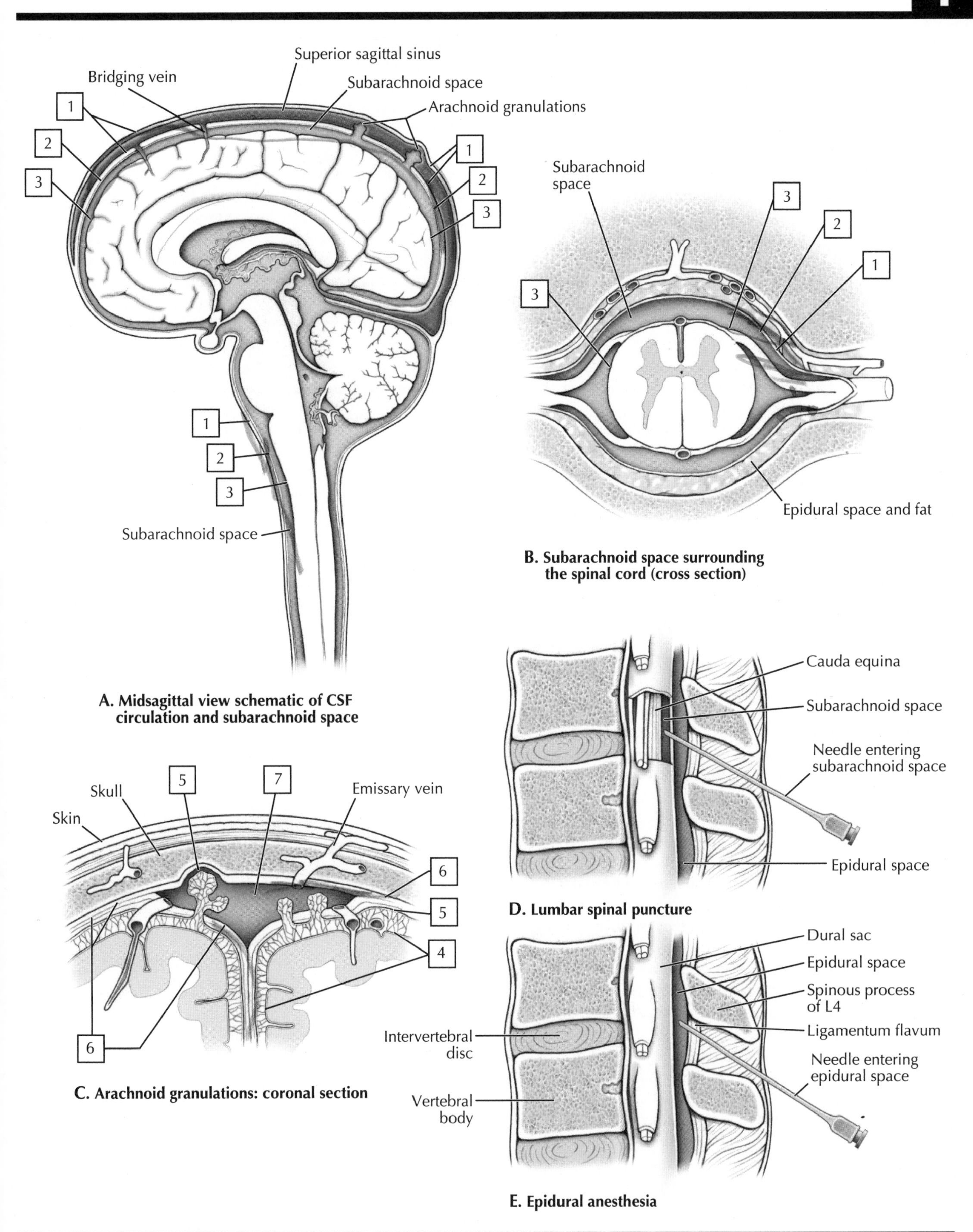


**A. Midsagittal view schematic of CSF circulation and subarachnoid space**

**B. Subarachnoid space surrounding the spinal cord (cross section)**

**C. Arachnoid granulations: coronal section**

**D. Lumbar spinal puncture**

**E. Epidural anesthesia**

# Sympathetic Division of the Autonomic Nervous System

The autonomic nervous system (ANS) is divided into the **sympathetic** and **parasympathetic divisions**. In contrast to the somatic division of the PNS, the ANS is a two-neuron system with a preganglionic neuron in the CNS that sends its axon into a peripheral nerve to synapse on a postganglionic neuron in a **peripheral autonomic ganglion**. The postganglionic neuron then sends its axon to the target (smooth muscle, cardiac muscle, and glands). The ANS is a visceral system because many of the body's organs are composed of smooth muscle walls and/or contain secretory glandular tissue.

The sympathetic division is also known as the **thoracolumbar division** because:

- Its preganglionic neurons are found only in the T1-L2 spinal cord levels
- Its preganglionic neurons lie within the intermediolateral gray matter of the spinal cord in the 14 segments defined above

Preganglionic axons exit the T1-L2 spinal cord in an anterior root, and enter a spinal nerve and then a white ramus communicans to enter the **sympathetic chain**. The sympathetic chain is a bilateral chain of ganglia just lateral to the vertebral bodies that runs from the base of the skull to the coccyx. Once in the sympathetic chain, the preganglionic axon may do one of three things:

- Synapse on a sympathetic chain postganglionic neuron at the T1-L2 level or ascend or descend to synapse on a sympathetic chain postganglionic neuron at any of the 31 spinal nerve levels
- Pass through the sympathetic chain, enter a **splanchnic (visceral) nerve**, and synapse in a **collateral ganglion** (celiac, superior mesenteric, inferior mesenteric) in the abdominopelvic cavity
- Pass through the sympathetic chain, enter a splanchnic nerve, pass through a collateral ganglion, and synapse on the cells of the **adrenal medulla**

Axons of the postganglionic sympathetic neurons may do one of four things:

- Those axons from sympathetic chain neurons reenter the spinal nerve via a gray ramus communicans and join any one of the 31 spinal nerves as they distribute widely throughout the body
- Do the same as in the previous option but course along blood vessels in the head or join cardiopulmonary or hypogastric plexuses of nerves to distribute to the head, thorax, and pelvic viscera
- Arise from postganglionic neurons in collateral ganglia and course with blood vessels to abdominopelvic viscera
- The postganglionic cells of the adrenal medulla are differentiated **endocrine cells (paraneurons)** that do not have axons but release their hormones (epinephrine and norepinephrine) directly into the bloodstream

**COLOR** the sympathetic preganglionic neuron and its axons red (solid lines), and color the postganglionic neuron and its axons green (dashed lines) in both figures.

Preganglionic axons release **acetylcholine (ACh)** at their synapses, whereas **norepinephrine (NE)** is the transmitter released by postganglionic axons (except on sweat glands, where it is ACh). The cells of the adrenal medulla (modified postganglionic sympathetic neurons) release **epinephrine** and some NE not as neurotransmitters but as hormones into the blood. The sympathetic system acts globally throughout the body to mobilize it in "fright-flight-fight" situations. The specific functions are summarized in the following table.

| STRUCTURE | EFFECTS |
|---|---|
| Eyes | Dilates the pupil |
| Lacrimal glands | Reduces secretion slightly (vasoconstriction) |
| Skin | Causes goose bumps (arrector pili muscle contraction) |
| Sweat glands | Increases secretion |
| Peripheral vessels | Causes vasoconstriction |
| Heart | Increases heart rate and force of contraction |
| Coronary arteries | Assists in vasodilation |
| Lungs | Assists in bronchodilation and reduced secretion |
| Digestive tract | Decreases peristalsis, contracts internal anal sphincter muscle, causes vasoconstriction to shunt blood elsewhere |
| Liver | Causes glycogen breakdown, glucose synthesis and release |
| Salivary glands | Reduces and thickens secretion via vasoconstriction |
| Genital system | Causes ejaculation (males) and orgasm (both sexes), and remission of erection (penis and clitoris) |
| | Constricts male internal urethral sphincter muscle |
| Urinary system | Decreases urine production via vasoconstriction |
| | Constricts male internal urethral sphincter muscle |
| Adrenal medulla | Increases secretion of epinephrine or norepinephrine |

***Clinical Note:***

While sympathetic activation provides an acute "fight or flight" response and is critical for our survival, it can be damaging if this acute response becomes chronic.

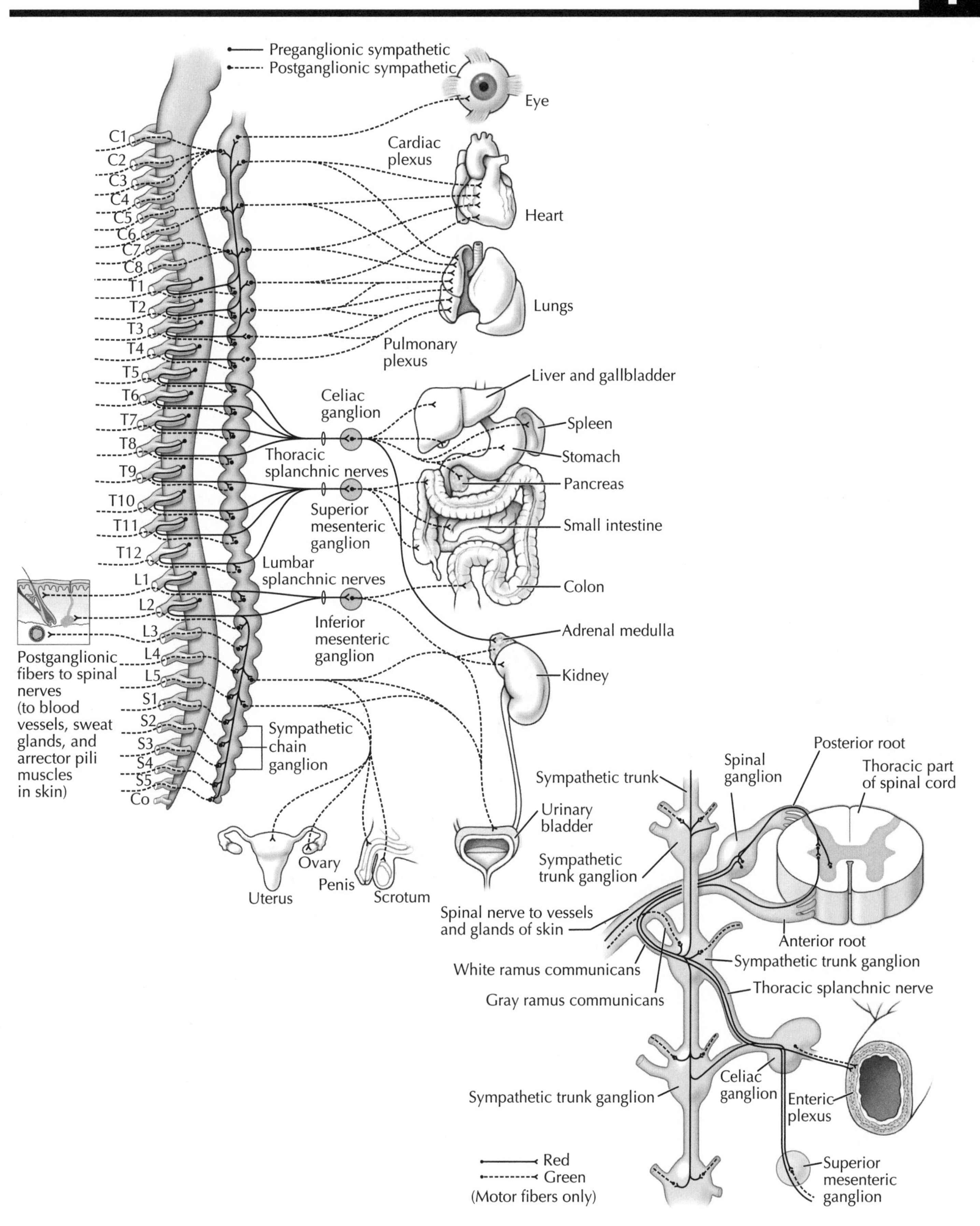
Preganglionic sympathetic
Postganglionic sympathetic
Eye
Cardiac plexus
Heart
Lungs
Pulmonary plexus
Liver and gallbladder
Celiac ganglion
Spleen
Stomach
Thoracic splanchnic nerves
Pancreas
Superior mesenteric ganglion
Small intestine
Lumbar splanchnic nerves
Colon
Inferior mesenteric ganglion
Adrenal medulla
Kidney
C1
C2
C3
C4
C5
C6
C7
C8
T1
T2
T3
T4
T5
T6
T7
T8
T9
T10
T11
T12
L1
L2
L3
L4
L5
S1
S2
S3
S4
S5
Co
Postganglionic fibers to spinal nerves (to blood vessels, sweat glands, and arrector pili muscles in skin)
Sympathetic chain ganglion
Uterus
Ovary
Penis
Scrotum
Urinary bladder
Sympathetic trunk
Spinal ganglion
Posterior root
Thoracic part of spinal cord
Sympathetic trunk ganglion
Spinal nerve to vessels and glands of skin
Anterior root
White ramus communicans
Sympathetic trunk ganglion
Gray ramus communicans
Thoracic splanchnic nerve
Celiac ganglion
Enteric plexus
Sympathetic trunk ganglion
Red
Green
(Motor fibers only)
Superior mesenteric ganglion

# Parasympathetic Division of the Autonomic Nervous System

The parasympathetic division of the ANS also is a two-neuron system with its preganglionic neuron in the CNS and postganglionic neuron in a peripheral ganglion. The parasympathetic division also is known as the **craniosacral division** because:

- Its preganglionic neurons are found in cranial nerves III, VII, IX, and X, and in the sacral spinal cord at levels S2-S4
- Its preganglionic neurons reside in the four cranial nuclei associated with the four cranial nerves listed previously, or in the lateral gray matter of the sacral spinal cord at levels S2-S4

Preganglionic parasympathetic axons may do one of two things:

- Exit the brainstem in the cranial nerve (except CN X, see below) and pass to a peripheral ganglion in the head (ciliary, pterygopalatine, submandibular, and otic ganglia) to synapse on the parasympathetic postganglionic neurons residing in these ganglia
- Exit the sacral spinal cord via an anterior root and then enter the **pelvic splanchnic nerves** to synapse on postganglionic neurons in **terminal ganglia** located in or near the viscera to be innervated

Axons of the postganglionic parasympathetic neurons may do one of two things:

- Pass from the **parasympathetic ganglion in the head** on existing nerves or blood vessels to innervate smooth muscle and glands of the head
- Pass from **terminal ganglia** in or near the viscera innervated and synapse on smooth muscle, cardiac muscle (thorax only), or glands in the neck, thorax, and abdominopelvic cavity

CN X (vagus nerve) is unique. Its preganglionic axons exit the brainstem and synapse on terminal ganglia in or near the targets in the neck, thorax (heart, lungs, glands, smooth muscle), and abdominal cavity (proximal two-thirds of the gastrointestinal tract and its accessory organs). Axons of the terminal ganglia neurons then synapse on their targets.

**COLOR** the preganglionic parasympathetic neurons and their axons (solid lines) arising from a cranial nerve or S2-S4 red, and color the postganglionic neuron and axon (dashed lines) in the peripheral or terminal ganglion green.

The sympathetic axons pass into the limbs, but the parasympathetic axons do not. Therefore the vascular smooth muscle, arrector pili muscles of the skin (attached to hair follicles), and sweat glands are all innervated only by the sympathetic system. ACh is the neurotransmitter at all parasympathetic synapses, both pre- and postganglionic. The parasympathetic system is concerned with feeding and sexual arousal and acts more slowly and focally than the sympathetic system. For example, CN X can slow the heart rate without affecting input to the stomach. In general, the sympathetic and parasympathetic systems maintain **homeostasis**, although as a protective measure, the body does maintain a low level of "sympathetic tone" and can activate this division on a moment's notice. This is an important survival mechanism. ANS function is regulated ultimately by the **hypothalamus**. The specific functions of the parasympathetic division of the ANS are summarized in the table below.

| STRUCTURE | EFFECTS |
|---|---|
| Eyes | Constricts pupil |
| Ciliary body | Constricts muscle for accommodation (near vision) |
| Lacrimal glands | Increases secretion |
| Heart | Decreases heart rate and force of contraction |
| Coronary arteries | Causes vasoconstriction with reduced metabolic demand |
| Lungs | Causes bronchoconstriction and increased secretion |
| Digestive tract | Increases peristalsis, increases secretion, inhibits internal anal sphincter for defecation |
| Liver | Aids glycogen synthesis and storage |
| Salivary glands | Increases secretion |
| Genital system | Promotes engorgement of erectile tissues (crus, corpus cavernosum, bulb of the penis, and corpus spongiosum of the penis in males, and bulb of the vestibule, crus, body, and glans of the clitoris in females) |
| Urinary system | Contracts bladder (detrusor muscle) for urination, inhibits contraction of internal urethral sphincter (only males have this sphincter), increases urine production |

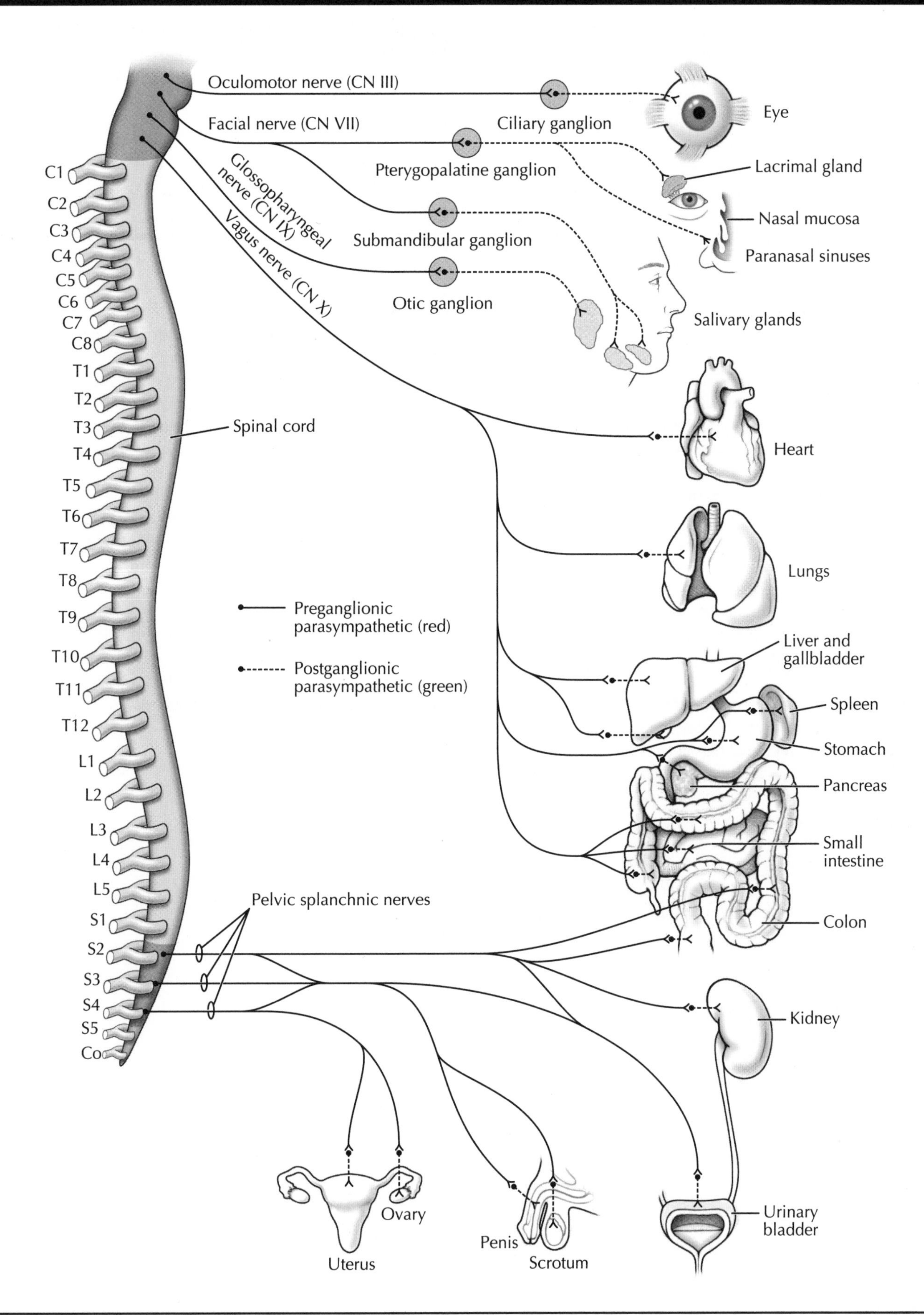

Oculomotor nerve (CN III)
Facial nerve (CN VII)
Glossopharyngeal nerve (CN IX)
Vagus nerve (CN X)
Ciliary ganglion
Pterygopalatine ganglion
Submandibular ganglion
Otic ganglion
Eye
Lacrimal gland
Nasal mucosa
Paranasal sinuses
Salivary glands
C1
C2
C3
C4
C5
C6
C7
C8
T1
T2
T3
T4
T5
T6
T7
T8
T9
T10
T11
T12
L1
L2
L3
L4
L5
S1
S2
S3
S4
S5
Co
Spinal cord
Heart
Lungs
Preganglionic parasympathetic (red)
Postganglionic parasympathetic (green)
Liver and gallbladder
Spleen
Stomach
Pancreas
Small intestine
Colon
Pelvic splanchnic nerves
Kidney
Ovary
Uterus
Penis
Scrotum
Urinary bladder

# Enteric Nervous System

Historically, the third division of the ANS was the enteric nervous system (the intrinsic neurons and nerve plexus found in the myenteric and submucosal layers of the bowel). Because the enteric neurons can function somewhat independently, the bowels were simplistically viewed as having a "brain of their own." However, the enteric nervous system is linked to the sympathetic and parasympathetic divisions of the ANS, and these are required for optimal regulation of bowel secretion, absorption, and motility. Some have characterized the enteric system as a "computer terminal" that has connections with the ANS and to the hypothalamus, which functions as the "master computer." Without the enteric nervous system, the sympathetic and parasympathetic inputs cannot coordinate control of the gastrointestinal (GI) system.

The neurons and nerve plexuses of the enteric nervous system use a variety of neurotransmitters and neuromodulators to communicate with one another and coordinate bowel function. More than 20 such substances have been identified, and it is estimated that the number of neurons in the gut is at least equivalent to the number found in the spinal cord! These neurons regulate peristaltic responses through the longitudinal and circular smooth muscle (which can function independently from autonomic control), although optimal functioning requires a coordinated effort between the ANS (see the ANS contributions below) and the enteric system. A **myenteric (smooth muscle) plexus** controls bowel motility and a deeper **submucosal plexus** controls fluid secretion and absorption.

ANS connections to the enteric nervous system include:

- Vagal parasympathetic input to the esophagus, stomach, small intestine, and proximal half of the colon
- S2-S4 parasympathetic input via pelvic splanchnic nerves to the distal half of the colon and to the rectum
- Sympathetic input from thoracic splanchnic nerves (T5-T12) to the stomach, small intestine, and proximal half of the colon
- Sympathetic input from lumbar splanchnic nerves (L1-L2) to the distal half of the colon and to the rectum

**COLOR** the following pathways from the ANS to the enteric nerve plexuses, using a different color for each pathway:

- ☐ **1. Vagus nerve**
- ☐ **2. Pelvic splanchnic nerves**
- ☐ **3. Thoracic splanchnic nerves**
- ☐ **4. Lumbar splanchnic nerves**

***Clinical Note:***

**Congenital megacolon** (distended large bowel) (also known as **Hirschsprung's disease**) results from a developmental defect that leads to an aganglionic segment of bowel that lacks both a submucosal and a myenteric plexus. Distention of the bowel proximal to the aganglionic region may occur shortly after birth or may cause symptoms only later, in early childhood.

Optimal GI function requires both the enteric nervous system and the ANS. A complex array of endocrine, paracrine, and neurocrine mediators must work in concert. A disruption to the extrinsic innervation to the GI tract from a neuropathy such as **diabetic neuropathy** will affect bowel motility and cause either diarrhea or constipation.

Additionally, some medications may affect bowel function, such as narcotic analgesics, which may block bowel constriction and peristalsis, also resulting in **constipation**.

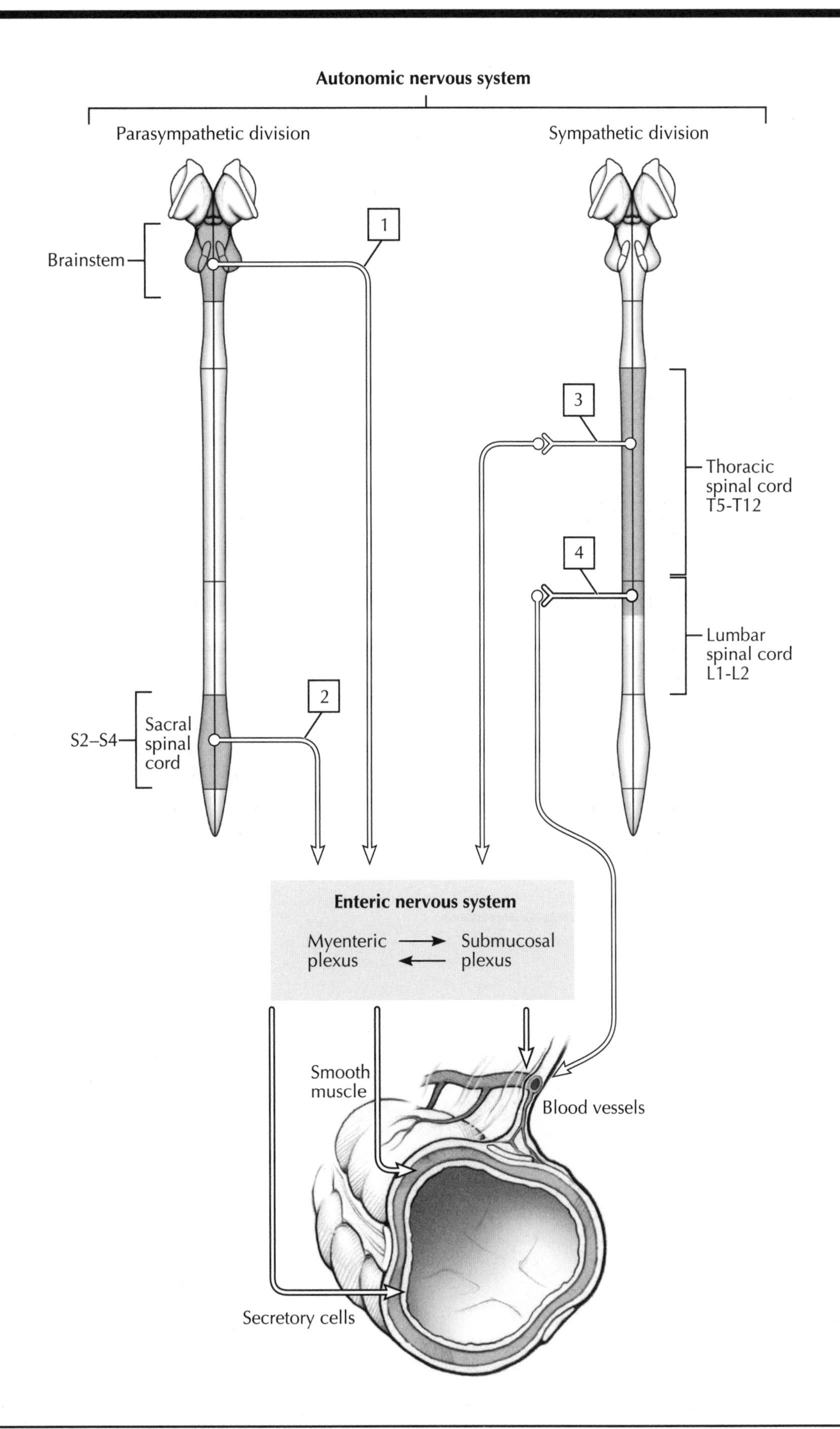
Autonomic nervous system
Parasympathetic division
Sympathetic division
Brainstem
1
2
3
4
Thoracic spinal cord T5-T12
Lumbar spinal cord L1-L2
S2–S4
Sacral spinal cord
Enteric nervous system
Myenteric plexus
Submucosal plexus
Smooth muscle
Blood vessels
Secretory cells

# 4 Cranial Nerves

In addition to the 31 pairs of spinal nerves, 12 pairs of cranial nerves arise from the brain, and they are identified both by their names and by **Roman numerals I through XII**. The cranial nerves are somewhat unique and can contain multiple functional components:

- **General**: same general functions as spinal nerves
- **Special**: functions found only in cranial nerves
- **Afferent and efferent**: sensory or motor functions, respectively
- **Somatic and visceral**: related to skin and skeletal muscle (somatic), or to smooth muscle and glands (visceral)

Hence, each cranial nerve may possess multiple functional components, such as general somatic afferents, meaning the nerve contains sensory fibers from the skin, not unlike those of the spinal nerve; general visceral efferents, meaning the nerve contains motor fibers to visceral structures (smooth muscle and/or glands), such as parasympathetic fibers from the sacral spinal cord (S2-S4 gives rise to parasympathetics); or special somatic afferents, meaning the nerve contains special sensory fibers, such as those for vision or hearing.

In general, CN I and II arise from the forebrain and are really tracts of the brain for the special senses of smell and sight. CN III, IV, and VI move the extraocular skeletal muscles of the eyeball. CN V has three divisions: $V_1$ and $V_2$ are sensory, and $V_3$ has both motor fibers to skeletal muscle and sensory fibers. CN VII, IX, and X have both motor and sensory fibers. CN VIII is involved in the special sense of hearing and in balance. CN XI and XII have motor fibers that innervate skeletal muscle. CN III, VII, IX, and X also contain parasympathetic fibers of origin (visceral fibers); many of the ANS fibers will "jump" onto the branches of CN V to reach their targets. The following table summarizes the types of fibers in each cranial nerve.

**COLOR** each cranial nerve as it arises from the brain or brainstem:

- ☐ 1. **CN I, olfactory nerve**
- ☐ 2. **CN II, optic nerve**
- ☐ 3. **CN III, oculomotor nerve**
- ☐ 4. **CN IV, trochlear nerve**
- ☐ 5. **CN V, trigeminal nerve**
- ☐ 6. **CN VI, abducens nerve**
- ☐ 7. **CN VII, facial nerve**
- ☐ 8. **CN VIII, vestibulocochlear nerve**
- ☐ 9. **CN IX, glossopharyngeal nerve**
- ☐ 10. **CN X, vagus nerve**
- ☐ 11. **CN XI, accessory nerve**
- ☐ 12. **CN XII, hypoglossal nerve**

| CRANIAL NERVE | FUNCTIONAL COMPONENT |
|---|---|
| CN I Olfactory nerve | Special sense of smell |
| CN II Optic nerve | Special sense of sight |
| CN III Oculomotor nerve | Motor to extraocular muscles |
| | Parasympathetic to smooth muscle in eye, constricts pupil and accommodates lens |
| CN IV Trochlear nerve | Motor to one extraocular muscle |
| CN V Trigeminal nerve | Sensory to face, orbit, nose, maxilla, mandible, teeth, anterior tongue |
| | Motor to skeletal muscles |
| CN VI Abducens nerve | Motor to one extraocular muscle |
| CN VII Facial nerve | Sensory to skin of ear |
| | Special sense of taste to anterior tongue |
| | Motor to glands—salivary, nasal, palatine, lacrimal |
| | Motor to scalp and facial muscles, stapedius muscle of middle ear, stylohyoid and posterior belly digastric muscles |
| CN VIII Vestibulocochlear nerve | Special sense of hearing and balance |
| CN IX Glossopharyngeal nerve | Sensory to posterior tongue |
| | Special sense of taste—posterior tongue |
| | Sensory from middle ear, pharynx, carotid body, and sinus |
| | Motor to parotid gland |
| | Motor to one muscle of pharynx, stylopharyngeus |
| CN X Vagus nerve | Sensory from external ear and dura of posterior cranial fossa |
| | Special sense of taste—epiglottis and palate |
| | Sensory from pharynx, larynx, and thoracic and abdominal organs |
| | Motor to smooth muscle of trachea, bronchi, thoracic and abdominal organs, and cardiac muscle of heart |
| | Motor to muscles of pharynx/larynx and striated muscle of esophagus |
| CN XI Accessory nerve | Motor to two muscles, sternocleidomastoid and trapezius |
| CN XII Hypoglossal nerve | Motor to tongue muscles |

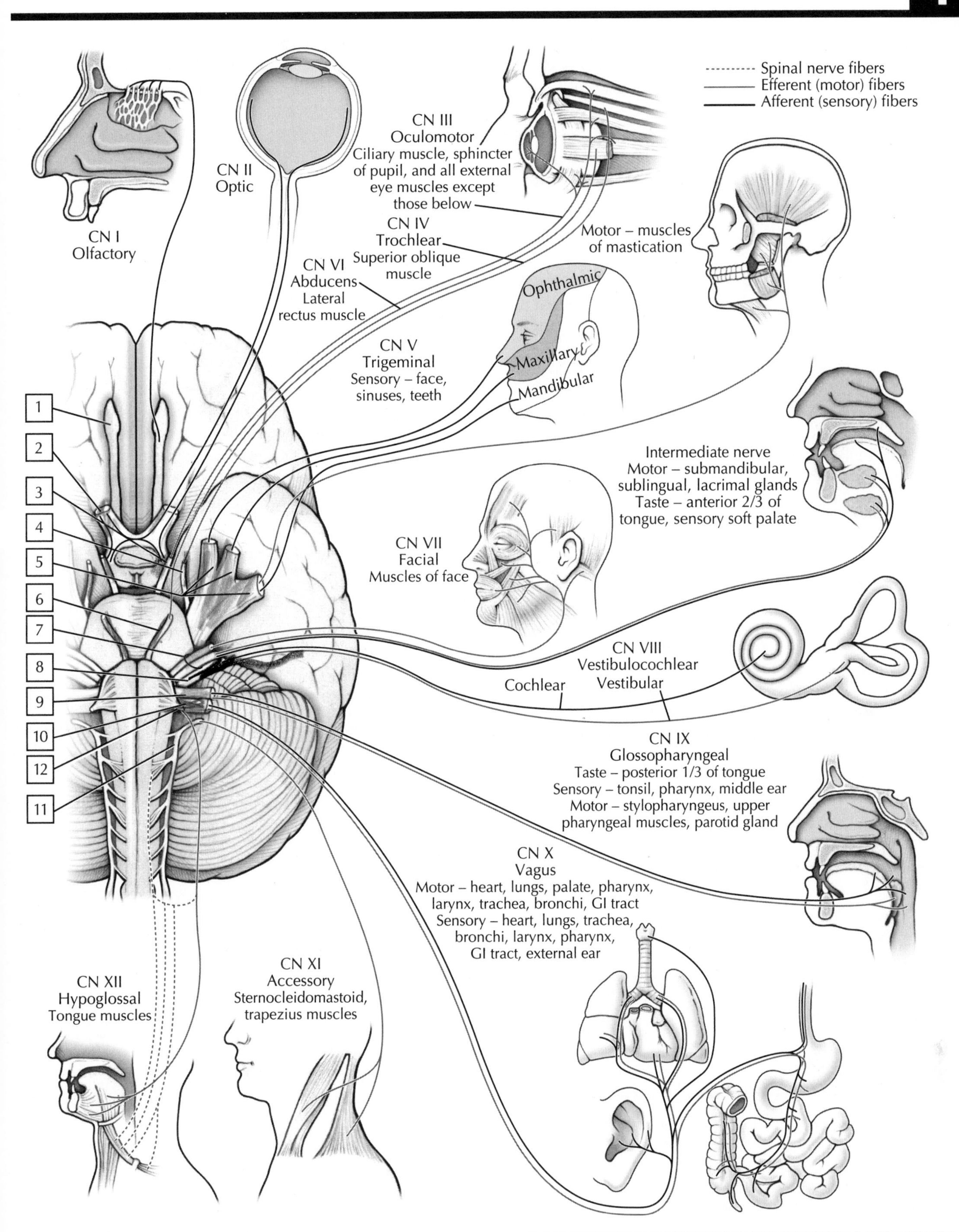
Spinal nerve fibers
Efferent (motor) fibers
Afferent (sensory) fibers
CN I
Olfactory
CN II
Optic
CN III
Oculomotor
Ciliary muscle, sphincter of pupil, and all external eye muscles except those below
CN IV
Trochlear
Superior oblique muscle
CN VI
Abducens
Lateral rectus muscle
Motor – muscles of mastication
CN V
Trigeminal
Sensory – face, sinuses, teeth
Ophthalmic
Maxillary
Mandibular
1
2
3
4
5
6
7
8
9
10
12
11
Intermediate nerve
Motor – submandibular, sublingual, lacrimal glands
Taste – anterior 2/3 of tongue, sensory soft palate
CN VII
Facial
Muscles of face
CN VIII
Vestibulocochlear
Cochlear
Vestibular
CN IX
Glossopharyngeal
Taste – posterior 1/3 of tongue
Sensory – tonsil, pharynx, middle ear
Motor – stylopharyngeus, upper pharyngeal muscles, parotid gland
CN X
Vagus
Motor – heart, lungs, palate, pharynx, larynx, trachea, bronchi, GI tract
Sensory – heart, lungs, trachea, bronchi, larynx, pharynx, GI tract, external ear
CN XII
Hypoglossal
Tongue muscles
CN XI
Accessory
Sternocleidomastoid, trapezius muscles

# 4 Visual System I

The eyeball (globe) is protected by the eyelid, which, in combination with the lacrimal apparatus, keeps the cornea moist by laying down a thin layer of tear film that coats the exposed surface of the eyeball (conjunctiva and cornea).

**COLOR** the following features of the lacrimal apparatus, using a different color for each feature:

- ☐ **1. Lacrimal gland: secretes tears under control of the parasympathetic fibers that originate in the facial nerve (CN VII)**
- ☐ **2. Lacrimal ducts: excretory ducts of the lacrimal gland**
- ☐ **3. Lacrimal sac: receives tears that are collected by the lacrimal canaliculi associated with the superior and inferior lacrimal punctum**
- ☐ **4. Nasolacrimal duct: conveys tears from the lacrimal sac into the nasal cavity**

Excessive irritation, pain, or emotional triggers can increase tear production (crying). Excess tears overwhelm the collecting system of the lacrimal ducts such that the tears will spill over the lower eyelid and run down the cheek. Likewise, copious tears collected in the lacrimal sacs will flow into the nasal cavity and cause a "runny" nose. Tears contain albumins, lactoferrin, lysozyme, lipids, metabolites, electrolytes, and serve an important antimicrobial function.

The human eyeball measures about 25 mm in diameter, is tethered in the bony orbit by six **extraocular muscles** that move the globe (see Plate 3-3), and is cushioned by fat that surrounds the posterior two-thirds of the globe. The eyeball is composed of three concentric layers:

- **Fibrous**: an outer layer that includes the cornea and sclera
- **Vascular**: the middle (uveal) layer that includes the choroid, and the stroma of the ciliary body and iris
- **Retina**: an outer, pigmented epithelium upon which the neural retina (photosensitive) lies

**COLOR** the following layers of the eyeball, using a different color for each layer:

- ☐ **5. Cornea**
- ☐ **6. Iris**
- ☐ **7. Ciliary body**
- ☐ **8. Retina**
- ☐ **9. Choroid**
- ☐ **10. Sclera**

The large chamber behind the lens is the **vitreous chamber** (body) and is filled with a gel-like substance called the **vitreous humor**, which helps cushion and protect the fragile retina during rapid movements of the eye. The chamber between the cornea and the iris is the **anterior chamber**, and the space between the iris and lens is the **posterior chamber**. Both of these chambers are filled with **aqueous humor**, which is produced by the **ciliary body** and circulates from the posterior chamber through the pupil (central opening in the iris) and into the anterior chamber, where it is absorbed by the trabecular meshwork into the **scleral venous sinus** at the angle of the cornea and iris.

The **ciliary body** contains smooth muscle that is arranged in a circular fashion like a sphincter muscle. When this muscle is relaxed, it pulls a set of zonular fibers attached to the elastic lens taut and flattens the lens for viewing objects at some distance from the eye. When focusing on near objects, the sphincter-like ciliary muscle contracts and constricts closer to the lens, relaxing the zonular fibers and allowing the elastic lens to round up for accommodation. This **accommodation reflex** is controlled by parasympathetic fibers that originate in the oculomotor nerve (CN III).

The iris also contains smooth muscle. Contraction of the circular **sphincter pupillae muscle**, under the control of CN III parasympathetic fibers, makes the pupil smaller, whereas contraction of the radially oriented **dilator pupillae muscle,** under sympathetic control, makes the pupil larger.

**COLOR** the following features of the anterior portion of the eyeball, using a different color for each feature:

- ☐ **11. Sphincter pupillae muscle of the iris**
- ☐ **12. Lens**
- ☐ **13. Dilator pupillae muscle of the iris**
- ☐ **14. Zonular fibers**

| FEATURE | DEFINITION |
|---|---|
| Sclera | Outer fibrous layer of eyeball |
| Cornea | Transparent part of outer layer; very sensitive to pain |
| Choroid | Vascular middle layer of eyeball |
| Ciliary body | Vascular and muscular extension of choroid anteriorly |
| Ciliary process | Radiating pigmented folds on ciliary body; secretes aqueous humor that fills posterior and anterior chambers |
| Iris | Contractile diaphragm with central aperture (pupil) |
| Lens | Transparent lens supported in capsule by zonular fibers |
| Retina | Optically receptive part of optic nerve (optic retina) |
| Macula lutea | Retinal area of most acute vision; in its center is the fovea centralis |
| Optic disc | Nonreceptive area where optic nerve axons leave retina for brain |

***Clinical Note:***

A **cataract** is an opacity, or cloudy area, in the crystalline lens.

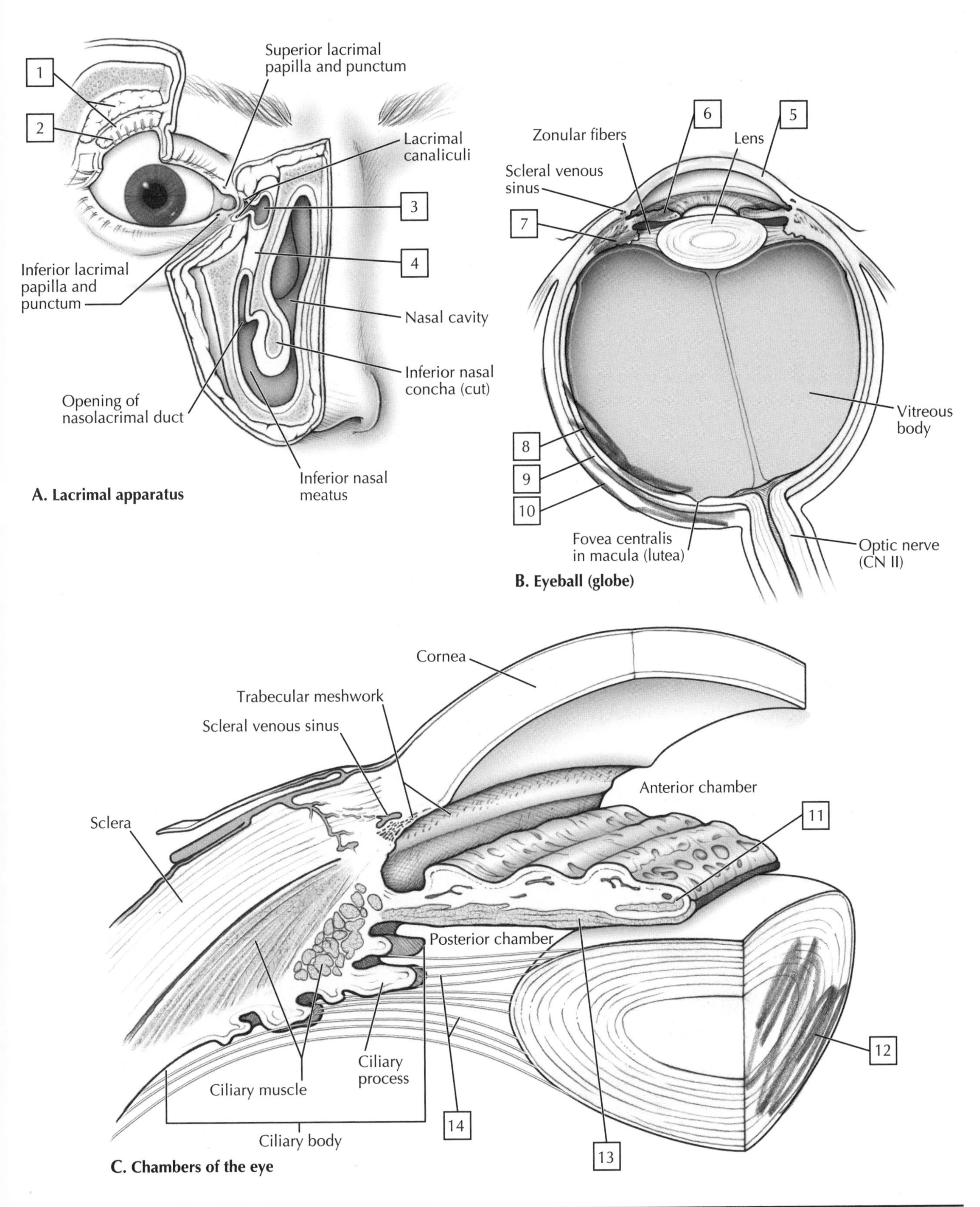


A. Lacrimal apparatus

B. Eyeball (globe)

C. Chambers of the eye

# 4 Visual System II

The **retina** is a very thin layer of tissue that is a direct extension of the brain, with most of its ganglion cell axons coursing back through the optic nerve to reach their first synapse in the lateral geniculate bodies of the thalamus. Light passes through the refractile media of the eye (cornea, aqueous humor, lens, and vitreous humor) to impinge on the neural retina, where it passes through the thickness of the retina to finally encounter the **photoreceptor cells** resting on a layer of **pigmented epithelium** (this epithelium prevents backscatter). The photoreceptors (rods and cones) synapse with bipolar cells, which synapse with the ganglion cells, whereas amacrine and horizontal cells provide interconnections. **Cones** are specialized for bright-light (color) vision and rods for low-light (night) vision. Each human retina contains about 7 million cones and about 120 million rods.

The portion of the retina directly in line with the focus of the lens and situated at the posterior pole of the globe is specialized. Here there is an area called the **macula lutea** with a very small pit, about the size of the head of a pin, called the **fovea centralis** (area of most acute focus) in the middle of the macula lutea. In the fovea, the retina is very thin and consists only of cones and ganglion cells, and it represents our area of greatest visual acuity. The macula lutea contains mostly cones and some rods, and outside of the macula lutea, the rods predominate over cones.

**COLOR** the cells of the neural retina, using the suggested colors for each cell:

- ☐ **1. Pigmented epithelium (brown)**
- ☐ **2. Ganglion cells and their axons (yellow)**
- ☐ **3. Bipolar cells (red)**
- ☐ **4. Rods (gray) (the thinner cells)**
- ☐ **5. Cones (blue) (the thicker cells)**

The visual pathway is organized topographically throughout its course to the occipital lobe. **Nasal** (medial side of the retina) ganglion cells send axons that cross the midline at the **optic chiasm,** whereas **temporal** (lateral side of the retina) ganglion cell axons remain ipsilateral (on the same side). Ganglion cell axons in the optic tracts:

- Largely terminate in the lateral geniculate body, which is organized in six layers
- Optic radiations from the geniculate body pass to the calcarine cortex of the occipital lobe, where conscious visual perception occurs
- From this region of the primary visual cortex, axons pass to the association visual cortex for processing of form, color, and movement
- Connections to the temporal lobe provide high-resolution object recognition (faces, and classification of objects)
- Connections to the parietal cortex provide analysis of motion of positional relationships of objects in the visual scene

***Clinical Note:***

**Ametropias** are conditions that occur when light rays focus in an aberrant manner on a site other than the macula lutea, which is the optimal retinal site. Optically, the cornea, lens, and axial length of the eyeball must be in precise balance to achieve sharp focus. Common disorders include:

- **Myopia (nearsightedness)**: 80% of ametropias, where the point of focus is in front of the retina
- **Hyperopia (farsightedness)**: age-related occurrence, where the point of focus is behind the retina
- **Astigmatism**: a nonspheroidal cornea causes focusing at multiple locations instead of at a single point; affects about 25% to 40% of the population
- **Presbyopia**: age-related progressive loss of the ability to accommodate the lens because of a loss of elasticity in the lens, requiring correction for seeing near objects or reading

The retina is attached to the choroid at the ora serrata, in the area adjacent to the nonpigmented epithelium of the ciliary body. At this location, the retina may separate from the choroid, leading to a **detached retina**. This will distort or impair normal vision. Retinal reattachment often can be achieved through a laser procedure.

**Diabetic retinopathy** develops in almost all patients with type 1 diabetes mellitus and in 50% to 80% of patients with type 2 diabetes mellitus of 20 years duration or more. Diabetic retinopathy is the number-one cause of blindness in middle-aged individuals and about the fourth leading cause of blindness overall in the United States.

**Glaucoma** is an optic neuropathy; the cause of glaucoma usually is an increase to resistance to outflow of aqueous humor in the anterior chamber, which leads to an increase in the intraocular pressure.

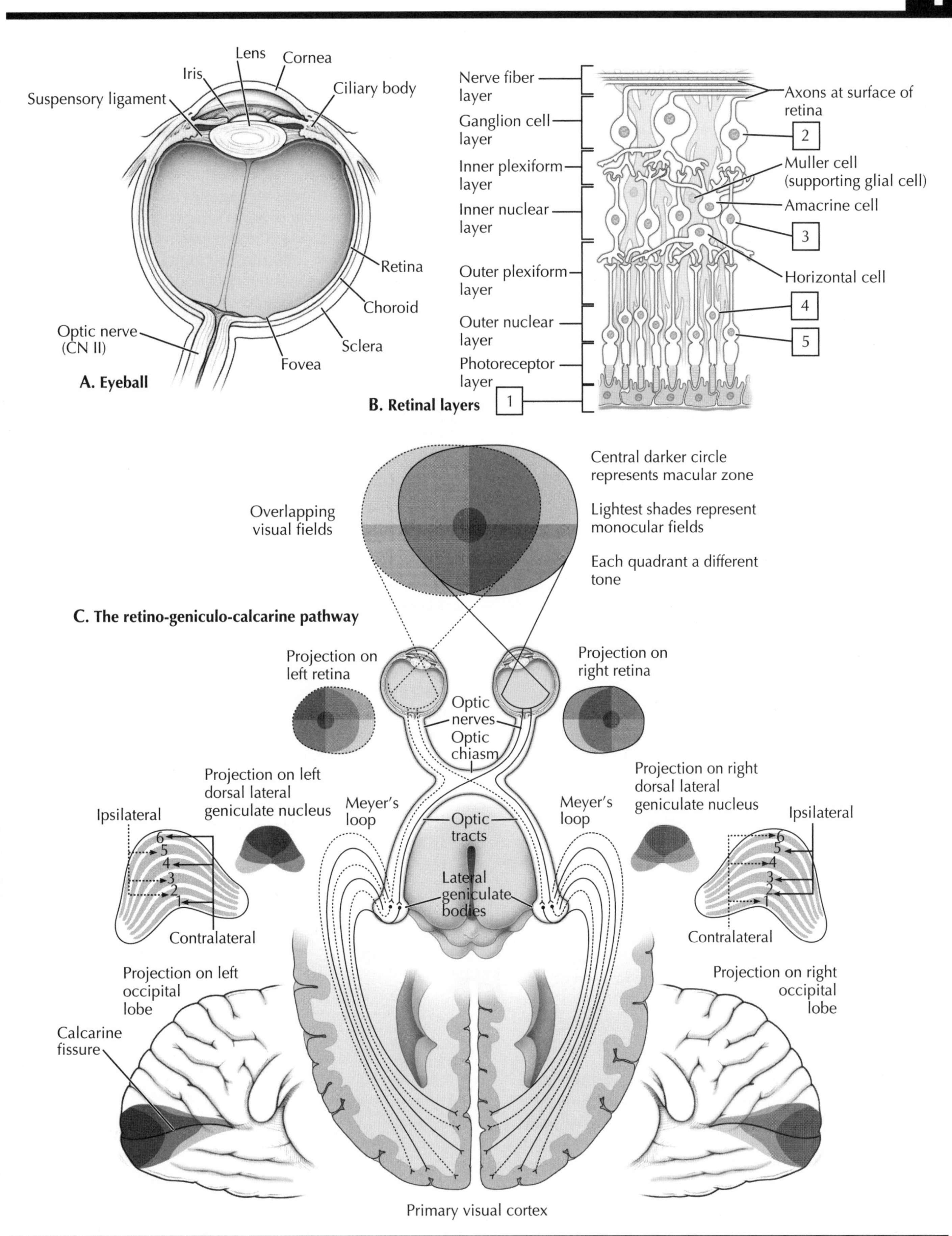
Lens
Cornea
Iris
Suspensory ligament
Ciliary body
Retina
Choroid
Optic nerve (CN II)
Fovea
Sclera
A. Eyeball
Nerve fiber layer
Ganglion cell layer
Inner plexiform layer
Inner nuclear layer
Outer plexiform layer
Outer nuclear layer
Photoreceptor layer
B. Retinal layers
1
Axons at surface of retina
2
Muller cell (supporting glial cell)
Amacrine cell
3
Horizontal cell
4
5
Central darker circle represents macular zone
Lightest shades represent monocular fields
Each quadrant a different tone
Overlapping visual fields
C. The retino-geniculo-calcarine pathway
Projection on left retina
Projection on right retina
Optic nerves
Optic chiasm
Projection on left dorsal lateral geniculate nucleus
Projection on right dorsal lateral geniculate nucleus
Meyer's loop
Meyer's loop
Optic tracts
Ipsilateral
Ipsilateral
6
5
4
3
2
1
Contralateral
Contralateral
Lateral geniculate bodies
Projection on left occipital lobe
Projection on right occipital lobe
Calcarine fissure
Primary visual cortex

# 4 Auditory and Vestibular Systems I

The transduction mechanism of the ear (hearing) and vestibular system (equilibrium) are closely aligned anatomically. The ear consists of three parts:

- **External**: the auricle (pinna), external acoustic meatus (canal), and tympanic membrane (eardrum)
- **Middle**: the tympanic cavity that contains the ear ossicles (malleus, incus, and stapes); this cavity communicates with the mastoid antrum posteriorly and with the auditory (eustachian) tube anteriorly
- **Inner (internal)**: the acoustic apparatus (cochlea) and vestibular apparatus (vestibule with the utricle and saccule, and semicircular canals)

**COLOR** the following features of the ear, using a different color for each feature:

- ☐ 1. **Middle ear ossicles (malleus, incus, and stapes) (also known as the hammer, anvil, and stirrup, respectively)**
- ☐ 2. **Cochlea**
- ☐ 3. **Tympanic membrane**
- ☐ 4. **External acoustic meatus**

Sound waves travel through the external ear and set up vibrations of the tympanic membrane. These vibrations, in turn, cause the middle ear ossicles to vibrate, causing the stapes to vibrate against the **oval (vestibular) window**, initiating a wave action within the fluid-filled (perilymph) **scala vestibuli** and **scala tympani of the cochlea** that causes deflection and depolarization of tiny hair cells within the **organ of Corti**. This stimulates action potentials in the afferent axons of the spiral ganglion cells, which are then conveyed centrally to the cochlear nuclei of the medulla oblongata. From this point, impulses are conveyed to higher brain centers for auditory processing, ending in the auditory cortex in the temporal lobe.

**COLOR** the following features of the bony and membranous labyrinths of the cochlea and vestibular apparatus, using a different color for each feature:

- ☐ 5. **Semicircular canals (anterior, lateral, and posterior): which are arranged at 90 degrees to each other and represent the x-, y-, and z-axes**
- ☐ 6. **Utricle**
- ☐ 7. **Saccule**
- ☐ 8. **Round (cochlear) window: closed by a secondary tympanic membrane, which dissipates the fluid wave initiated at the oval window by the vibratory action of the stapes**

The final step in the auditory transduction pathway from mechanical vibrations to neuronal action potentials, which then are conveyed to the brain, occurs at the level of the organ of Corti within the cochlea. Cochlear hair cells (inner and outer rows) rest on a basilar membrane and are arranged functionally. Traveling pressure waves in the scala vestibuli are transmitted through the **vestibular membrane** to the endolymph-filled cochlear duct. These traveling pressure waves displace the **basilar membrane** (louder sounds cause more displacement), and **tectorial membrane**. The hair cells on the basilar membrane have their tufts attached to the tectorial membrane, and the different displacements of these two membranes cause a shearing effect of the hair cells. This shearing effect deflects the hairs, depolarizes the hair cells, causes the release of neurotransmitters, and initiates an action potential in the afferent axons of the spiral ganglion cells.

**COLOR** the following features of the organ of Corti, using a different color for each feature:

- ☐ 9. **Cochlear nerve, spiral ganglion, and axons**
- ☐ 10. **Inner hair cells**
- ☐ 11. **Outer hair cells**
- ☐ 12. **Basilar membrane**
- ☐ 13. **Tectorial membrane**

***Clinical Note:***

Several kinds of **hearing loss** can occur:

- **Conductive loss**: usually due to a disorder or damage to the tympanic membrane and/or the middle ear ossicles
- **Sensorineural loss**: disorder of the inner ear or cochlear division of the vestibulocochlear nerve (CN VIII), which may include such causes as infection, exposure to loud noises, tumors, e.g., an acoustic Schwannoma, or adverse reactions to certain administered drugs. Loud noises above 85 dB, such as loud rock music, may damage hair cells responsive to the offending frequencies. Other loud, repetitive noises, say from jet engines, industrial sources, gunshots, and explosions can cause permanent damage. Also, some viral infections, e.g., mumps, may damage hair cells.

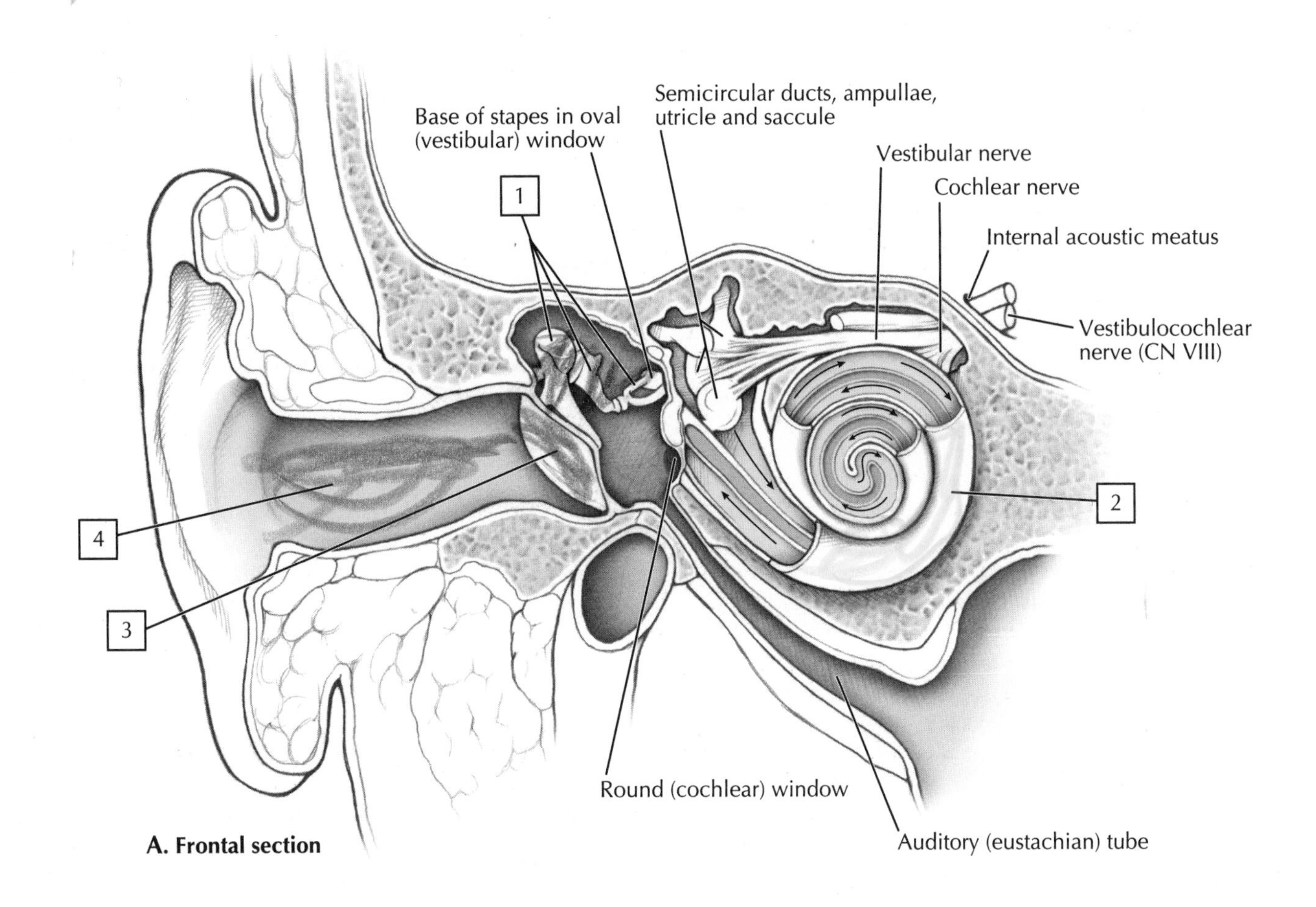


A. Frontal section

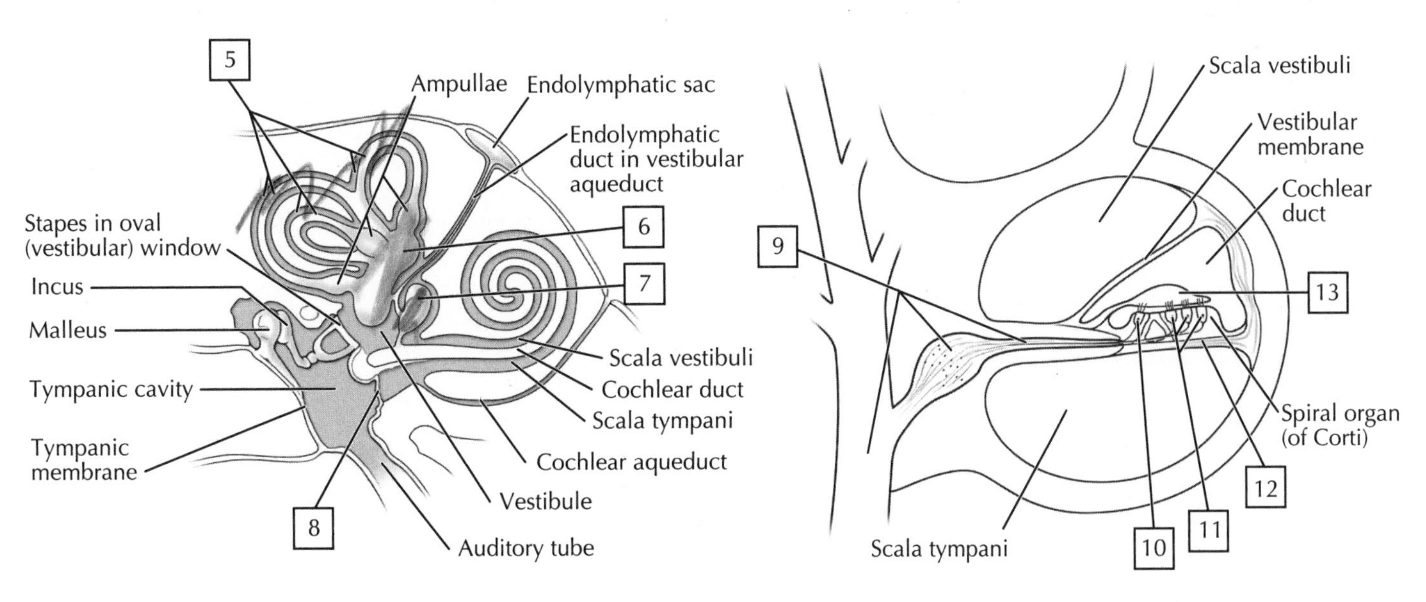


B. Bony and membranous labyrinths: schema

C. Section through turn of cochlea

# 4 Auditory and Vestibular Systems II

While half of the vestibulocochlear nerve (CN VIII) is concerned with hearing, the other half conveys sensory information that is important in maintaining the special sense of equilibrium. Receptors for equilibrium involve two functional components:

- **Static**: a special receptor called the macula resides in each utricle and saccule and is concerned with the position of the head and linear acceleration, as well as gravity and low-frequency vibrations (saccule only)
- **Dynamic**: special receptors called the crista ampullaris reside in the ampulla of each semicircular canal and are concerned with angular (rotational) movements of the head

The maculae also have hair cells (like the organ of Corti), but a single kinocilium also exists at the edge of each bundle of hairlike stereocilia (really long microvilli). The "hair" tufts are embedded in a gelatinous polysaccharide mass called the **otolithic membrane,** which is capped by very small **otoliths** (calcium carbonate crystals), giving the mass a rigidity that resists a change in motion. During linear acceleration, the hairs are displaced and will increase their release of neurotransmitters onto the primary sensory axons of the vestibular ganglion cells. This occurs as the hairs are bent toward the kinocilium, thus depolarizing the hair cells. Movement of the hairs away from the kinocilium hyperpolarizes the hair cells, decreasing their release of neurotransmitters. Finally, the utricle's macula senses acceleration on a horizontal plane, whereas the saccule's macula is better at sensing vertical acceleration, the sensation one feels when one starts to ascend in an elevator.

**COLOR** the following features of the vestibular system (see part ***A***) and maculae (see part ***B***), using a different color for each feature:

- ☐ **1. Maculae of the saccule and utricle**
- ☐ **2. Vestibular ganglion and its afferent axons**
- ☐ **3. Cristae within the ampulla of the semicircular canals**
- ☐ **4. Otoliths (on the surface of the otolithic membrane)**
- ☐ **5. Gelatinous otolithic membrane**
- ☐ **6. Hair cells and "hair" tufts extending into the otolithic membrane**

The **cristae** in the ampullae of the semicircular canals also have hair cells and a kinocilium just like the maculae. However, the gelatinous protein-polysaccharide mass is called a **cupula** (pointed cap) and it projects into the endolymph of the semicircular canal. During rotational movements, the cupula is swayed by the movement of the endolymph and the deflection of the hair cells causes depolarization and release of a neurotransmitter on the sensory nerve endings.

**COLOR** the following features of the crista, using a different color for each feature:

- ☐ **7. Gelatinous cupula**
- ☐ **8. Hair cells and "hair" tufts extending into the cupula**

Vestibular afferent axons terminate in the vestibular nuclei in the brainstem or directly in the cerebellum, to modulate and coordinate muscle movement, tone, and posture. Descending axons from the vestibular nuclei course to the spinal cord to regulate head and neck movements, while other projections coordinate eye movements (CN III, IV, and VI). Finally, some axons ascend to the thalamus and then to the insular, temporal, and parietal cortex.

***Clinical Note:***

**Vertigo** is the sensation of movement or rotation with a loss of equilibrium (dizziness). It can be produced by excessive stimulation of the vestibular system, as occurs in seasickness, carsickness, or carnival rides. Viral infections, certain medications, and tumors also can lead to vertigo.

The auditory nerve may be damaged by **trauma** e.g., to the petrous portion of the temporal bone, or by tumors, e.g., **acoustic Schwannoma**, and infections. Hair cells and the auditory nerve can be damaged by **irritative lesions**, e.g., increased endolymph pressure on hair cells (in Ménière's disease) which can produce **tinnitus** (ringing, buzzing, or clicking noises in the ear).

An **acoustic Schwannoma** on the vestibular portion of CN VIII can irritate the auditory apparatus leading to tinnitus and vestibular symptoms of vertigo, dizziness, nystagmus, and balance problems.

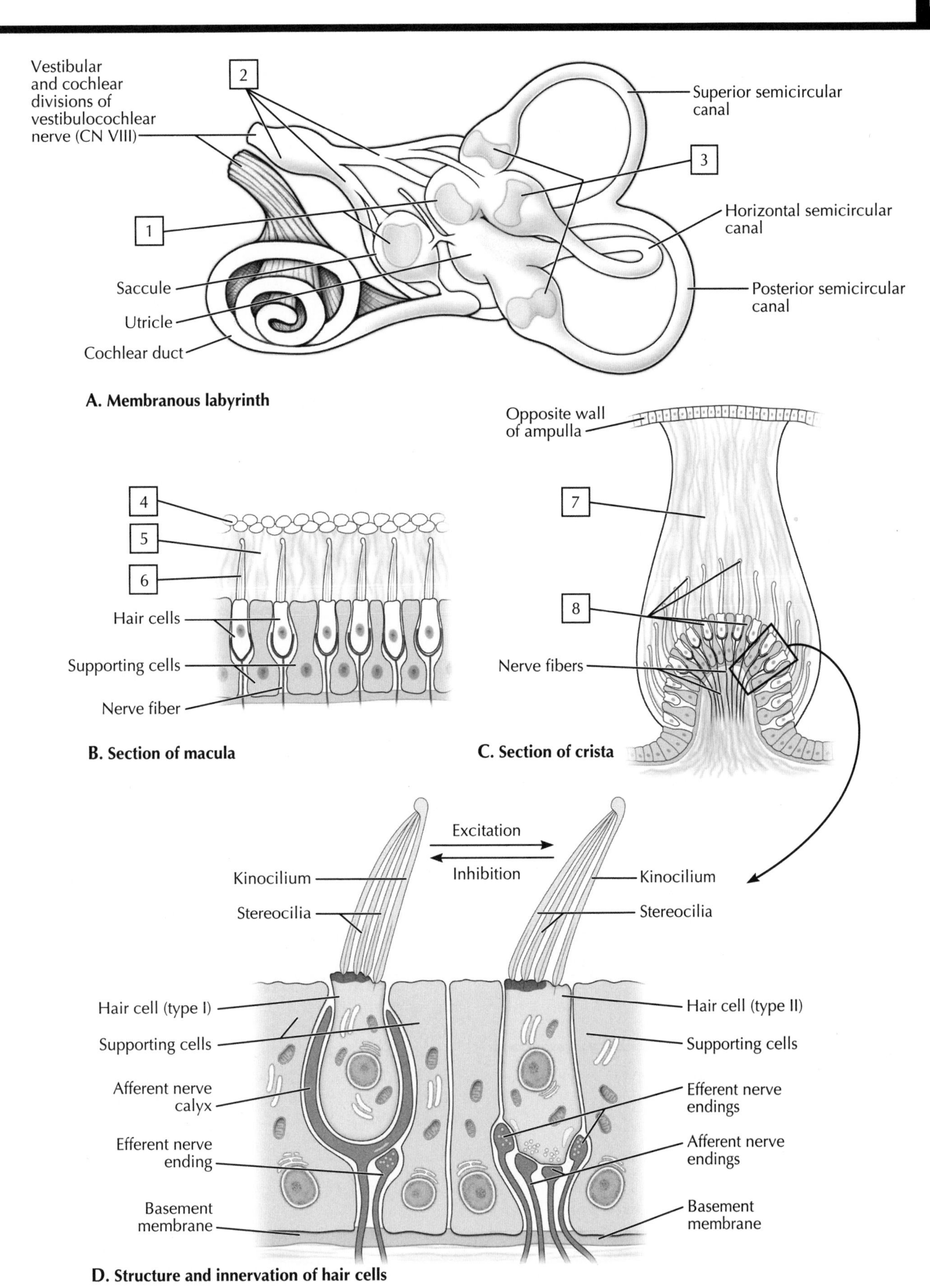


**A. Membranous labyrinth**

**B. Section of macula**

**C. Section of crista**

**D. Structure and innervation of hair cells**

# 4 Taste and Olfaction

**Taste buds** are chemoreceptors that transduce chemical "tastes" into electrical signals that are conveyed to the CNS for higher processing. We have about 2,000 to 5,000 taste buds (each with 50-150 taste receptor cells) mostly located on the dorsal tongue (also present on the epiglottis and palate), which can distinguish the following taste sensations:

- **Salty**: inorganic salts
- **Sweet**: organic molecules, such as sugar, alcohol, saccharin, and some amino acids
- **Sour**: acids and protons (hydrogen ions)
- **Bitter**: alkaloids and poisons
- **Umami (Japanese for "delicious")**: glutamate (the taste of MSG)

On the tongue, various mucosal specializations called **lingual papillae** are evident and include four types, three of which possess taste buds:

- **Filiform**: small and the most numerous papillae, they serve only a mechanical function and do not possess taste buds
- **Fungiform**: mushroom-shaped papillae that are more numerous near the tip of the tongue and possess taste buds
- **Foliate**: parallel rows of papillae concentrated near the lateral edge of the tongue; poorly developed in humans
- **Vallate**: large papillae (about 8-12) close to the back of the body of the tongue that possess taste buds

Most taste buds respond to multiple "tastes," and our gustatory and olfactory receptors function in parallel; most flavors are enhanced by both taste and smell. Pinching your nose shut while eating will significantly diminish your sensation of taste! Molecules, dissolved in saliva, contact the gustatory microvilli in the taste pore and depolarize the taste cells, causing the release of a neurotransmitter onto afferent nerve endings. Nerve impulses are conveyed to the CNS via the facial (from the anterior two-thirds of the tongue), glossopharyngeal (posterior one-third of the tongue), and vagus (epiglottis and palate) nerves to the pontine taste area (parabrachial nucleus in the pons). Axons then project to the thalamus, hypothalamus, and amygdala, and to the gustatory cortex. These limbic/hypothalamic projections may account for the emotional, motivational, and behavioral aspect of a gustatory experience. Input from the trigeminal system and olfactory system also integrate into the enjoyable experience of eating a meal.

**COLOR** the following features of the tongue and taste bud, using a different color for each feature:

- ☐ 1. **Vallate papillae**
- ☐ 2. **Foliate papillae**
- ☐ 3. **Filiform papillae**
- ☐ 4. **Microvilli of the taste cells in the taste pore**
- ☐ 5. **Taste cells**

Olfactory chemoreceptors lie in the olfactory epithelium at the roof of the nasal cavity. The receptors are bipolar neurons whose dendritic end projects into the nasal cavity and ends in a tuft of microvilli in a mucous film. Odors, dissolved in the mucous film, bind to specific odorant-binding proteins and interact with the microvilli, depolarizing the olfactory neuron. Impulses then are conveyed along the neuron's central process through the cribriform plate to neurons in the olfactory bulb. The olfactory tract (CN I) projects centrally, bypassing the thalamus and distributing to the anterior olfactory nucleus, nucleus accumbens, the uncus (primary olfactory cortex) and periamygdaloid cortex, and the amygdala, and the lateral entorhinal cortex (see part ***E***).

It is estimated that we can sense thousands of substances, but most can be reduced to the following six categories: floral, ethereal (pears), musky, camphor (eucalyptus), putrid, and pungent (vinegar, peppermint).

**COLOR** the following features of the olfactory receptors, using a different color for each feature:

- ☐ 6. **Region of olfactory epithelium distribution in the nose**
- ☐ 7. **Olfactory receptor cells: their dendrites and microvilli projecting into the nasal cavity, and their axons coursing through the cribriform plate**

***Clinical Note:***

The olfactory axons are very fragile and can be easily injured by trauma. If they are permanently damaged, a person can lose the sense of smell, a condition known as **anosmia**. The olfactory receptor cells survive for about one month and then are replaced (bipolar neurons); they are one of a few types of nerve cells that can be replaced throughout life.

Olfactory information goes directly to limbic regions associated with emotional interpretation, behavioral responsiveness, hypothalamic visceral activity for which odors are important, and appropriate autonomic responses. **Olfactory responses** are rapid, specific, and long lasting.

The olfactory bulbs and the olfactory tracts can be damaged by **tumors** (gliomas), by **aneurysms** at the anterior vascular circle of Willis (see the blood supply to the brain, Plate 5-10), and by **meningitis**.

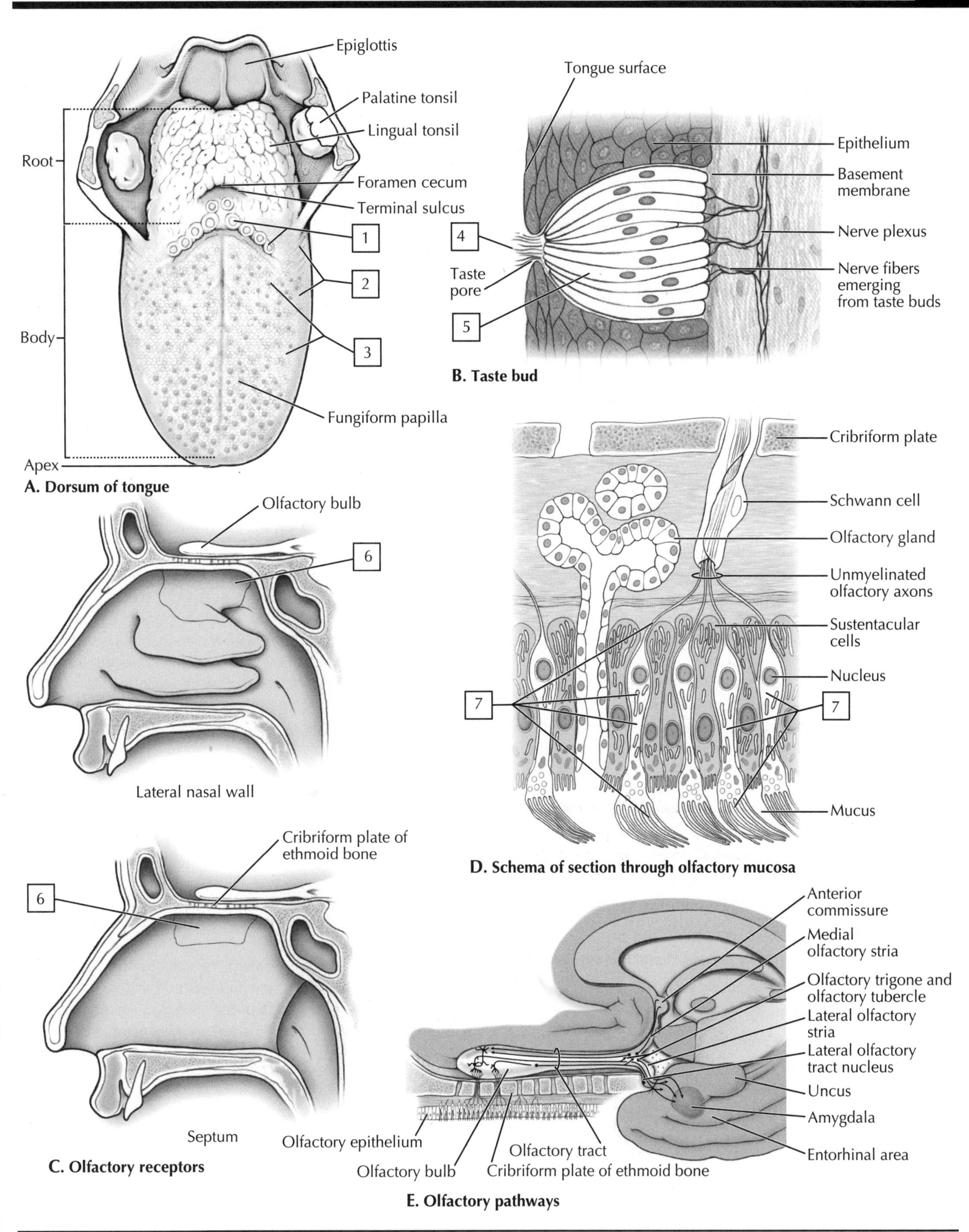


**A. Dorsum of tongue**

**B. Taste bud**

**C. Olfactory receptors**

**D. Schema of section through olfactory mucosa**

**E. Olfactory pathways**

# 4 Cervical Plexus

The anterior rami of the 31 pairs of spinal nerves often join one another shortly after they branch from the spinal nerve and form a network or **plexus of nerves**. A plexus is not unlike a vast network of different railway tracks that interconnect in a major rail terminal or switching yard. The nerve plexus is a mixing of nerve fibers from several adjacent spinal cord levels that ultimately gives rise to several "terminal" nerve branches, which then pass further into the periphery and innervate skeletal muscle, joints, and the skin. Although a muscle may be innervated by a single nerve, that nerve usually contains fibers from several spinal cord levels.

The first and most rostral of the nerve plexuses is the **cervical (neck) plexus**, composed of the anterior rami of the first four cervical nerves. The motor branches of the plexus, as is typical of all spinal nerves, contain hundreds or thousands of three types of nerve fibers (***somatic motor fibers*** to innervate skeletal muscle; ***postganglionic sympathetic fibers*** to innervate smooth muscle of the hair follicles, vasculature, and sweat glands; and ***sensory fibers***).

The major motor branches include the:

- **Ansa cervicalis**: innervates the infrahyoid or "strap" muscles of the anterior neck
- **Phrenic nerve**: from C3, C4, and C5, this nerve "keeps the diaphragm alive"; it innervates the respiratory diaphragm, which is critical for our breathing
- **Minor branches**: several smaller motor branches innervate individual neck muscles

The remaining branches of the cervical plexus are largely sensory, and innervate the skin of the neck and even send sensory branches superiorly to the skin around the ear and back of the scalp. The table summarizes the branches of the cervical plexus.

**COLOR** the following branches of the cervical plexus. Color the motor branches one color and the sensory branches a different color:

- ☐ 1. **Nerves to the geniohyoid and thyrohyoid muscles**
- ☐ 2. **Transverse cervical: sensory**
- ☐ 3. **Ansa cervicalis (*ansa* means "loop"): motor branch**
- ☐ 4. **Supraclavicular nerves: sensory**
- ☐ 5. **Phrenic nerve: motor branch**
- ☐ 6. **Lesser occipital: sensory**
- ☐ 7. **Great auricular: sensory**

| NERVE | INNERVATION |
|---|---|
| C1 | Travels with CN XII to innervate geniohyoid and thyrohyoid muscles |
| Ansa cervicalis | Is C1-C3 loop that sends motor branches to infrahyoid muscles |
| Lesser occipital | From C2, is sensory to neck and scalp posterior to ear |
| Great auricular | From C2 to C3, is sensory over parotid gland and posterior ear |
| Transverse cervical | From C2 to C3, is sensory to anterior triangle of neck |
| Supraclavicular | From C3 to C4, are anterior, middle, and posterior sensory branches to skin over clavicle and shoulder region |
| Phrenic | From C3 to C5, is motor and sensory nerve to diaphragm |
| Motor branches | Are small twigs that supply scalene, levator scapulae, and prevertebral muscles |

***Clinical Note:***

The **phrenic nerve** (C3-C5) receives two of its three nerve segment contributions from the cervical plexus and is an important nerve, because it innervates the respiratory diaphragm. This nerve passes through the thorax in close association with the heart and its pericardial sac, so any surgeon operating in the thorax must identify this nerve and be certain to preserve it. Likewise, a person with cervical spine injuries above the level of C3 that severely damage the spinal cord will need mechanical ventilation, because the nerve fibers to the phrenic nerve will degenerate. In fact, most or all motor function below the level of a severe spinal cord injury will be lost.

A **phrenic nerve block** may be performed to achieve paralysis of the diaphragm on one side, for example when surgery on the lung is required. The anesthetic agent is injected around the phrenic nerve where it lies on the anterior aspect of the anterior scalene muscle.

A **cervical plexus block** may be indicated when surgery is required in the neck. The anesthetic agent is injected in several locations along the posterior margin of the sternocleidomastoid muscle near the junction of the superior and middle thirds of this muscle, close to the sites of the emerging lesser occipital and greater auricular nerves (6 and 7 in the illustrations).

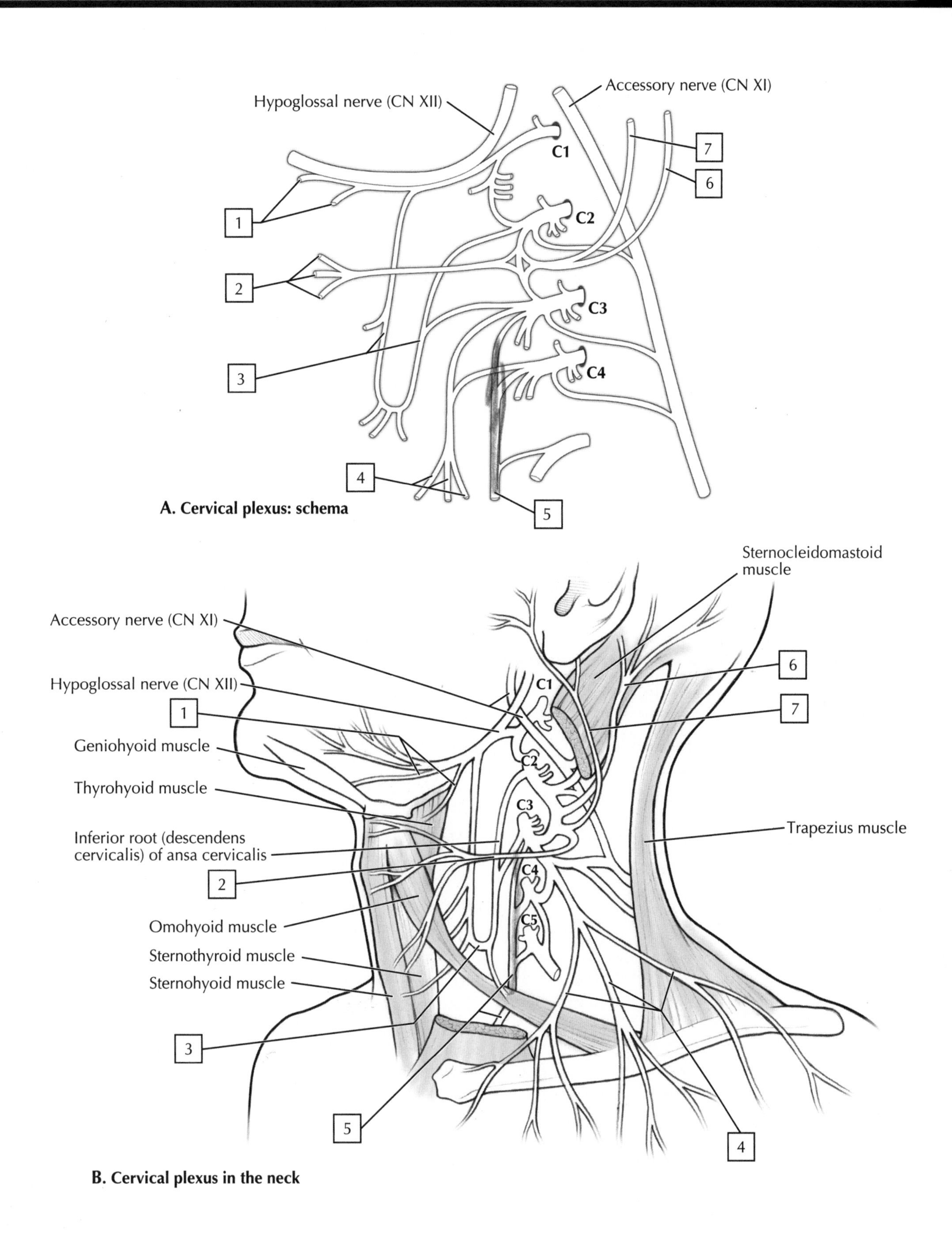


**A. Cervical plexus: schema**

**B. Cervical plexus in the neck**

# 4 Brachial Plexus

The brachial plexus is formed by the anterior rami of spinal nerves C5-T1. This plexus consists of the following descriptive components:

- **Roots**: five anterior rami of C5-T1 form the "roots" of the plexus
- **Trunks**: the five roots sort into three trunks, called superior, middle, and inferior, all lying beneath the clavicle and above the first rib
- **Divisions**: each trunk divides into an anterior and a posterior division, forming six divisions
- **Cords**: all the posterior divisions combine to form the posterior cord; the lateral and medial cords are formed by combinations of the anterior divisions
- **Terminal branches**: the plexus gives rise to five large terminal branches that innervate the muscles of the shoulder, arm, forearm, and hand

The three cords of the plexus are named for their relationship to the axillary artery, because they wrap around this artery in the axilla (armpit) and its accompanying vein(s), with the entire neurovascular bundle being wrapped in a fascial sheath called the **axillary sheath**. A number of other, smaller nerves also arise from components of the brachial plexus to innervate some muscles of the back and lateral and anterior chest wall. The following table summarizes some of the more important nerves of the brachial plexus and the muscles they innervate (see individual muscle tables for more detail).

| ARISE FROM | NERVE | MUSCLES INNERVATED |
|---|---|---|
| Roots | Posterior scapular | Levator scapulae and rhomboids |
| | Long thoracic | Serratus anterior |
| Upper trunk | Suprascapular | Supraspinatus and infraspinatus |
| | Subclavius | Subclavius |
| Lateral cord | Lateral pectoral | Pectoralis major |
| | Musculocutaneous | Anterior compartment muscles of arm |
| Medial cord | Medial pectoral | Pectoralis minor and major |
| | Ulnar | Some medial forearm and most hand muscles |
| Medial and lateral cords | Median | Most forearm and some hand muscles |
| Posterior cord | Upper subscapular | Subscapularis |
| | Thoracodorsal | Latissimus dorsi |
| | Lower subscapular | Subscapularis and teres major |
| | Axillary | Deltoid and teres minor |
| | Radial | Posterior compartment muscles of the arm and forearm |

**COLOR** the five roots, three trunks, six divisions, three cords, and five terminal branches of the brachial plexus (see part ***A***), using a different color for each component (e.g., red for roots, blue for trunks). Also, **color** the five terminal branches of the cord as they pass into the upper limb (see part ***B***), using a different color for each nerve:

- ☐ **1. Axillary**
- ☐ **2. Musculocutaneous**
- ☐ **3. Median**
- ☐ **4. Radial**
- ☐ **5. Ulnar**

***Clinical Note:***

Various injuries to the upper limb can result in damage to one or more of the terminal branches of the brachial plexus. As you note each of these nerve lesions described below, return to Chapter 3 and review the muscles of the upper limb, whose actions will be affected by damage to the nerve(s) that innervate them.

***Musculocutaneous nerve***: because this nerve runs through the arm and is protected by overlying muscles, it is not frequently injured.

***Axillary nerve***: damage results in weakened ability to abduct the limb at the shoulder. A dislocated shoulder injury could stretch this nerve and damage its axons.

***Radial nerve***: because this nerve innervates all extensors, a proximal injury would result in a weakened ability to extend at the elbow, wrist, and fingers. A somewhat lower injury might only result in "**wrist drop**" (inability to extend the wrist and fingers).

***Median nerve***: damage results in weakness in flexing the wrist and weakened flexion of the thumb, index finger, and middle finger when one is asked to make a fist. Compression of the nerve at the wrist (**carpal tunnel syndrome**) would not affect wrist movement but would weaken the function of the thenar muscles of the hand.

***Ulnar nerve***: damage results in weakness in flexing the wrist and the little and ring fingers; when this damage is combined with hyperextended metacarpophalangeal (MP) joints of these same fingers, a "**claw hand**" results, which is indicative of an ulnar nerve injury. Atrophy of the hypothenar eminence may also occur. The ulnar nerve is the most frequently injured nerve of the upper limb.

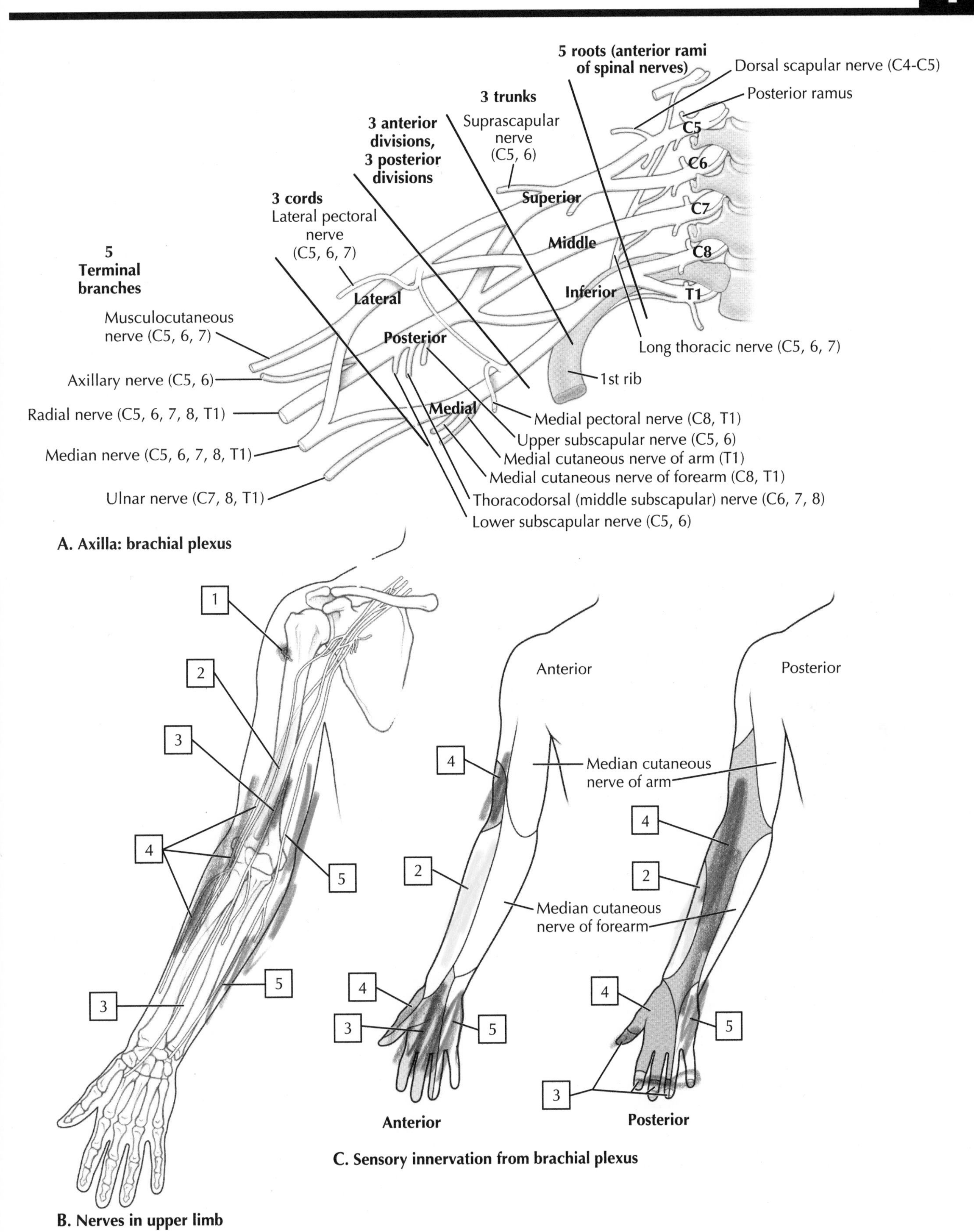


A. Axilla: brachial plexus

B. Nerves in upper limb

C. Sensory innervation from brachial plexus

# 4 Lumbar Plexus

The lumbar plexus is formed by the anterior rami of spinal nerves L1-L4. The major motor components of this plexus are included in the following nerves:

- **Femoral nerve**: from L2-L4, this nerve innervates the anterior thigh muscles (largely extensors of the knee)
- **Obturator nerve**: from L2-L4, this nerve innervates the medial thigh muscles (largely adductors of the hip)
- **Genitofemoral nerve**: from L1-L2, motor fibers to the cremaster muscle (a covering of the spermatic cord) in males, and sensory fibers to the anteromedial thigh in both sexes

A large nerve trunk from the inferior portion of the lumbar plexus, called the **lumbosacral trunk**, will continue into the pelvis and join anterior rami of sacral nerves to form the **sacral plexus** (L4-S4). Nerves from these two plexuses innervate muscles of the pelvis and perineum, muscles of the gluteal (buttocks) region, muscles of the posterior thigh (hamstrings), and all of the muscles of the leg and foot.

The **sensory components** of the lumbar plexus innervate the inguinal region; groin; medial, anterior, and lateral aspects of the thigh; and anteromedial leg and ankle (see individual muscle tables in Chapter 3 for more detail).

**COLOR the following nerves of the lumbar plexus, using a different color for each nerve:**

- ☐ **1. Iliohypogastric: largely sensory fibers to the inguinal region but does provide some motor fibers to several abdominal wall muscles (internal oblique and transversus abdominis) (L1)**
- ☐ **2. Ilioinguinal: largely sensory fibers to the inguinal region and external genitalia but does supply some motor fibers to the same abdominal muscles listed above (L1)**
- ☐ **3. Genitofemoral: motor fibers to the cremaster muscle in males and sensory fibers to the anteromedial thigh in both sexes (L1-L2)**
- ☐ **4. Lateral cutaneous nerve of the thigh: largely sensory fibers to the lateral thigh (L2-L3)**
- ☐ **5. Femoral: motor fibers to the anterior compartment thigh muscles and sensory fibers over the anterior thigh, and the anteromedial leg and ankle (via the saphenous nerve); passes deep to the inguinal ligament (L2-L4)**
- ☐ **6. Obturator: motor fibers to the medial compartment thigh muscles and sensory fibers to the medial thigh; passes through the obturator foramen to enter the medial thigh (L2-L4)**

***Clinical Note:***

Various injuries to the lower limb can result in damage to one or more of the large nerves innervating the muscles of the thigh. (The resulting conditions will make more sense if you also review the muscle compartments of the lower limb.) Some examples include:

***Femoral nerve***: damage results in a weakened ability to extend the knee. A patient may have to push on the anterior thigh when placing the affected limb on the ground during walking to "force" the knee into an extended position.

***Obturator nerve***: damage results in a weakened ability to adduct the hip. The obturator nerve lies beneath several layers of muscle and is well protected in the thigh, unless it is cut by a deep laceration. Most injuries of the nerve occur as it passes through the pelvis (e.g., pelvic trauma from automobile accidents).

A **regional nerve block** using an anesthetic agent may be used when clinically indicated. The femoral nerve (L2-L4), for example, may be blocked (motor and sensory) about 2 cm inferior to the inguinal ligament and a finger's breadth lateral to the femoral artery (which must be avoided!).

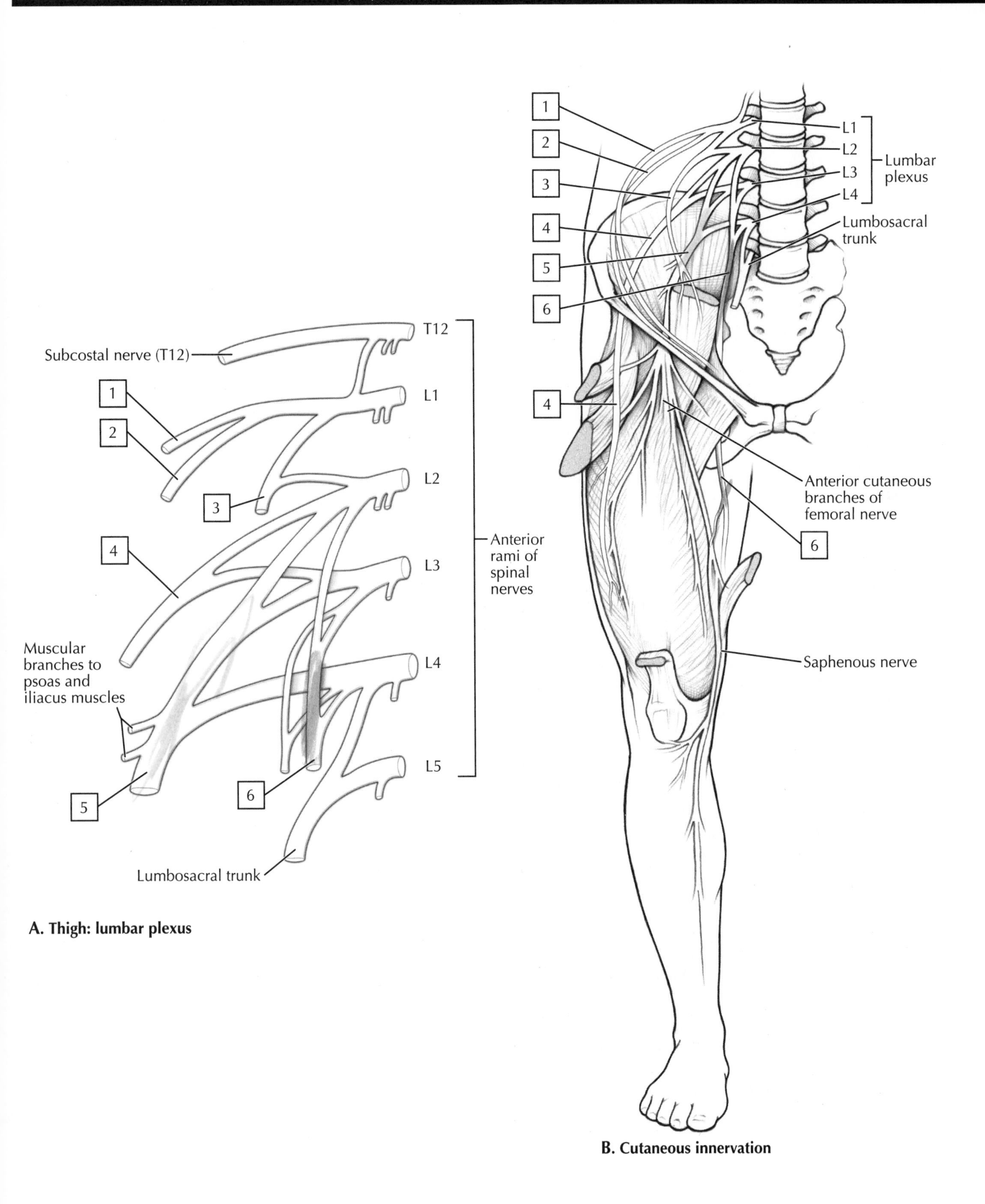


A. Thigh: lumbar plexus

B. Cutaneous innervation

# 4 Sacral Plexus

The sacral plexus is formed by the anterior rami of spinal nerves L4-S4. The major motor components of the sacral plexus are summarized in the table. In general, the sacral plexus innervates:

- Muscles forming the walls and floor of the pelvic cavity
- Muscles of the gluteal (buttocks) region
- Muscles of the perineum
- Muscles of the posterior thigh (hamstrings)
- All the muscles of the leg and foot

The largest nerve in the body, the **sciatic nerve**, arises from the sacral (sometimes referred to as the lumbosacral) plexus, with nerve fiber contributions from L4-S3. The lumbar contribution comes from the union of the lumbosacral trunk (L4-L5), which joins the first four sacral nerves to form the plexus. The sciatic nerve is really two nerves combined to form one:

- **Tibial nerve**: innervates the three hamstring muscles of the posterior thigh, the muscles of the posterior compartment of the leg, and all of the muscles of the foot (via plantar branches)
- **Common fibular nerve**: innervates the short head of the biceps femoris muscle in the posterior thigh and the muscles of the lateral and anterior compartments of the leg (see individual muscle tables in Chapter 3 for more detail)

Sensory fibers of the sacral plexus innervate skin and muscles of the perineum, gluteal region, posterior thigh, and posterolateral leg and ankle, and all of the muscles of the foot.

| DIVISION AND NERVE | INNERVATION |
|---|---|
| | **Anterior** |
| Pudendal (S2-S4) | Supplies motor and sensory innervation to perineum |
| Tibial (L4-S3) | Innervates muscles of the posterior thigh, posterior leg, and foot; with common fibular nerve, it forms sciatic nerve (largest nerve in body) |
| | **Posterior** |
| Superior gluteal (L4-S1) | Innervates several gluteal muscles and the tensor fasciae latae |
| Inferior gluteal (L5-S2) | Innervates gluteus maximus muscle |
| Common fibular (L4-S2) | Portion of sciatic nerve (with tibial nerve) that innervates lateral and anterior muscle compartments of leg |

**COLOR** the following nerves of the sacral plexus, using a different color for each nerve:

☐ 1. **Superior gluteal: motor and sensory fibers to two of the three gluteal muscles and tensor fasciae latae muscle**

☐ 2. **Inferior gluteal: motor and sensory fibers to the gluteus maximus muscle**

☐ 3. **Sciatic (combination of the tibial and common fibular nerves): motor fibers to the posterior thigh muscles and all of the muscles below the knee; sensory fibers to the posterior thigh muscles, posterolateral leg and ankle muscles, and all of the muscles of the foot**

☐ 4. **Pudendal (means "shameful"): motor and sensory fibers to the perineum and external genitalia**

***Clinical Note:***

Athletically active individuals may report pain (sciatica) related to the lumbar spine (**herniated disc** impinging on L4, L5, or S1 nerve roots), buttocks (**bursitis** or **hamstring muscle pulls**), or pelvic region (intrapelvic disorders). **Sciatica** is the pain associated with the large sciatic nerve and is often felt in the buttocks and/or radiating down the posterior thigh and into the posterolateral leg. As noted above, it can be due to multiple problems (disc herniation, direct trauma, inflammation, compression).

A **stab wound** to the sciatic nerve may result in weakness or paralysis of the muscles that are important in thigh extension and leg flexion.

**Intragluteal (intramuscular) injections** are given in the gluteus maximus muscle, but only in the superolateral quadrant of the buttocks, to avoid injury to the large gluteal nerve.

The common fibular nerve is the nerve most often injured in the lower limb. It is most vulnerable to trauma where it passes laterally around the fibular head. Weakness of the muscles of the anterior and lateral compartments of the leg leads to "**footdrop**" (an inability to adequately dorsiflex the foot) and weakened eversion of the foot.

**Tibial fractures** (the most common fracture of a long bone) may be severe enough to damage the tibial and/or the fibular nerves, leading to a compromised innervation of multiple leg muscles.

Injury to the common fibular or deep fibular nerve may lead to **footdrop**, resulting in a "steppage" gait where, at the end of the swing phase of walking, the foot slaps down to the ground.

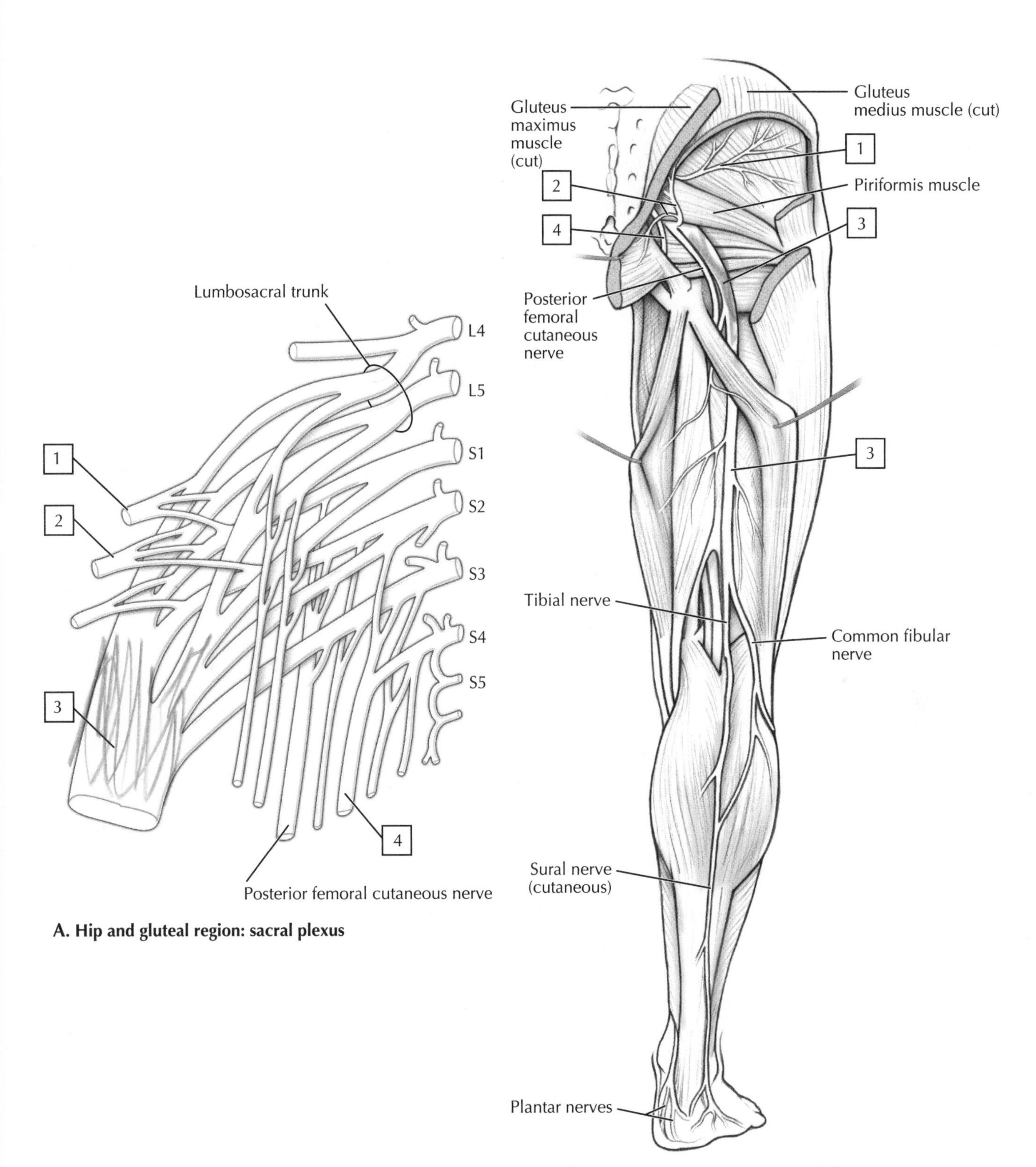


**A. Hip and gluteal region: sacral plexus**

**B. Distribution of sacral plexus in lower limb**

## REVIEW QUESTIONS

For each description below (1-3), color the relevant structure in the image.

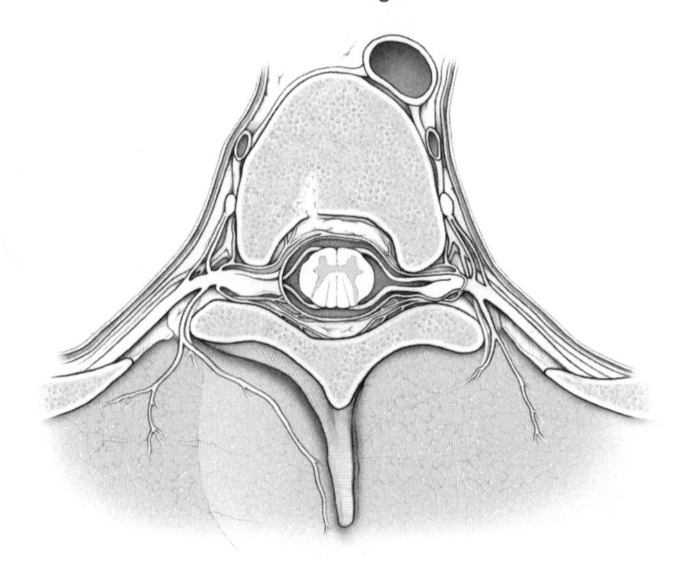

1. The nerve cell bodies (soma) of afferent (sensory) nerves are found in this structure (color it red).
2. This structure innervates the intrinsic back muscles (color it blue).
3. This structure contains somatic efferent and preganglionic sympathetic fibers (color it green).
4. In multiple sclerosis, the CNS myelin is progressively destroyed. Which of the following cells myelinates CNS axons?
   A. Astrocytes
   B. Microglia
   C. Oligodendrocytes
   D. Schwann cells
   E. Tanycytes
5. The primary motor cortex is located in which of the following structures?
   A. Cingulate gyrus
   B. Corpus callosum
   C. Insula
   D. Precentral gyrus
   E. Thalamus
6. Bradykinesia (slow movements) and resting tremor suggest that a patient has Parkinson's disease and a reduction of dopamine release from the substantia nigra. Which of the following brain regions is the target area for these dopamine-secreting neurons?
   A. Amygdaloid body
   B. Cingulate gyrus
   C. Hippocampus
   D. Striatum
   E. Thalamus
7. A patient presents with a fractured humerus and wrist drop. Which of the following nerves is most likely injured?
   A. Axillary
   B. Median
   C. Musculocutaneous
   D. Radial
   E. Ulnar

8. Which is the largest nerve in the human body (hint: innervates most of the muscles of the lower limb)? ______________________

9. Which cranial nerve has three large divisions? ______________________

10. Which cranial nerve innervates the submandibular salivary glands? ______________________

## ANSWER KEY

1. Spinal ganglion
2. Posterior ramus (of a spinal nerve)
3. Anterior root

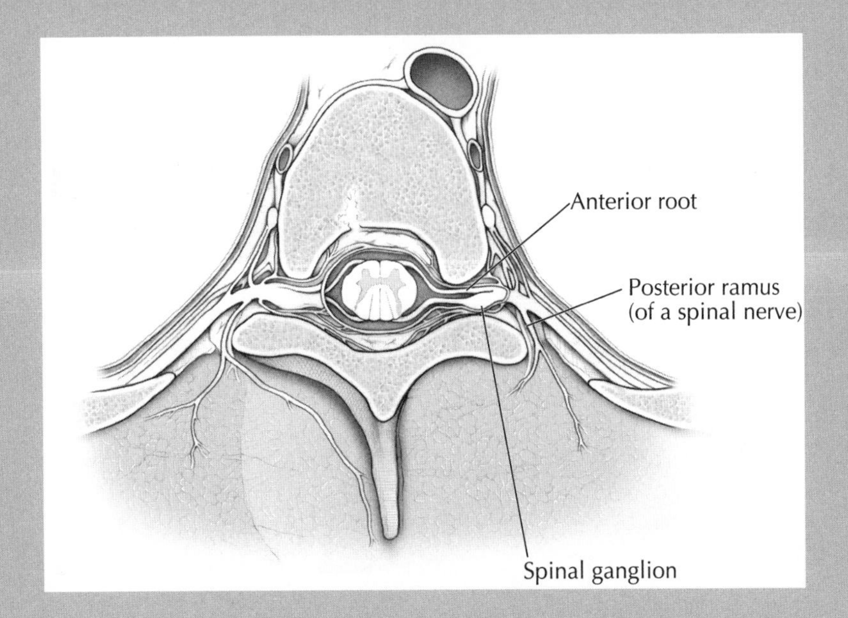


4. C
5. D
6. D
7. D
8. Sciatic nerve (combined tibial and common fibular nerves)
9. Trigeminal nerve (CN V)
10. Facial nerve (CN VII) via its parasympathetic fibers

# Chapter 5 Cardiovascular System

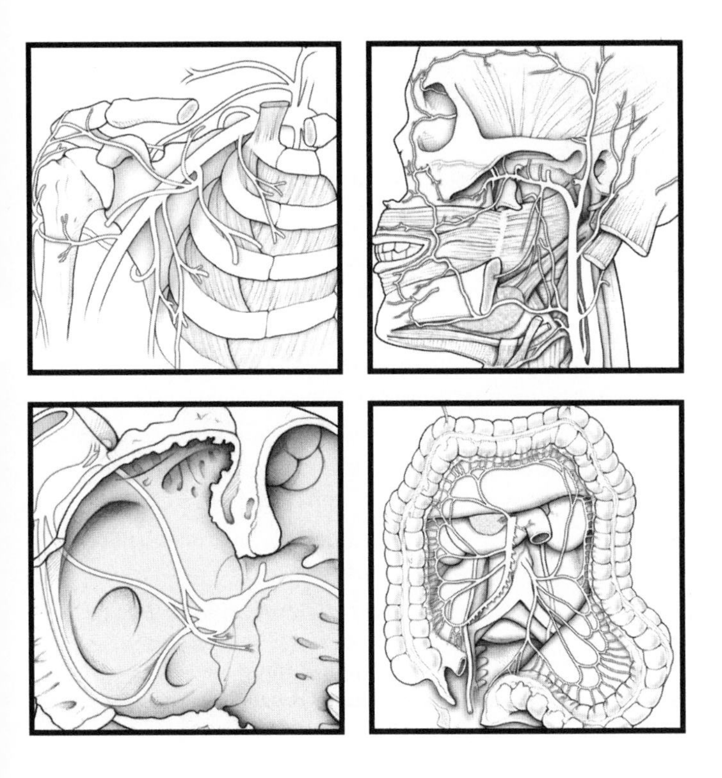

# 5 Composition of Blood

The blood consists of the following formed elements:
- Platelets, cell fragments that form clots to stop bleeding
- White blood cells (WBCs), also called leukocytes
- Red blood cells (RBCs), also called erythrocytes
- Plasma, a nonliving fluid matrix; about 90% water and over 100 different solutes, including albumin, globulins, clotting proteins, and metabolic enzymes, antibacterial proteins, hormones, gases, wastes, ions, and products of cell activity

Blood is a **fluid connective tissue** that circulates through the arteries to reach the body's tissues and returns to the heart through the veins. The functions of blood include:
- Transport of dissolved gases, nutrients, metabolic waste products, and hormones to and from tissues
- Prevention of fluid loss via clotting mechanisms
- Immune defense activities
- Regulation of pH and electrolyte balance
- Thermoregulation via blood vessel constriction and dilation

When blood is "spun down" in a centrifuge, the RBCs precipitate to the bottom of the tube; they make up about 45% of the blood volume. The next layer is a "buffy coat," which makes up slightly less than 1% of the blood volume and includes the WBCs (leukocytes) and platelets. The remaining 55% of the blood volume is the **plasma** (**serum** is plasma with the clotting factors removed), which includes:
- Water
- Plasma proteins
- Other solutes (electrolytes, organic nutrients, organic wastes, and others as listed above in the first paragraph)

The volume of the packed RBCs represents the **hematocrit**, which normally ranges from about 40% to 50% in males and 35% to 45% in females. The "buffy coat" includes platelets and the WBCs. The WBCs include the following types of leukocytes (see Plate 6-2):
- **Neutrophils**: the most numerous of the granular WBCs, accounting for about 50% to 70% of the WBC population (and all WBCs, granular and agranular), they possess a multilobed nucleus, function as phagocytes at sites of inflammation, and live 8 to 12 hours in the blood and about 1 to 2 days in the extravascular compartment
- **Eosinophils**: account for about 2% to 4% of all WBCs and are granular WBCs that respond to allergic reactions, participate in immune responses, phagocytose antigen-antibody complexes, and live about 8 to 12 hours in the blood and for an unknown period of time in the connective tissues (an average of about 8 to 12 days)
- **Lymphocytes**: the most common type of agranular WBCs, and account for about 20% to 30% of white cells; there are three types of lymphocytes: (1) B cells, which are derived from the bone marrow and produce circulating antibodies; (2) T cells, which are derived from the bone marrow but complete their differentiation in the thymus; they are either cytotoxic, helper, or suppressor cell-mediated immune cells; and (3) natural killer (NK) cells, which kill virus-infected cells
- **Basophils**: least numerous WBCs (about 0.5% to 1%), they are granular, function in immune, allergic, and inflammatory reactions, release vasoactive substances that can lead to hypersensitivity or allergic reactions, and live in the blood for about 8 hours and for an unknown period of time in connective tissues
- **Monocytes**: the largest of the WBCs and account for about 2% to 8% of white cells; they are agranular, travel from the bone marrow into the connective tissue, where they differentiate into macrophages, and live as monocytes in the blood for about 16 hours and for an unknown period of time in connective tissues as macrophages

**COLOR** the following blood cells, using the colors suggested:

- ☐ 1. **Red blood cells: do not possess a nucleus as mature cells (red)**
- ☐ 2. **Platelets (yellow)**
- ☐ 3. **Neutrophil (color the multilobed nucleus purple or dark blue and the cytoplasm light blue)**
- ☐ 4. **Monocyte (color the crescent-shaped nucleus purple or dark blue and the cytoplasm light blue)**
- ☐ 5. **Eosinophil (color the nucleus dark blue or purple, the small cytoplasmic granules red, and the surrounding cytoplasm light blue)**
- ☐ 6. **Lymphocyte (color the nucleus blue or purple and the cytoplasm light blue)**
- ☐ 7. **Basophil (color the nucleus dark blue or purple, the cytoplasmic granules dark blue, and the surrounding cytoplasm light blue)**

***Clinical Note:***

**Blood doping** is practiced by some athletes who participate in aerobic athletic events to artificially induce polycythemia, an increase in RBCs and their production (via erythropoiesis). Generally, this is considered unethical and can lead to problems by dramatically increasing the viscosity of the blood and blood pressure.

Overproduction of abnormal leukocytes occurs in **leukemia** and **infectious mononucleosis. Leukemia** refers to a group of cancerous conditions of the white blood cells, and **mononucleosis** is a highly contagious viral disease. It is caused by the Epstein-Barr virus and results in an excessive number of atypical agranulocytes, leading to a fever, sore throat, and an achy, tired feeling.

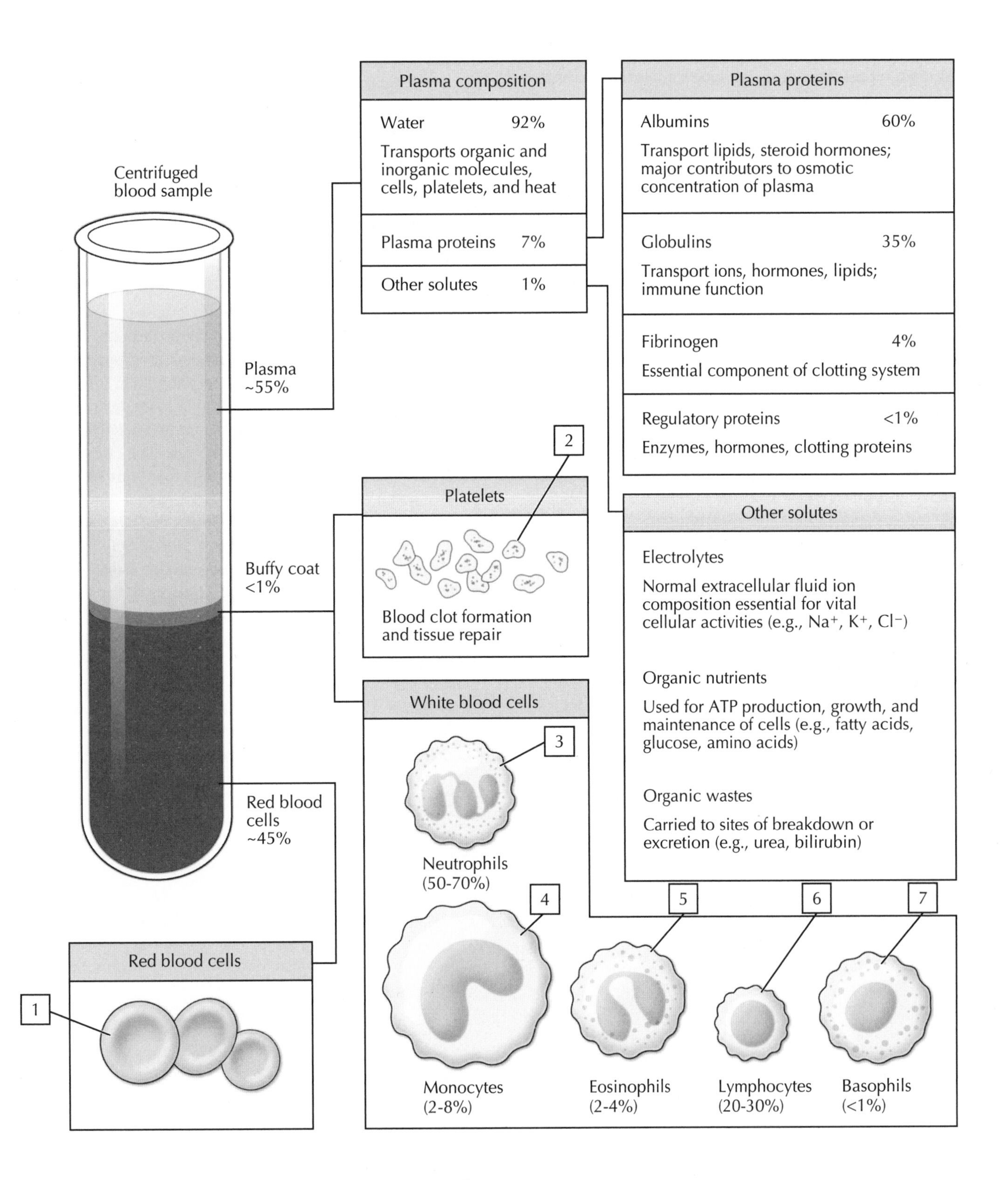

Centrifuged blood sample
Plasma ~55%
Buffy coat <1%
Red blood cells ~45%
Plasma composition
Water 92%
Transports organic and inorganic molecules, cells, platelets, and heat
Plasma proteins 7%
Other solutes 1%
Plasma proteins
Albumins 60%
Transport lipids, steroid hormones; major contributors to osmotic concentration of plasma
Globulins 35%
Transport ions, hormones, lipids; immune function
Fibrinogen 4%
Essential component of clotting system
Regulatory proteins <1%
Enzymes, hormones, clotting proteins
Platelets
Blood clot formation and tissue repair
Other solutes
Electrolytes
Normal extracellular fluid ion composition essential for vital cellular activities (e.g., $Na^+$, $K^+$, $Cl^-$)
Organic nutrients
Used for ATP production, growth, and maintenance of cells (e.g., fatty acids, glucose, amino acids)
Organic wastes
Carried to sites of breakdown or excretion (e.g., urea, bilirubin)
White blood cells
Neutrophils (50-70%)
Monocytes (2-8%)
Eosinophils (2-4%)
Lymphocytes (20-30%)
Basophils (<1%)
Red blood cells
1
2
3
4
5
6
7

The cardiovascular system consists of the following components:

- **Heart**: pumps the blood throughout the circulation
- **Pulmonary circulation**: a closed loop circulation between the heart and lungs for gas exchange
- **Systemic circulation**: a closed-loop circulation between the heart and all of the tissues of the body

The circulatory system's vessels include the following:

- **Arteries**: any vessel that carries blood away from the heart (including vessels carrying blood low in oxygen and high in carbon dioxide from the right ventricle of the heart to the lungs via the pulmonary arteries)
- **Veins**: any vessel that returns blood to the heart (including oxygenated blood returning from the lungs to the left atrium of the heart via pulmonary veins)

At rest, the cardiac output is about 5 L/min in both the pulmonary and systemic circulations. The amount of blood flow per minute ($\dot{Q}$), as a percentage of cardiac output, and relative to the percentage of oxygen used per minute ($\dot{V}O_2$) in various organ systems, is shown for the resting state in the illustration. Note that the brain uses over 20% of the available oxygen. At any point, most of the blood (64%) resides in the veins (a low-pressure system) and is returned to the right side of the heart. The arterial side of the systemic circulation (a high-pressure system) possesses significant amounts of smooth muscle in the vessel walls, and the small arteries and arterioles are largely responsible for most of the vascular resistance in the circulatory system.

## COLOR

- ☐ **1. The arterial side (right shaded side) of the central schematic figure red**
- ☐ **2. The venous side (left side) blue. Note that the vessels passing from the right ventricle (RV) to the lungs are the pulmonary arteries (even though the blood is less saturated with oxygen) and that the vessels from the lungs to the left atrium (LA) are called pulmonary veins (fully saturated with oxygen).**

***Clinical Note:***

**Hypertension** (high blood pressure) is a major risk factor for atherogenesis, atherosclerotic cardiovascular disease, stroke, coronary artery disease, and renal failure. Hypertension can result from an unknown cause (idiopathic, or essential hypertension) or secondary causes (e.g., medications, hormone imbalance, tumors). Hypertension is defined as two or more blood pressure readings of systolic pressure higher than 140 mm Hg or a diastolic pressure higher than 90 mm Hg. One reading of over 210 mm Hg systolic or over 120 mm Hg diastolic also indicates hypertension.

**Blood group designations** include groups based on the presence or absence of two **agglutinogens**, Type A and Type B. A person inherits their ABO blood group, which will be one of the following: Type A, Type B, Type AB, or Type O. In the United States, the type O blood group has neither agglutinogen and is the most common ABO group in white, black, Asian, and Native Americans. Type A is the next most common blood type, type B is next, and type AB is the least common. Type O groups can be "universal donors" and type AB individuals can be "universal recipients," but other factors, such as the Rh factor (Rh agglutinogens), must be considered.

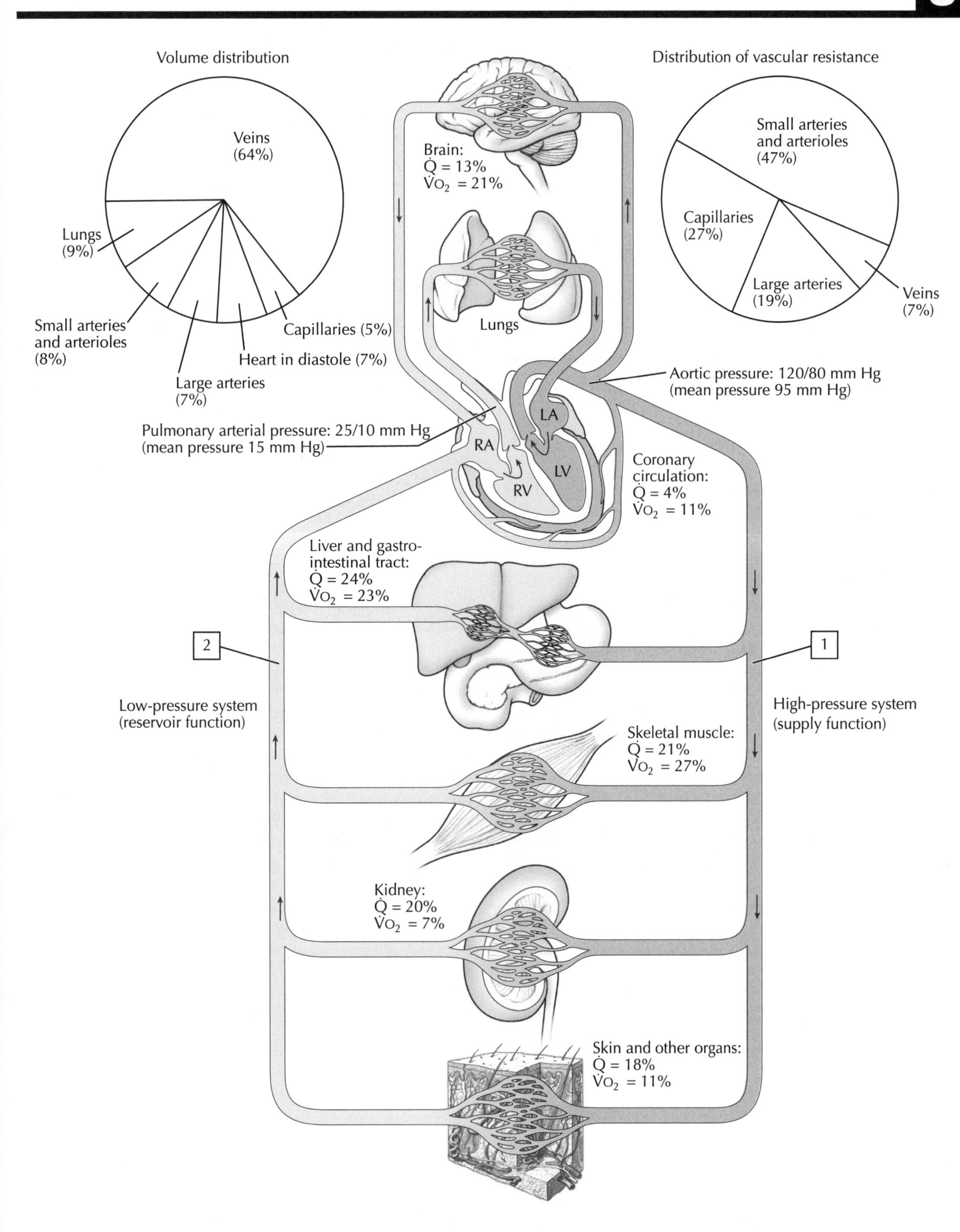
Volume distribution
Veins (64%)
Lungs (9%)
Small arteries and arterioles (8%)
Large arteries (7%)
Heart in diastole (7%)
Capillaries (5%)
Distribution of vascular resistance
Small arteries and arterioles (47%)
Capillaries (27%)
Large arteries (19%)
Veins (7%)
Brain: Q̇ = 13% V̇O2 = 21%
Lungs
Aortic pressure: 120/80 mm Hg (mean pressure 95 mm Hg)
Pulmonary arterial pressure: 25/10 mm Hg (mean pressure 15 mm Hg)
RA
LA
LV
RV
Coronary circulation: Q̇ = 4% V̇O2 = 11%
Liver and gastro-intestinal tract: Q̇ = 24% V̇O2 = 23%
2
1
Low-pressure system (reservoir function)
High-pressure system (supply function)
Skeletal muscle: Q̇ = 21% V̇O2 = 27%
Kidney: Q̇ = 20% V̇O2 = 7%
Skin and other organs: Q̇ = 18% V̇O2 = 11%

# 5 Heart I

The thoracic cavity is divided into left and right pleural sacs, which contain the lungs, and a "middle space" called the **mediastinum**. The mediastinum is further subdivided into the following regions:

- **Superior**: lies deep to the manubrium of the sternum and contains the great vessels (superior vena cava and aorta)
- **Inferior**: has three subdivisions of its own:
  - **Anterior**: lies deep (posterior) to the body of the sternum and contains some fat and connective tissue
  - **Middle**: lies deep to the anterior mediastinum and contains the heart encased in its pericardial sac
  - **Posterior**: lies deep to the heart and contains the descending thoracic aorta, thoracic lymphatic duct, and esophagus

**COLOR** the following subdivisions of the mediastinum, using a different color for each subdivision:

☐ **1. Middle mediastinum**
☐ **2. Anterior mediastinum**
☐ **3. Superior mediastinum**
☐ **4. Posterior mediastinum**

The heart lies in the middle mediastinum and is encased within a tough fibrous sac called the **pericardium**. The pericardium has a tough outer layer called the **fibrous pericardium**, which reflects onto the great vessels in the superior mediastinum. A parietal layer of the serous pericardium lines the inner aspect of the fibrous pericardium and then reflects onto the heart itself as the visceral serous pericardium (**epicardium**). The serous layers secrete a thin film of serous fluid that lubricates the walls of the pericardium and reduces the friction created by the beating of the heart. The features of the pericardium are summarized in the table below.

| FEATURE | DEFINITION |
| --- | --- |
| Fibrous pericardium | Tough outer layer that reflects onto great vessels |
| Serous pericardium | Layer that lines inner aspect of fibrous pericardium (parietal layer); reflects onto heart as epicardium (visceral layer) |
| Innervation | Phrenic nerve (C3-C5) for conveying pain; vasomotor innervation via sympathetic nerves |
| Transverse sinus | Space posterior to aorta and pulmonary trunk; can clamp vessels with fingers in this sinus (important for some surgical procedures) |
| Oblique sinus | Pericardial space posterior to heart |

**COLOR** the components of the pericardium, using a different color for each component:

☐ **5. Fibrous pericardium**
☐ **6. Parietal layer of the serous pericardium**
☐ **7. Visceral layer of the serous pericardium (epicardium)**

Note that when viewed in situ, the heart cannot be seen because it is encased within the pericardial sac. The great vessels in the superior mediastinum are visible superior to the pericardium, and the fatty thymus gland can be seen overlying the upper portion of the pericardium. The base of the pericardium and heart lies upon the abdominal diaphragm, with the lungs in their pleural sacs bordering the pericardium on each side (note that in part ***C*** the anterior portion of the pleural sacs has been opened to reveal the lungs).

**COLOR** the following features of the pericardium in situ, using the colors suggested:

☐ **8. Arch of the aorta (red)**
☐ **9. Thymus gland (yellow)**
☐ **10. Superior vena cava (blue)**
☐ **11. Pericardium (gray or tan)**

***Clinical Note:***
Diseases of the pericardium involve inflammatory conditions (**pericarditis**) and effusions (fluid accumulation in the pericardial sac). Additionally, bleeding into the pericardial cavity can cause **cardiac tamponade** (compression of the heart from bleeding into this enclosed space). The bleeding may be from a ruptured aortic aneurysm, ruptured myocardial infarct, or a penetrating injury (stab wound). The collection of blood in the pericardial cavity is called **hemopericardium** and it compromises the beating of the heart, decreases venous return to the heart, and affects cardiac output. The accumulating blood needs to be drawn out of the pericardial cavity using a needle and syringe, and the appropriate repair initiated, because this is often a life-threatening condition.

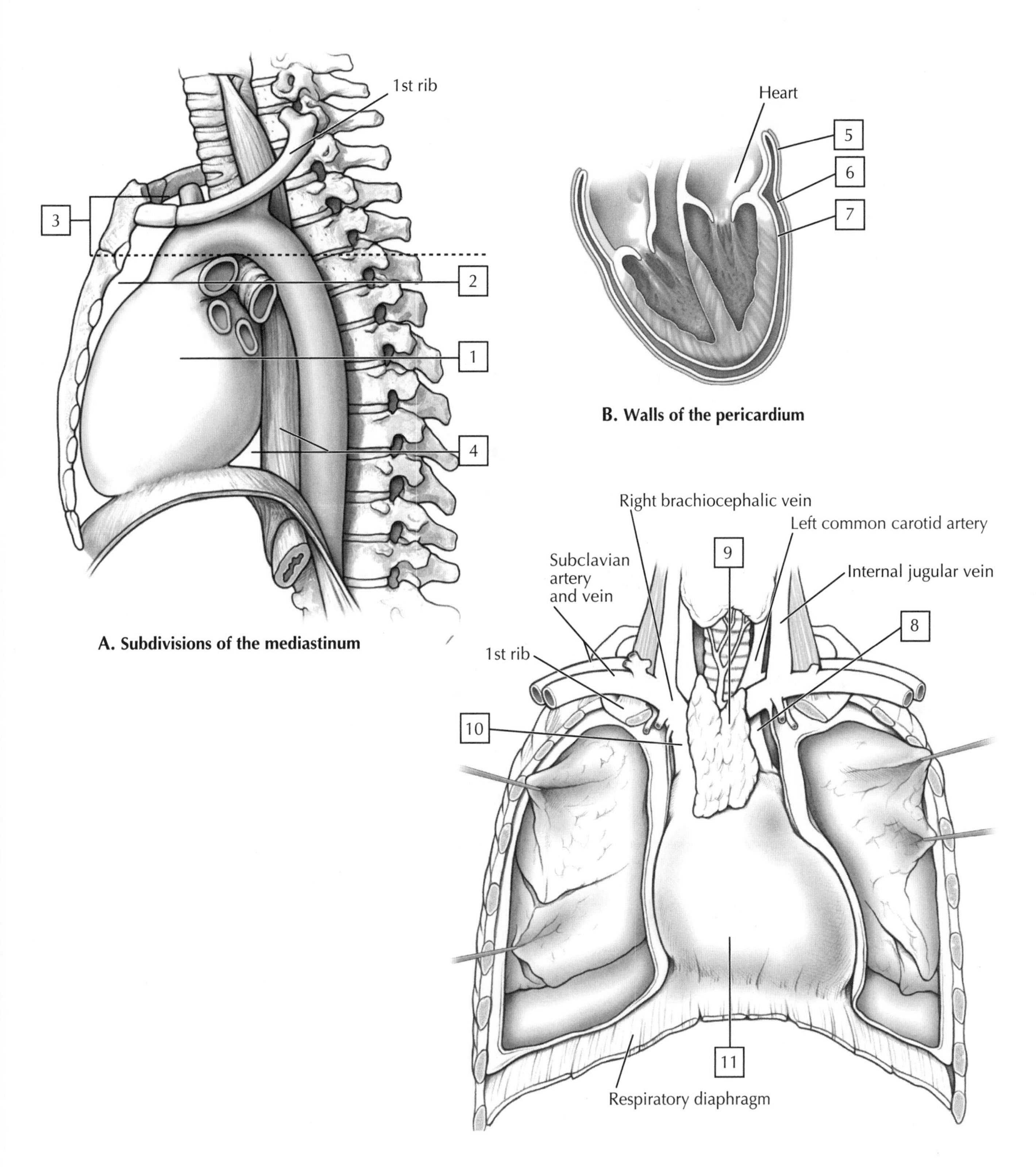


A. Subdivisions of the mediastinum

B. Walls of the pericardium

C. Pericardium and heart: heart in situ

# 5 Heart II

The human heart has four chambers: two atria and two ventricles. Blood returning from the **systemic circulation** (via the superior and inferior vena cava) enters the right atrium and right ventricle and is pumped into the **pulmonary circulation** (via pulmonary arteries) to the lungs for gas exchange. Blood returning from the lungs via the pulmonary circulation (pulmonary veins) enters the left atrium and left ventricle and then is pumped into the systemic circulation (via the aorta).

The atria and ventricles are separated by atrioventricular valves (**tricuspid valve** on the right side and **mitral valve** on the left side), which prevent blood from refluxing into the atria when the ventricles contract. Likewise, the two major outflow vessels, the **pulmonary trunk** from the right ventricle and the **ascending aorta** from the left ventricle, also possess valves called **semilunar valves** (pulmonary and aortic valves). Each semilunar valve has three leaflets that look like a crescent moon; hence, their designation as "semilunar valves." Details of the features of each heart chamber are summarized in the table below.

| FEATURE | DEFINITION |
|---|---|
| **Right Atrium** | |
| Auricle | Pouchlike appendage of atrium; embryonic heart tube derivative |
| Pectinate muscles | Ridges of myocardium inside auricle |
| Crista terminalis | Ridge that runs from the inferior vena cava (IVC) to superior vena cava (SVC) openings; its superior extent marks the site of the SA node |
| Fossa ovalis | Depression in interatrial septum; former site of foramen ovale |
| Atrial openings | One each for SVC, IVC, and coronary sinus (venous return from cardiac veins); the drainage derived from the coronary arteries that supply the cardiac muscle (see Plate 5-6) |
| **Right Ventricle** | |
| Trabeculae carneae | Irregular ridges of ventricular myocardium |
| Papillary muscles | Superoposterior (anterior), inferior, and septal projections of myocardium extending into ventricular cavity; prevent valve leaflet prolapse |
| Chordae tendineae | Fibrous cords that connect papillary muscles to valve leaflets |
| Moderator band (septomarginal trabecula) | Muscular band that conveys AV bundle from septum to base of ventricle at site of superoposterior (anterior) papillary muscle |
| Ventricular openings | One to pulmonary trunk through pulmonary valve; one to receive blood from right atrium through tricuspid valve |
| **Left Atrium** | |
| Auricle | Small appendage representing primitive embryonic atrium whose wall has pectinate muscle |
| Atrial wall | Wall slightly thicker than thin-walled right atrium |
| Atrial openings | Usually four openings for four pulmonary valve |
| **Left Ventricle** | |
| Papillary muscles | Superoposterior (anterior) and inferior (posterior) muscles, larger than those of right ventricle |
| Chordae tendineae | Fibrous cords that connect papillary muscles to valve leaflets |
| Ventricular wall | Wall much thicker than that of right ventricle |
| Membranous septum | Very thin superior portion of interventricular septum and site of most ventricular septal defects (VSDs); this defect allows blood from the higher pressure left ventricle to pass through the defect into the lower pressure right ventricle |
| Ventricular openings | One to aorta through aortic valve; one to receive blood from left atrium through mitral valve |

**COLOR** the following features of the heart chambers, using a different color for each feature, except where a color is suggested:

- ☐ 1. **Outflow to the pulmonary trunk (blue)**
- ☐ 2. **Left atrium**
- ☐ 3. **Pulmonary veins (usually two from each side) (light red)**
- ☐ 4. **Mitral leaflet (posterior cusp)**
- ☐ 5. **Ascending aorta and aortic arch (red)**
- ☐ 6. **Superior vena cava (blue)**
- ☐ 7. **Aortic valve**
- ☐ 8. **Right atrium**
- ☐ 9. **Right atrioventricular (tricuspid) valve**
- ☐ 10. **Right ventricle**
- ☐ 11. **Inferior vena cava (blue)**
- ☐ 12. **Papillary muscles**
- ☐ 13. **Left ventricle**
- ☐ 14. **Pulmonic valve**

***Clinical Note:***
Typically, the **heart sounds** are described as "lub-dub," signifying the sounds made by the closing of the atrioventricular valves followed rapidly by the closing of the semilunar valves. Two additional sounds occur with the filling of the ventricles but are more difficult to discern. Using a stethoscope, one can listen to the four valves to determine if they are functioning properly. To do so, it is best to place the stethoscope over the chest wall and heart at a point where the blood has passed through the valve, to the heart chamber or vessel downstream, because the sound is better carried in the fluid medium. The gray dots in part ***C*** show the approximate placement of a stethoscope to auscultate each valve.

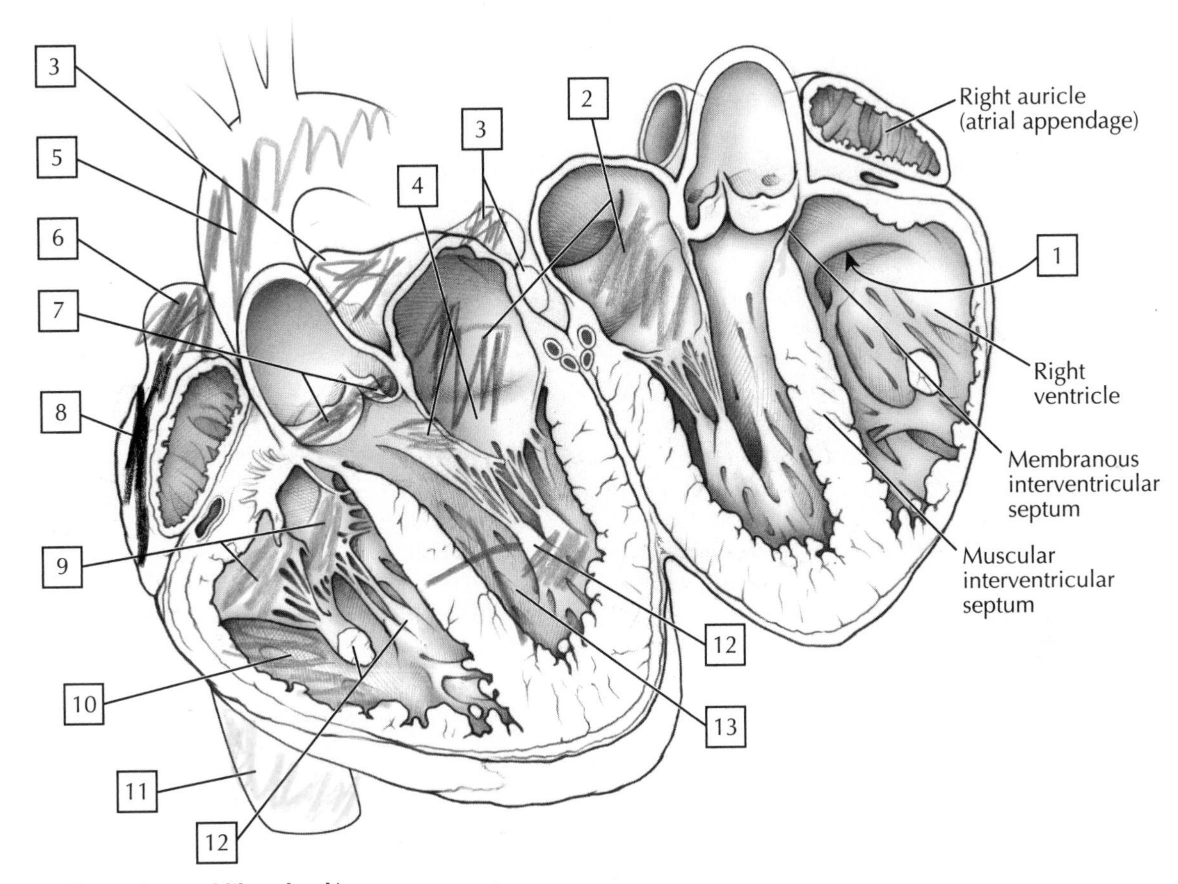


A. Sectioned heart (opened like a book)

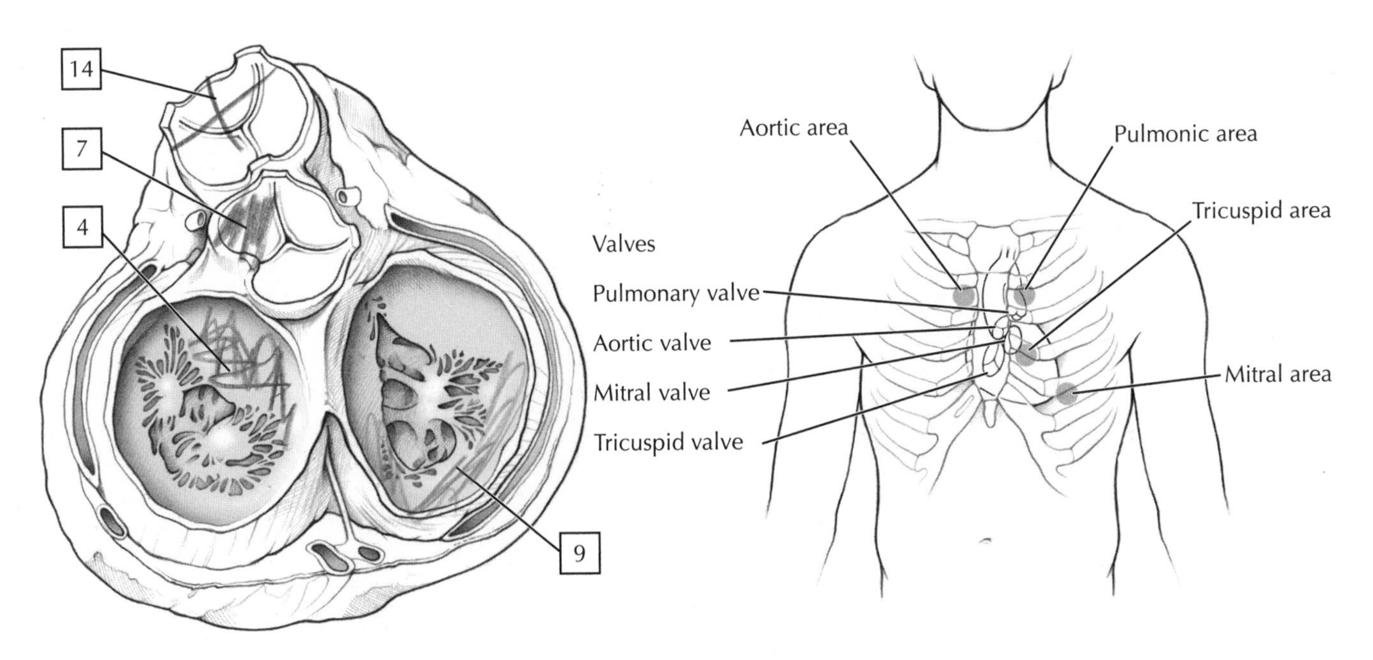


B. Heart in diastole: viewed from base with atria removed

C. Precordial areas of auscultation

# 5 Heart III

The pericardium is innervated by somatic pain fibers that course in the phrenic nerves (C3-C5), whereas the heart itself is innervated by the autonomic nervous system. The chief components of this innervation pattern include:

- **Parasympathetics**: derived from the vagus nerve (CN X), which courses to the cardiac plexus; parasympathetic stimulation slows the heart rate, decreases the force of contraction, and vasodilates coronary resistance vessels (although most vagal effects are restricted directly to the sinoatrial [SA] nodal region [pacemaker]
- **Sympathetics**: derived from upper thoracic cardiac nerves originating in the T1-T4/T5 intermediolateral cell column, which then enter the sympathetic trunk. These fibers synapse in upper cervical and thoracic ganglia and send postganglionic sympathetic fibers to the cardiac plexus. Sympathetic stimulation increases the heart rate, the force of contraction, and minimally vasoconstricts the coronary resistance vessels (however, this vasoconstriction is often masked by a powerful metabolic coronary vasodilation, which is important because coronary arteries must dilate to supply blood to the heart muscle as it increases its workload under sympathetic stimulation)
- **Afferents**: sensory nerve fibers course from the heart in the sympathetic nerves to spinal ganglia associated with the T1-T4 spinal cord levels; these fibers convey pain associated with myocardial ischemia

The heart maintains an intrinsic spontaneous rhythm of about 90-100 beats/min, but the normal parasympathetic tone overrides this intrinsic rate and maintains the resting heart rate at about 72 beats/min. The cardiac muscle of the heart exists in two forms:

- Contractile myocardium
- Specialized conducting myocardium

The specialized conducting myocardium does not contract but does spread the wave of depolarization rapidly throughout the chambers of the heart. Impulses are initiated in the sinuatrial (SA) node and are conveyed to the atrioventricular (AV) node. From here, the impulses pass through the common AV bundle (of His) and then spread through the ventricles via the right and left bundle branches and Purkinje fiber system. Components of this intrinsic conduction system are summarized in the table below.

| FEATURE | DEFINITION |
|---|---|
| SA node | Pacemaker of heart; site where action potential is initiated |
| AV node | Node that receives impulses from SA node and conveys them to the common AV bundle (of His) |
| Bundle branches | Right and left bundles that convey impulses down either side of interventricular septum to subendocardial Purkinje system |

The wave of depolarization, beginning at the SA node, and the repolarization of the myocardium generate the familiar electrocardiographic pattern (P, QRS, and T waves; see the ECG tracing in part ***A***) used clinically to assess the heart's conduction system.

**COLOR** the features of the heart's intrinsic conduction pathway and the elements (action potential wave forms) of the electrocardiogram (ECG) listed below, using the colors suggested:

- ☐ **1. SA node (blue)**
- ☐ **2. AV node (yellow)**
- ☐ **3. Common AV bundle (of His)**
- ☐ **4. Ventricular bundle branches (Purkinje system)**

***Clinical Note:***

**Atrial fibrillation** is the most common arrhythmia (although uncommon in children) and affects about 4% of people older than 60 years. **Ventricular tachycardia** is a dysrhythmia originating from a ventricular focus with a heart rate typically greater than 120 beats/min. It is usually associated with coronary artery disease, because myocardial ischemia often affects the ventricular endocardium where the Purkinje conduction system is localized.

**Valvular heart disease** most often affects the mitral and aortic valves (they are working against a greater pressure). Major problems can include stenosis (narrowing of the valve) or insufficiency (compromised valve function), often leading to regurgitation.

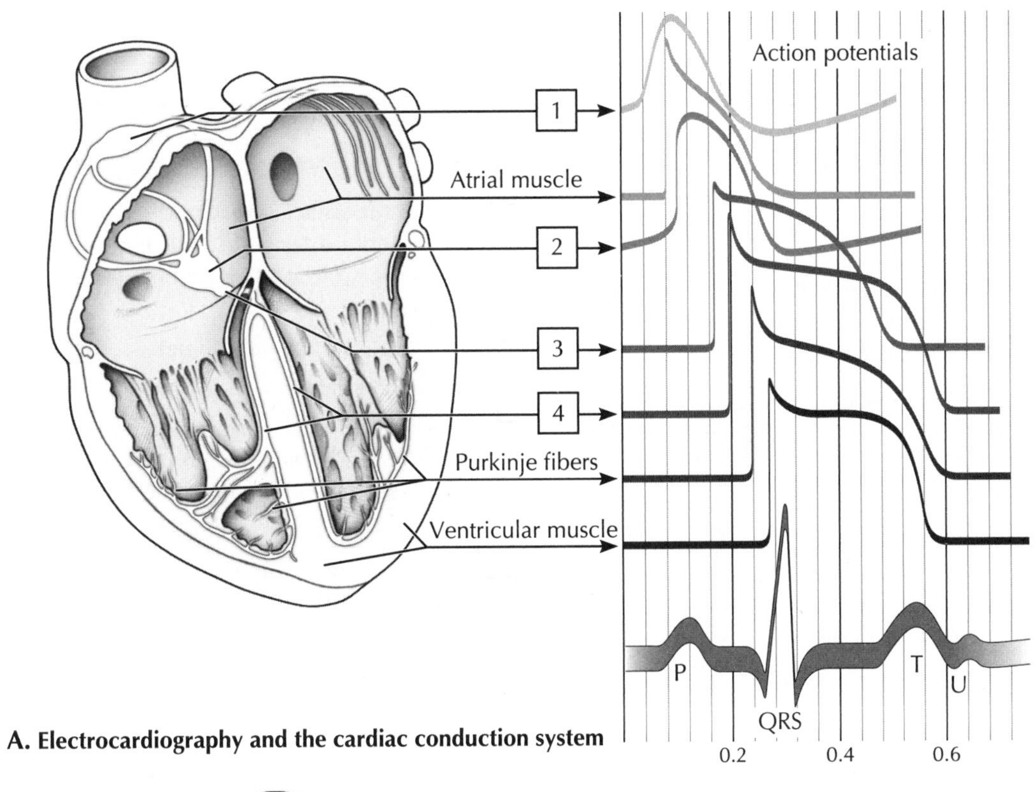


**A. Electrocardiography and the cardiac conduction system**

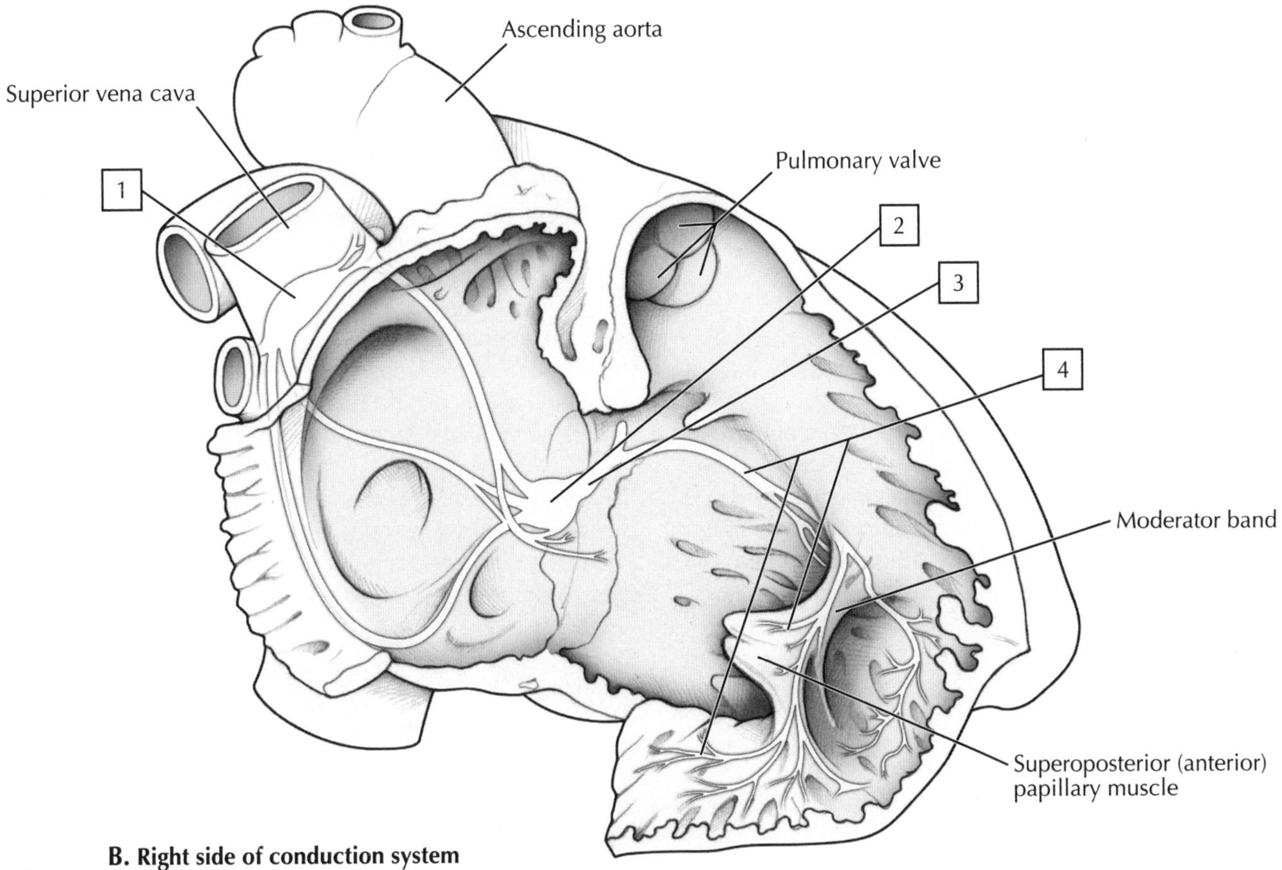


**B. Right side of conduction system**

# Heart IV

The first set of arteries to arise from the ascending aorta as it leaves the heart are the **coronary arteries**, which are named for their location, crowning the heart. The blood that they supply to the heart has the highest oxygen saturation of any blood in the body and therefore can meet the heart's high metabolic needs. There are two coronary arteries, left and right, and three major **cardiac veins**, great, middle, and small. These veins return most of the blood to the **coronary sinus** and the right atrium, although several small anterior cardiac veins drain directly into the right atrium and several of the smallest cardiac veins drain through the cardiac wall directly into all four heart chambers, but mostly the right atrium. The vascular supply to the heart is summarized in the following table.

| VESSEL | COURSE |
| --- | --- |
| Right coronary artery | Consists of major branches: sinuatrial (SA) nodal, right marginal, inferior (posterior) interventricular (also called posterior descending) branch of right coronary artery, atrioventricular (AV) nodal |
| Left coronary artery | Consists of major branches: circumflex, anterior interventricular branch (left anterior descending) (LAD), left marginal |
| Great cardiac vein | Parallels anterior interventricular branch (LAD) artery and drains into coronary sinus |
| Middle cardiac vein | Parallels inferior (posterior) interventricular branch of right coronary artery and drains into coronary sinus |
| Small cardiac vein | Parallels right marginal artery and drains into coronary sinus |
| Anterior cardiac veins | Are several small veins that drain directly into right atrium |
| Smallest cardiac veins | Drain through the cardiac wall directly into all four heart chambers |

Coronary blood flow varies with the aortic pressure but also is influenced by physical factors, such as compression of the vessels during contraction of the heart chambers (coronary flow is significantly decreased as the contracting ventricular myocardium compresses the coronary arteries), and metabolic factors released from the myocytes. A number of metabolic factors have been implicated in the regulation of coronary blood flow:

- $H^+$
- $CO_2$
- Decreased $O_2$
- $K^+$
- Lactic acid
- Nitric oxide
- Adenosine (probably the most important factor)

When cardiac work demand increases, adenosine is released by the myocytes and this leads to vasodilation and increased blood flow in the coronary arteries.

**COLOR** each of the coronary arteries and cardiac veins listed below, using the colors suggested:

- ☐ 1. **Left coronary artery and its major branches (anterior interventricular [anterior descending] branch, circumflex branch, left marginal branch) (orange)**
- ☐ 2. **Great cardiac vein (blue)**
- ☐ 3. **Small cardiac vein (brown)**
- ☐ 4. **Right coronary artery and its major branches, the SA nodal branch, right marginal branch, and inferior (posterior) interventricular branch (red)**
- ☐ 5. **Coronary sinus (purple)**
- ☐ 6. **Middle cardiac vein (green)**

***Clinical Note:***

**Angina pectoris** is the sensation caused by myocardial ischemia and is usually described as pressure, discomfort, or a feeling of choking in the left chest or substernal region that radiates to the left shoulder and arm as well as the neck, jaw and teeth, abdomen, and back. This radiating pattern is an example of **"referred pain,"** in which visceral afferents from the heart enter the upper thoracic spinal cord along with somatic afferents, both converging in the spinal cord's dorsal horn. Interpretation of the visceral pain may initially be confused with somatic sensations from the same cord levels. Myocardial ischemia due to atherosclerosis and coronary artery thrombosis is the major cause of **myocardial infarction** (MI), which affects more than one million Americans each year. If the ischemia is severe enough, necrosis (tissue death) of the myocardium can occur; it usually begins in the subendocardium, because this region is the most poorly perfused region of the ventricular wall.

A **coronary artery bypass** graft offers a surgical approach for revascularization. Veins or arteries from elsewhere in the patient's body are grafted to the coronary arteries to improve the blood supply to the cardiac muscle.

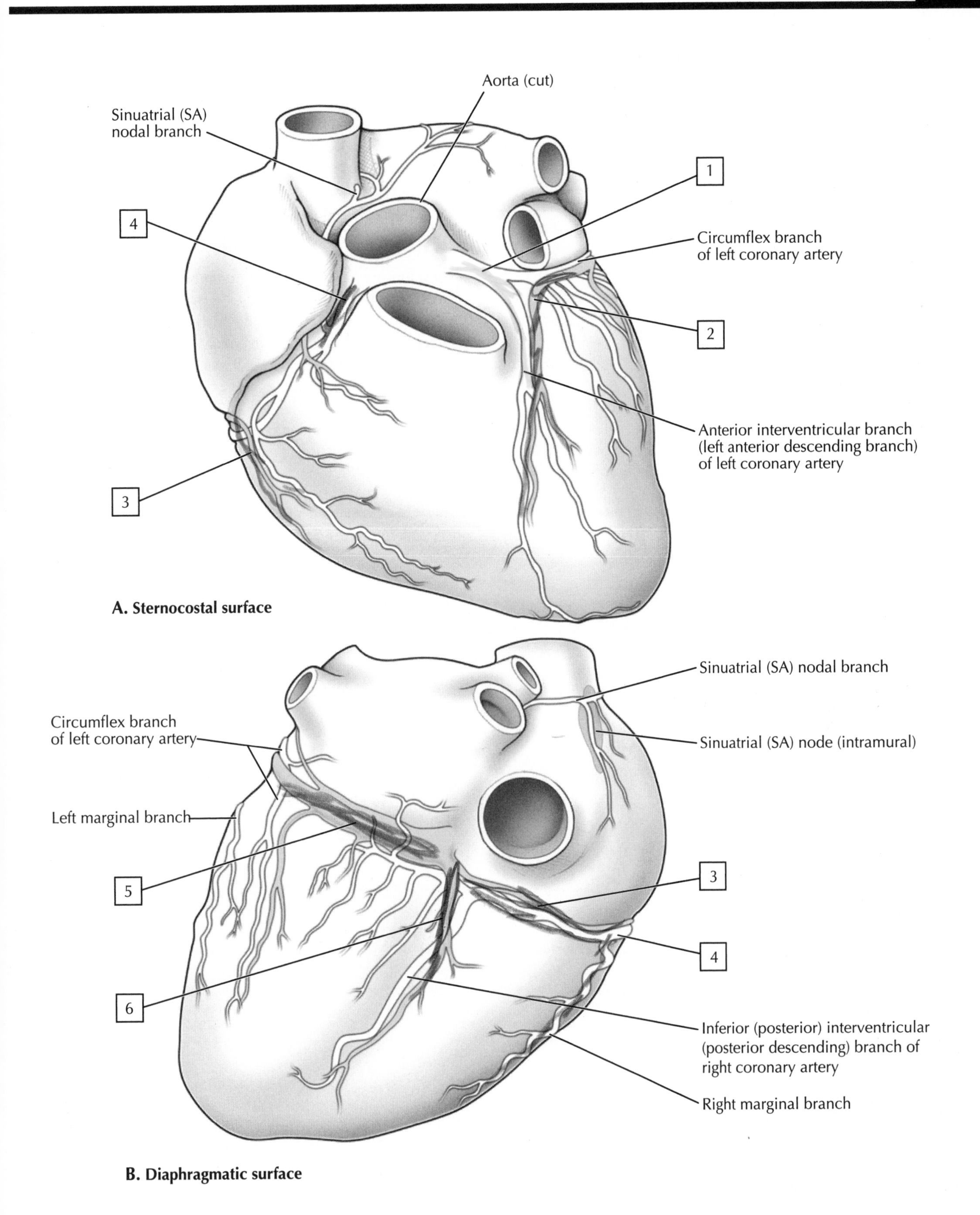


**A. Sternocostal surface**

**B. Diaphragmatic surface**

# 5 Features of Arteries, Capillaries, and Veins

Most arteries and veins have three essential layers, or tunicae (capillaries and postcapillary venules are the exceptions). The three layers are the:

- **Tunica intima**: an inner layer of simple squamous epithelium, called the endothelium, that lines all arteries, veins, and capillaries
- **Tunica media**: a middle layer of concentrically oriented layers of smooth muscle; in large arteries (aorta), elastic lamellae are interspersed between the smooth muscle layers
- **Tunica adventitia**: an outer layer of connective tissue, composed primarily of collagen and a few elastic fibers

**Arteries** can be classified into four different types based upon their size and the relative thickness of the tunicae, or the lack of them:

- **Large (elastic) arteries**: aorta and proximal portions of the subclavian and common carotid arteries
- **Medium (muscular) arteries**: most of the commonly "named" arteries in the body
- **Small arteries and arterioles**: responsible for most of the vascular resistance; arterioles regulate the blood flow into the capillary beds
- **Capillaries**: composed of endothelium only; they are functionally responsible for the exchange of gases and metabolites between the tissues and the blood

**Veins** can be classified into three different types based on their size and the relative thickness of the tunica media:

- **Venules and small veins**: venules include postcapillary venules (composed of endothelium and pericytes only) and muscular venules (in which the tunica media has one or two layers of smooth muscle); small veins have two to three smooth muscle layers
- **Medium veins**: most of the commonly "named" veins in the body; these veins in the extremities contain valves that assist in the venous return against gravity
- **Large veins**: the tunica adventitia is much thicker than the tunica media in these veins; they include the subclavian veins and venae cavae

The human body contains over 50,000 miles of blood vessels. The key distinguishing features of these vessels are summarized in the table below.

| VESSEL | DIAMETER | INNER LAYER (TUNICA INTIMA) | MIDDLE LAYER (TUNICA MEDIA) | OUTER LAYER (TUNICA ADVENTITIA) |
|---|---|---|---|---|
| | | **Arteries** | | |
| Large artery (elastic artery) | >1 cm | Endothelium; connective tissue; smooth muscle | Smooth muscle; elastic lamellae | Connective tissue; elastic fibers; thinner than tunica media |
| Medium artery (muscular artery) | 2-10 mm | Endothelium; connective tissue; smooth muscle | Smooth muscle; collagen fibers; little elastic tissue | Connective tissue; some elastic fibers; thinner than tunica media |
| Small artery | 0.1-2 cm | Endothelium; connective tissue; smooth muscle | Smooth muscle (8-10 cell layers); collagen fibers | Connective tissue; some elastic fibers; thinner than tunica media |
| Arteriole | 10-100 μm | Endothelium; connective tissue; smooth muscle | Smooth muscle (1-2 cell layers) | Thin, ill-defined |
| Capillary | 4-10 μm | Endothelium | None | None |
| | | **Veins** | | |
| Postcapillary venule | 10-50 μm | Endothelium; pericytes | None | None |
| Muscular venule | 50-100 μm | Endothelium; pericytes | Smooth muscle (1-2 cell layers) | Connective tissue; some elastic fibers; thicker than tunica media |
| Small vein | 0.1-1 mm | Endothelium; connective tissue; smooth muscle (2-3 layers) | Smooth muscle (2-3 layers continuous with tunica intima) | Connective tissue; some elastic fibers; thicker than tunica media |
| Medium vein | 1-10 mm | Endothelium; connective tissue; smooth muscle; some have valves | Smooth muscle; collagen fibers | Connective tissue; some elastic fibers; thicker than tunica media |
| Large vein | >1 cm | Endothelium; connective tissue; smooth muscle | Smooth muscle (2-15 layers); collagen fibers | Connective tissue; some elastic fibers, longitudinal smooth muscles; much thicker than tunica media |

**COLOR** the following features of the blood vessels, using a different color for each feature:

- ☐ 1. **Tunica intima (endothelium)**
- ☐ 2. **Tunica media (smooth muscle)**
- ☐ 3. **Tunica adventitia (connective tissue)**

***Clinical Note:***

A thickening and narrowing of the arterial wall and eventual deposition of lipid into the wall can lead to one form of **atherosclerosis**. The narrowed artery may not be able to meet the metabolic needs of the adjacent tissues, with the danger that they may become ischemic (lack of oxygen). Multiple factors, including focal inflammation of the arterial wall, may result in this condition.

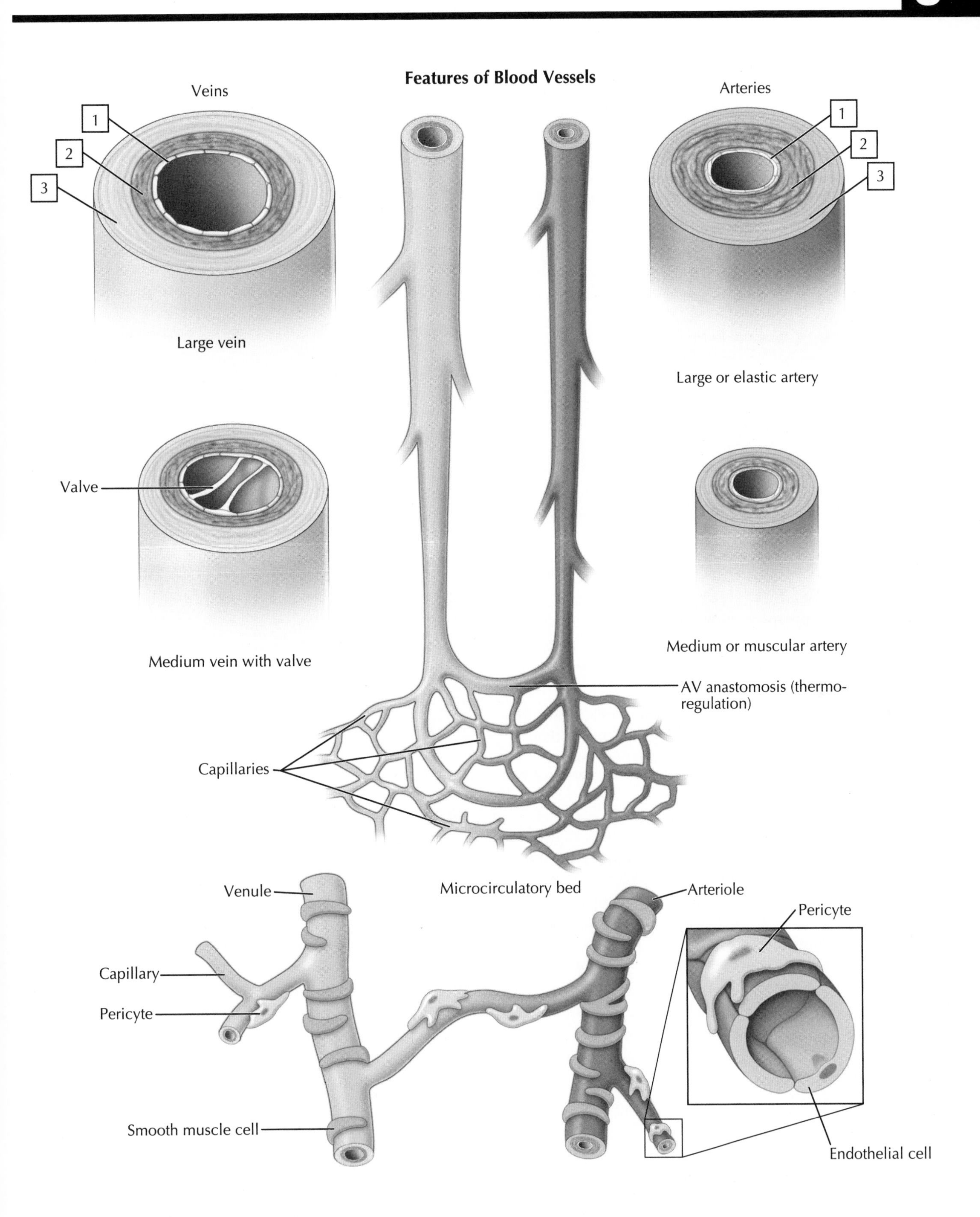
Features of Blood Vessels
Veins
Arteries
1
2
3
1
2
3
Large vein
Large or elastic artery
Valve
Medium vein with valve
Medium or muscular artery
AV anastomosis (thermo-
regulation)
Capillaries
Microcirculatory bed
Venule
Arteriole
Pericyte
Capillary
Pericyte
Smooth muscle cell
Endothelial cell

# Head and Neck Arteries

Arteries supplying the head and neck region arise principally from the **subclavian** and **common carotid arteries**. The subclavian artery is divided into three parts by the anterior scalene muscle. Part 1 lies medial, part 2 posterior, and part 3 lateral to the anterior scalene muscle. Branches of the subclavian are summarized in the following table.

| BRANCH | COURSE |
|---|---|
| | **Part 1** |
| Vertebral | Ascends through C6-C1 foramen transversarium and enters foramen magnum |
| Internal thoracic | Descends parasternally to anastomose with superior epigastric artery |
| Thyrocervical trunk | Gives rise to inferior thyroid, transverse cervical, and suprascapular arteries |
| | **Part 2** |
| Costocervical trunk | Gives rise to deep cervical and superior intercostal arteries |
| | **Part 3** |
| Dorsal scapular | Is inconstant; may also arise from transverse cervical artery |

**COLOR** the following branches of the subclavian artery, using a different color for each branch:

- ☐ 1. **Vertebral: provides blood to the posterior portion of the brain**
- ☐ 2. **Costocervical trunk: its deep cervical branch supplies the deep lateral neck**
- ☐ 3. **Thyrocervical trunk: its transverse cervical and inferior thyroid branches supply portions of the neck, and the thyroid and parathyroid glands**

The **common carotid artery** ascends in the carotid sheath, which also contains the internal jugular vein and vagus nerve, and divides into the internal and external carotid branches. The **internal carotid artery** essentially does not give off any branches in the neck (it does, but they are very small and seldom mentioned) but does pass into the intracranial carotid canal to supply the middle and anterior portions of the brain, and orbit. The **external carotid artery** gives rise to eight branches that supply the neck, face, scalp, dura mater, nasal and paranasal regions, and oral cavity. Its branches are summarized in the following table.

| BRANCH | COURSE AND STRUCTURES SUPPLIED |
|---|---|
| Superior thyroid | Supplies thyroid gland, larynx, and infrahyoid muscles |
| Ascending pharyngeal | Supplies pharyngeal region, middle ear, meninges, and prevertebral muscles |
| Lingual | Passes deep to hyoglossus muscle to supply the tongue |
| Facial | Courses over the mandible and supplies the face |
| Occipital | Supplies sternocleidomastoid muscle and anastomoses with costocervical trunk |
| Posterior auricular | Supplies region posterior to ear |
| Maxillary | Passes into infratemporal fossa (described later) |
| Superficial temporal | Supplies face, temporalis muscle, and lateral scalp |

**COLOR** the following branches of the external carotid artery, using a different color for each branch:

- ☐ 4. **Maxillary**
- ☐ 5. **Facial**
- ☐ 6. **Lingual**
- ☐ 7. **Superior thyroid**
- ☐ 8. **Superficial temporal**

***Clinical Note:***

A **carotid pulse** in the neck is felt where it lies between the trachea and infrahyoid muscles, usually just deep to the anterior border of the sternocleidomastoid muscle.

Obstruction of the internal carotid artery resulting from **atherosclerosis** (a thickening of the tunica intima) may cause carotid occlusion and stenosis of this important artery. In such cases, a **carotid endarterectomy** may be performed to strip off the offending atherosclerotic plaque from the intima.

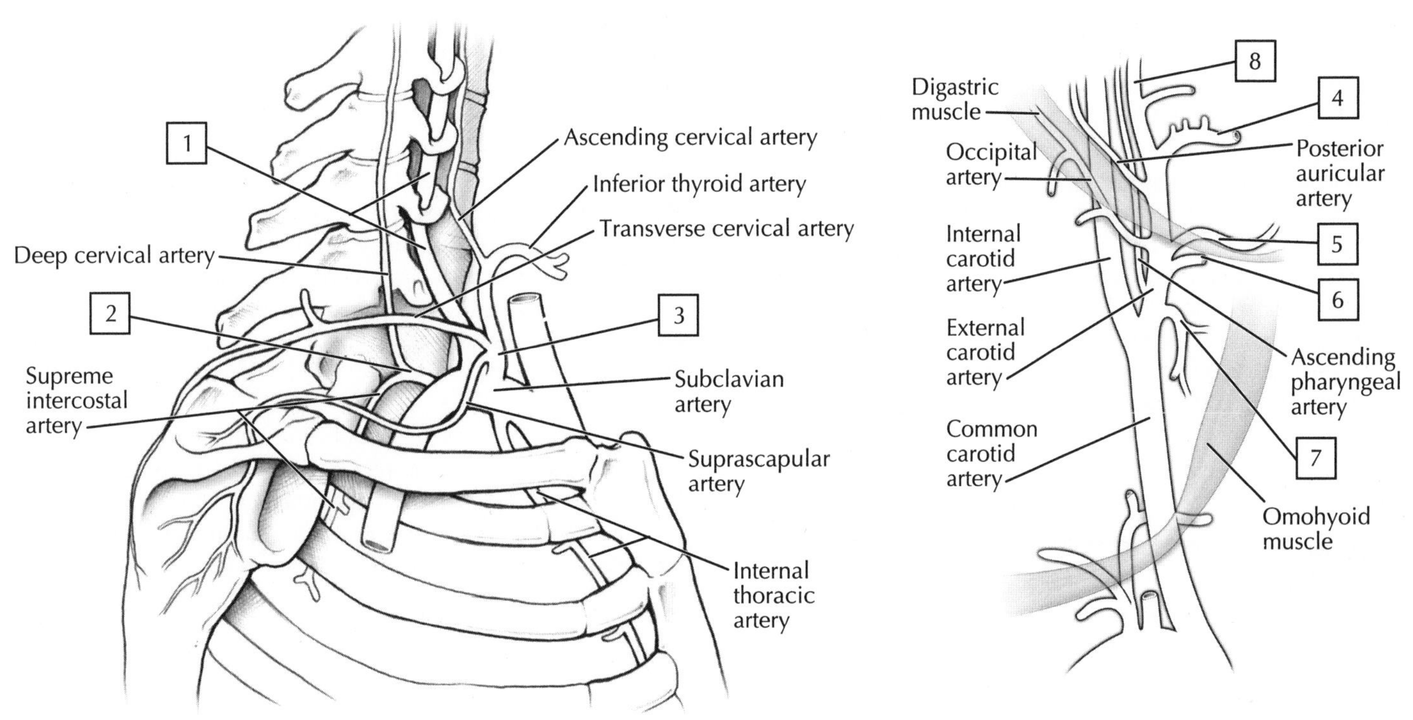


A. Neck: subclavian artery

B. Right external carotid branches: schema

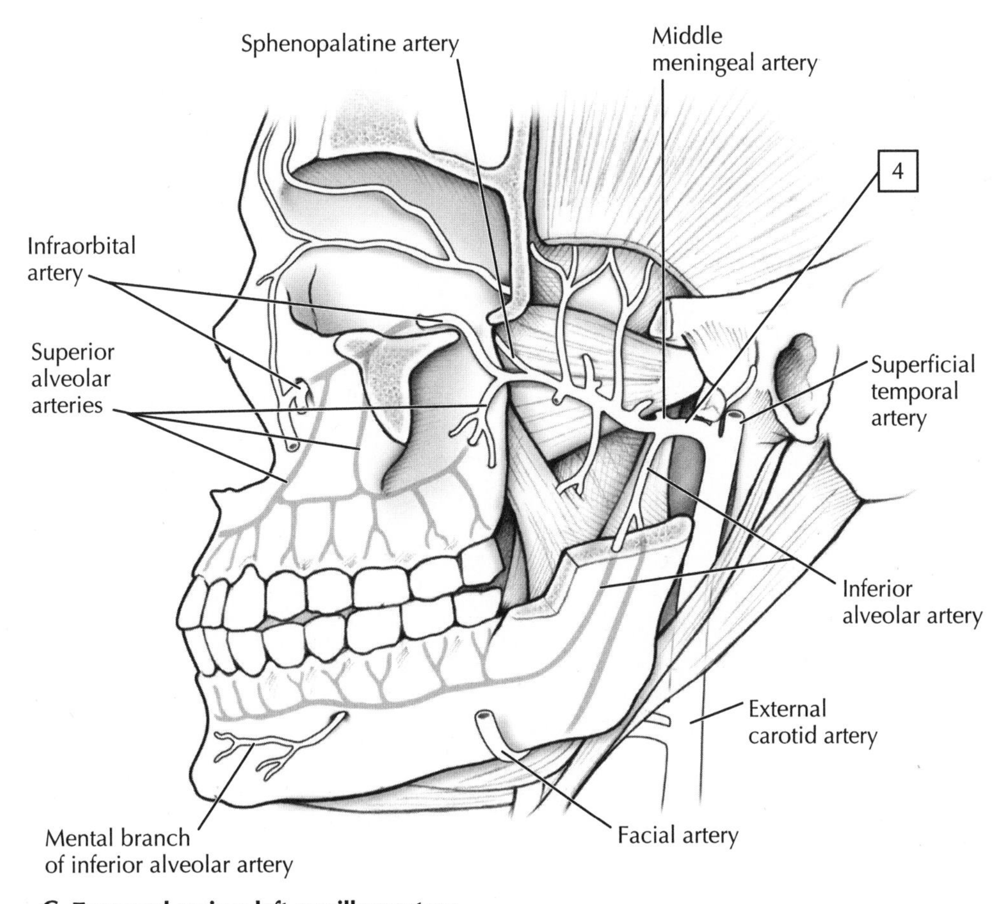


C. Temporal region: left maxillary artery

# Head Arteries

The **maxillary artery** supplies the infratemporal region, dura mater, nasal region, and a portion of the oral cavity. It is the largest branch of the external carotid artery and has the most extensive distribution of its branches. It gives rise to 15 or more branches of its own but, for descriptive purposes, is divided into three parts:

- **Retromandibular**: arteries enter foramina of the skull or jaw and supply the dura mater, mandibular teeth and gums, ear, and chin
- **Pterygoid**: branches supply the muscles of mastication and buccinator muscle
- **Pterygopalatine**: branches enter foramina of the skull and supply maxillary teeth and gums, orbital floor, nose, paranasal sinuses, palate, auditory tube, and superior pharynx

**COLOR** the following major branches of the maxillary artery using the color red:

☐ 1. **Maxillary artery and only these major branches arising from it:**

☐ 2. **Inferior alveolar (to mandibular teeth and gums)**

☐ 3. **Middle meningeal (to dura mater covering the brain)**

☐ 4. **Posterior superior alveolar (its branches to maxillary teeth and gums)**

☐ 5. **Infraorbital (to floor of orbit)**

☐ 6. **Sphenopalatine (to nose, paranasal sinuses, palate, and superior pharynx)**

The maxillary artery passes through the infratemporal fossa, enmeshed within the medial and lateral pterygoid muscles and a large venous plexus called the pterygoid plexus of veins (see Plate 5-11).

***Clinical Note:***

Because of the extensive arterial supply and venous drainage in the infratemporal fossa region, trauma to this area of the face and head can cause significant bleeding. Numerous nerves, muscles, and other structures lie within this region, and hemostasis and infection control must be a priority for the healthcare team.

If the internal carotid artery should become narrowed or occluded, the branches of the external carotid artery, especially its larger facial, maxillary, and superficial temporal branches may provide **collateral routes of circulation**. These routes are more likely to develop when the occlusion is gradual, as in atherosclerosis, rather than acute, as in an embolic obstruction.

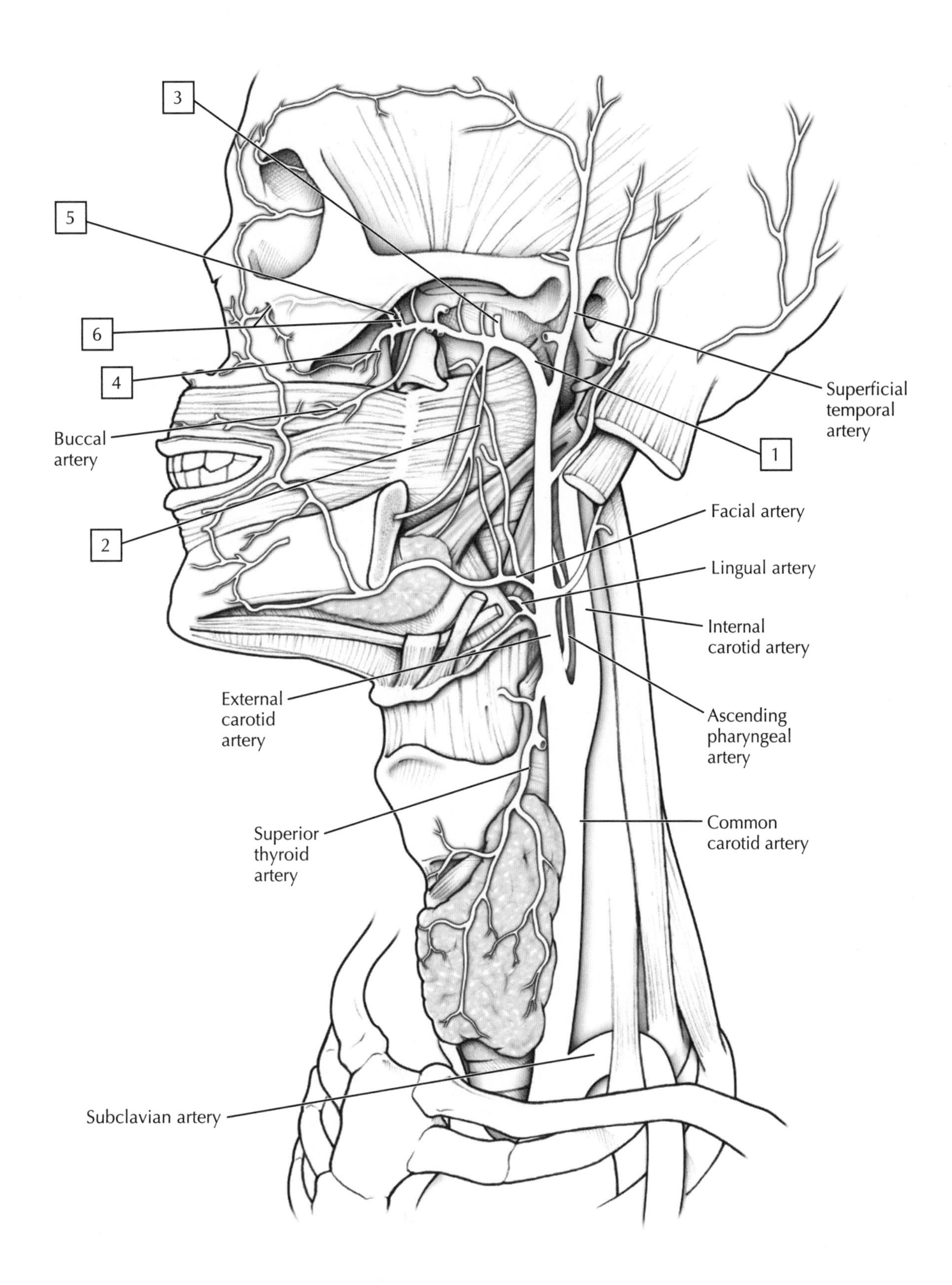
3
5
6
4
Buccal artery
2
Superficial temporal artery
1
Facial artery
Lingual artery
Internal carotid artery
External carotid artery
Ascending pharyngeal artery
Superior thyroid artery
Common carotid artery
Subclavian artery

# 5 Arteries of the Brain

The arteries supplying the brain arise largely from two pairs of arteries:

- **Vertebrals**: arise from the two subclavian arteries in the neck, ascend through the foramen transversarium of the cervical vertebrae, and enter the foramen magnum of the skull to supply the posterior portion of the brain
- **Internal carotids**: arise from the two common carotid arteries in the neck (see Plate 5-9), ascend in the neck and enter the carotid canal. They then traverse the foramen lacerum to terminate as the middle and anterior cerebral arteries, which anastomose with the circle of Willis around the optic chiasm, hypophysis, and basal hypothalamus

| ARTERY | COURSE AND STRUCTURES SUPPLIED |
|---|---|
| Vertebral | From subclavian artery, supplies cerebellum |
| Posterior inferior cerebellar | From vertebral artery, goes to posteroinferior cerebellum |
| Basilar | From both vertebrals, goes to brainstem, cerebellum, cerebrum |
| Anterior inferior cerebellar | From basilar, supplies inferior cerebellum |
| Superior cerebellar | From basilar, supplies superior cerebellum |
| Posterior cerebral | From basilar, supplies inferior cerebrum, occipital lobe |
| Posterior communicating | Cerebral arterial circle (of Willis) |
| Internal carotid | From common carotid, supplies cerebral lobes and eye |
| Middle cerebral | From internal carotid, goes to lateral aspect of cerebral hemispheres |
| Anterior communicating | Cerebral arterial circle (of Willis) |
| Anterior cerebral | From internal carotid, goes to cerebral hemispheres (except occipital lobe) |

**COLOR** the following arteries that supply the brain, using a different color for each artery:

- ☐ 1. **Anterior communicating**
- ☐ 2. **Anterior cerebral**
- ☐ 3. **Middle cerebral**
- ☐ 4. **Posterior communicating**
- ☐ 5. **Posterior cerebral**
- ☐ 6. **Basilar**
- ☐ 7. **Anterior inferior cerebellar**
- ☐ 8. **Vertebral**

***Clinical Note:***

Bleeding from an artery supplying the dura mater results in arterial blood collecting in the space between the dura mater and the skull and is called an **epidural hematoma**. This often occurs from blunt trauma to the head and involves bleeding from the middle meningeal artery (from the maxillary artery) or one of its branches. If not immediately treated, this bleeding may begin to press on the brain.

A **subarachnoid hemorrhage** usually occurs from the rupture of a saccular, or berry, aneurysm (a ballooning of an artery) involving one of the branches of the vertebral, internal carotid, or circle of Willis arteries.

**Occlusion** (by an atherosclerotic plaque or thrombus) of the:

- Anterior cerebral artery can disrupt sensory and motor functions on the contralateral lower extremity
- Middle cerebral artery can disrupt sensory and motor functions on the contralateral upper extremity or, if the internal capsule is affected, the entire contralateral body
- Posterior cerebral artery can disrupt visual functions from the contralateral visual field

A **subdural hematoma** is often caused by an acute venous hemorrhage of the cortical bridging veins draining cortical blood into the superior sagittal sinus. The blood collects between the meningeal dura mater and the arachnoid mater.

A **cerebrovascular accident** (CVA), or stroke, is a localized brain injury caused by a vascular episode that lasts more than 24 hours, whereas a **transient ischemic attack** (TIA) is a focal ischemic episode lasting less than 24 hours. Stroke is classified as ischemic (about 70% to 80%), as a result of infarction (thrombotic or embolic) from atherosclerosis of the extracranial (usually carotid) and intracranial arteries, or from underlying heart disease. Hemorrhagic stroke occurs when a cerebral vessel weakens and ruptures (subarachnoid or intracerebral hemorrhage), which causes intracranial bleeding, usually affecting a larger brain area.

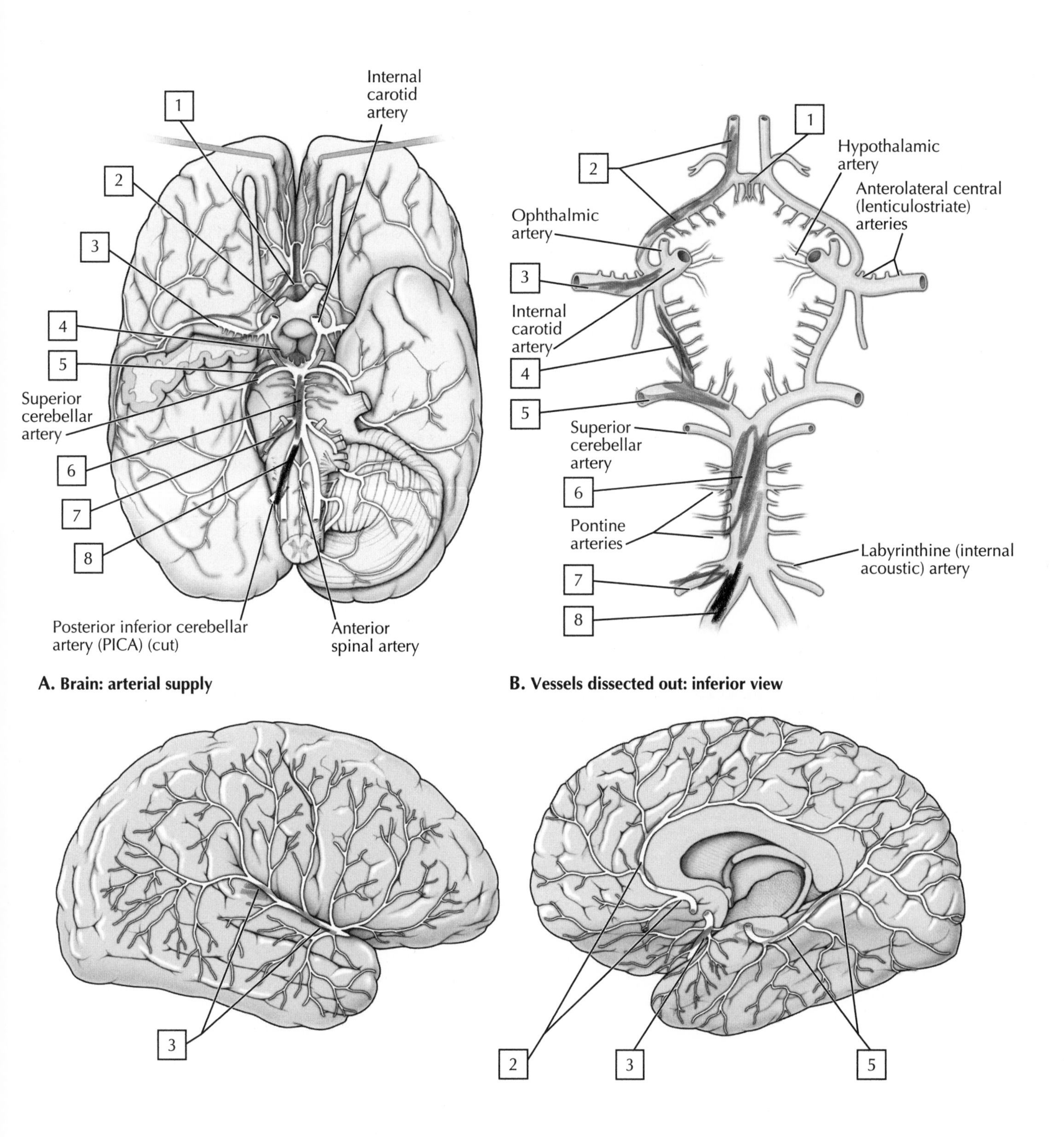


**A. Brain: arterial supply**

**B. Vessels dissected out: inferior view**

# 5 Veins of the Head and Neck

Much of the blood drained from the brain collects into various **dural venous sinuses** (the dural layers separate to form a large vein or sinus) (see Plates 4-17 and 4-18), which tend to direct the flow of venous blood posteriorly along the superior and inferior sagittal sinuses to the confluence of sinuses. Cerebral veins empty into the dural sinuses by traversing the subarachnoid space. The dural sinuses are thick-walled channels formed from a split of the inner and outer layers of dura. From here, blood flows in the right and left transverse and sigmoid sinuses to collect and form the origin of the **internal jugular veins**.

Venous blood also can drain from the **cavernous sinus** (see part ***A***, #1) into the ophthalmic veins and thence into facial veins to reach the internal jugular veins (IJV), or drain inferiorly into the **pterygoid plexus of veins** (see part B), and then into the facial and/or retromandibular veins, ultimately finding their way to the internal jugular veins.

**COLOR** the following venous sinuses, using a different color for each sinus:

- ☐ **1. Cavernous**
- ☐ **2. Sigmoid**
- ☐ **3. Transverse**
- ☐ **4. Superior sagittal**
- ☐ **5. Straight**
- ☐ **6. Superior petrosal**

The venous drainage of the head and neck ultimately causes blood to collect into the following major paired veins (numerous anastomoses exist between these veins):

- **Retromandibular**: receives tributaries from the temporal and infratemporal regions (pterygoid plexus), nasal cavity, pharynx, and oral cavity
- **Internal jugular**: drains the brain, face, thyroid gland, and neck
- **External jugular**: drains the superficial neck, lower neck and shoulder, and upper back (often communicates with the retromandibular vein)

**COLOR** the following veins, using a different color for each vein:

- ☐ **7. Facial**
- ☐ **8. Superior, middle, and inferior thyroid**
- ☐ **9. Retromandibular**
- ☐ **10. Internal jugular**

***Clinical Note:***

The cavernous sinus surrounds the pituitary gland and has connections to ophthalmic veins, the pterygoid plexus of veins, the basilar plexus, and the superior and inferior petrosal sinuses. Venous blood flow through this sinus is stagnant because the interior of the sinus is filled with a trabecular web of connective tissue fibers that impede blood flow. Consequently, blood-borne infections can "seed" themselves in this sinus and cause a **cavernous sinus thrombosis**. Additionally, pituitary tumors can expand laterally into this sinus and stretch its dural wall, potentially placing pressure on a number of cranial nerves (CN III, IV, $V_1$, $V_2$, and VI) related to the sinus.

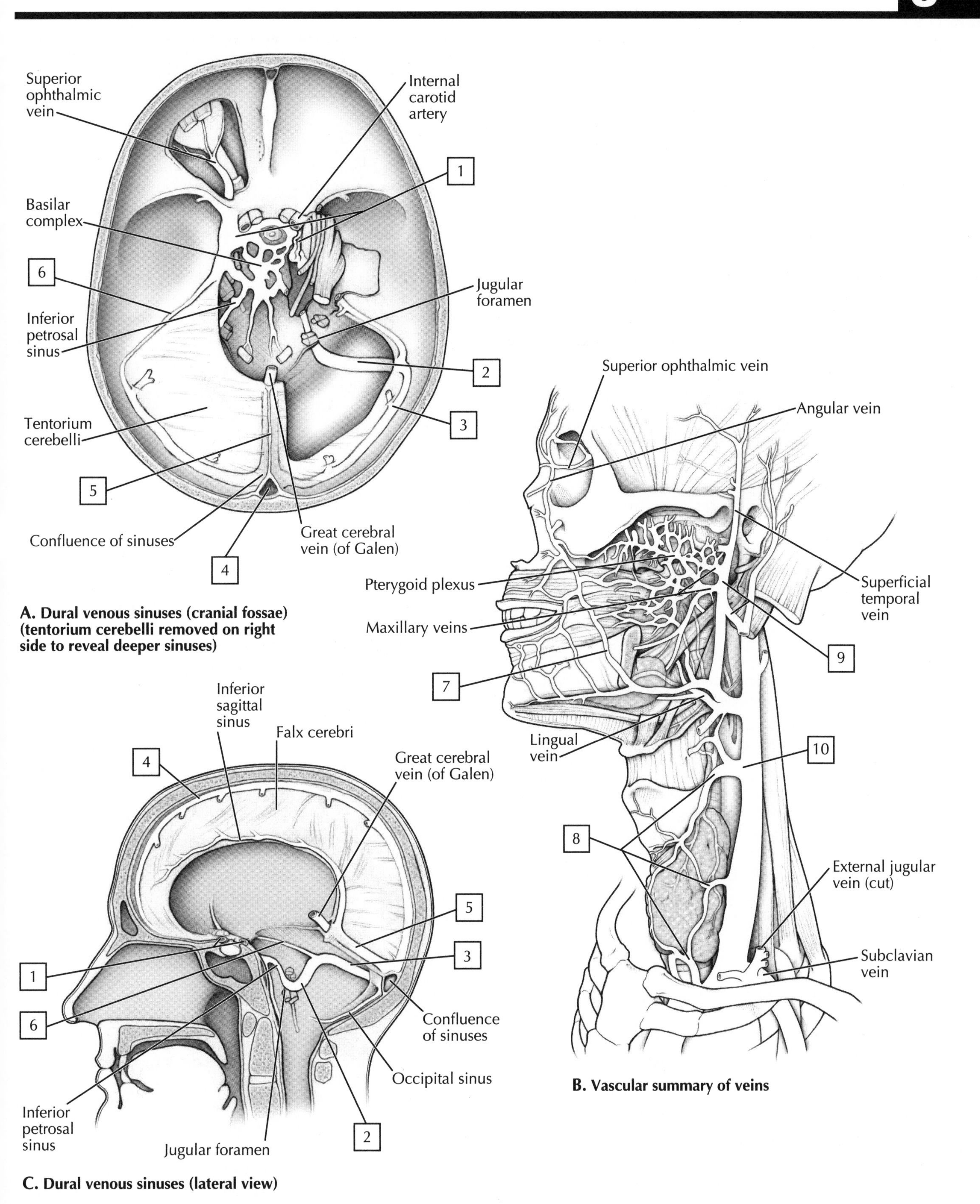


**A. Dural venous sinuses (cranial fossae) (tentorium cerebelli removed on right side to reveal deeper sinuses)**

**B. Vascular summary of veins**

**C. Dural venous sinuses (lateral view)**

# 5 Arteries of the Upper Limb

Arteries of the upper limb arise from a continuation of the **subclavian arteries.** Once the subclavian artery emerges from beneath the clavicle and crosses the first rib, its name changes to the **axillary artery**; it continues on to course through the axillary region (armpit). Once the axillary artery reaches the inferior border of the teres major muscle, it becomes the **brachial artery**, which itself divides into the **ulnar** and **radial arteries** in the cubital fossa (region anterior to the elbow).

The **axillary artery** begins at the first rib and descriptively is divided into three parts by the presence of the overlying pectoralis minor muscle. Branches of the subclavian and axillary arteries form a rich anastomosis around the scapula, supplying the muscles acting on the shoulder joint.

| PART OF AXILLARY ARTERY | BRANCH | COURSE AND STRUCTURES SUPPLIED |
|---|---|---|
| 1st | Superior thoracic | Supplies first two intercostal spaces |
| 2nd | Thoracoacromial | Has clavicular, pectoral, deltoid, and acromial branches |
| | Lateral thoracic | Runs with long thoracic nerve and supplies muscles that it traverses |
| 3rd | Subscapular | Divides into thoracodorsal and circumflex scapular branches |
| | Anterior humeral circumflex | Passes around surgical neck of humerus |
| | Posterior humeral circumflex | Runs with axillary nerve through the quadrangular space to anastomose with anterior circumflex branch |

The **brachial artery** is a direct continuation of the axillary artery inferior to the teres major muscle.

| ARTERY | COURSE |
|---|---|
| Brachial | Begins at inferior border of teres major and ends at its bifurcation in cubital fossa as the radial and ulnar arteries |
| Deep brachial (profunda brachii) | Runs with radial nerve around humeral shaft |
| Superior ulnar collateral | Runs with ulnar nerve |
| Inferior ulnar collateral | Passes anterior to medial epicondyle of humerus |
| Radial | Is smaller lateral branch of brachial artery |
| Ulnar | Is larger medial branch of brachial artery |

The brachial artery divides into the ulnar and radial arteries in the cubital fossa.

| ARTERY | COURSE |
|---|---|
| **Radial** | Arises from brachial artery in cubital fossa |
| Radial recurrent branch | Anastomoses with radial collateral artery in arm |
| Palmar carpal branch | Anastomoses with carpal branch of ulnar artery |
| **Ulnar** | Arises from brachial artery in cubital fossa |
| Anterior ulnar recurrent | Anastomoses with inferior ulnar collateral in arm |
| Posterior ulnar recurrent | Anastomoses with superior ulnar collateral in arm |
| Common interosseous | Gives rise to anterior and posterior interosseous arteries |
| Palmar carpal branch | Anastomoses with carpal branch of radial artery |

The **ulnar** and **radial arteries** anastomose in the palm of the hand by forming **two palmar arches**. Common digital and proper digital branches arise from the superficial palmar arch to supply the fingers. The ulnar and radial arteries are summarized in the following table.

| ARTERY | COURSE |
|---|---|
| | **Radial** |
| Superficial palmar branch | Forms superficial palmar arch with ulnar artery |
| Princeps pollicis | Passes under flexor pollicis longus tendon and divides into two proper digital arteries to thumb |
| Radialis indicis | Passes to index finger on its lateral side |
| Deep palmar arch | Is formed by terminal part of radial artery |
| | **Ulnar** |
| Deep palmar branch | Forms deep palmar arch with radial artery |
| Superficial palmar arch | Is the termination of ulnar artery; gives rise to three common digital arteries, each of which gives rise to two proper digital arteries |

**COLOR** the following arteries, using a different color for each artery:

☐ 1. **Subclavian**
☐ 2. **Axillary**
☐ 3. **Brachial**
☐ 4. **Deep brachial**
☐ 5. **Radial**
☐ 6. **Ulnar**
☐ 7. **Deep palmar arch**
☐ 8. **Superficial palmar arch**

***Clinical Note:***

**Pulse points** of the upper limb include:

- **Brachial**: in the proximal third of the medial arm, where the brachial artery can be pressed against the humerus
- **Cubital**: brachial artery in the cubital fossa, medial to the biceps muscle tendon and just before it divides into the ulnar and radial branches
- **Radial**: common site for taking a pulse, felt just lateral to the flexor carpi radialis muscle tendon in the distal forearm (at the wrist)
- **Ulnar**: in the distal forearm (wrist), just lateral to the flexor carpi ulnaris muscle tendon

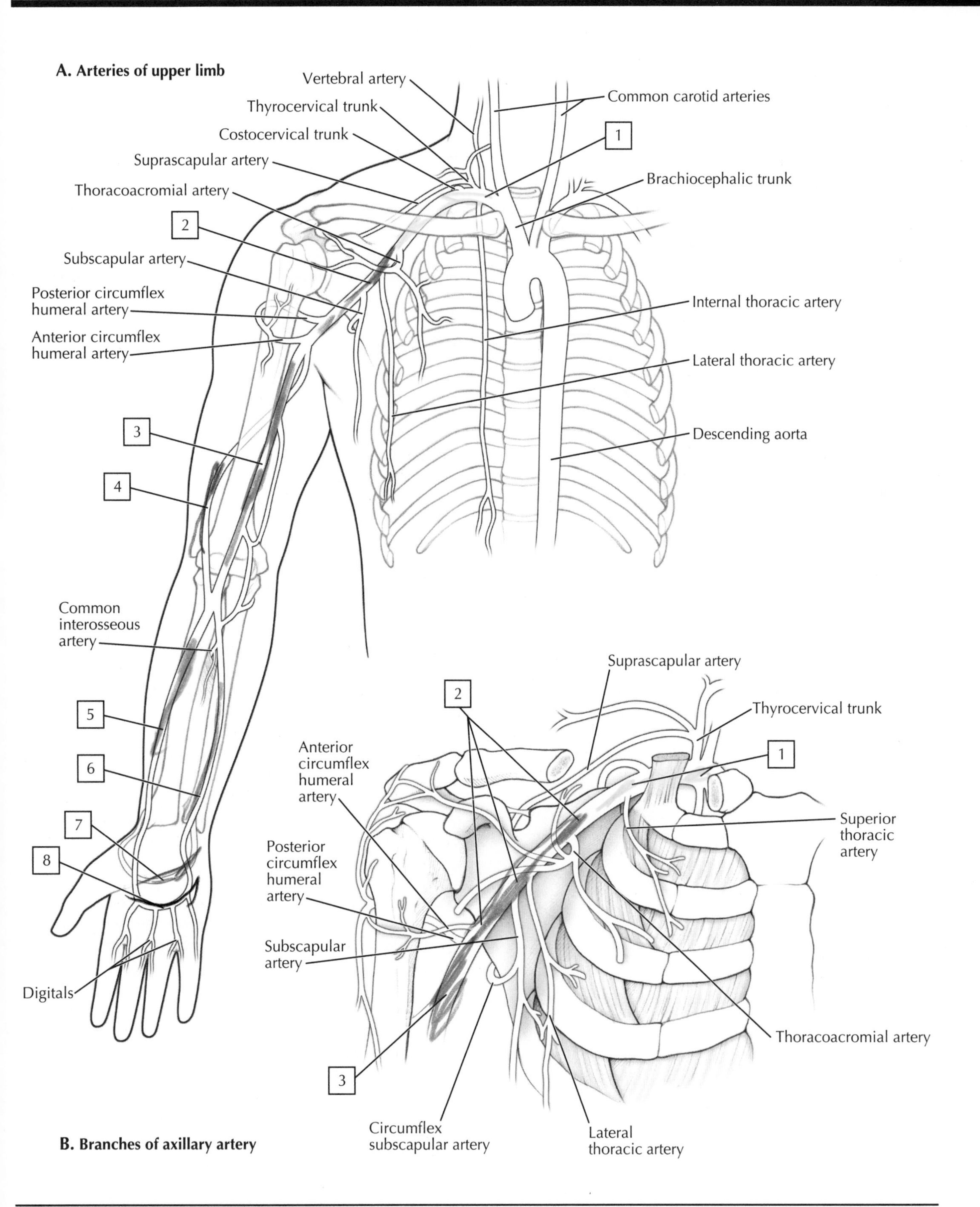
A. Arteries of upper limb
Vertebral artery
Thyrocervical trunk
Costocervical trunk
Suprascapular artery
Thoracoacromial artery
2
Subscapular artery
Posterior circumflex humeral artery
Anterior circumflex humeral artery
3
4
Common interosseous artery
5
6
7
8
Digitals
Common carotid arteries
1
Brachiocephalic trunk
Internal thoracic artery
Lateral thoracic artery
Descending aorta
B. Branches of axillary artery
Suprascapular artery
2
Thyrocervical trunk
1
Anterior circumflex humeral artery
Posterior circumflex humeral artery
Subscapular artery
Superior thoracic artery
Thoracoacromial artery
3
Circumflex subscapular artery
Lateral thoracic artery

# 5 Arteries of the Lower Limb

Arteries of the lower limb arise from the pelvis. The **obturator artery** arises from the internal iliac artery and supplies the medial compartment of the thigh. The much larger **femoral artery** arises as a direct continuation of the external iliac artery as it passes beneath the inguinal ligament. These two arteries and the deep artery of the thigh are summarized in the following table.

| ARTERY | COURSE AND STRUCTURES SUPPLIED |
|---|---|
| Obturator | Arises from internal iliac artery (pelvis); has anterior and posterior branches; passes through obturator foramen with obturator nerve and enters medial thigh |
| Femoral | Continuation of external iliac artery with numerous branches to perineum, hip, thigh, and knee |
| Deep femoral | Arises from femoral artery; supplies hip and thigh |

In the distal thigh, the **femoral artery** passes through the adductor hiatus of the adductor magnus muscle to reach the posterior aspect of the knee, where it becomes the **popliteal artery**. Just inferior to the knee, the popliteal artery divides into the anterior and **posterior tibial arteries**, which course down the leg in the anterior and posterior muscle compartments, respectively. The posterior tibial artery also gives rise to a small **fibular artery**, which courses in the lateral compartment of the leg.

In the foot, the anterior tibial artery forms an anastomosis around the ankle joint and continues on the dorsum of the foot as the **dorsalis pedis artery**. The major blood supply to the muscles of the sole of the foot arises from the **posterior tibial artery**, which passes inferior to the medial malleolus at the ankle and divides into the **medial and lateral plantar arteries**. The medial plantar artery divides into superficial and deep branches, whereas the lateral plantar artery forms a deep plantar arch and anastomoses with arteries on the dorsum of the foot.

**COLOR** the following arteries of the lower limb, using a different color for each artery:

- ☐ **1. Femoral**
- ☐ **2. Popliteal**
- ☐ **3. Anterior tibial**
- ☐ **4. Posterior tibial**
- ☐ **5. Dorsalis pedis**
- ☐ **6. Medial plantar**
- ☐ **7. Lateral plantar**

***Clinical Note:***

**Pulse points** in the lower limb include:

- **Femoral**: just inferior to the midpoint of the inguinal ligament at the point where the femoral artery is lying superficially
- **Popliteal**: behind the knee
- **Posterior tibial**: just superior to the medial malleolus as the artery begins to descend into the foot
- **Dorsalis pedis**: on the dorsum of the foot; this pulse point is the one lying farthest from the heart

The femoral artery and its medially positioned femoral vein may be used to gain access to major vessels of the limbs, abdominopelvic cavity, and thorax, e.g., a catheter may be threaded through the femoral artery and into the aorta for **coronary artery angiography and angioplasty**. Similarly, access to the larger veins of the inferior vena cava and the right side of the heart and pulmonary veins may be obtained through the **femoral vein**.

An ankle fracture involving the talus (see Plate 2-19) usually occurs at the talar neck. This kind of fracture often occurs from direct trauma or from landing on the foot after a fall from a great height. Such a fracture may lead to **avascular necrosis of the talus body** because most of the blood supply to the talus passes through the talar neck.

Diabetes mellitus may lead to **microvascular disease**, resulting in a decreased cutaneous blood flow. Additionally, hyperglycemia predisposes the extremity to bacterial and fungal infection.

Atheroscerosis can also affect the lower limb vasculature. Arterial stenosis (narrowing) or occlusion in the leg leads to **peripheral vascular disease** (PVD), a disorder largely associated with increasing age. PVD produces symptoms of claudication, which results from decreased blood flow at times of greater demand. Signs may include pain, hair loss on the leg, ulceration on the foot or great toe, and diminished peripheral pulses (dorsalis pedis pulse).

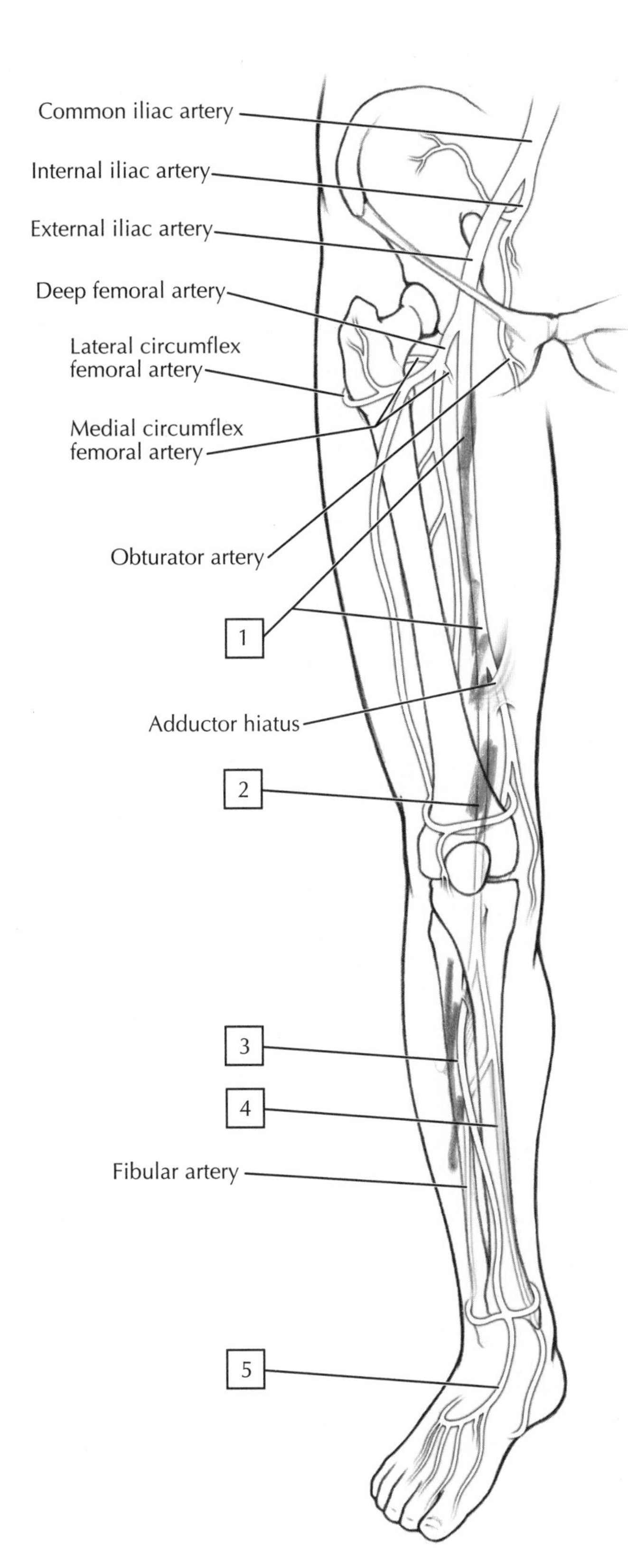


A. Arteries of lower limb: anterior view

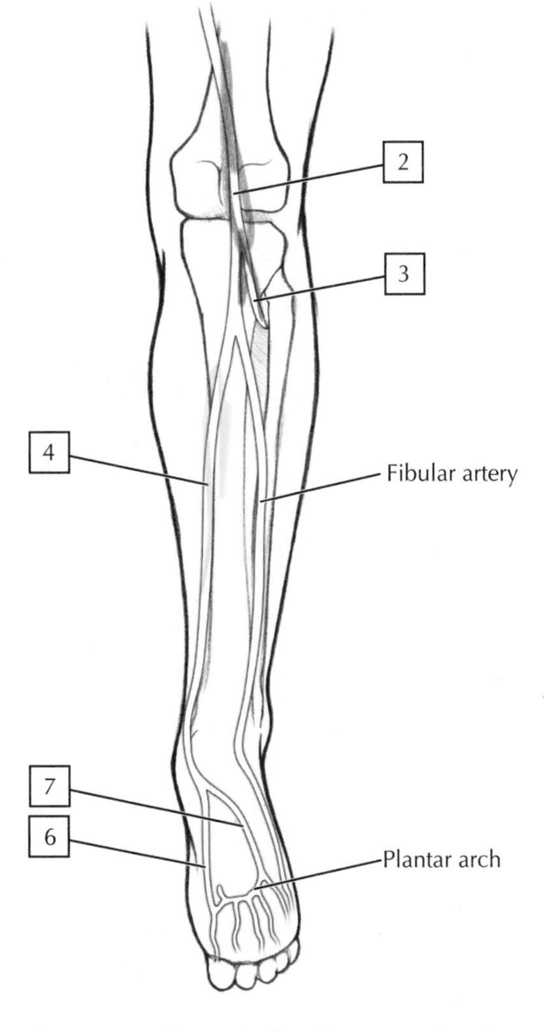


B. Arteries of leg and sole of foot: posterior view

# 5 Thoracic and Abdominal Aorta

The **thoracic aorta** descends alongside and slightly to the left of the esophagus and gives rise to the following major arteries:

- Right and left coronary arteries to the heart
- Brachiocephalic trunk (divides into the right common carotid and right subclavian arteries)
- Left common carotid artery
- Left subclavian artery
- Right and left bronchial arteries, variable in number, to the primary bronchi and lungs
- Pericardial branches (small and variable in number)
- Posterior intercostal arteries (course along the inferior margins of each rib) and their small collateral branches
- Esophageal arteries to supply the esophagus
- Mediastinal arteries that are small and supply the lymph nodes, nerves, and connective tissue of the posterior mediastinum
- Superior phrenic arteries to the diaphragm
- Small subcostal arteries that lie below the last rib

The **abdominal aorta** enters the abdomen via the aortic hiatus (T12 vertebral level) of the respiratory diaphragm, and divides into the common iliac arteries anterior to the L4 vertebra. Unpaired arteries to the gastrointestinal tract include the celiac artery and superior and inferior mesenteric arteries. Paired arteries to the other viscera include the suprarenal, renal, and gonadal (ovarian or testicular) arteries. Arteries to the musculoskeletal structures include paired inferior phrenic arteries, four to five pairs of lumbar arteries, and the unpaired median sacral artery. These arteries are summarized in the following table.

**COLOR** the following arteries arising from the aorta, using a different color for each artery:

- ☐ 1. **Brachiocephalic trunk**
- ☐ 2. **Celiac trunk (artery)**
- ☐ 3. **Superior mesenteric**
- ☐ 4. **Gonadals (ovarian or testicular)**
- ☐ 5. **Common iliacs**
- ☐ 6. **Inferior mesenteric**
- ☐ 7. **Aorta (color the entire aorta red)**
- ☐ 8. **Renals**
- ☐ 9. **Left subclavian**
- ☐ 10. **Left common carotid**

**BRANCHES OF THE ABDOMINAL AORTA**

| ARTERY | ARISES FROM AORTA | SITE OF ORIGIN | SUPPLIES |
|---|---|---|---|
| Celiac trunk | Anterior | Just inferior to aortic hiatus of respiratory diaphragm | Abdominal embryonic foregut derivatives* |
| Superior mesenteric artery | Anterior | Just inferior to celiac trunk | Abdominal embryonic midgut derivatives* |
| Inferior mesenteric artery | Anterior | Inferior to gonadal arteries | Abdominal embryonic hindgut derivatives* |
| Middle suprarenal arteries | Lateral | Just superior to renal arteries | Suprarenal glands |
| Renal arteries | Lateral | Just inferior to superior mesenteric artery | Kidneys |
| Gonadal (testicular or ovarian) arteries | Paired anterolateral | Inferior to the renal arteries | Testes or ovaries |
| Inferior phrenic arteries | Paired anterolateral | Just inferior to aortic hiatus | Diaphragm |
| Lumbar arteries | Posterior | Four pairs | Posterior abdominal wall and spinal cord |
| Median sacral artery | Posterior | Just superior to aortic bifurcation | Remnant of our caudal (tail) artery |
| Common iliac arteries | Terminal | Bifurcation of L4 vertebra | Pelvis, perineum, gluteal region, and lower limb |

*Foregut embryonic derivatives include the stomach, liver, gallbladder, pancreas, spleen, and the first half of the duodenum. Midgut embryonic derivatives include the second half of the duodenum, jejunum, ileum, cecum, ascending colon, and two-thirds of the transverse colon. Hindgut embryonic derivatives include the left one-third of the transverse colon, descending colon, sigmoid colon, and rectum.

***Clinical Note:***

**Aneurysms** (bulges in the arterial wall) usually involve larger arteries. The etiology includes a family history, hypertension, a breakdown of collagen and/or elastin within the vessel wall that leads to inflammation and weakening of the wall, and atherosclerosis. The abdominal aorta (below the level of the renal arteries) and iliac arteries are most often involved. Surgical repair for large aneurysms is important, because a ruptured aneurysm can be fatal.

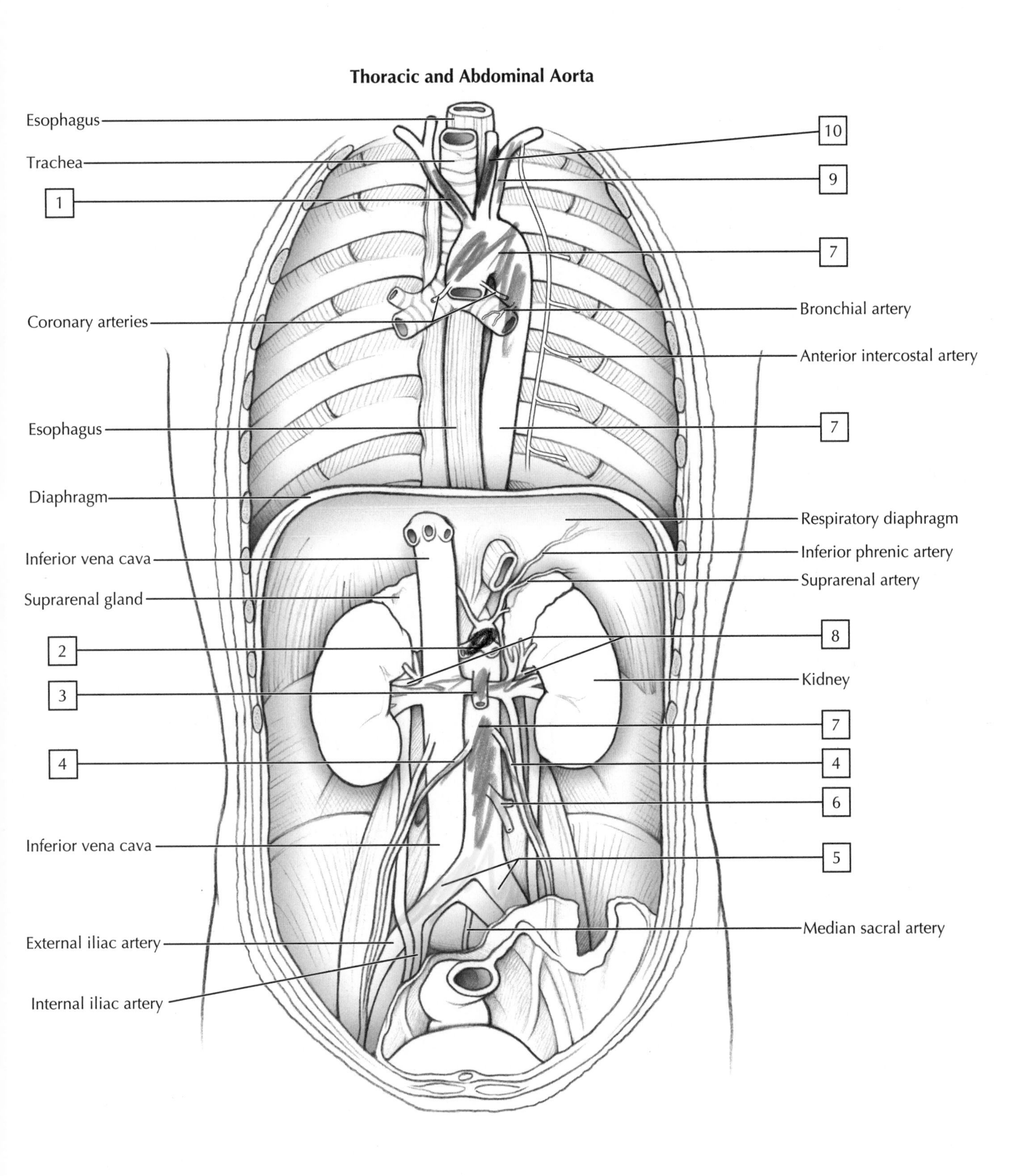
Thoracic and Abdominal Aorta
Esophagus
Trachea
1
Coronary arteries
Esophagus
Diaphragm
Inferior vena cava
Suprarenal gland
2
3
4
Inferior vena cava
External iliac artery
Internal iliac artery
10
9
7
Bronchial artery
Anterior intercostal artery
7
Respiratory diaphragm
Inferior phrenic artery
Suprarenal artery
8
Kidney
7
4
6
5
Median sacral artery

# Arteries of the Gastrointestinal Tract

Arteries that supply the gastrointestinal (GI) tract include the three unpaired arteries that arise from the anterior aspect of the abdominal aorta and include the:

- **Celiac trunk**: supplies visceral structures derived from the embryonic foregut and the spleen
- **Superior mesenteric (SMA)**: supplies visceral structures derived from the embryonic midgut
- **Inferior mesenteric (IMA)**: supplies visceral structures derived from the embryonic hindgut

These three GI tract arteries and their major branches are summarized in the tables below.

| ARTERY | STRUCTURES SUPPLIED |
|---|---|
| Celiac | Supplies stomach, spleen, liver, gallbladder, portions of pancreas, and first half of duodenum |
| Left gastric | Supplies proximal stomach and distal esophagus |
| Splenic | Supplies pancreas (dorsal branch), stomach (short gastrics and left gastroepiploic), and spleen |
| Common hepatic | Divides into proper hepatic artery and gastroduodenal artery, which supply liver, gallbladder, stomach, duodenum, and pancreas |
| SMA | Supplies small intestine and proximal half of colon; arises from aorta posterior to neck of pancreas |
| Inferior pancreaticoduodenal | Supplies second half of duodenum and pancreas |
| Middle colic | Supplies transverse colon |
| Intestinal | About 15 branches supply jejunum and ileum (jejunal and ileal branches) |
| Ileocolic | Supplies ileum, cecum, and appendix |
| Right colic | Supplies ascending colon and proximal portion of the transverse colon |
| IMA | Supplies distal colon; arises from aorta about 2 cm superior to its bifurcation |

| ARTERY | STRUCTURES SUPPLIED |
|---|---|
| Left colic | Supplies distal transverse and entire descending colon |
| Sigmoid arteries | Three or four branches supply sigmoid colon |
| Superior rectal | Supplies proximal rectum (anastomoses with other rectal arteries) |

The arterial supply of the GI tract in some senses mirrors the autonomic innervation of the GI tract. Thus, if one is familiar with the foregut, midgut, and hindgut embryonic derivatives of the GI tract, one can correlate the blood supply with the parasympathetic and sympathetic innervation of the same bowel regions. This relationship between the GI tract, its blood supply, and its innervation is summarized in the bottom table, arranged around the derivatives of the embryonic foregut, midgut, and hindgut.

**COLOR** the following arteries supplying the GI tract, using a different color for each artery:

- ☐ **1. Common hepatic branch of the celiac trunk**
- ☐ **2. Left gastric branch of the celiac trunk**
- ☐ **3. Splenic branch of the celiac trunk**
- ☐ **4. Main portion of the superior mesenteric artery (SMA)**
- ☐ **5. Middle colic branch of the SMA**
- ☐ **6. Right colic branch of the SMA**
- ☐ **7. Ileocolic branch of the SMA**
- ☐ **8. Left colic branch of the inferior mesenteric artery (IMA)**
- ☐ **9. Sigmoid branches of the IMA**
- ☐ **10. Superior rectal branch of the IMA**

| | FOREGUT | MIDGUT | HINDGUT |
|---|---|---|---|
| Organs | Stomach<br>Liver<br>Gallbladder<br>Pancreas<br>Spleen<br>First half of duodenum | Second half of duodenum<br>Jejunum<br>Ileum<br>Cecum<br>Ascending colon<br>Two-thirds of transverse colon | Left one-third of transverse colon<br>Descending colon<br>Sigmoid colon<br>Rectum |
| Arteries | Celiac trunk:<br>Splenic<br>Left gastric<br>Common hepatic | Superior mesenteric:<br>Ileocolic<br>Right colic<br>Middle colic | Inferior mesenteric:<br>Left colic<br>Sigmoid branches<br>Superior rectal |
| Ventral mesentery | Lesser omentum<br>Falciform ligament<br>Coronary/triangular ligaments | None | None |
| Dorsal mesentery | Gastrosplenic ligament<br>Splenorenal ligament<br>Gastrocolic ligament<br>Greater omentum | Mesointestine<br>Mesoappendix<br>Transverse mesocolon | Sigmoid mesocolon |
| *Nerve supply:*<br>Parasympathetic<br>Sympathetic | <br>Vagus nerve (CN X)<br>Thoracic splanchnic nerves (T5-T10) | <br>Vagus nerve (CN X)<br>Thoracic splanchnic nerves (T11-T12) | <br>Pelvic splanchnic nerves (S2-S4)<br>Lumbar splanchnic nerves (L1-L2) |

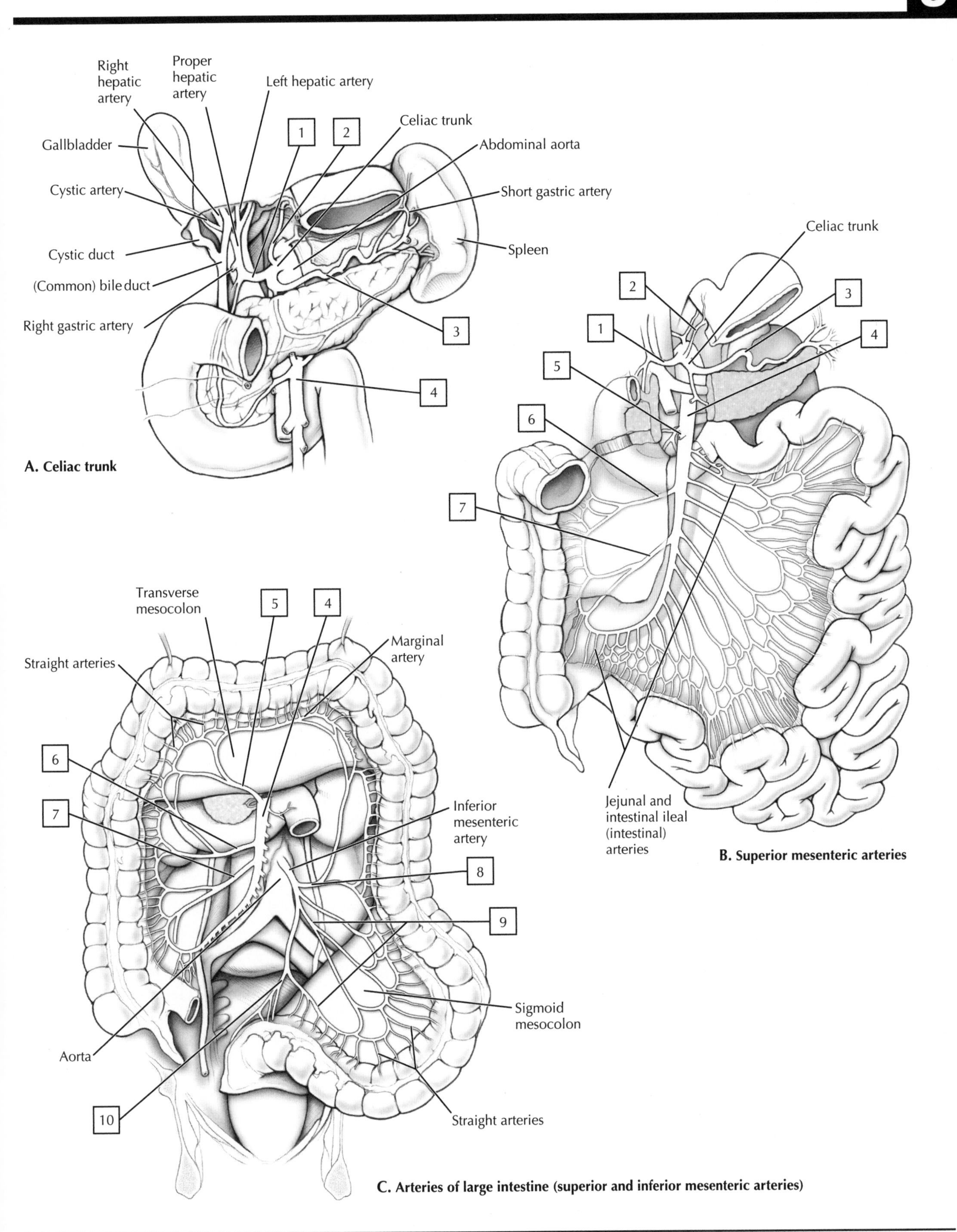


A. Celiac trunk

B. Superior mesenteric arteries

C. Arteries of large intestine (superior and inferior mesenteric arteries)

# 5 Arteries of the Pelvis and Perineum

The abdominal aorta divides at the level of the L4 vertebra into the **right** and **left common iliac arteries**. The common iliac arteries then divide into the external iliac arteries, each of which passes forward and beneath the inguinal ligament to enter the thigh as the femoral arteries and the **internal iliac arteries**. The internal iliac arteries supply the pelvic viscera, its muscular walls, the muscles of the gluteal (buttock) region, and the perineum and external genitalia. The major branches of the pelvic arteries are summarized in the following table (note these are for the female).

| ARTERY (DIVISION)* | COURSE AND STRUCTURES SUPPLIED |
|---|---|
| Common iliac | Divides into external (to thigh) and internal (to pelvis) iliac |
| Internal iliac | Divides into posterior division (P) and anterior division (A) |
| *Iliolumbar (P)* | To iliacus (iliac artery), psoas, and quadratus lumborum muscles and spine (lumbar artery) |
| *Lateral sacral (P)* | Piriformis muscle and sacrum (meninges and nerves) |
| *Superior gluteal (P)* | Between lumbosacral trunk and S1 nerves, through greater sciatic foramen and into gluteal region |
| *Inferior gluteal (A)* | Between S1 or S2 and S2 or S3 to gluteal region |
| *Internal pudendal (A)* | To perineal structures, passing through the greater sciatic foramen and then back through the lesser sciatic foramen to the perineum |
| *Umbilical (A)* | Gives rise to superior vesical artery to bladder and becomes medial umbilical ligament when it reaches anterior abdominal wall |
| *Obturator (A)* | Passes into medial thigh via obturator foramen (with nerve) |
| *Uterine (A)* | Runs over levator ani and under the ureter to reach uterus |
| *Vaginal (A)* | From internal iliac or uterine, passes to vagina |
| *Middle rectal (A)* | To lower rectum and superior part of anal canal |
| Ovarian | From abdominal aorta, runs in suspensory ligament of ovary |
| Superior rectal | Continuation of inferior mesenteric artery (IMA) to rectum |
| Median sacral | From aortic bifurcation, unpaired artery to sacrum and coccyx |

**A*, branch of anterior trunk; *P*, branch of posterior trunk.

Arteries for the male are similar, except that the uterine, vaginal, and ovarian branches are replaced by arteries to the ductus deferens (from a vesical branch), prostate (from the inferior branch), and testis (from the aorta). Significant variability exists for these arteries, so they are best identified by naming them for the structure they supply.

The blood supply to the perineum is via the **internal pudendal artery** from the internal iliac artery. The internal pudendal (pudendal means "shameful") artery gives rise to the following branches:

- **Inferior anorectal**: to the external anal sphincter
- **Perineal**: arises from the pudendal and provides branches to the labia (scrotum in males)
- **Terminal portion of the pudendal**: terminates by providing branches to the erectile tissues (bulb of the vestibule in females and bulb of the penis in males) and branches to the crus of the clitoris (crus of the penis in male)

**COLOR** the following branches of the internal iliac artery, using a different color for each artery:

- ☐ 1. **Superior gluteal**
- ☐ 2. **Umbilical**
- ☐ 3. **Inferior gluteal**
- ☐ 4. **Internal pudendal**
- ☐ 5. **Inferior anorectal**
- ☐ 6. **Superior vesical**
- ☐ 7. **Uterine**
- ☐ 8. **Obturator**
- ☐ 9. **Perineal**

***Clinical Note:***
**Erectile dysfunction** (ED) is an inability to achieve and/or maintain penile erection sufficient for sexual intercourse. Its occurrence increases with age. Normal erectile function occurs when a sexual stimulus causes the release of nitric oxide from nerve endings and the endothelial cells of the corpus cavernosum. This relaxes the vascular smooth muscle tone and increases the blood flow that simultaneously engorges the erectile tissues and compresses the veins that might otherwise drain the blood away. Available drugs to treat ED in males aid in relaxing the vascular smooth muscle of the small arteries that supply the penile erectile tissues (these arteries are branches of the internal pudendal). This same mechanism also functions in females, and is responsible for erectile tissue engorgement of the bulb of the vestibule and crus of the clitoris.

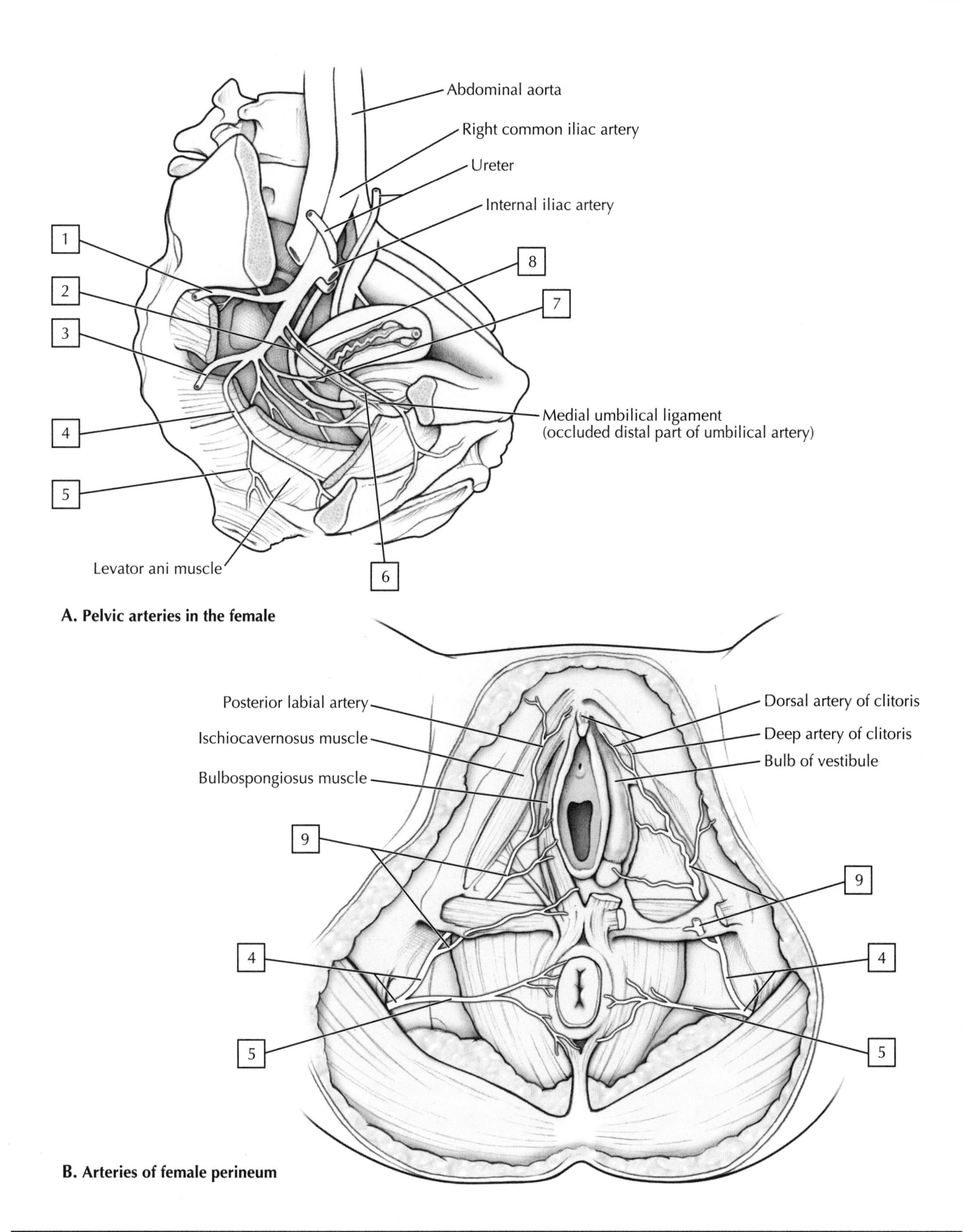


**A. Pelvic arteries in the female**

**B. Arteries of female perineum**

# 5 Veins of the Thorax

The venous system of the body cavities (thorax and abdominopelvic cavities) is composed of the:

- Caval system: superior and inferior vena cava and their tributaries
- Hepatic portal system: hepatic portal vein and its tributaries

The **caval system** drains:

- Body walls, including the musculoskeletal components and the overlying skin
- Head and neck regions, via the dural venous sinuses (brain) and the internal and external jugular system of veins
- Upper and lower limbs, via a set of deep and superficial veins that ultimately drain into the superior vena cava (upper limb) or into the inferior vena cava (lower limb)
- Inferior vena cava, which drains the abdominopelvic cavity, and receives venous return from the lower limbs, but does not drain the GI tract or its accessory organs (liver, gallbladder, and pancreas) or the spleen

The **hepatic portal system** drains the:

- GI tract in the abdominopelvic cavity and its accessory organs (liver, gallbladder, pancreas) via its superior and inferior mesenteric veins and its branches and their tributaries
- Spleen, an organ of the lymphoid system, via the splenic vein

In the thorax, the thoracic walls and visceral structures (lungs, esophagus, thymus) are drained by the **azygos system** of veins (the heart is drained by its own system of cardiac veins). The azygos venous blood ultimately drains into the superior vena cava (SVC) just before the SVC enters the right atrium of the heart. The azygos tributaries include the:

- Intercostal veins (posterior)
- Hemiazygos vein (drains into the azygos vein and often the accessory hemiazygos vein)
- Accessory hemiazygos vein (ultimately drains into the left brachiocephalic vein)
- Lumbar veins (ascending connections with azygos vein)
- Esophageal veins
- Mediastinal veins
- Pericardial veins
- Bronchial veins
- Right superior intercostal vein

The azygos system forms an important venous conduit between the inferior vena cava and the SVC. It is part of the deep venous drainage system but has connections with superficial veins that course in the subcutaneous tissues. The azygos system of veins generally does not possess valves (so the direction of blood flow is pressure dependent) and its branches can be variable, as is typical of the venous system in general.

Remember, veins are variable and more numerous than arteries; at rest, about two-thirds of the blood resides in the venous system.

**COLOR** the following veins, using a different color for each vein:

- ☐ 1. **Right brachiocephalic**
- ☐ 2. **Superior vena cava (SVC)**
- ☐ 3. **Azygos**
- ☐ 4. **Left brachiocephalic**
- ☐ 5. **Accessory hemiazygos**
- ☐ 6. **Hemiazygos**

***Clinical Note:***

Be aware of the variability in the venous drainage, especially in the thoracic and abdominal regions, and in the distribution of superficial veins in the upper and lower extremities. A good rule of thumb is to remember that veins are variable. At rest, about 65% of the blood resides in the venous system; veins are variable, are capacitance vessels, and in most regions of the body veins occur as a superficial set of veins in the subcutaneous tissues that connect with a deeper set of veins that more closely parallel the major arteries.

Clinically, some of the more prominent superficial veins may be accessed for **venipuncture** to withdraw a blood sample for clinical examination, or to administer fluids into the bloodstream.

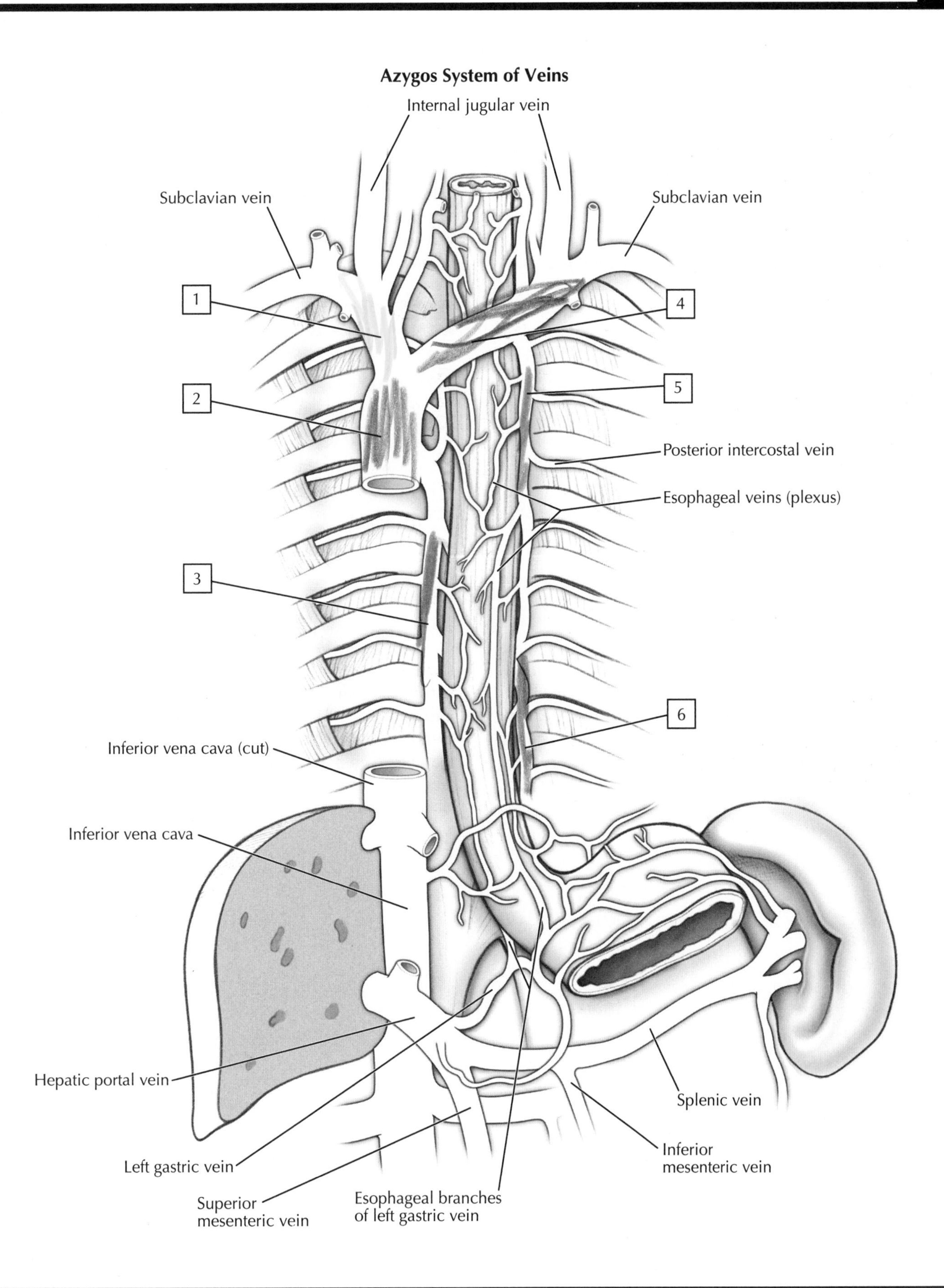
Azygos System of Veins
Internal jugular vein
Subclavian vein
Subclavian vein
1
4
2
5
Posterior intercostal vein
Esophageal veins (plexus)
3
6
Inferior vena cava (cut)
Inferior vena cava
Hepatic portal vein
Splenic vein
Left gastric vein
Inferior mesenteric vein
Superior mesenteric vein
Esophageal branches of left gastric vein

# 5 Veins of the Abdominopelvic Cavity

Veins that drain everything in the abdominopelvic cavity except the GI tract, its accessory organs (liver, gallbladder, pancreas), and the spleen are tributaries that primarily drain into the **inferior vena cava** (IVC).

Venous drainage from the pelvis occurs primarily into tributaries that correspond to the arterial branches of the internal iliac artery, and are correspondingly given the same names. Ultimately, this venous blood collects in the common iliac veins, which then drain into the IVC. The perineum and external genitalia are largely drained by the **internal pudendal vein**, which corresponds to the artery of the same name that supplies this region. The IVC runs superiorly and pierces the dome of the respiratory diaphragm anterior to the T8 vertebra to drain directly into the right atrium of the heart.

The major tributaries of the IVC include the:

- Common iliac veins
- Lumbar veins (the upper lumbar veins usually form connections to the azygos system of veins via ascending lumbar veins)
- Right gonadal (ovarian or testicular) vein (the left gonadal vein drains into the left renal vein)
- Renal veins
- Right suprarenal vein (left suprarenal drains into the left renal vein)
- Inferior phrenic veins
- Hepatic veins

These abdominopelvic veins do not possess valves, so blood flow direction is dependent upon the pressure gradient in the vessels. As with the azygos system in the thorax, connections with superficial veins in the subcutaneous tissues occur with the veins draining the interior body walls.

A set of **superficial veins** drain the anterolateral abdominal wall, the superficial inguinal region, rectus sheath, and lateral thoracic wall. Most of its connections ultimately drain into the axillary vein, then into the subclavian veins, and then into the two brachiocephalic veins, which unite to form the superior vena cava.

**Inferior epigastric veins** from the external iliac veins enter the posterior rectus sheath, course cranially above the umbilicus as the superior epigastric veins, and then anastomose with the internal thoracic veins that drain into the subclavian veins.

**COLOR** the following veins, using a different color for each vein:

- ☐ 1. **IVC**
- ☐ 2. **External iliac**
- ☐ 3. **Internal iliac**
- ☐ 4. **Inferior rectal**
- ☐ 5. **Hepatic**
- ☐ 6. **Renal**
- ☐ 7. **Right and left gonadal (ovarian or testicular veins)***

*Note that the left gonadal vein drains into the left renal vein, not the IVC.

***Clinical Note:***

If venous return through the liver is obstructed, say, by **cirrhosis**, then these superficial veins (the superficial epigastric and paraumbilical veins) can become engorged and so dilated that they form a **caput medusae** (tortuous enlarged veins around the umbilicus) (see Plate 5-19).

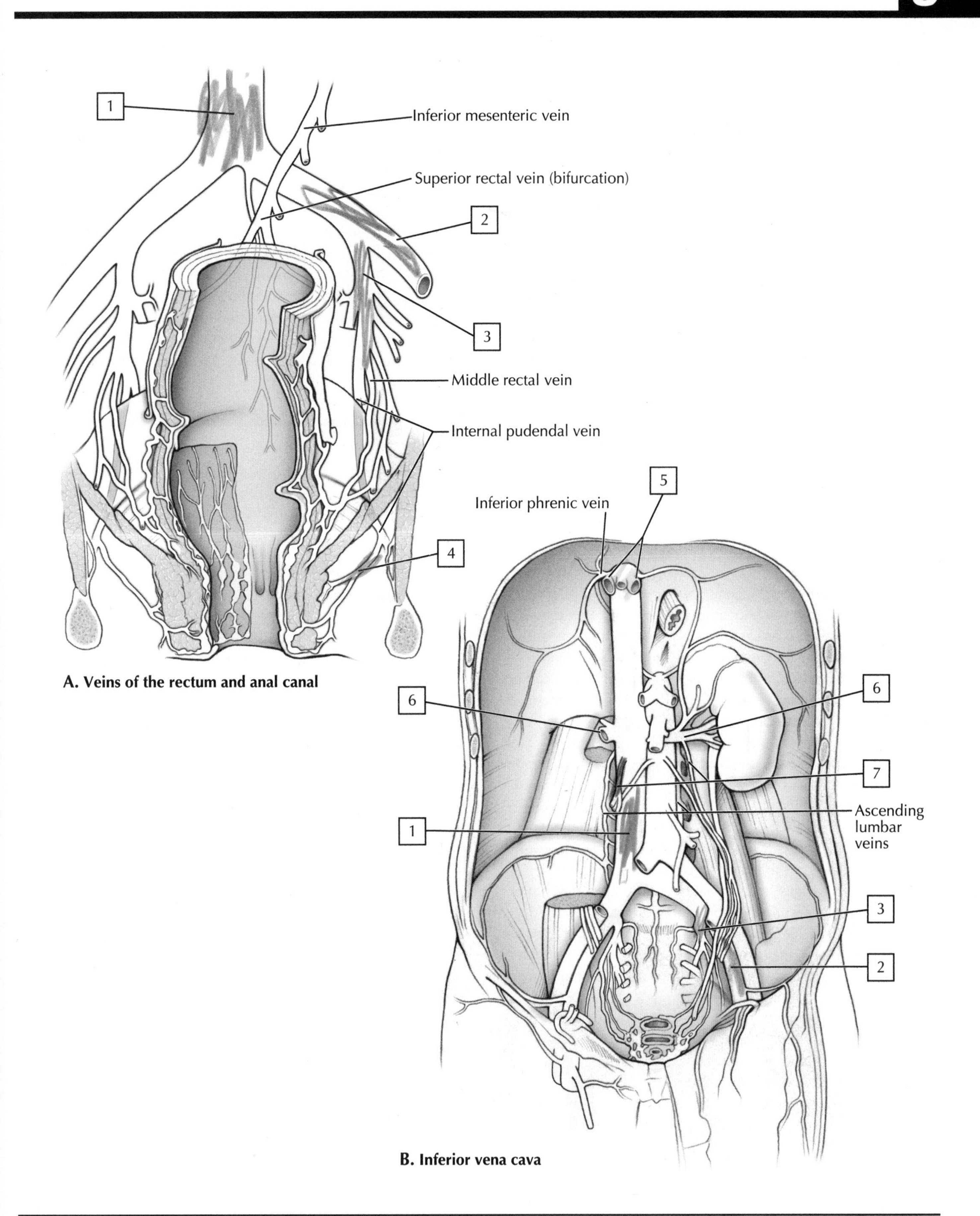


A. Veins of the rectum and anal canal

B. Inferior vena cava

# Portosystemic Anastomoses

The GI tract, its accessory organs (gallbladder, liver, pancreas), and the spleen are drained by the hepatic portal system of veins. Four major veins make up this system:

- **Inferior mesenteric vein (IMV)**: drains the hindgut derivatives of the GI tract, including the distal transverse colon, descending colon, sigmoid colon, and proximal rectum; it may drain into the junction of the superior mesenteric vein and splenic vein, drain only into the splenic vein, or drain directly into the superior mesenteric vein (remember, veins are variable)
- **Superior mesenteric vein (SMV)**: drains the midgut derivatives of the GI tract, including the distal duodenum, small intestine, ascending colon, and proximal transverse colon, as well as the pancreas
- **Splenic vein**: drains the spleen, a portion of the stomach, and pancreas; the IMV often drains into the splenic vein, as shown in the image, but as noted previously, the IMV may also drain directly into the SMV or the junction of the SMV and the splenic vein
- **Hepatic portal**: formed by the union of the **splenic** and **superior mesenteric veins**, this large vein drains the stomach (via left and right gastric veins) and gallbladder, and receives all the venous drainage from the GI tract via the superior and inferior mesenteric veins and the splenic vein; occasionally, the gallbladder will drain directly into the liver

All of the blood from the visceral structures listed previously drains ultimately into the hepatic portal vein and then into the liver. The liver processes important products and energy sources (glucose, fat, protein, vitamins) from the GI tract, produces cellular fuels, produces plasma proteins and clotting factors, metabolizes toxins and drugs, excretes substances like bilirubin, and produces bile acids. From the liver, the venous blood flows into several hepatic veins, which immediately drain into the IVC just before it pierces the respiratory diaphragm and enters the right atrium of the heart.

Various conditions, such as cirrhosis, can damage the liver and impede venous blood flow through this vital organ. However, blood must return to the heart for gas exchange in the lungs, so it will bypass the liver via important **portosystemic anastomoses** to gain access to the caval system (SVC, IVC, and azygos veins) and its tributaries, which can then return the blood to the heart. The impeded venous return raises the blood pressure in the portal system, causing **portal hypertension**; because the veins of the portal system lack valves, the venous blood can reverse flow and seek alternative routes back to the heart. Clinically, these portosystemic anastomoses are lifesaving and include the following major routes:

- **Esophageal**: blood will shunt from the hepatic portal and splenic veins into gastric veins of the stomach and then into esophageal veins that are connected to the azygos system of veins, ultimately draining into the SVC and the heart (see part ***A***) (next to the labels in the image)
- **Rectal**: blood will drain inferiorly in the inferior mesenteric vein to the superior rectal vein and then into the middle and inferior rectal veins (anastomosis around the rectum) to reach the IVC and the heart (see part ***B*** next to the labels in the image)
- **Paraumbilical**: blood from the portal vein will drain into the paraumbilical veins and fill the subcutaneous veins of the abdominal wall (forms a tortuous tangle of veins visible on the abdominal surface called the caput medusae), which then may drain into tributaries of the SVC, IVC, and azygos system (see part ***C*** label in the image)
- **Retroperitoneal**: least important of the pathways; some blood will drain from retroperitoneal GI viscera into parietal veins in the body wall to reach the caval tributaries (not shown)

**COLOR** the following veins that contribute to the portacaval anastomotic system, using the colors suggested for each vein:

- ☐ **1. Portal (dark blue)**
- ☐ **2. Superior mesenteric (purple)**
- ☐ **3. Splenic (dark red)**
- ☐ **4. Inferior mesenteric (light blue)**

***Clinical Note:***

**Cirrhosis**, a largely irreversible disease, is characterized by diffuse fibrosis, parenchymal nodular regeneration, and a disturbed hepatic architecture that progressively disrupts portal blood flow through the liver (leading to portal hypertension). Major causes of cirrhosis include:

- Alcoholic liver disease: 60% to 70%
- Viral hepatitis: 10%
- Biliary diseases: 5% to 10%
- Other: 5% to 15%

**Portal hypertension,** a result of the increased resistance to venous blood flow through the diseased liver, has the following clinical consequences:

- Ascites (abnormal accumulation of fluid in the abdominal cavity)
- Formation of portosystemic venous shunts via the anastomoses previously noted
- Congestive splenomegaly (engorgement of the spleen with blood)
- Hepatic encephalopathy (toxins in the blood, not removed by the diseased liver, cause brain disease)

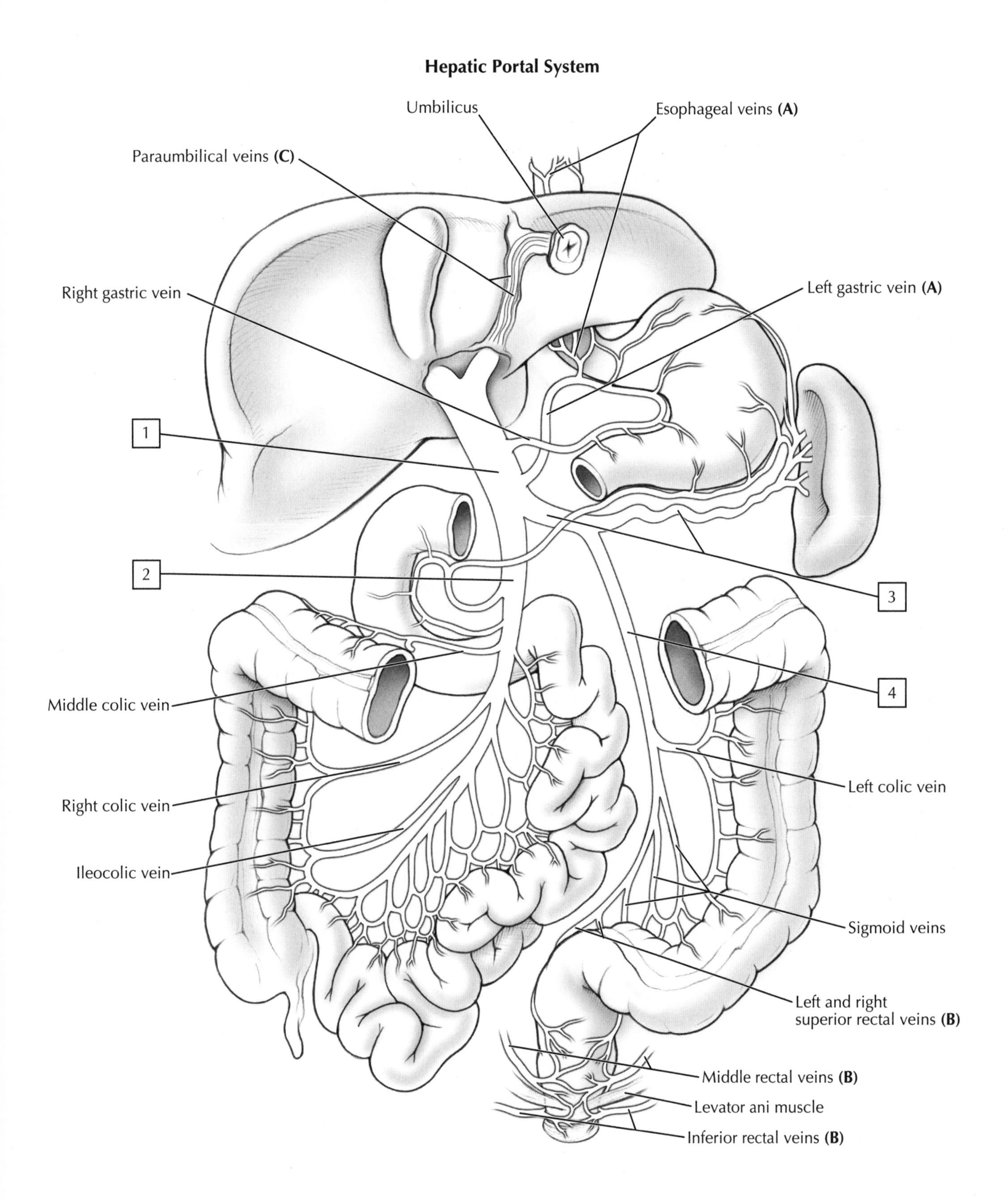
Hepatic Portal System
Umbilicus
Esophageal veins (A)
Paraumbilical veins (C)
Right gastric vein
Left gastric vein (A)
1
2
3
4
Middle colic vein
Right colic vein
Ileocolic vein
Left colic vein
Sigmoid veins
Left and right superior rectal veins (B)
Middle rectal veins (B)
Levator ani muscle
Inferior rectal veins (B)

# 5 Veins of the Upper Limb

Like the rest of the body, the upper limb is drained by a set of deep and superficial veins. However, the veins of the upper (and lower) limb contain valves, which help (largely by the action of adjacent muscle contraction) to return venous blood to the heart against gravity.

The **deep set of veins** of the upper limb parallels the arteries and includes the following major veins:

- **Radial**: parallels the radial artery in the lateral forearm
- **Ulnar**: parallels the ulnar artery in the medial forearm
- **Brachial**: formed by the union of the radial and ulnar veins in the cubital fossa; this vein parallels the brachial artery in the medial aspect of the arm
- **Axillary**: in the armpit, it parallels the axillary artery in the axillary sheath (surrounded by the cords of the brachial nerve plexus)
- **Subclavian**: parallels the subclavian artery but passes anterior to the anterior scalene muscle rather than posterior to it (artery lies posterior)

The **superficial set of veins** of the upper limb is connected by communicating veins to the deep set of veins and provides an additional route for venous return to the heart. These veins can vary considerably from person to person and have extensive tributaries. The veins also have valves to assist in venous return and include the following major veins:

- **Dorsal venous network**: most of the blood from the palm will drain into these veins (especially when the hand is squeezed)
- **Cephalic**: runs in the subcutaneous tissue along the lateral forearm and arm to ultimately drain into the axillary vein
- **Basilic**: runs in the subcutaneous tissue along the medial forearm and distal arm to ultimately dive deep into the medial arm and drain into the axillary vein
- **Median cubital**: passes from the cephalic vein to the basilic vein in the cubital fossa and is a common site for **venipuncture** to withdraw a blood sample or administer fluids intravenously

**COLOR** the following veins of the upper limb, using a different color for each vein:

☐ 1. **Subclavian**
☐ 2. **Axillary**
☐ 3. **Cephalic (superficial)**
☐ 4. **Brachial**
☐ 5. **Median cubital (superficial)**
☐ 6. **Radial**
☐ 7. **Ulnar**
☐ 8. **Basilic (superficial)**

***Clinical Note:***

In general, veins are more numerous than arteries and more variable in their location. They often parallel arteries, especially deep within the body or extremities. Veins of the limbs and those in the lower neck (internal jugular veins) contain valves, whereas most other veins in the body are valveless. Often, when a vein such as the brachial or axillary vein parallels the artery of the same name, the vein actually forms "**venae comitantes**" (accompanying veins), or a network of veins that entwines the parallel artery as vines might entwine a tree trunk. With several major exceptions, many veins may be sacrificed during surgery because so many alternative venous channels exist to return blood from a region back to the heart (of course, if venous repair is feasible, it is preferred). Additionally, the body will usually "sprout" new veins from adjacent tributaries to drain an area denuded of its venous drainage.

The median cubital vein is often used for **venipuncture** (withdrawal of blood or injection of solutions into the venous system). Tying a tourniquet around the arm will constrict the venous return, and the forearm superficial veins distal to the tourniquet will become distended and become palpable and often visible. These engorged veins (usually the median cubital vein) can then be used to withdraw a blood sample or used for intravenous injections, the administration of fluids, or for intravenous feeding. Superficial veins on the dorsal aspect of the hand also may be used for venipuncture.

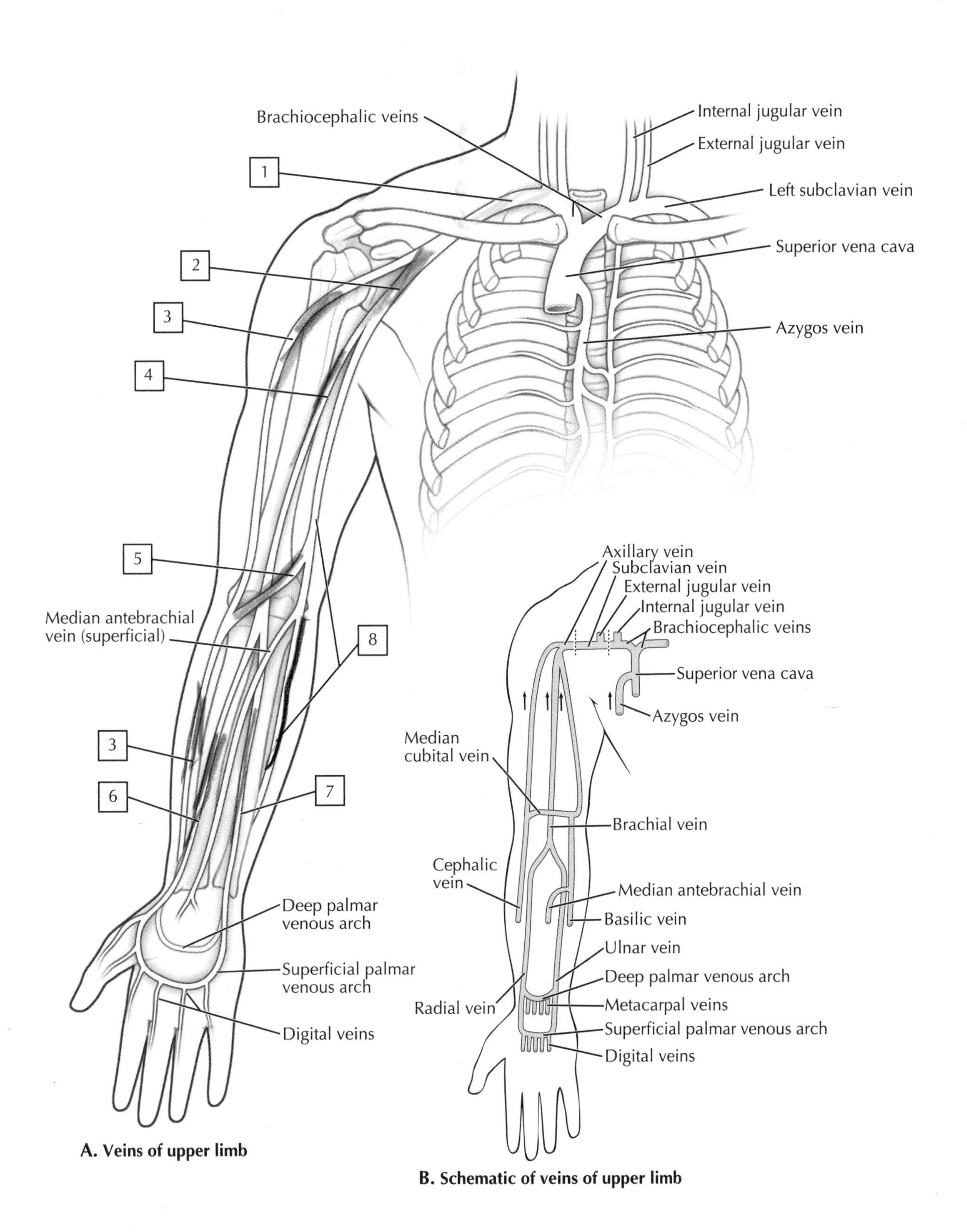


**A. Veins of upper limb**

**B. Schematic of veins of upper limb**

# 5 Veins of the Lower Limb

Like the rest of the body, the lower limb is drained by a set of deep and superficial veins. However, the veins of the lower (and upper) limb contain valves, which help (largely by the action of adjacent muscle contraction) to return venous blood to the heart against gravity.

The **deep set of veins** of the lower limb parallels the arteries and includes the following major veins:

- **Posterior tibial**: drains from the sole of the foot and medial ankle superiorly up the leg, paralleling the posterior tibial artery in the posterior compartment of the leg
- **Anterior tibial**: begins as the dorsalis pedis vein on the dorsum of the foot and parallels the anterior tibial artery in the anterior compartment of the leg
- **Fibular**: small vein that parallels the artery of the same name in the lateral compartment of the leg and drains into the posterior tibial vein
- **Popliteal**: lies behind the knee and is formed by the anterior and posterior tibial veins
- **Femoral**: the popliteal vein becomes the femoral vein in the distal thigh, and then the femoral vein drains deep to the inguinal ligament to become the external iliac vein in the pelvis

The **superficial set of veins** of the lower limb is connected by communicating veins to the deep set of veins and provides an additional route for venous return to the heart. These veins can vary considerably from person to person and have extensive tributaries. The veins also have valves to assist in venous return and include the following major veins:

- **Dorsal venous arch**: drains blood from the foot into the small and large saphenous veins at the lateral and medial aspects of the ankle, respectively
- **Small saphenous**: courses superiorly in the subcutaneous tissue of the calf (posterior aspect of the leg) and then dives deeply to drain into the popliteal vein behind the knee
- **Great saphenous**: courses superiorly from the medial side of the ankle to run up the medial leg and thigh, draining into the femoral vein just inferior to the inguinal ligament

Note that the great saphenous vein of the lower limb and the cephalic vein of the upper limb are analogous veins, as are the small saphenous vein of the lower limb and the basilic vein of the upper limb (both dive deeply to join a deeper vein).

**COLOR** the following veins of the lower limb, using a different color for each vein:

- ☐ 1. **Femoral**
- ☐ 2. **Great saphenous (superficial)**
- ☐ 3. **Anterior tibial**
- ☐ 4. **Popliteal**
- ☐ 5. **Small saphenous (superficial)**
- ☐ 6. **Posterior tibial**

***Clinical Note:***

Veins of the extremities and those in the lower neck contain valves. The valves are an extension of the tunica intima of the venous wall, project into the vein's lumen, and are similar in appearance to the semilunar valves of the heart. Venous valves assist in venous return against gravity by preventing the backflow of blood. The blood in the veins of the extremities is propelled along, in part, by the contraction of adjacent skeletal muscles. The walls of veins adjacent to a valve can become weakened and distended, thus compromising the ability of the valve to work properly and affecting venous return. Such veins are called **varicose** (enlarged and tortuous) veins, and this condition is most common in the veins of the lower limb, especially the saphenous vein and its tributaries.

**Deep venous thrombosis** (DVT) to the deep veins of the lower limb results in swelling, warmth, and inflammation leading to infection. **Venous stasis** (stagnation of the blood), caused by external pressure on the veins (tight bandages, prolonged periods of being bedridden, inactivity common on long airline flights), is a common cause of DVT. Sometimes, a large thrombus may develop, break free, and travel via the venous system proximally to the heart, and become lodged in the pulmonary artery, a life-threatening event.

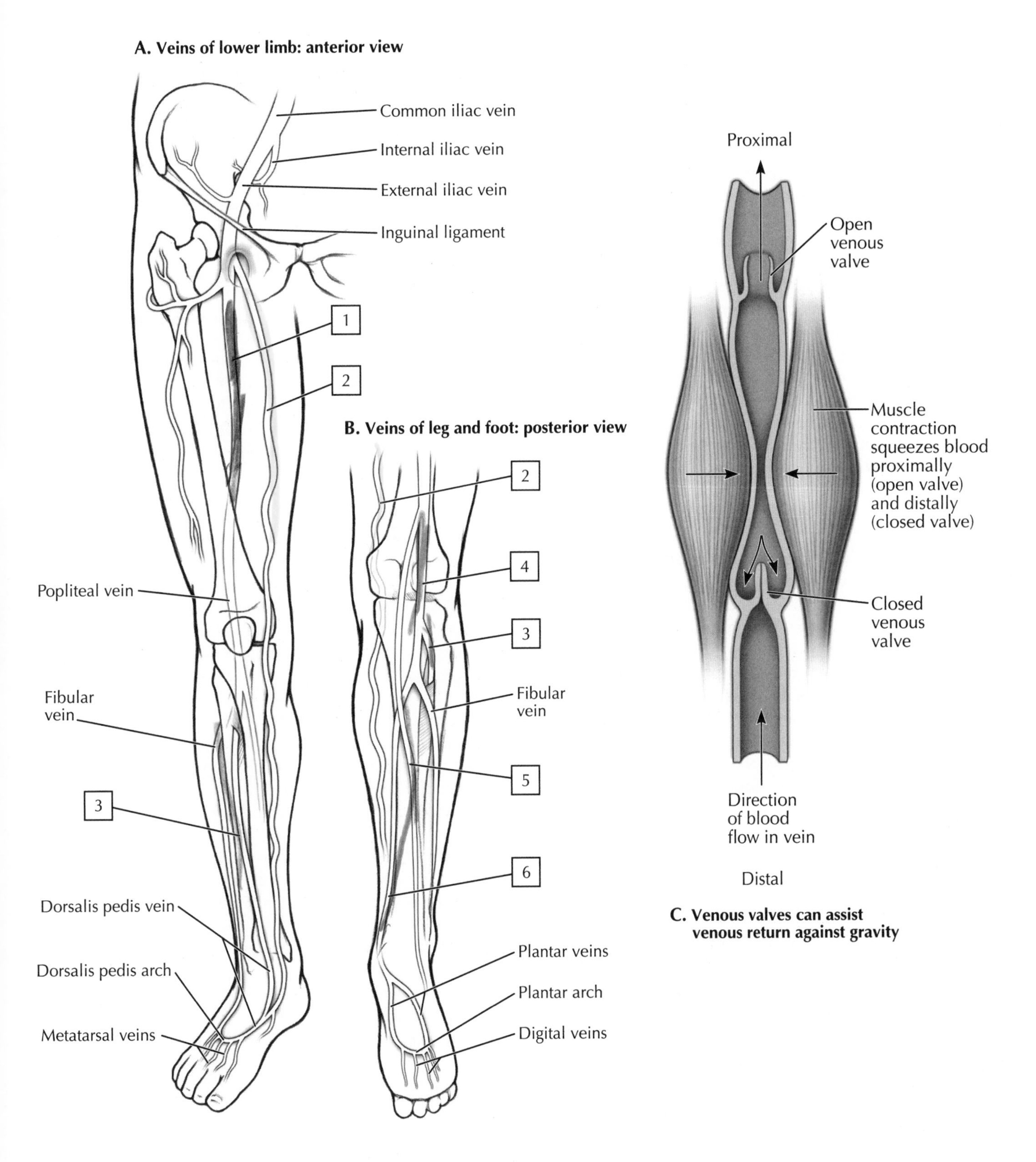
A. Veins of lower limb: anterior view
Common iliac vein
Internal iliac vein
External iliac vein
Inguinal ligament
1
2
Popliteal vein
Fibular vein
3
Dorsalis pedis vein
Dorsalis pedis arch
Metatarsal veins
B. Veins of leg and foot: posterior view
2
4
3
Fibular vein
5
6
Plantar veins
Plantar arch
Digital veins
Proximal
Open venous valve
Muscle contraction squeezes blood proximally (open valve) and distally (closed valve)
Closed venous valve
Direction of blood flow in vein
Distal
C. Venous valves can assist venous return against gravity

# 5 Prenatal and Postnatal Circulation

The pattern of fetal circulation is one of gas exchange and nutrient/metabolic waste exchange across the placenta with the maternal blood (but not the exchange of blood cells) and distribution of oxygen and nutrient-rich blood to the tissues of the fetus. Various shunts allow fetal blood to largely bypass the liver, which is not needed for metabolic processing in utero, and the lungs, also not needed in utero for gas exchange. In fact, in the fetus the lungs are filled with fluid and provide a high resistance to significant blood flow. The mother takes care of oxygenation for the fetus. Therefore, the blood in the fetus needs to bypass the liver and lungs and gain direct access to the left side of the heart so that it may be pumped into the fetal systemic circulation. Several fetal shunts allow this to happen and include the following:

- **Ductus venosus** (bypasses the liver)
- **Foramen ovale** (shunts blood from the right atrium to the left atrium, thus bypassing the lungs)
- **Ductus arteriosus** (shunts any blood that doesn't cross through the foramen ovale and thus gains entrance into the right ventricle and, via ventricular contraction, passes into the pulmonary trunk and through the ductus arteriosus into the aorta, thereby bypassing the lungs)
- **Umbilical arteries and vein** (placental vessels that return blood to the placenta or convey blood from the placenta to the heart)

These shunts close at birth or shortly thereafter, and the newborn infant begins gas exchange through his or her own lungs and processes ingested liquids, and ultimately solid food, through his or her own liver. These changes at birth include the following:

- Ductus venosus becomes a ligament (ligamentum venosum)
- Foramen ovale becomes the fossa ovalis, the thin superior portion of the interatrial septum
- Ductus arteriosus becomes the ligamentum arteriosum between the pulmonary trunk and arch of the aorta
- Umbilical arteries and veins become ligaments

**COLOR** the following features of the prenatal and postnatal circulation (note the direction of blood flow indicated by the arrows):

- ☐ 1. **Umbilical arteries (carry blood, low in oxygen and nutrients, and waste products from fetus to placenta) (color blue as blood is low in oxygen)**
- ☐ 2. **Umbilical vein (carries oxygenated and nutrient-rich blood from placenta to fetal heart) (color red as blood from placenta to fetus is rich in oxygen)**
- ☐ 3. **Ductus venosus (shunt to largely but not completely bypass the fetal liver)**
- ☐ 4. **Foramen ovale (shunt from fetal right atrium to left atrium to bypass the fetal lungs)**
- ☐ 5. **Ductus arteriosus (shunt from pulmonary trunk to aorta to largely bypass the fetal lungs)**
- ☐ 6. **Ligamentum arteriosum (obliterated ductus arteriosus)**
- ☐ 7. **Fossa ovalis (obliterated foramen ovale in the superior part of the interatrial septum)**
- ☐ 8. **Ligamentum venosum (obliterated ductus venosus that once largely bypassed the fetal liver to deliver oxygen- and nutrient-rich blood to the fetal heart)**
- ☐ 9. **Ligamentum teres of liver (obliterated umbilical vein that carried oxygen- and nutrient-rich blood to the fetus from the placenta)**
- ☐ 10. **Medial umbilical ligaments (occluded part of umbilical arteries that carried oxygen- and nutrient-poor blood along with its waste products, back to the placenta)**

Note that while the liver and lungs were largely "off-line" during fetal development, these important organs still needed oxygenated and nutrient-rich blood from the placenta to continue to develop and grow as the fetus grew in utero. They also had to off-load deoxygenated blood and waste products, and return these to the placenta and ultimately the maternal circulation.

***Clinical Note:***

Sometimes the various shunts that exist in the fetal circulation fail to close postnatally, and must be repaired surgically. **Atrial septal defects** account for about 10% to 15% of congenital cardiac anomalies. Most of these defects are ostium secundum defects from the incomplete closure of the foramen ovale (label #4 in part ***A***).

A **patent ductus arteriosus** (PDA) is a failure of the ductus arteriosus to close shortly after birth. This results in a postnatal shunt of blood from the aorta into the pulmonary trunk, which may lead to congestive heart failure. PDA accounts for about 10% of congenital heart defects.

The **Tetralogy of Fallot** usually results from a maldevelopment of the spiral septum that normally divides the truncus arteriosus into the pulmonary trunk and ascending aorta. This defect involves the following:

- Pulmonary stenosis or a narrowing of the right ventricular outflow tract
- An overriding (transposed) aorta
- Right ventricular hypertrophy
- Ventricular septal defect (VSD)

Surgical repair is done on cardiopulmonary bypass to close the VSD and provide unobstructed flow into the pulmonary trunk. The stenotic pulmonary outflow tract is widened by inserting a patch into the wall of the vessel, thus increasing the volume of the subpulmonic stenosis and/or the pulmonary artery stenosis.

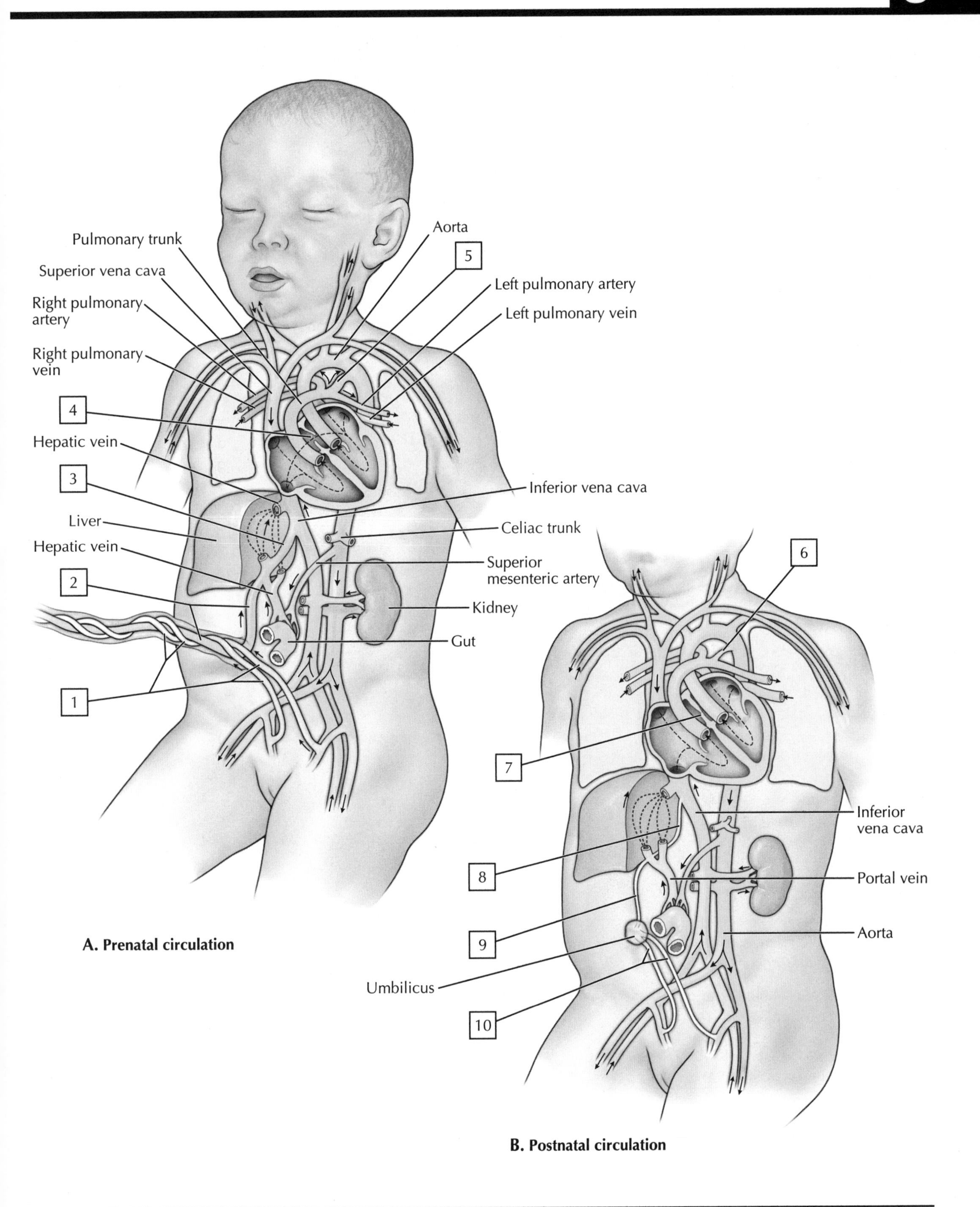


A. Prenatal circulation

B. Postnatal circulation

## REVIEW QUESTIONS

**For each description below, color that feature in the image of the sectioned heart.**

1. This muscle extends into the ventricles and prevents prolapse of the valve leaflets.
2. Blood from the left ventricle passes through this valve.
3. Blood from the left atrium passes through this valve to enter the left ventricle.
4. Blood returning from the lower portion of the body enters the right atrium via this vein.

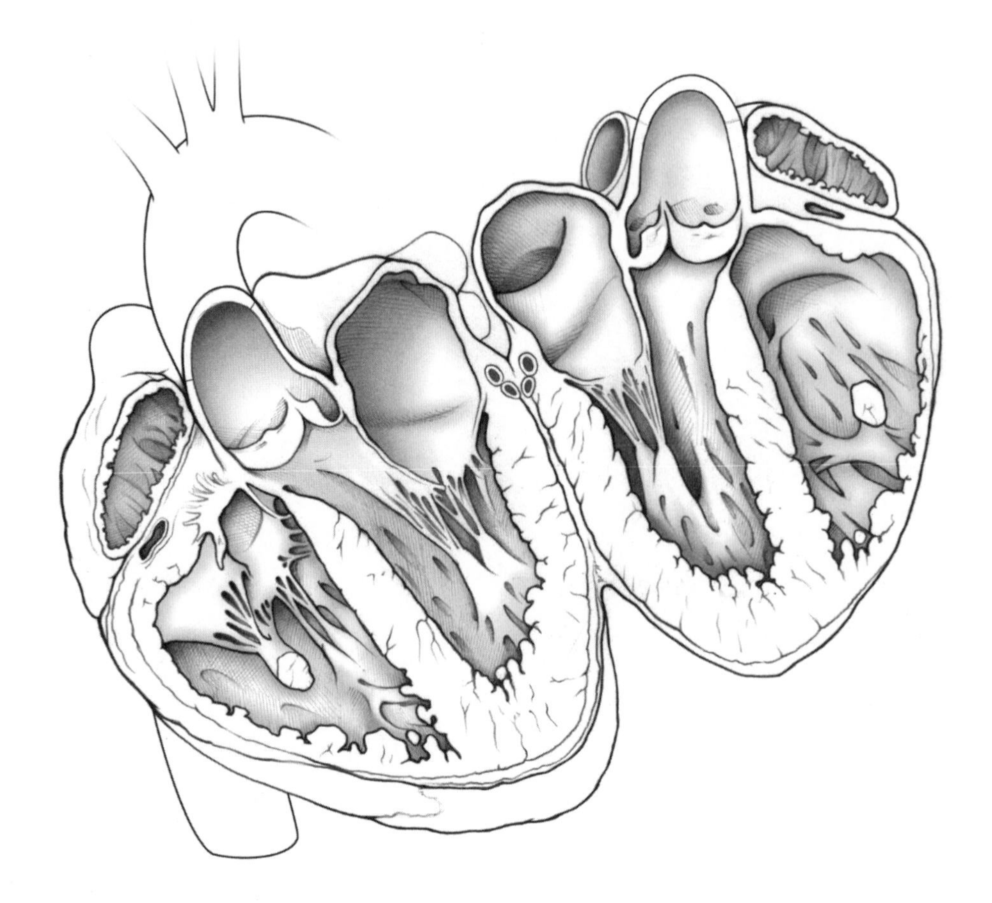

5. An atrial septal defect often occurs at the site of this interatrial septal shunt in the fetal heart. Which of the following structures or features of the fetal heart are involved in this defect?
   A. Ductus arteriosus
   B. Ductus venosus
   C. Foramen ovale
   D. Ligamentum arteriosum
   E. Ligamentum venosum
6. A gunshot wound to the anterior shoulder region traumatizes the cords of the brachial plexus and most likely would also involve damage to which of the following arteries?
   A. Axillary
   B. Brachial
   C. Brachiocephalic
   D. Common carotid
   E. Subclavian

7. The left ovarian vein drains into which of the following veins?
   A. Inferior mesenteric
   B. Inferior vena cava
   C. Left external iliac
   D. Left renal
   E. Portal

8. What three unpaired arteries provide the major blood supply to the abdominal gastrointestinal tract?
   A. ___
   B. ___
   C. ___

9. A laceration to the perineal region would most likely involve bleeding from branches of which major artery supplying this region?
   ___

10. Which artery is responsible for the most distal pulse in the body that is frequently assessed by clinicians? ___
   ___

## ANSWER KEY

1. Papillary muscle(s) in either ventricle
2. Aortic valve
3. Mitral valve
4. Inferior vena cava

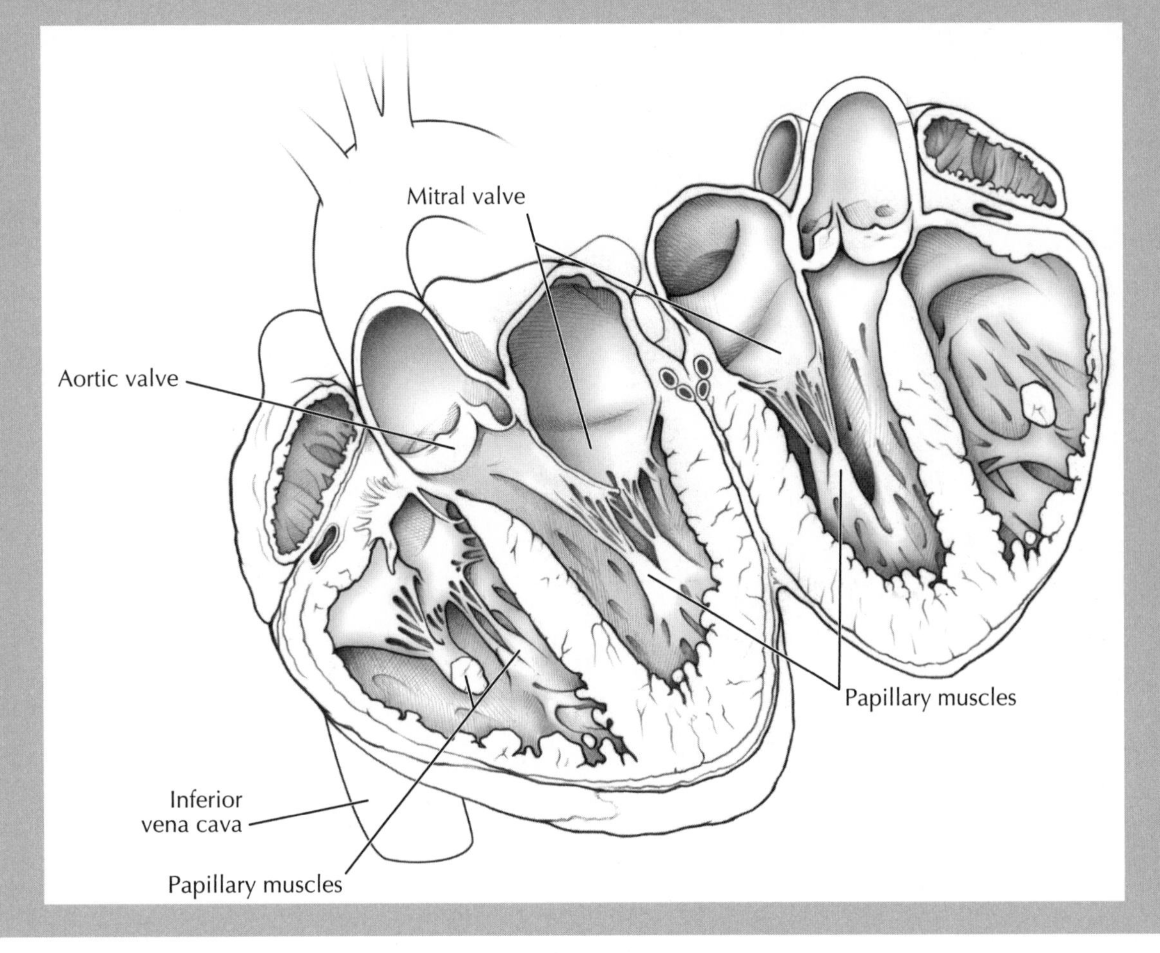

5. C
6. A
7. D
8. Celiac trunk, superior mesenteric artery, and inferior mesenteric artery
9. Internal pudendal artery
10. Dorsalis pedis pulse on the dorsum of the foot

# Chapter 6 Lymphatic Vessels and Lymphoid Organs

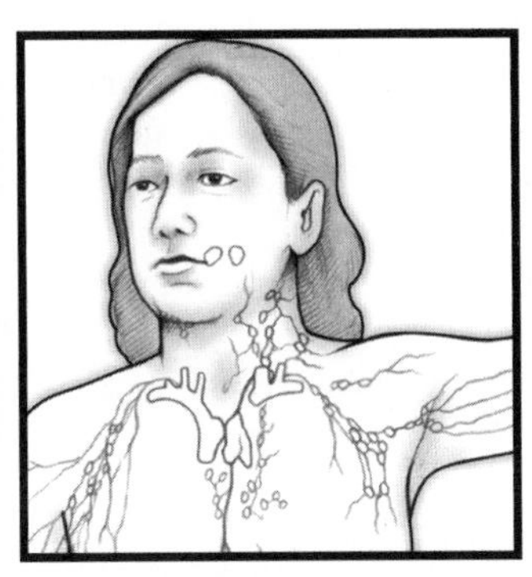
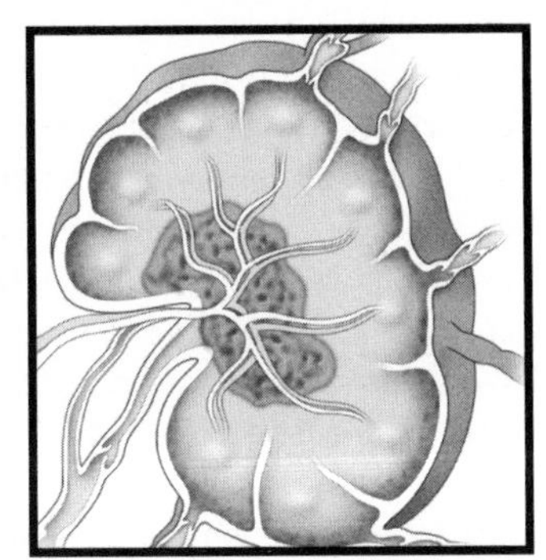
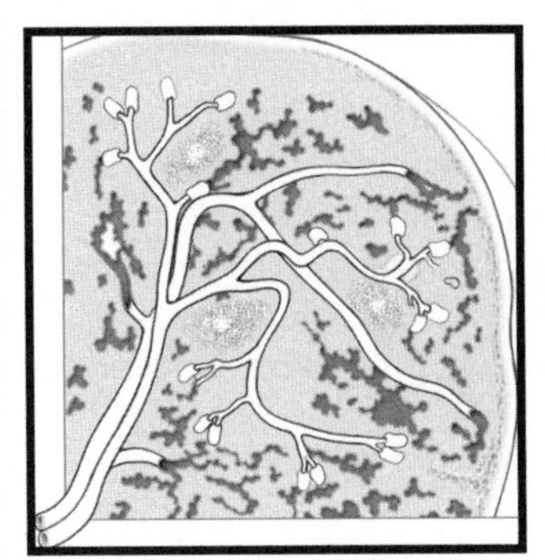
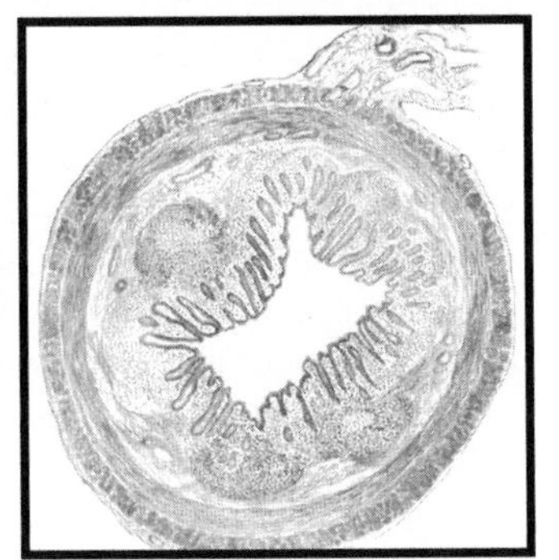

# 6 General Organization of the Lymphatic System

The lymphatic system is intimately associated with the cardiovascular system, both in the development of its lymphatic vessels and in its immune function. The lymphatic system functions to:

- Protect the body against infection by activating defense mechanisms that make up our immune system
- Collect tissue fluids, solutes, hormones, and plasma proteins and return them to the circulatory system (bloodstream)
- Absorb fat (chylomicrons) from the small intestine into the lymphatic lacteals

Components of the lymphatic system include:

- **Lymph**: a watery fluid that resembles plasma but contains fewer proteins and may contain fat together with cells (mainly lymphocytes and a few red blood cells)
- **Lymphocytes**: cellular components of lymph, which include T cells and B cells, and NK cells
- **Lymph vessels**: an extensive network of vessels and capillaries in peripheral tissues that transport lymph and lymphocytes
- **Lymphoid organs**: collections of lymphoid tissue that include lymph nodes, aggregates of lymphoid tissue along the respiratory and gastrointestinal (GI) passageways, tonsils, thymus, spleen, and bone marrow

**COLOR** the lymphoid organs, using a different color for each organ:

- ☐ 1. **Tonsils**
- ☐ 2. **Thymus gland**
- ☐ 3. **Spleen**
- ☐ 4. **Bone marrow**

The body is about 60% fluid by weight, with 40% of it intracellular fluid and 20% of it extracellular fluid (ECF). The lymphatics are essential for returning ECF, solutes, and protein (lost via the capillaries into the ECF compartment) back to the bloodstream. The lymphatics return about 3.5 to 4.0 L of fluid per day back to the bloodstream and also distribute hormones, nutrients (fats from the bowel and proteins from the interstitium), and waste products from the ECF to the bloodstream.

**Lymphatic vessels** transport lymph from everywhere in the body, including the central nervous system via small meningeal lymphatics (our glymphatic system), with the majority of the lymph ultimately collecting in the **thoracic lymphatic duct** (joins the veins at the union of the left internal jugular and left subclavian veins). A much smaller **right lymphatic duct** drains lymphatics from the right upper quadrant of the body to a similar site on the right side. Encapsulated **lymph nodes** are strategically placed to act as "filters" of the lymph as it moves toward the venous system.

**COLOR** the following features of a lymph node, using the colors indicated for each feature:

- ☐ 5. **Vein (blue)**
- ☐ 6. **Artery (red)**
- ☐ 7. **Efferent lymph vessel (yellow)**
- ☐ 8. **Afferent lymph vessels (green)**

Cells associated with the lymphatic system and its immune responses include:

- **Lymphocytes**: **B cells** (bone marrow–derived cells, which make up 10% to 15% of the circulating lymphocytes; can differentiate into **plasma cells**, which secrete antibodies that can bind to foreign antigens); **T cells** (thymus-dependent cells, which make up about 80% of circulating lymphocytes; they attack foreign cells and virus-infected cells, and can be cytotoxic, helper, or suppressor T cells); and **NK cells** (natural killer cells, which make up 5% to 10% of circulating lymphocytes; they attack foreign cells, cancer cells, or virus-infected cells, and constantly provide immunologic surveillance of the body)
- **White blood cells**: monocytes, neutrophils, basophils, and eosinophils (see Plate 5-1).
- **Macrophages**: phagocytic cells that act as scavengers and are antigen-presenting cells, which initiate immune responses
- **Reticular cells**: similar to fibroblasts, these cells can attract T and B cells and dendritic cells
- **Dendritic cells**: bone marrow–derived cells that are potent antigen-presenting cells to T cells and are found mainly in the skin, nose, lungs, stomach, and intestines
- **Follicular dendritic cells**: highly branching cells that mingle with B cells in the germinal center of the lymph node and contain antigen-antibody complexes for months or years, but are not antigen-presenting cells

***Clinical Note:***

Lymph vessels may become secondarily infected (**lymphangitis**) along with lymph nodes (**lymphadenitis**). These processes may result from the lymphatic spread of cancer cells (**metastasis**) to distant sites. Enlarged lymph nodes may be diagnostic of a localized infection and/or metastasis of cancer from a local or distant site. If cancer is suspected, enlarged lymph nodes may be biopsied to determine if they are malignant.

Because of the relatively high incidence of **breast cancer** in women (although it may occur in males as well, but is far less frequent), the lymphatic dissemination (metastases) of cancer cells via the lymphatics can be extensive. However, most of the lymphatic drainage pattern of the breast passes initially to the **axillary lymph nodes**. Routine self-examination of the breast and axillary region, along with routine mammograms, is highly recommended for early detection of breast cancer.

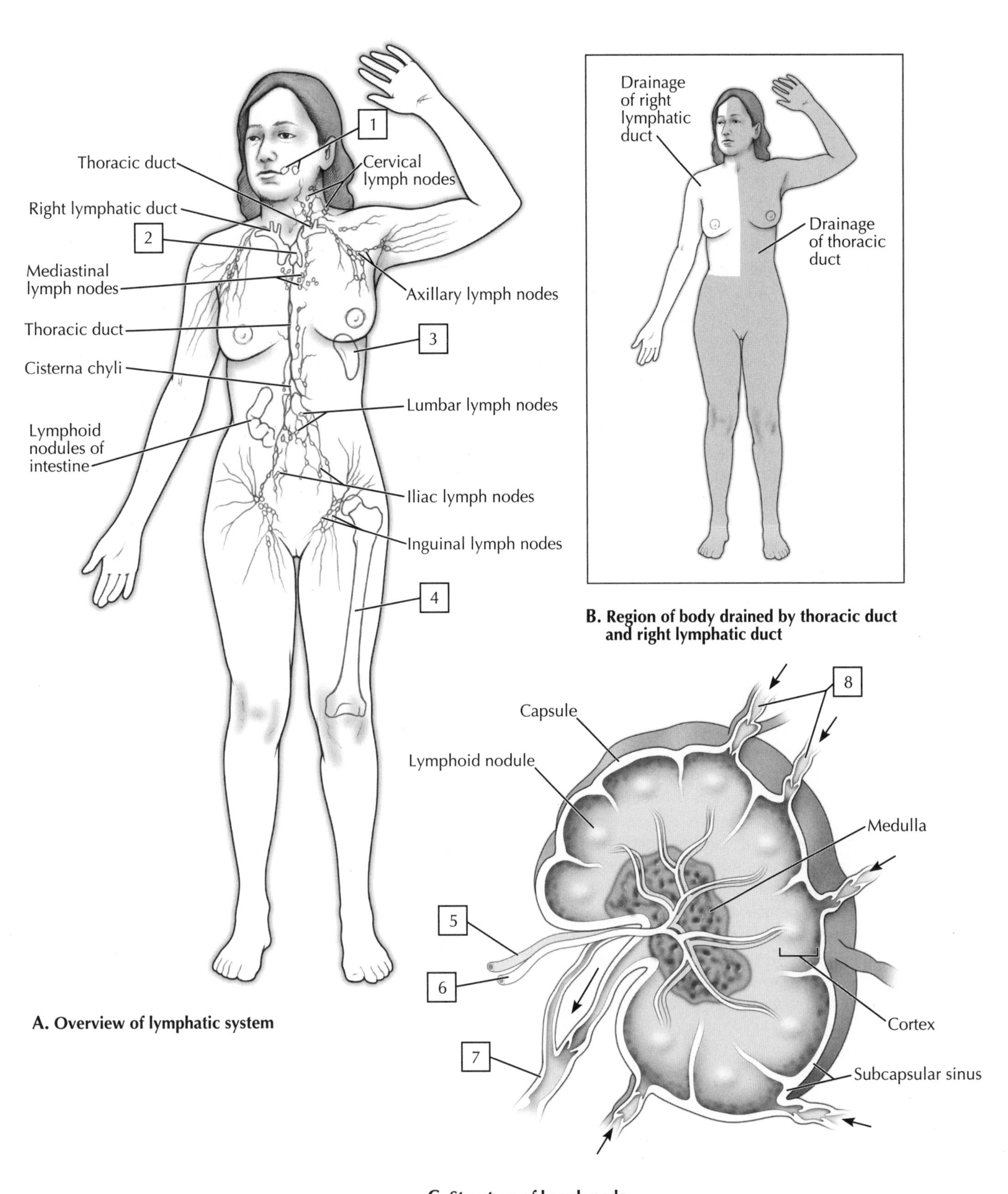


A. Overview of lymphatic system

B. Region of body drained by thoracic duct and right lymphatic duct

C. Structure of lymph node

# Innate Immunity

When a foreign microorganism, virus-infected cell, or cancer cell is detected within the body, the lymphatic system mounts what is called an **immune response**. The detected pathogens are distinguished from the body's own normal cells, and then a response is initiated to neutralize the pathogen. The human body has evolved two major responses to protect against foreign invaders:

- **Nonspecific innate immunity**: a first line of defense composed of **physical barriers** to invasion that include the skin and mucous membranes that line the body's exterior (skin) or its respiratory, GI, urinary, and reproductive systems; additional barriers include mucosae and their secretions, which may include enzymes and acidic secretions (lysozymes interferons, fibronectin, serum complement); saliva containing thiocyanate; and physical coughing and sneezing to remove pathogens and irritants; inflammation and fever also are part of this defense system

Thus the hallmark of innate immunity is **inflammation**, a relatively nonspecific response with symptoms of redness, heat, swelling, and pain. The key elements of inflammation include:

- **Tissue injury**: the physical nonspecific barriers are breached by a pathogen
- **Leukocytosis**: significant increase in white blood cells in the bloodstream, primarily neutrophils, which flow to and migrate from the vasculature (diapedesis) into the site of inflammation; also an increase in phagocytic cells such as macrophages, monocytes, and natural killer (NK) cells
- **Release of chemical inflammatory mediators**: histamine (mast cells and basophils), kinins (neutrophils and other sources), prostaglandins (neutrophils and other cells), cytokines (leukocytes, fibroblasts, endothelial cells, lymphocytes), and complement (normally inactive, circulating plasma proteins—the humoral component of the innate immune response) are released by various cells that cause vasodilation, increased capillary permeability, and chemotaxis
- **Phagocytosis**: pathogens, dead cells, and debris are phagocytosed, usually forming pus at the site of injury
- **Healing**: the area is walled off, clots may form, and debris is removed as the healing process begins
- **Natural killer cells**: these cells are part of the nonspecific or innate immunity, and develop from the common lymphoid progenitor cells; they are genetically programed to recognize transformed cells (virus-infected cells or tumor cells) and kill them

The inflammation associated with the innate immune response is **genetically determined**; it does not involve prior exposure to antigens, but it does involve cells and various chemical inflammatory mediators.

Moreover, it appears that the innate response activates the elements of the specific adaptive immune response, which is the second form of immune defense in our bodies (see Plate 6-3).

**COLOR** the following elements of the innate immune response that lead to inflammation, using the suggested colors for each element:

- ☐ **1. Pathogens (yellow)**
- ☐ **2. Dendritic cell and its cytokines and inflammatory mediators (green)**
- ☐ **3. Macrophages (blue)**
- ☐ **4. Neutrophils (purple)**
- ☐ **5. Blood vessel (red)**
- ☐ **6. Monocytes (light blue)**

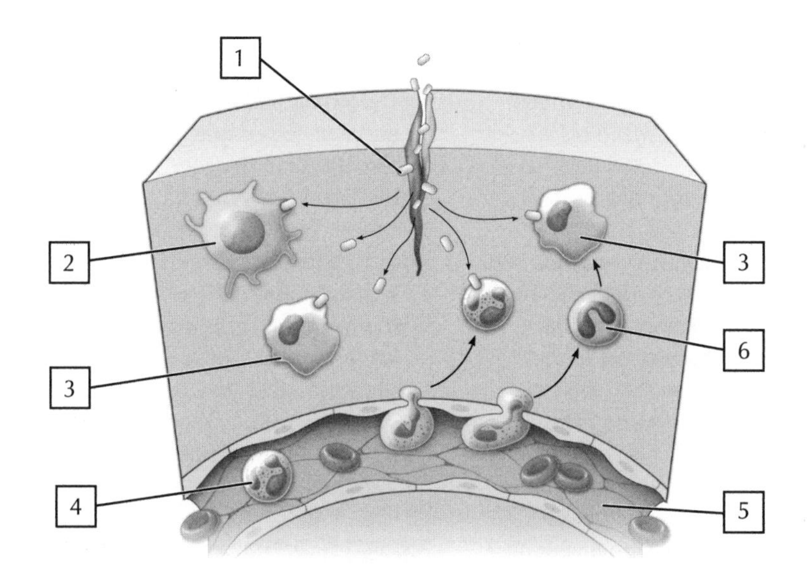

**Physical barriers**

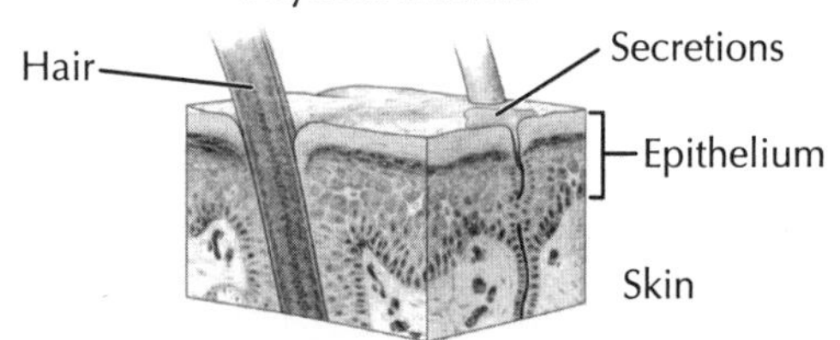


**Phagocytes**

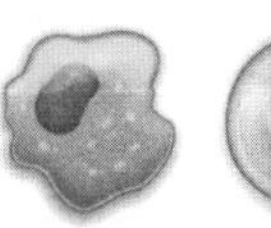
Fixed macrophage

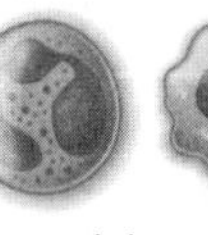
Neutrophil

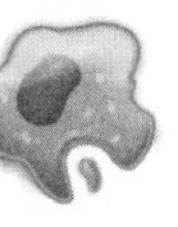
Free macrophage

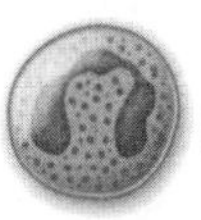
Eosinophil

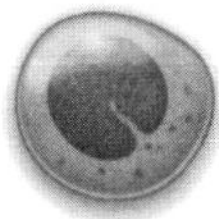
Monocyte

**Immunological surveillance**
*Killer cells: destroy abnormal cells*

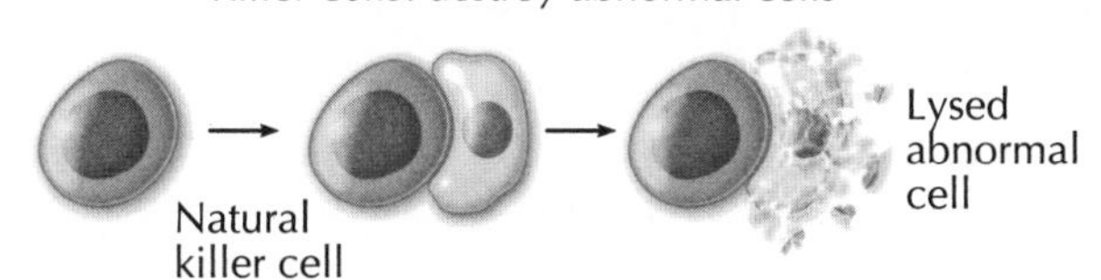


**Interferons**
*Protect cells by increasing their resistance to disease*

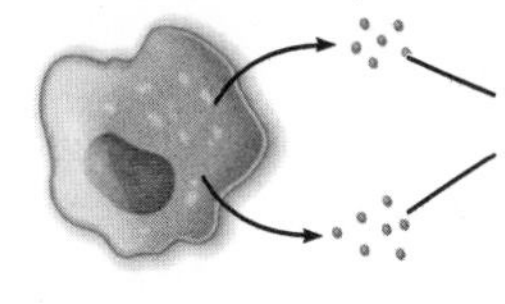


**Complement system**
*Lyses cells: stimulates inflammatory response*

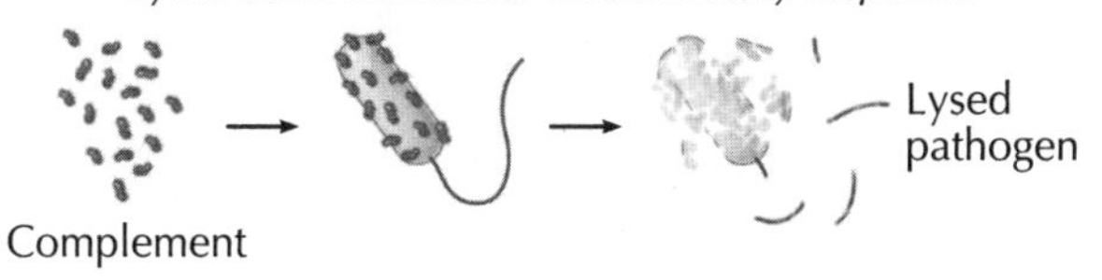


**Inflammatory response**

Mast cells

- Increases blood flow
- Activates phagocytes
- Increases capillary permeability
- Activates the complement system
- Infected region is sequestered by clotting
- Fever
- Systemic defenses activated

**Fever**
*Reduces pathogens, facilitates tissue repair, activates defenses*

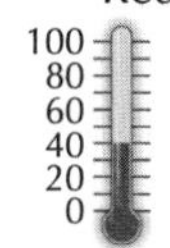


Body temperature rises above 37° C in response to pyrogens

# Adaptive Immunity

**Specific (adaptive) immunity**: a second line of defense is characterized by **specific pathogen recognition**, immunological memory (acquired resistance), amplification of immune responses (humoral responses producing antibodies), and a rapid response against pathogens that reinvade; adaptive immunity induces acquired resistance. The adaptive immune response is a specific response that is characterized by the following features:

- **Specificity**: a response that is directed toward a specific pathogen
- **Passive or active forms**: immunity that can be passed from another individual via antibodies (passive) or produced by antibodies that develop in response to antigens (active)
- **Systemic**: a response that is not confined simply to the site of inflammation; it is a slower response than the innate response but lasts much longer and induces acquired resistance
- **Memory**: once antibodies are developed in response to a foreign antigen, the body "remembers" the response and can mount an even stronger response upon a second exposure to the same antigen

The cells of the adaptive response are **lymphocytes** (B and T and NK cells), derived from the pluripotent hemopoietic stem cells of the bone marrow. B cells are involved in the **humoral (chemical attack) response**, which can be summarized as follows:

- B cell recognizes a pathogen by the binding of its surface antibodies to a foreign antigen and becomes sensitized
- B cells then become activated when an inactive helper T cell recognizes the same antigens, binds to the B cell, and secretes lymphokines that cause activated B cells to divide
- B cell division yields millions of B cells, which then become plasma cells that secrete antibodies (immunoglobulins, Ig) to the antigen into the circulating blood and lymph
- These circulating antibodies bind to the specific antigens on pathogens and label them for destruction by phagocytes; the antibodies also may bind directly to bacterial toxins or receptors used by bacteria and viruses such that they directly neutralize the invader
- B cell division also yields memory B cells, which remain in reserve should the body be reexposed to the same foreign antigen

The T cells are of several types and are involved in **cell-mediated responses**:

- **Helper T cells**: although not directly involved in killing pathogens or infected cells, these T cells (that express CD4 markers) control the immune response by directing the activities of other cells of the immune system; they recognize antigens presented by B cells, become activated, and secrete cytokines that promote humoral- and cell-mediated immunity
- **Memory T cells**: derived from helper and killer (cytotoxic) T cells, they remain in reserve in case of reinfection
- **Regulatory (suppressor) T cells**: activated later than other B and T cells, they suppress the immune response, thus limiting the overall intensity of any single response
- **Cytotoxic (killer) T cells**: respond to an antigen on cell surfaces (other than B cells), become activated and divide, and produce memory T cells and killer T cells, which travel throughout the body to find and destroy virus-infected cells, cancer cells, bacteria, fungi, protozoa, and foreign cells (e.g., from tissue transplants)
- **Natural killer (NK) cells**: these are cells that develop from the common lymphocyte progenitor cell as B and T cells, but are named for their ability to selectively kill certain types of target cells; they do not mature in the thymus but are genetically programed to "police" the body in blood and lymph and recognize transformed cells (cells with a virus or tumor cells), which they lyse and kill

**COLOR** the following cells involved in the adaptive immune response, using the colors recommended for each cell type:

- ☐ 1. **Antigen (yellow)**
- ☐ 2. **Infected cell displaying antigen (brown)**
- ☐ 3. **B cell (blue)**
- ☐ 4. **Dying infected cell (gray/light black)**
- ☐ 5. **Antibodies (red)**
- ☐ 6. **Memory B cell (light blue)**
- ☐ 7. **Memory T cell (light green)**
- ☐ 8. **Killer T cell (orange)**
- ☐ 9. **Activated T cell (green)**

***Clinical Note:***

**B lymphocytes** (bursa of Fabricius in birds and bursa-equivalent organs in mammals) are the GALT (gut-associated lymphoid tissue) and bone marrow derived cells that participate in humeral immunity and account for about 20% to 30% of the circulating lymphocytes.

**T lymphocytes** are named for the thymus, where they differentiate. They account for about 60% to 80% of circulating lymphocytes, have a long life span, and are involved in cell-mediated immunity.

**NK cells** kill cancer cells and virus-infected cells before the adaptive immune system is activated. They constitute about 5% to 10% of circulating lymphocytes.

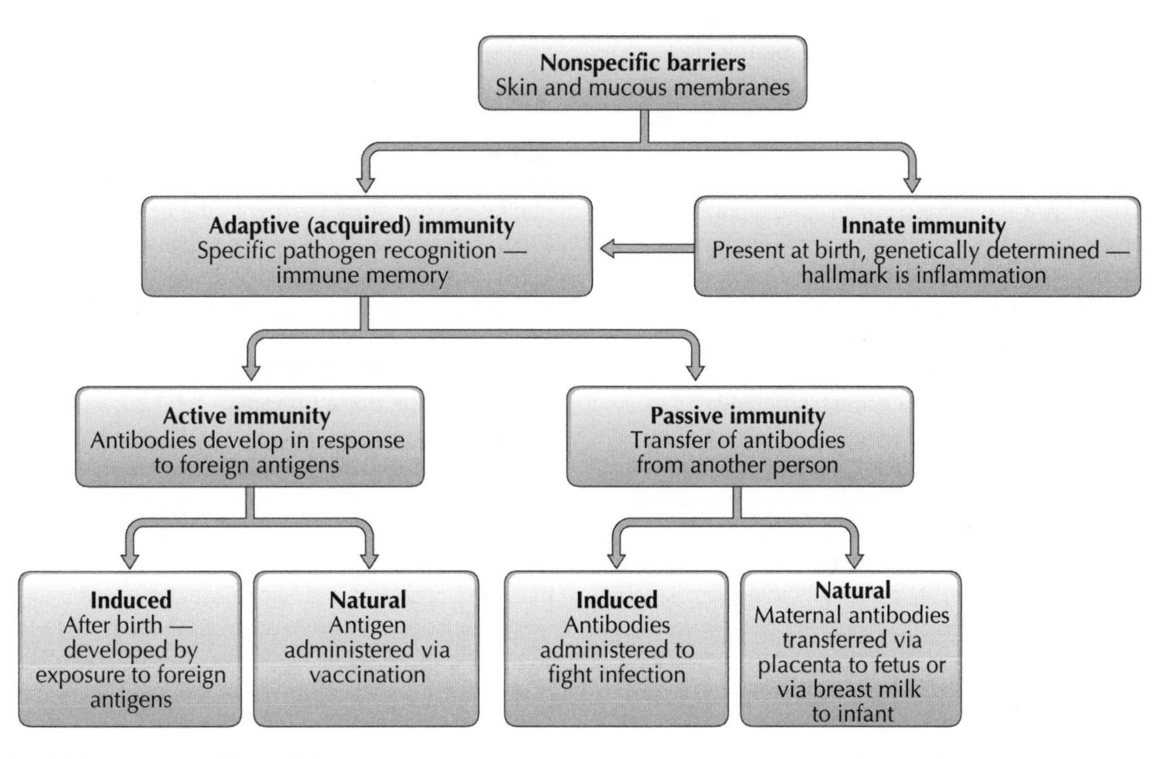


A. Different types of immunity

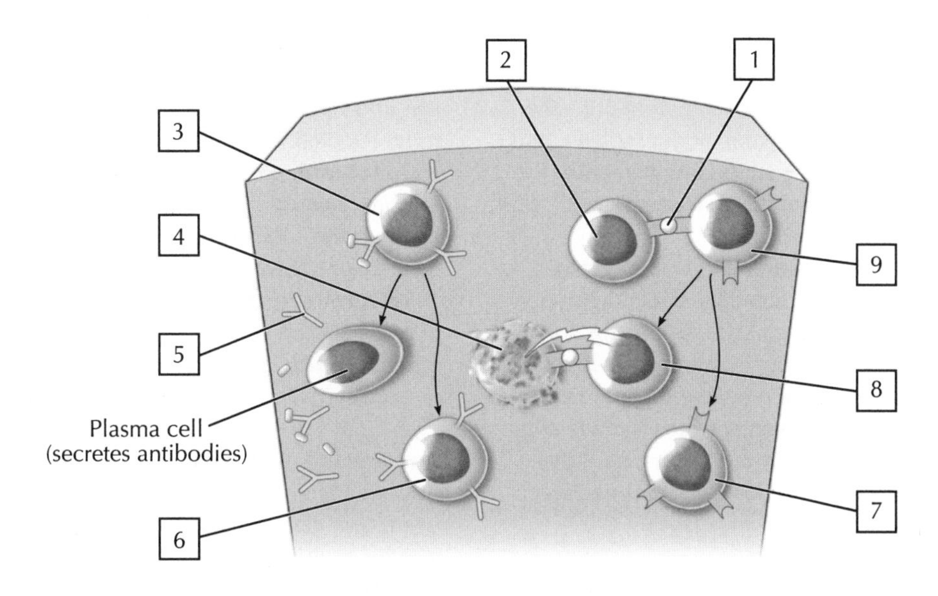


B. Adaptive immune system

# Thymus and Bone Marrow

Lymphocytes are derived from the **pluripotent hemopoietic stem cells** of the bone marrow, but at this stage, all are immature cells (neither B nor T cells). This distinction occurs as part of the lymphocyte maturation process; B cells, so named for the bone marrow, mature in the red bone marrow and become immunocompetent (can recognize a specific antigen) and self-tolerant (can recognize the body's own antigens as "self" and not "foreign"). However, these lymphocytes are immature and must "seed" the lymph nodes, spleen, and other lymphoid tissues where the antigen challenge occurs. Once mature (antigen-activated), immunocompetent lymphocytes circulate continuously in the bloodstream and lymph, and throughout the lymphoid organs of the body.

T cells, on the other hand, leave the **bone marrow** and travel to the **thymus,** where they become immunocompetent. The thymus is a bilobed organ in the superior mediastinum that is quite large in neonates but involutes after puberty. In the thymus, T cells undergo rapid cell division, greatly increasing their numbers before their "education" as T cells. Positive selection occurs in the thymic cortex, where T cells recognize self-MHC (major histocompatibility complex) molecules; T cells that cannot recognize self-MHC molecules are destroyed. Next, the surviving T cells must "learn" to recognize self-antigens by not binding too vigorously to self-MHC or self-MHC–bound self-peptides; if they do, they are destroyed as a safety measure to ensure that T cells do not attack the body's own antigens. It is estimated that only about 2% of T cells survive this education process. While T cells undergo their education, they are sequestered from circulating antigens by the **blood-thymus barrier**, so they won't be "distracted" by any circulating antigens. Eventually, immunocompetent but naive T cells migrate via the blood to the lymph nodes, spleen, and other lymphoid organs where they mature and then enter the bloodstream and lymph as activated immunocompetent T cells.

Lymphocytes are immunocompetent before encountering foreign antigens, and this process is entirely dependent upon our genes; our "endowed" genetic makeup for recognizing all of the possible antigens in our surrounding environment has been acquired through a process of natural selection during evolution. Many of the possible foreign antigens that we might encounter in our lifetime never invade our bodies, so those lymphocytes specifically selected to deal with those antigens will lie dormant.

Despite the fact that T and B cells are immunocompetent and have survived the rigorous "weeding out" process, they are still not mature until they have traveled to the spleen, lymph nodes, or other secondary lymphoid tissues and encountered their specific antigens, at which time they become antigen activated and are ready to initiate a response. Most of the T cells become helper and killer T cells once they reach the secondary lymphoid tissues; 60% to 80% of all circulating lymphocytes are T cells.

**COLOR** the following elements related to lymphocyte traffic from the bone marrow to the thymus and secondary lymph organs, using the colors suggested:

- ☐ **1. Thymus (yellow)**
- ☐ **2. Bone marrow (red)**
- ☐ **3. Immature lymphocytes (blue with pink nuclei)**
- ☐ **4. Lymph node (green)**
- ☐ **5. Spleen (dark red)**

***Clinical Note:***
**Vaccines** provide an "artificial" way to acquire immunity. Many vaccines contain dead or attenuated (alive but extremely weakened) pathogens. Generally speaking, vaccines may cause us some discomfort, or none, but do provide antigenic determinants that afford us immunity.

**Organ transplants** are a viable alternative for some patients suffering organ failure, but immunologic rejection is a problem. Therefore, careful screening involving blood typing, major histocompatibility complex (MHC) screening, and postoperative immunosuppressive therapy is vitally important if the graft is not to be rejected. Basically, four types of grafts are most common:

- **Autografts**: tissue grafts from one site to another in the same person
- **Isografts**: donated graft to a patient by a genetically identical individual (must be identical twin)
- **Allografts**: graft from a person that is not genetically identical but belongs to the same species
- **Xenografts**: a graft from another animal species into a human

**Autoimmune diseases** are not uncommon (occur in about 5% of adults), and autoantibodies are produced that attack one's own tissues. Some more common autoimmune diseases include multiple sclerosis, myasthenia gravis, Graves' disease, juvenile type I diabetes mellitus, and rheumatoid arthritis.

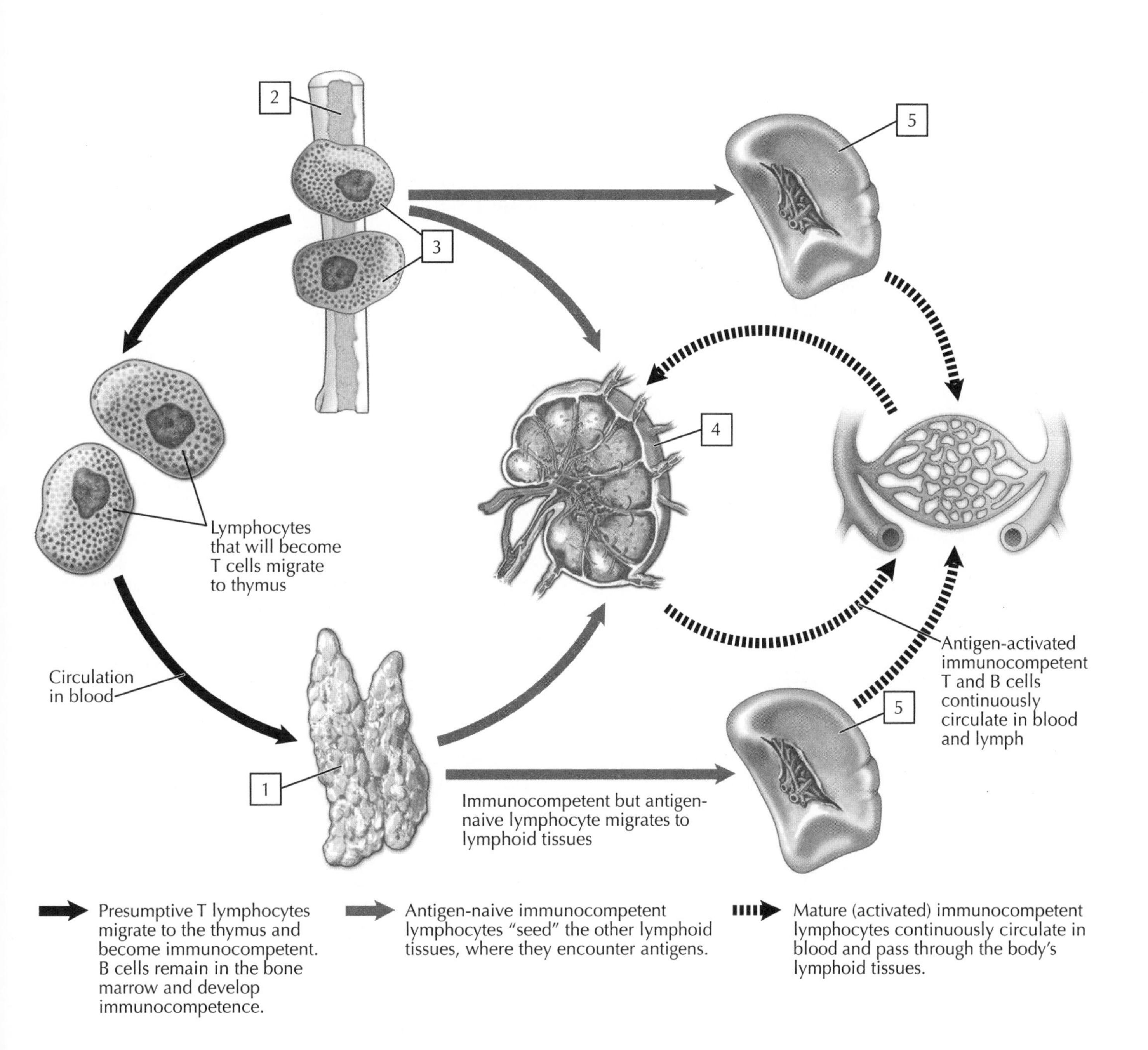


Presumptive T lymphocytes migrate to the thymus and become immunocompetent. B cells remain in the bone marrow and develop immunocompetence.

Antigen-naive immunocompetent lymphocytes "seed" the other lymphoid tissues, where they encounter antigens.

Mature (activated) immunocompetent lymphocytes continuously circulate in blood and pass through the body's lymphoid tissues.

# 6 Spleen

The spleen is about the size of your clenched fist and lies in the upper left quadrant of the abdomen, tucked in posterolaterally to the stomach under the protection of the lower left rib cage. Simplistically, it is a large lymph node (and can become quite large during infections), although functionally it is much more and is involved in the following functions:

- Lymphocyte proliferation (B and T cells)
- Immune surveillance and response
- Blood filtration
- Destruction of old or damaged red blood cells (RBCs)
- Destruction of damaged platelets
- Recycles iron and globin
- Blood reservoir
- Production of RBCs in early fetal life

The spleen is an encapsulated organ with an extensive infrastructure composed of a trabecular network of connective tissue, which supports concentrations of lymphocytes in regions called the "**white pulp**." There are also regions of venous sinusoids rich in macrophages and RBCs called the "**red pulp**."

The **white pulp** is organized as an aggregation of lymphocytes that surround a **central artery**, forming a periarterial lymphatic sheath (PALS). The PALS gives the appearance of lymph nodules consisting largely of B cells surrounded by a more diffuse collection of T cells. The nodules contain a germinal center where B cells proliferate and become activated. The immune functions of the spleen include:

- Antigen presentation by macrophages and dendritic cells
- Proliferation and activation of B and T cells
- Production of antibodies directed against circulating antigens
- Removal of antigens from the blood

The **red pulp** is organized into regions of **splenic (venous) sinuses** separated by splenic cords (of Billroth) that consist of a meshwork of reticular fibers and cells, including:

- RBCs
- Macrophages
- Dendritic cells
- Lymphocytes
- Plasma cells
- Granulocytes

Macrophages associated with the splenic sinuses phagocytose damaged RBCs, break down the hemoglobin (heme is broken down to bilirubin), and recycle the iron (stored as ferritin or hemosiderin for recycling). Blood from the central artery flows into the white pulp and the splenic sinuses, with the blood cells percolating through the splenic cords before squeezing back into the collecting splenic venous sinuses. This "open circulatory" pattern exposes the RBCs to macrophages, which remove old or damaged cells from the circulation. Thus the primary function of the red pulp is to filter the blood (the removal of particulate material, macromolecular antigens, and aged, abnormal, or damaged blood cells and platelets).

**COLOR** the features of the splenic architecture, using a different color for each feature:

- ☐ 1. **Lymph vessel in the splenic capsule**
- ☐ 2. **Central artery**
- ☐ 3. **Splenic venous sinuses of the red pulp**
- ☐ 4. **White pulp (splenic nodule)**

***Clinical Note:***

The spleen, despite its protective position under the lower left rib cage, is the most commonly injured abdominal organ. Trauma to the abdominal wall (playground accidents in children, automobile accidents, and falls) can lacerate or **rupture the spleen**. This is serious because the rich blood supply to the spleen means that intraperitoneal hemorrhage and possible shock can result if the spleen's capsule and parenchyma are damaged. Surgical removal of the spleen generally is not problematic, because we can live without our spleen. Other lymphatic tissues, the liver, and the bone marrow can take over the functions of the spleen.

If diseased, e.g., granulocytic leukemia, the spleen may enlarge up to 10 times its normal size and weight (**splenomegaly**). The spleen can also increase moderately in size in patients with portal hypertension.

**Infectious mononucleosis** (the "kissing disease") is caused by the Epstein-Barr virus; as a result of this infection the immune system is activated, the lymph nodes enlarge, and the spleen swells, sometimes quite significantly.

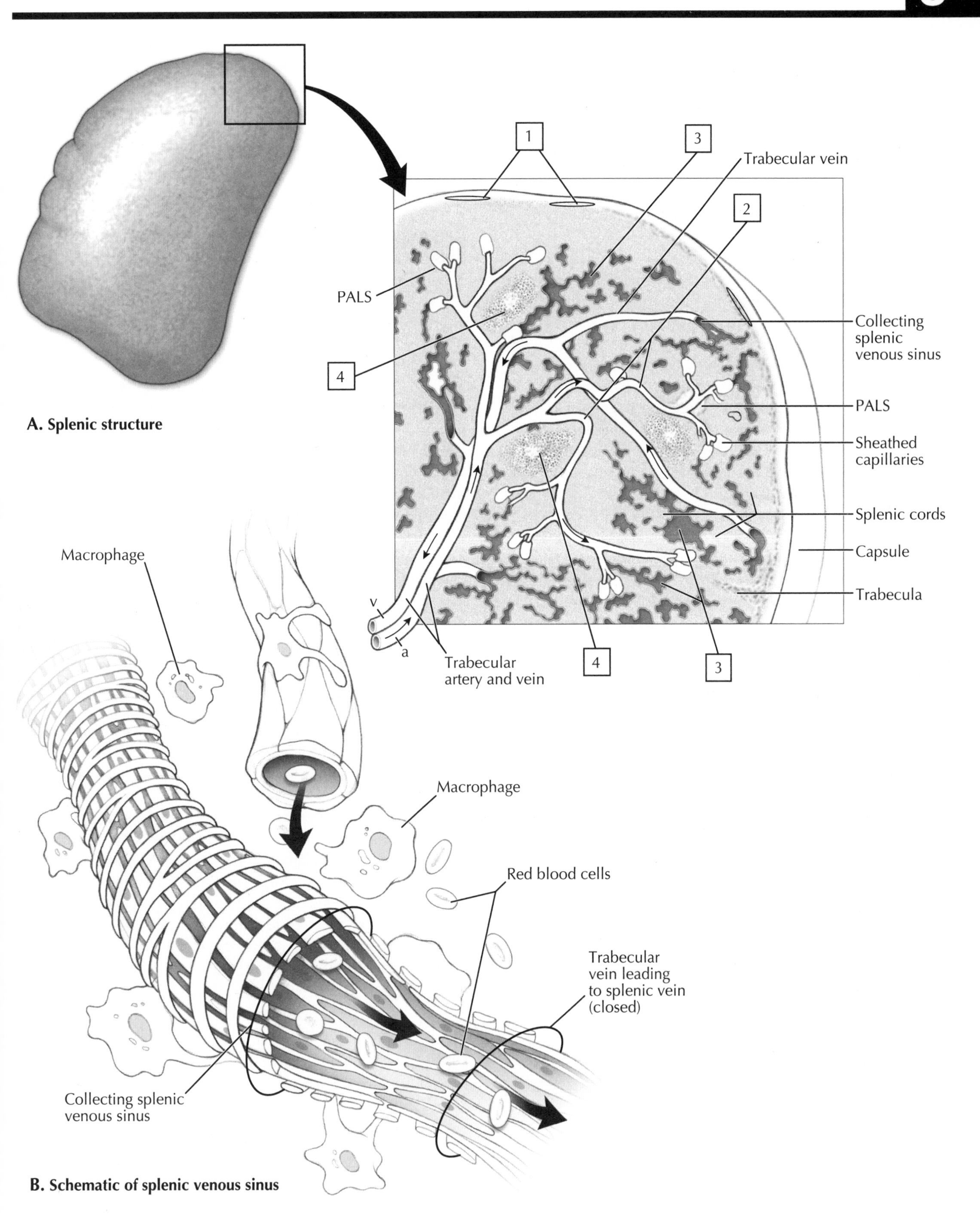


**A.** **Splenic structure**

**B.** **Schematic of splenic venous sinus**

# 6 Tonsils, BALT, GALT, and MALT

In addition to the lymph nodes and vessels, bone marrow, thymus, and spleen, a number of other **diffuse lymphatic tissues** exist in the body and play a regional and systemic role in immune function. These accumulations include the:

- Tonsils
- Bronchus-associated lymphoid tissue (BALT)
- Vermiform appendix and gut-associated lymphoid tissue (GALT)
- Mucus-associated lymphoid tissue (MALT)

## Tonsils

The tonsils include collections of lymphatic tissue in the oral cavity (palatine tonsils, visible when you open your mouth and say "ah"), lingual tonsils on the base of the tongue, pharyngeal tonsils (when enlarged and inflamed they are called adenoids) in the roof of the nasopharynx, and tubal tonsils around the opening of the auditory (eustachian) tube. Together, these lymphatic aggregations form the "**Waldeyer's lymphatic ring**." They play an important immune role by protecting the nasal and oral passages from invading pathogens, especially during childhood. Some of these tissues atrophy with advancing age and become less important.

## BALT

Accumulations of lymphoepithelial cells are diffusely located around the bronchi and the bronchial tree as they pass into the lung. BALT appears similar to the Peyer's patches that line the GI tract and provides immune responses against pathogens that may enter the airways and lungs.

## Vermiform Appendix and GALT

The vermiform (wormlike) appendix is attached to the cecum (first portion of the colon) and contains a small lumen lined with mucosa and rich in lymphatic nodules. The amount of lymphatic tissue tends to decrease with advancing age.

Likewise, numerous aggregations of lymphatic tissue containing B and T cells reside in the lamina propria and submucosa of the ileum, which are called **Peyer's patches**. Diffuse lymphatic tissue (lymphocytes and plasma cells) also resides in the lamina propria, and together these accumulations are referred to as the GALT. As one proceeds from proximal to distal in the bowel (including the colon), one tends to encounter a greater accumulation of lymphatic cells and nodules associated with the lamina propria; their primary function is to protect against pathogens and antigenic molecules that might invade the body.

## MALT

The term MALT really refers to all of the mucosal-associated lymph nodules and diffuse lymphatic cells that encompass the BALT and GALT, but also includes lymphatic accumulations in other organ systems, such as the female reproductive tract. Essentially the lymphatic tissues of the lamina propria of the digestive system, the respiratory lymphatics, and the genitourinary tract would be included as MALT.

**COLOR** the tissues associated with the lymphatic accumulations listed below:

- ☐ **1. Tonsils**
- ☐ **2. BALT**
- ☐ **3. GALT and Peyer's patches of the ileum**
- ☐ **4. Lymph nodules of the vermiform appendix**

***Clinical Note:***

**Acquired immunodeficiency syndrome** (AIDS) is caused by the human immunodeficiency virus (HIV), which is an RNA retrovirus. Unfortunately, most HIV-infected individuals will eventually develop AIDS. HIV gains entry into helper T cells and then "injects" its own genetic information into the cell cytoplasm. The T cell then makes copies of the virus which are released by exocytosis, where the HIV particles can infect other helper T cells. The major strategy to combat the virus is by anti-HIV treatment, which includes a combination of several chemotherapeutic agents.

As noted above, the vermiform (L.worm-like) appendix is a small, blind intestinal diverticulum, which can vary in length and possess a short mesentery (mesoappendix). Acute inflammation of the appendix (**appendicitis**) is one of the more common phenomenon leading to an acute abdomen. The pain of appendicitis usually begins as a vague periumbilical pain that, over time, radiates to the lower right quadrant of the abdomen. In this location, the inflamed appendix contacts and irritates the parietal peritoneum of the posterior abdominal wall, and the severe pain then becomes well-localized. An **appendectomy** is the usual procedure used to remove the inflamed appendix and is performed by a small incision and direct resection or more commonly, by laparoscopic surgery.

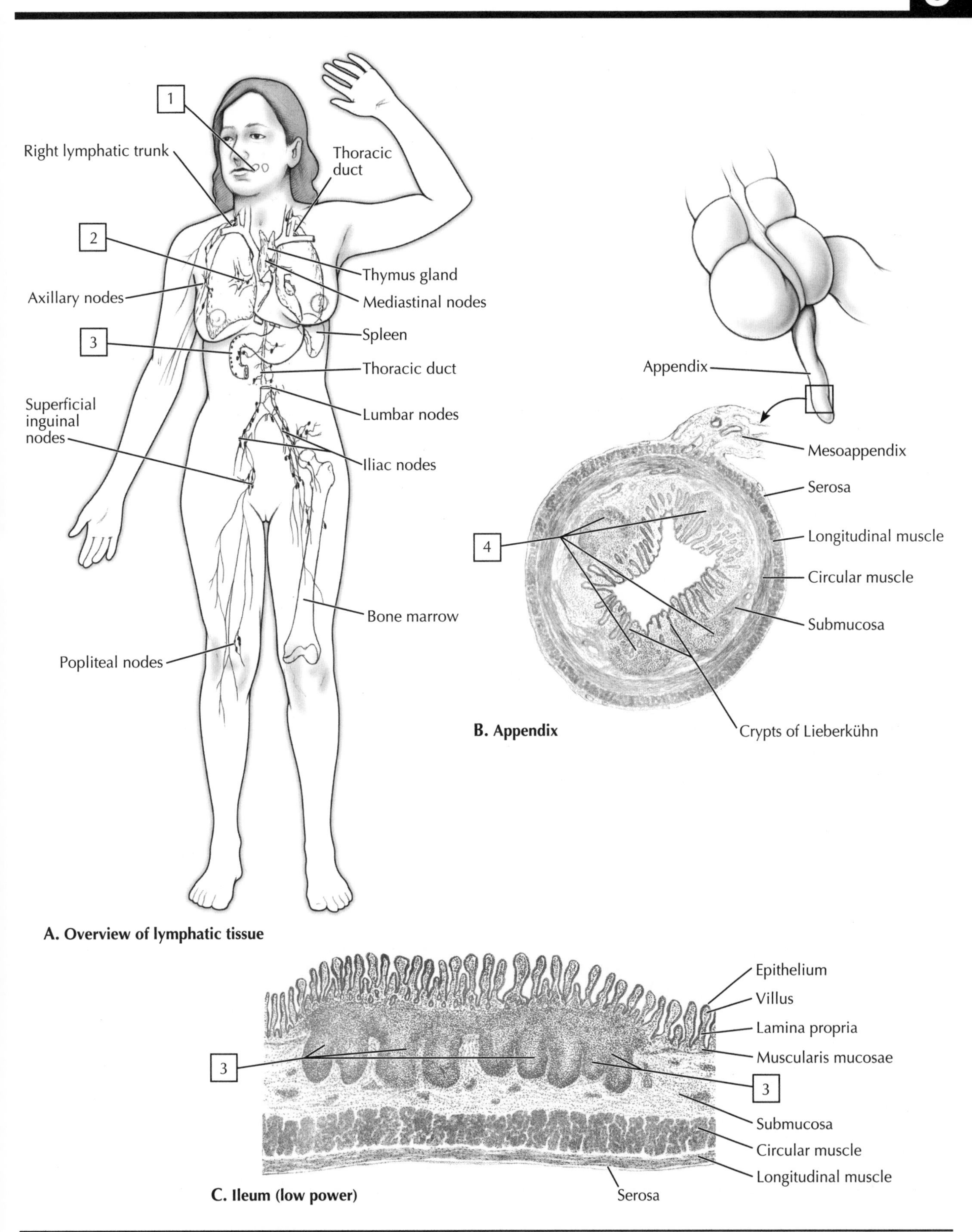


**A. Overview of lymphatic tissue**

**B. Appendix**

**C. Ileum (low power)**

# Clinical Aspects of the Lymphatic System

The combined lymphatic and immune systems are involved in a number of clinical disorders, coincident with the importance of the systems in fighting pathogens and cancer.

## Lymphatic Metastases

Cancer spreads from its primary site in one of three ways:
- Direct contact with adjacent tissues
- Via the venous system
- Via the lymphatic vessels

The lymphatics are especially important because cancer cells can easily gain access to the lymphatic system. Once in the lymphatic vessels, the cancer cells encounter lymph nodes, where they are filtered from the lymph, may seed the nodes, and may grow. This process may cause the nodes to enlarge and become fixed (immobile to palpation) but nontender. Nodes that are inflamed and enlarged due to noncancerous diseases are mobile and tender to the touch. Because of the predictable lymphatic drainage pattern, clinicians can usually trace the spread of cancer from one set of lymph nodes to the next set up the line. The first major lymph node that is enlarged because of metastasis is referred to as the **sentinel node**. Major lymph node accumulations exist in humans and include both a paired set of palpable nodes near the surface of the body and a deeper set of nodes that cannot be palpated but can be accessed by selective imaging techniques.

**COLOR** the major lymph node accumulations, using the suggested colors for each set of nodes:

☐ 1. **Jugulodigastric nodes of the deep cervical chain: lie along the internal jugular vein, drain the head and neck, and are palpable when enlarged (orange)**

☐ 2. **Axillary nodes: drain the upper limb, shoulder, and chest region and are palpable when enlarged (red)**

☐ 3. **Mediastinal nodes: clustered around the tracheal bifurcation and hilum of the lungs, drain the lungs and thorax, and are deep nodes that cannot be palpated when enlarged (purple)**

☐ 4. **Paraaortic (lumbar) nodes: receive lymph from the abdominal cavity and lower half of the body, are clustered around the aorta near to the renal arteries, and are not palpable when enlarged; they drain into the cisterna chyli and thoracic duct (brown)**

☐ 5. **Iliac nodes: lie along the iliac vessels, receive lymph from the lower limbs and pelvic viscera, and drain toward the paraaortic nodes; they are deep and cannot be palpated when enlarged (blue)**

☐ 6. **Superficial inguinal nodes: drain the lower limb and external genitalia and are palpable when enlarged (yellow)**

## Vaccination (Immunization)

Immunity can be artificially induced through the process of vaccination. This is done by injecting an antigen from the pathogen being immunized against that will stimulate the body's immune system. Most bacterial vaccines are designed to expose the body to antigens derived from acellular components of the bacterium or one of its harmless toxins. These antigens often produce a weak response in the body, so **adjuvants** are coinjected with the antigens to further activate the cells of the immune system. Most viral vaccines are **live attenuated** (diminished virulence) **viruses** that activate an immune response without infection.

## Autoimmunity

When the immune system cannot distinguish self from nonself, it can mount an immune reaction against the body's own cells. Some autoimmune disorders include:
- **Systemic lupus erythematosus**, which largely affects the skin, kidneys, lungs, and heart
- **Multiple sclerosis**, which affects the normal myelination in the CNS
- **Myasthenia gravis**, which affects communication between nerves and skeletal muscle
- **Type I diabetes mellitus**, which affects the insulin-producing cells of the pancreatic islets
- **Rheumatoid arthritis**, which affects many of the body's joints

## Immunodeficiencies

Immunodeficiencies occur when components of the immune system do not respond to pathogens and remain inactive. Common causes are genetic (congenital) or acquired (e.g., HIV), but can also include poor nutrition, alcoholism, and illicit drug use.

## Hypersensitivity

Hypersensitivity occurs when the body's immune system battles a pathogen in such an aggressive manner that it damages its own tissues. Four types are recognized:
- **Type I**: acute, such as an anaphylactic reaction; allergy is a good example
- **Type II**: antibodies bind to antigens on the body's own cells (called antibody-dependent or cytotoxic hypersensitivity); a reaction to transfusion with the wrong blood type is an example
- **Type III**: an abundance of antibody-antigen complexes in the body causes an inflammatory reaction, initiating a robust hypersensitivity reaction; chronic infection or allergic reactions are examples
- **Type IV**: cell-mediated or delayed hypersensitivity reactions that usually take several days to develop and include allergic skin reactions (poison ivy and contact dermatitis) as well as protective reactions to infections, cancer cells, or the rejection of foreign tissue grafts

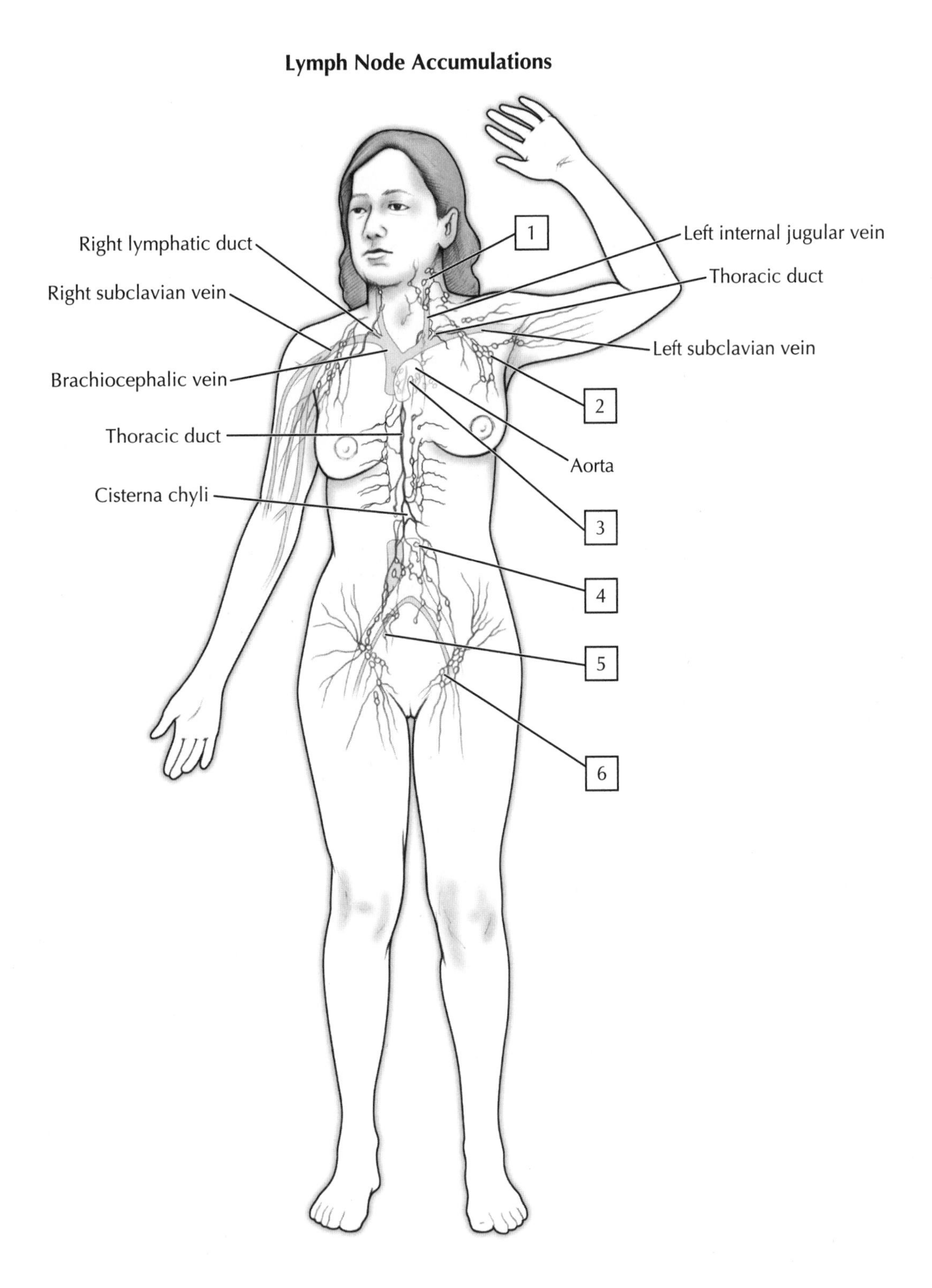
Lymph Node Accumulations
1
Right lymphatic duct
Left internal jugular vein
Thoracic duct
Right subclavian vein
Left subclavian vein
Brachiocephalic vein
2
Thoracic duct
Aorta
Cisterna chyli
3
4
5
6

## REVIEW QUESTIONS

1. T cells are part of the adaptive immune response and come in several different types. Which type of T cell responds to antigen on the cell surface and can become activated, destroy many virus- and bacteria-infected cells, and divide to produce other types of T cells?
   A. Helper T cells
   B. Killer T cells
   C. Memory T cells
   D. Suppressor T cells

2. When a T cell leaves the bone marrow, it travels to which organ to become immunocompetent?
   A. Lymph nodes
   B. Spleen
   C. Thymus
   D. Thyroid
   E. Tonsil

3. Which organ is important in recycling iron and globin?
   A. Colon
   B. Gallbladder
   C. Kidney
   D. Spleen
   E. Thymus

4. Many cells of the immune system are phagocytic. Which immune cells are especially important in the allergic response (hint: see Plate 5-1)?
   A. Eosinophils
   B. Fixed macrophages
   C. Free macrophages
   D. Monocytes
   E. Neutrophils

**For each description below (5-8), color the appropriate area of the spleen.**

5. This splenic region is important in phagocytosis of damaged red blood cells.
6. This region is organized around a central artery.
7. This feature of the spleen is thin and fragile, and damage to it can result in significant blood loss.
8. This is the site of the splenic red pulp and splenic sinuses.

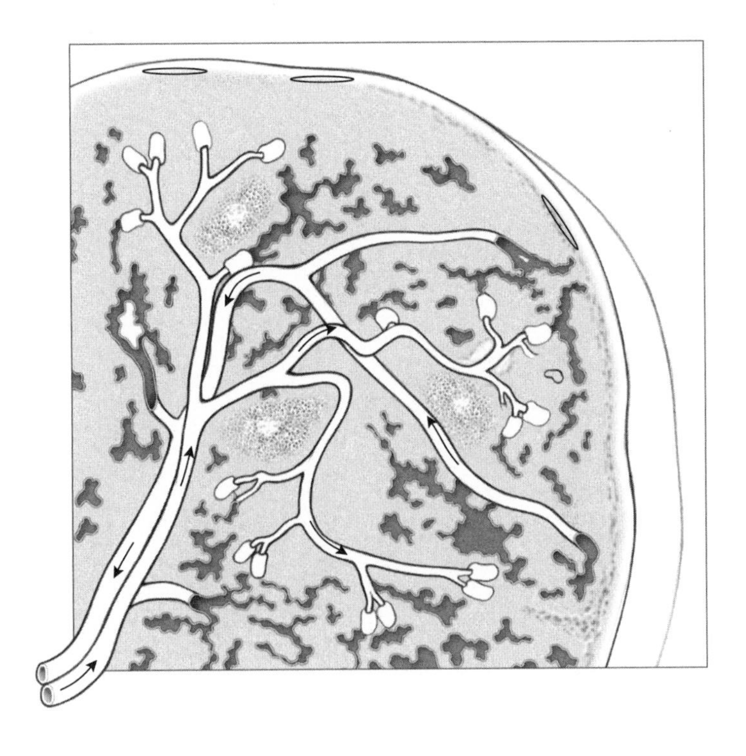

9. The thoracic duct begins in the upper abdomen, where numerous lymphatic vessels coalesce to form the beginning of the duct. What is this feature called? ______________________________

10. Where does the thoracic duct ultimately end?______________________________

## ANSWER KEY

1. B
2. C
3. D
4. A
5. Red pulp
6. Periarterial lymphatic sheath (PALS)
7. Splenic capsule enclosing the entire spleen
8. Splenic red pulp

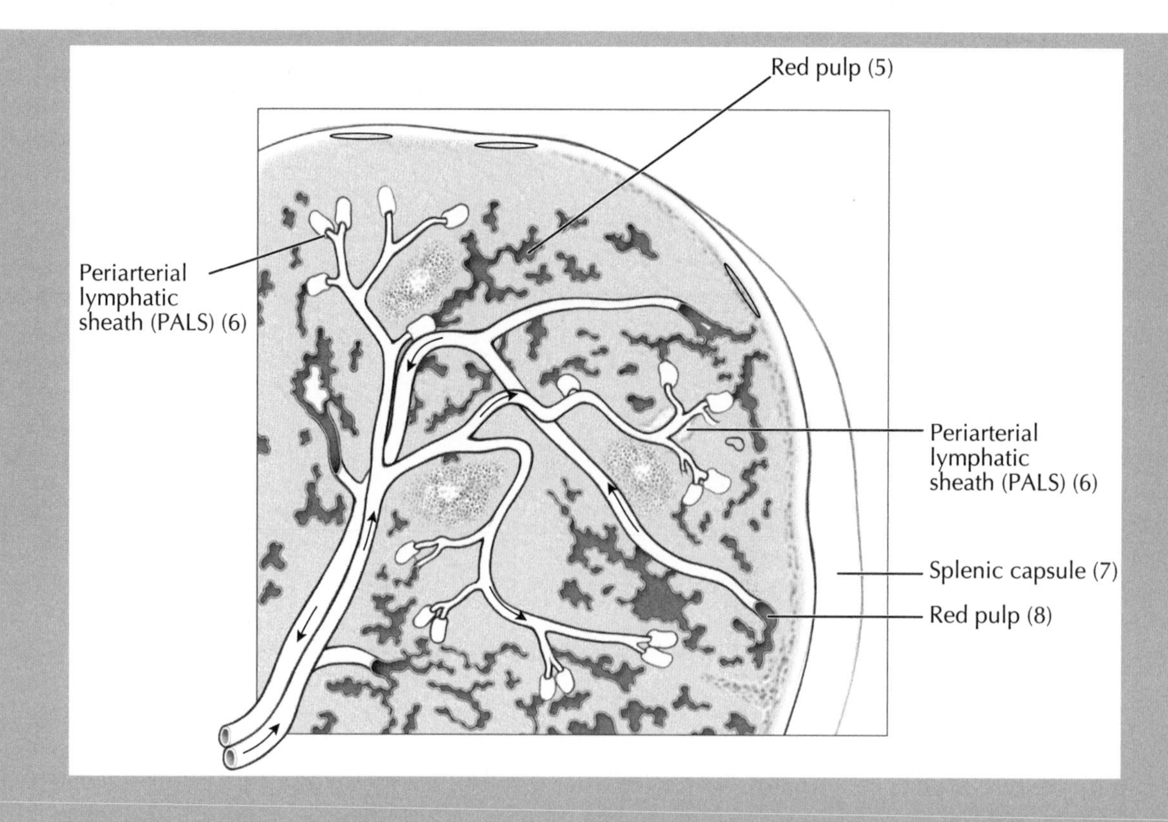


9. Cisterna chyli
10. In the venous system at the junction of the left subclavian and left internal jugular veins

# Chapter 7 Respiratory System

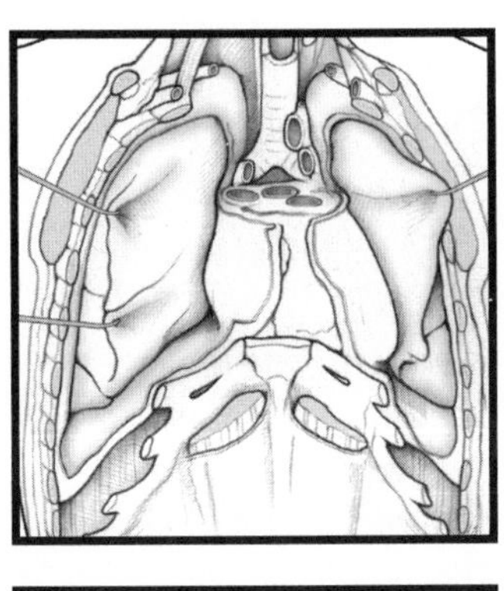
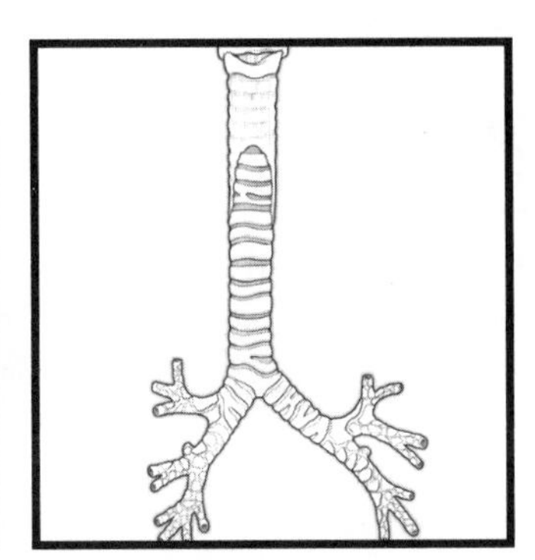
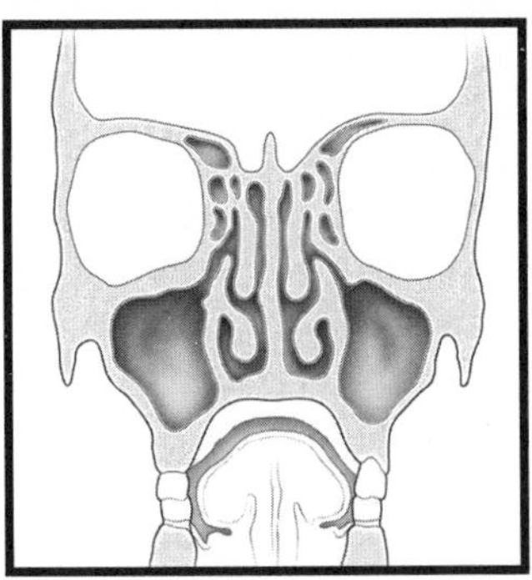
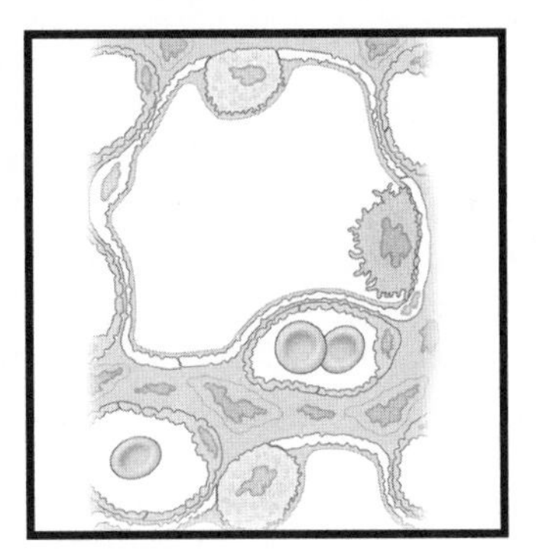

# 7 Overview

The human body is dependent upon the external environment for a source of food and oxygen for survival, and for a catch basin for its wastes. The cells of the human body require oxygen to carry out the vital functions necessary for survival. Respiration involves four basic processes:

- **Pulmonary ventilation**: the movement of air into and out of the lungs; necessary for the exchange of gases (fresh air into the lungs, especially oxygen, and expired gases, especially carbon dioxide)
- **External respiration**: involves the movement of oxygen from the lungs into the bloodstream and the movement of carbon dioxide from the blood into the lungs
- **Transport of respiratory gases**: the role of the cardiovascular system to carry oxygen to the tissues and carbon dioxide from the tissue cells to the lungs
- **Internal respiration (cellular respiration)**: the transport of oxygen from the blood to the cells of the body, and the transport of carbon dioxide from the cells to the blood

As described above, the respiratory system provides the body with oxygen for its metabolic needs and eliminates carbon dioxide. Structurally, the respiratory system includes the:

- Nose and paranasal sinuses (produce mucus, filter, warm, and moisten air; resonance chamber for speech, detect smell)
- Pharynx and its subdivisions, the nasopharynx, oropharynx, and laryngopharynx (passageway for air and food, facilitate exposure of immune system)
- Larynx (air passageway and prevent food from entering lower respiratory tract; voice production)
- Trachea (air passageway, cleans, warms, and moistens inspired air)
- Bronchi, bronchioles, alveolar ducts and sacs, and alveoli (air passageways that filter air and connect trachea with alveoli, which is the main site for gas exchange)
- Lungs (paired organs that contain the respiratory passages smaller than the primary bronchi)
- Pleurae (provide lubrication fluid and compartmentalize lung lobes)

Functionally, the respiratory system performs five basic functions:

- Filters and humidifies the air and moves air into and out of the lungs
- Provides a large surface area for gas exchange with the blood
- Helps to regulate the pH of the body fluids
- Participates in vocalization and provides a resonance chamber for speech
- Assists the olfactory system with the detection of smells

Histologically, most of the respiratory epithelium is ciliated, pseudostratified columnar epithelium. There are a few exceptions, however: The vocal folds and epiglottis are stratified squamous epithelium, and areas of transition to small bronchioles go from respiratory epithelium to simple cuboidal epithelium. Alveoli are lined with thin squamous cells (**type I pneumocytes**) and simple cuboidal cells (**type II pneumocytes** that secrete surfactant).

The epithelial lining of the respiratory tract is important in warming, humidifying, and filtering the air before it reaches the sensitive lung alveoli. A rich vascular network helps to warm the air, and the ciliated epithelium and presence of mucous cells (goblet cells) helps to humidify the air and capture particulate material, which is then swept away by the cilia, to be swallowed or expectorated.

**COLOR** each of the following features of the respiratory system, using a different color for each feature:

☐ 1. **Laryngopharynx**
☐ 2. **Oropharynx**
☐ 3. **Nasopharynx**
☐ 4. **Nasal cavity**
☐ 5. **Larynx**
☐ 6. **Trachea**
☐ 7. **Lungs**

***Clinical Note:***

**Asthma** can be intrinsic (no clearly defined environmental trigger) or extrinsic (has a defined trigger). Asthma usually results from a hypersensitivity reaction to an allergen (dust, pollen, mold), which leads to irritation of the respiratory passages, smooth muscle contraction (narrowing of the passages), swelling (edema) of the epithelium, and increased production of mucus. Presenting symptoms are often wheezing, shortness of breath, coughing, tachycardia, and feelings of chest tightness. Asthma is a pathological inflammation of the airways and occurs in both children and adults.

**Dyspnea** (trouble breathing) may occur for a variety of reasons, including asthma and emphysema.

**Emphysema** occurs when the walls of the alveoli of the lungs are damaged and break through, resulting in enlarged spaces (alveolar chambers) reducing the surface area for gas exchange. Gas exchange also is compromised when tumors, mucus, or inflammation obstruct gas exchange in the alveoli.

Inadequate delivery of oxygen to the body's tissues is called **hypoxia. Hypocapnia** happens when the $CO_2$ levels in the blood are low. For example, someone experiencing an anxiety attack may hyperventilate, thus blowing off $CO_2$ to such a level that it may trigger the constriction of blood vessels, reducing blood flow to the brain, and potentially resulting in cerebral ischemia.

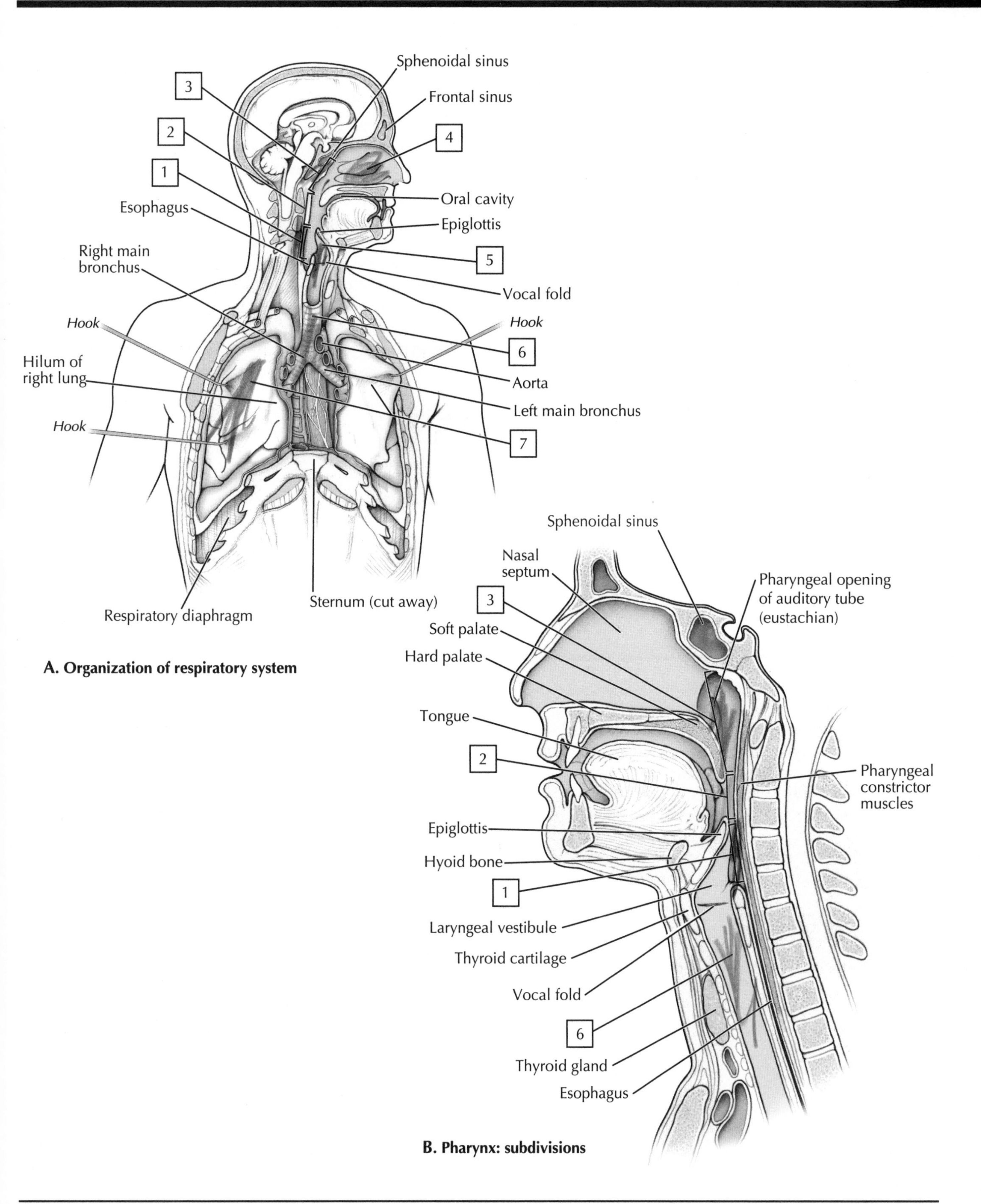


**A. Organization of respiratory system**

**B. Pharynx: subdivisions**

# 7 Nasal Cavity and Nasopharynx

The nose is composed primarily of cartilages, except at the "bridge" of the nose, where the nasal bone resides. In the anterior region, air enters or leaves the nose via the nares, which open into the nasal vestibule; in the posterior region, the nasal cavity communicates with the nasopharynx via paired apertures called the **choanae**.

**COLOR** the following cartilages contributing to the nose, using a different color for each cartilage:

- ☐ **1. Lateral processes of nasal septal cartilage**
- ☐ **2. Major alar cartilages**
- ☐ **3. Nasal septal cartilage**

The nasal cavity is separated from the cranial cavity by portions of the frontal, ethmoid, and sphenoid bones, and from the oral cavity inferiorly by the **hard palate**. A nasal septum, usually deviated slightly to one side or the other, divides the nasal cavity into right and left chambers. The anterior third of the nasal septum is cartilaginous and the posterior two-thirds is bony.

**COLOR** the following features of the nasal septum, using a different color for each feature:

- ☐ **4. Nasal septal cartilage**
- ☐ **5. Perpendicular plate of the ethmoid bone**
- ☐ **6. Vomer**

The lateral wall of the nasal cavity is characterized by three shelflike conchae, or turbinates (conchae covered with respiratory epithelium also are often referred to as turbinates), which protrude into the cavity. These structures, which have a covering of nasal respiratory epithelium, greatly increase the surface area for warming, humidifying, and filtering the air. The space beneath each shelflike concha is called a **meatus**. At the most superior aspect of the nasal cavity the olfactory region is found, with its olfactory epithelium and specialized sensory cells for the detection of smells via the first cranial nerve.

**COLOR** the following features of the lateral wall of the nasal cavity, using a different color for each feature:

- ☐ **7. Superior nasal concha (turbinate)**
- ☐ **8. Middle nasal concha (turbinate)**
- ☐ **9. Inferior nasal concha (turbinate)**

The innervation of the nasal cavities includes:

- **Olfactory nerve (CN I)**: olfaction (smell)
- **Trigeminal nerve (CN $V_1$ and $V_2$)**: sensory fibers via the maxillary division of the trigeminal nerve, except for the anterior part of the nose (CN $V_1$)
- **Facial nerve (CN VII)**: secretomotor preganglionic parasympathetic fibers course from the facial nerve to the pterygopalatine ganglion, synapse at that point, and course with branches of CN $V_2$ to innervate the nasal mucous glands
- Postganglionic sympathetic fibers from the superior cervical ganglion innervate the blood vessels

Most of the blood supply to the nasal cavities is from branches of the maxillary and facial arteries, with some contributions from the ethmoidal branches of the ophthalmic artery.

In the posterior region, the nasal cavities communicate via the choanae with the most superior portion of the pharynx, called the **nasopharynx**. In its lateral wall, the opening of the pharyngotympanic (auditory, eustachian) tube is visible and represents a direct conduit to the middle ear cavity.

***Clinical Note:***

**Acute otitis media**, an inflammation of the middle ear, is a common disorder in children under the age of 15 years. In part this disorder is prevalent because of the horizontal nature of the pharyngotympanic, or auditory, tube in children (the tube is slightly more vertical in adults), so that the normal gravitational drainage toward the nasopharynx is compromised. Infections may be bacterial or viral.

Viruses, bacteria, and various allergens can cause **rhinitis** and inflammation of the nasal mucosa. The mucosal inflammation reaction produces excessive mucus, nasal congestion, and postnasal drip.

The nasal mucosa contains a rich supply of sensory nerve endings, and exposure to irritating particles such as dust and pollens can trigger a **sneeze reflex**, providing a way of expelling these irritants.

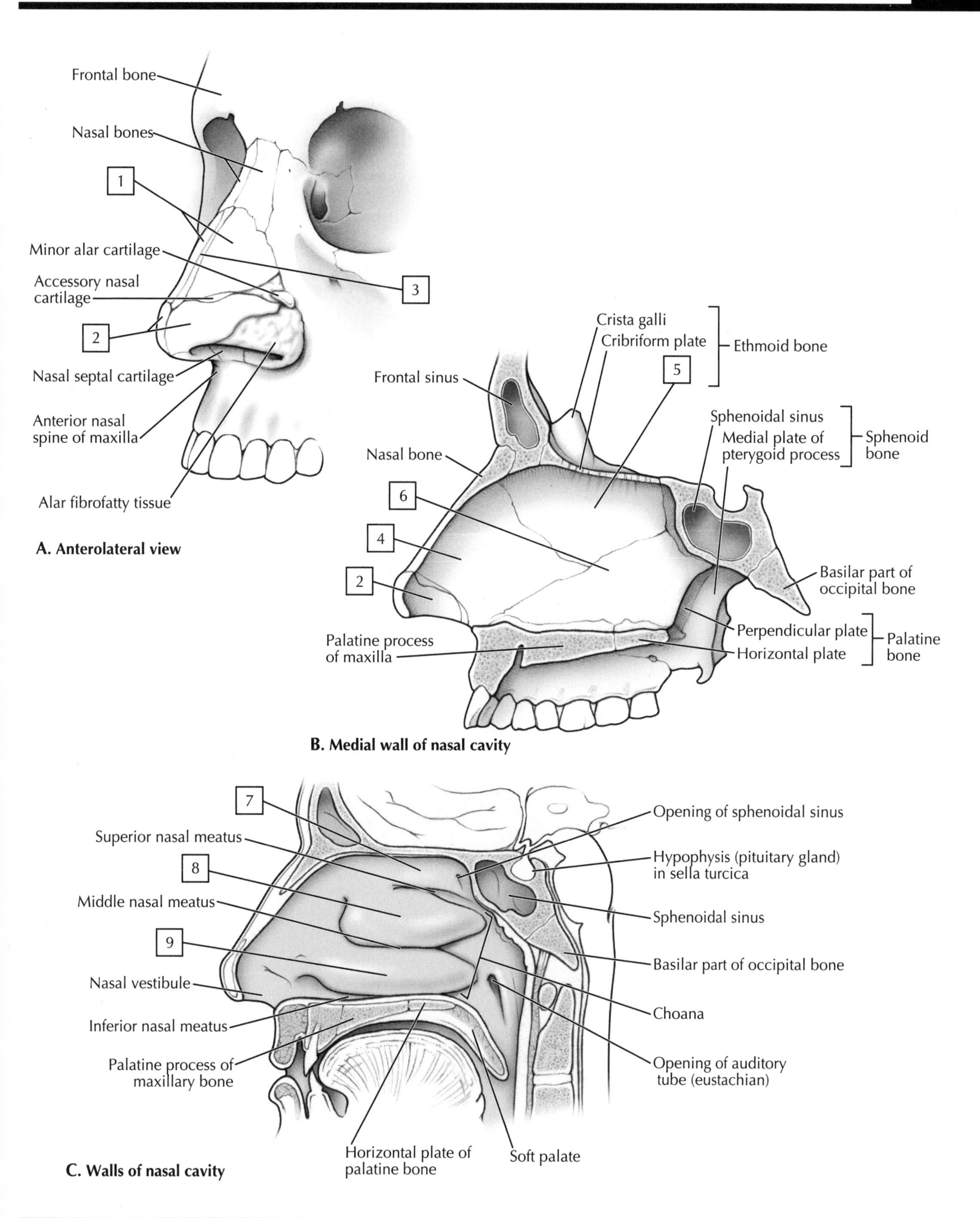


**A. Anterolateral view**

**B. Medial wall of nasal cavity**

**C. Walls of nasal cavity**

# Paranasal Sinuses

There are four pairs of air-filled paranasal sinuses, which are open chambers within several of the bones surrounding the nose and orbits. They are lined with respiratory epithelium, assist in warming and humidifying the inspired air, and drain their mucus secretions into the nasal cavities. Sneezing and blowing the nose clears the nasal cavity and sinuses of excess secretions. The paranasal sinuses and their features are summarized in the following table.

| SINUS | DESCRIPTION |
|---|---|
| Frontal | Paired sinuses, lying anteriorly in frontal bone and draining into semilunar hiatus of middle meatus |
| Ethmoid | Paired anterior, middle, and posterior sinuses in ethmoid bone; anterior and middle draining into middle meatus (hiatus semilunaris and ethmoid bulla, respectively), and posterior, into superior nasal meatus |
| Sphenoidal | Paired sinuses, in sphenoid bone, draining into sphenoethmoidal recess |
| Maxillary | Paired sinuses, in maxilla, draining into middle meatus (semilunar hiatus); it is the largest paranasal sinus (20-30 ml) |

The mucosa of the paranasal sinuses is innervated by sensory branches from the trigeminal nerve (CN V) (ophthalmic and maxillary divisions).

**COLOR** the following paranasal sinuses, using a different color for each sinus:

- ☐ **1. Frontal sinus**
- ☐ **2. Ethmoidal air cells (sinuses)**
- ☐ **3. Sphenoidal sinus**
- ☐ **4. Maxillary sinus**

***Clinical Note:***
**Rhinosinusitis** is an inflammation of the paranasal sinuses, most commonly the ethmoid and maxillary sinuses and nasal cavity. Usually, this condition begins as a viral infection followed by a secondary bacterial infection that obstructs the discharge of normal sinus mucus secretions and compromises the sterility of the sinuses. Infections of the nasal cavities can spread to the anterior cranial fossa via the cribriform plate, to the nasopharynx and soft tissues of the retropharyngeal region, to the middle ear via the auditory (pharyngotympanic or eustachian tube), paranasal sinuses, and even to the lacrimal apparatus and conjunctiva of the eye. Excessive **deviation of the nasal septum**, either congenital or as a result of injury by a fist fight or athletic injury, can usually be repaired surgically. **Epistaxis** (nosebleed) is fairly common because of the rich blood supply to the nasal mucosa and its susceptibility to irritants and trauma. The anterior third of the nose is the most common site for nosebleeds (*Kiesselbach's area*).

The **maxillary sinuses** are the most commonly infected of the paranasal sinuses. This may be due to the small ostia (opening for drainage) and from swelling of the rich sinus mucosa. If necessary, the maxillary sinus can be cannulated and drained.

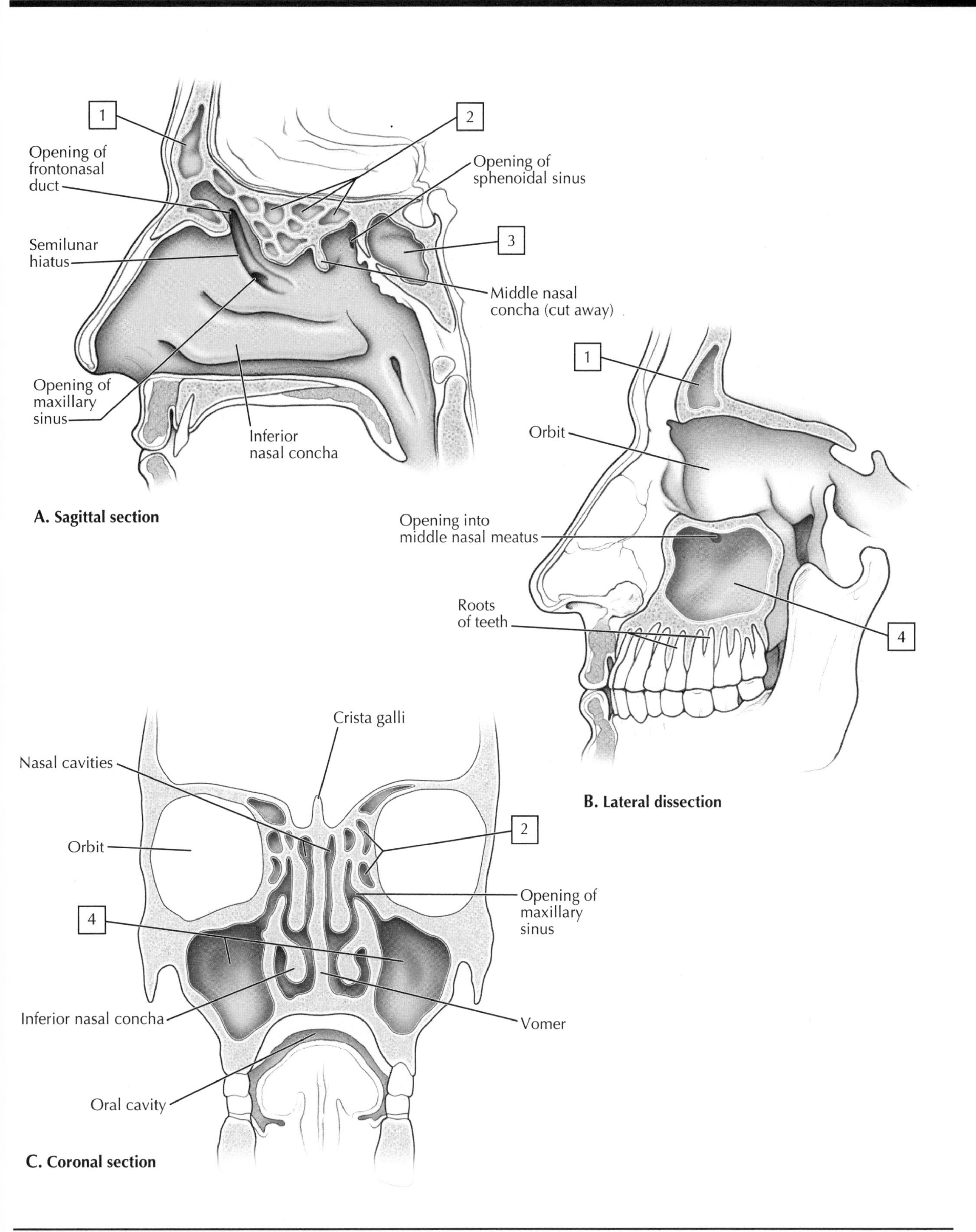


A. Sagittal section

B. Lateral dissection

C. Coronal section

# 7 Oropharynx, Laryngopharynx, and Larynx

The pharynx (throat) is subdivided into three regions:

- **Nasopharynx**: lies posterior to the nasal cavities and above the soft palate (already discussed)
- **Oropharynx**: extends from the soft palate to the superior tip of the epiglottis, and lies posterior to the oral cavity
- **Laryngopharynx**: extends from the tip of the epiglottis to the inferior aspect of the cricoid cartilage (often referred to by clinicians as the "hypopharynx"), lying posterior to the larynx

The oropharynx and laryngopharynx provide a passageway for both air and food (solids and liquids) and are essentially fibromuscular tubes lined with stratified squamous epithelium to protect the lining from abrasion. The muscular walls of the pharynx are formed largely by the three pharyngeal constrictors discussed previously (see Plate 3-5). **Waldeyer's lymphatic ring**, composed of the tubal tonsils, nasopharyngeal tonsils, lingual tonsils, and palatine tonsils, "guards" the openings into the pharynx and provides an important lymphatic immunologic defense mechanism, especially in children and adolescents.

The larynx lies anterior to the laryngopharynx and proximal esophagus, at about the level of the C3-C6 vertebrae and superior to the trachea. Structurally, the larynx consists of nine cartilages joined by ligaments and membranes.

| CARTILAGE | DESCRIPTION |
|---|---|
| Thyroid | Two hyaline laminae and the laryngeal prominence (Adam's apple) |
| Cricoid | Signet ring–shaped hyaline cartilage just inferior to thyroid |
| Epiglottis | Spoon-shaped elastic cartilage plate attached to thyroid |
| Arytenoid | Paired pyramidal cartilages that rotate on cricoid cartilage |
| Corniculate | Paired cartilages that lie on apex of arytenoid cartilages |
| Cuneiform | Paired cartilages in aryepiglottic folds that have no articulations |

The laryngeal cavity includes the following subdivisions:

- **Vestibule**: lies between the laryngeal inlet (just posterior to the epiglottis) and the vestibular folds
- **Rima glottidis**: the space or "slit" between the vocal folds
- **Ventricle**: the recesses that extend laterally between the vestibular and vocal folds
- **Infraglottic cavity**: the space below the vocal folds to the level of the cricoid cartilage; below the cricoid cartilage the infraglottic cavity becomes the proximal trachea

The **vestibular (false) folds** are protective in nature, but the **vocal (true) folds** control phonation much as a reed does in a reed instrument. Vibrations of the folds produce sounds as air passes through the rima glottidis; the pitch produced by these vibrations is dependent upon the diameter, length, thickness, and tension of the vocal folds. The size of the rima glottidis and tension on the folds are determined by the laryngeal muscles, (see Plate 3-6), but the amplification, resonance, and quality of the sound are products of the shape and size of the pharynx, oral cavity, nasal and paranasal cavities, and movements of the tongue, lips, cheeks, and soft palate.

**COLOR** the following features of the larynx, using a different color for each feature:

- ☐ 1. **Epiglottis**
- ☐ 2. **Thyroid cartilage**
- ☐ 3. **Infraglottic cavity**
- ☐ 4. **Cricoid cartilage**
- ☐ 5. **Trachea**
- ☐ 6. **Vestibular folds**
- ☐ 7. **Vocal folds**
- ☐ 8. **Laryngeal vestibule**
- ☐ 9. **Ventricle**

***Clinical Note:***

Inflammation of the pharyngeal tonsils (sometimes called adenoids), is a condition called **adenoiditis**, and this can obstruct the passageway of air from the nasal cavity to the nasopharynx. This infection may spread to the tubal tonsils and partially or completely obstruct the pharyngotympanic (eustachian) tube.

**Hoarseness** can be due to any condition that results in improper vibration or coaptation of the vocal folds. **Acute laryngitis** is an inflammation of the vocal folds that results in edema (swelling) of the vocal fold mucosa; it is usually a result of smoking, gastroesophageal reflux disease, chronic rhinosinusitis, cough, overuse of the voice (loud yelling, talking, or singing for extended periods), myxedema, or infections.

When other methods of establishing an airway have been exhausted or determined to be unsuitable, an incision can be made through the skin and the underlying cricothyroid membrane to gain access to the trachea (**cricothyrotomy**). The site of the incision can be judged by locating the space between the thyroid notch and sliding your finger inferiorly until the space between the thyroid and cricoid cartilages is palpated (about one fingerbreadth inferior to the thyroid notch).

Caution: if the patient has a midline thyroid pyramidal lobe, this procedure may lacerate that tissue and cause significant bleeding!

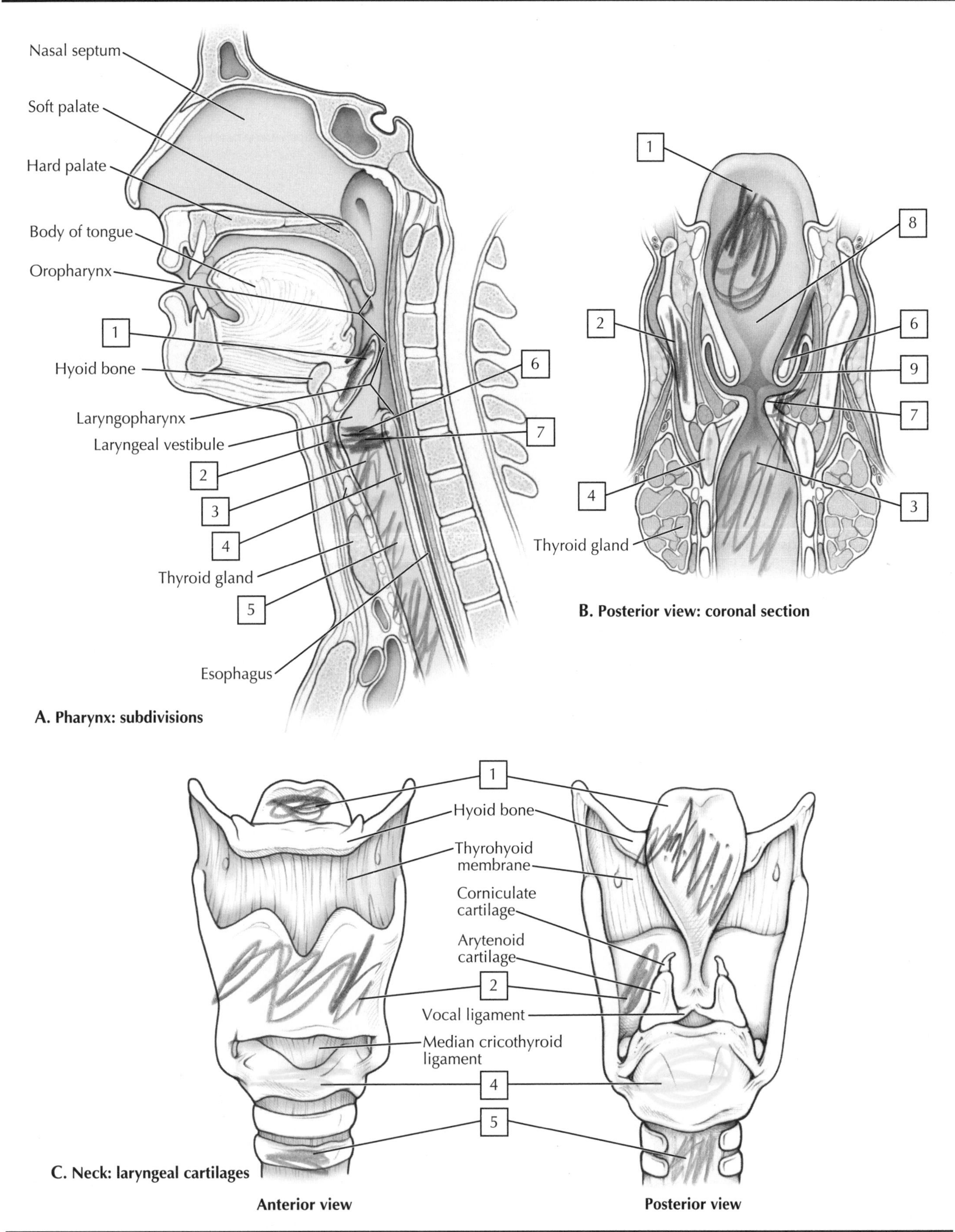


A. Pharynx: subdivisions

B. Posterior view: coronal section

C. Neck: laryngeal cartilages

# Trachea and Lungs

## Trachea and Bronchi

The trachea (windpipe) and bronchi (primary, secondary, and tertiary) convey air to and from the lungs, and their features are summarized in the table below.

| FEATURE | CHARACTERISTICS |
|---|---|
| Trachea | Is about 5 inches long and 1 inch in diameter; courses inferiorly anterior to esophagus and posterior to aortic arch |
| Cartilaginous rings | Are 16-20 C-shaped rings |
| Bronchus | Divides into right and left main (primary) bronchi at the level of the sternal angle of Louis |
| Right bronchus | Is shorter, wider, and more vertical than left bronchus; aspirated foreign objects more likely to pass into this bronchus |
| Carina | Is internal, keel-like cartilage at bifurcation of trachea |
| Secondary bronchi | Supply lobes of each lung (three on right side, two on left side) |
| Tertiary bronchi | Supply bronchopulmonary segments (10 for each lung) |

Beyond the tertiary bronchi, the passageways narrow considerably and eventually lose their cartilaginous support, thus forming the **bronchioles**, with a terminal bronchiole supplying a lobule of the lung. Within the lobules, the respiratory bronchioles divide into alveolar ducts, sacs, and alveoli.

**COLOR** the trachea and major bronchi, using a different color for each component:

☐ 1. **Trachea**

☐ 2. **Primary (main) bronchi (right side and left side)**

☐ 3. **Secondary bronchi (superior, middle, and inferior on the right side; superior and inferior on the left side)**

☐ 4. **Tertiary bronchi to the 10 bronchopulmonary segments in each lung**

## Lungs

Each lung is invested in a layer of **visceral pleura**, which reflects off of the lung surface and then forms an outer layer of **parietal pleura** that lines the inner aspect of the thoracic cage. The pleural cavities thus are potential spaces, like the pericardial sac, which normally contain a small amount of serous fluid that lubricates the surfaces and reduces the friction during respiration. The parietal pleura is sensitive to pain (the visceral pleura is not), and the two pleural cavities are separated from one another by the mediastinum. Features of the pleura are summarized in the table below.

| FEATURE | DEFINITION |
|---|---|
| Cupula pleura | Dome of cervical parietal pleura extending above the 1st rib |
| Parietal pleura | Membrane that in descriptive terms includes costal, mediastinal, diaphragmatic, and cervical (cupula) pleura |
| Pleural reflections | Points at which parietal pleura reflects off one surface and extends onto another (e.g., costal to diaphragmatic) |
| Pleural recesses | Reflection points at which lung does not fully extend into the pleural space (e.g., costodiaphragmatic and costomediastinal) |

The right lung has three lobes and the left lung two lobes. On the medial surface of each lung is the **hilum**, which is the region where vessels, bronchi, nerves, and lymphatics enter and leave the lungs. Features of each lung are summarized in the following table.

| FEATURE | CHARACTERISTICS |
|---|---|
| Lobes | Three lobes (superior, middle, inferior) in right lung; two in left |
| Horizontal fissure | Only on right lung, extends along line of 4th rib |
| Oblique fissure | On both lungs, extends from T2 vertebra to 6th costal cartilage |
| Impressions | Made by adjacent structures, in fixed lungs |
| Hilum | Points at which structures (bronchi, vessels, nerves, lymphatics) enter or leave lungs |
| Lingula | Tongue-shaped feature of left lung |
| Cardiac notch | Indentation for the heart, in left lung |
| Pulmonary ligament | Double layer of parietal pleura hanging from the hilum that marks reflection of visceral pleura to parietal pleura |
| Bronchopulmonary segment | 10 functional segments in each lung supplied by a segmental bronchus and a segmental artery from the pulmonary artery |

**COLOR** the following features of the lungs, using the colors recommended for each feature:

☐ 5. **Pulmonary arteries: carry blood from the right ventricle of the heart to the lungs for oxygenation (blue)**

☐ 6. **Bronchus (yellow)**

☐ 7. **Pulmonary veins: return oxygenated blood to the left atrium of the heart (red)**

***Clinical Note:***

**Lung cancer** is the leading cause of cancer-related deaths and arises either from the alveolar lining cells or from the epithelium of the tracheobronchial tree.

**Aspiration** of small objects (peanuts, marbles) into the lungs can block the bronchi. Usually, the object is aspirated into the right main bronchus because it is shorter, wider, and more vertical than the left main bronchus.

Generally, **chronic lung disease** can be lumped into chronic obstructive pulmonary disease (COPD) or chronic restrictive lung disease. Obstructive diseases include chronic bronchitis, asthma, and emphysema, and make it more difficult to exhale the air residing in the lung. Restrictive diseases (fibrosis) usually reduce the compliance of the lungs, making it more difficult to inflate the stiffened lungs.

**Pneumonia** accounts for about one-sixth of all deaths in the United States. Children and elderly adults are especially vulnerable to pneumococcal pneumonia, as are individuals with congestive heart failure, COPD, diabetes, or alcoholism.

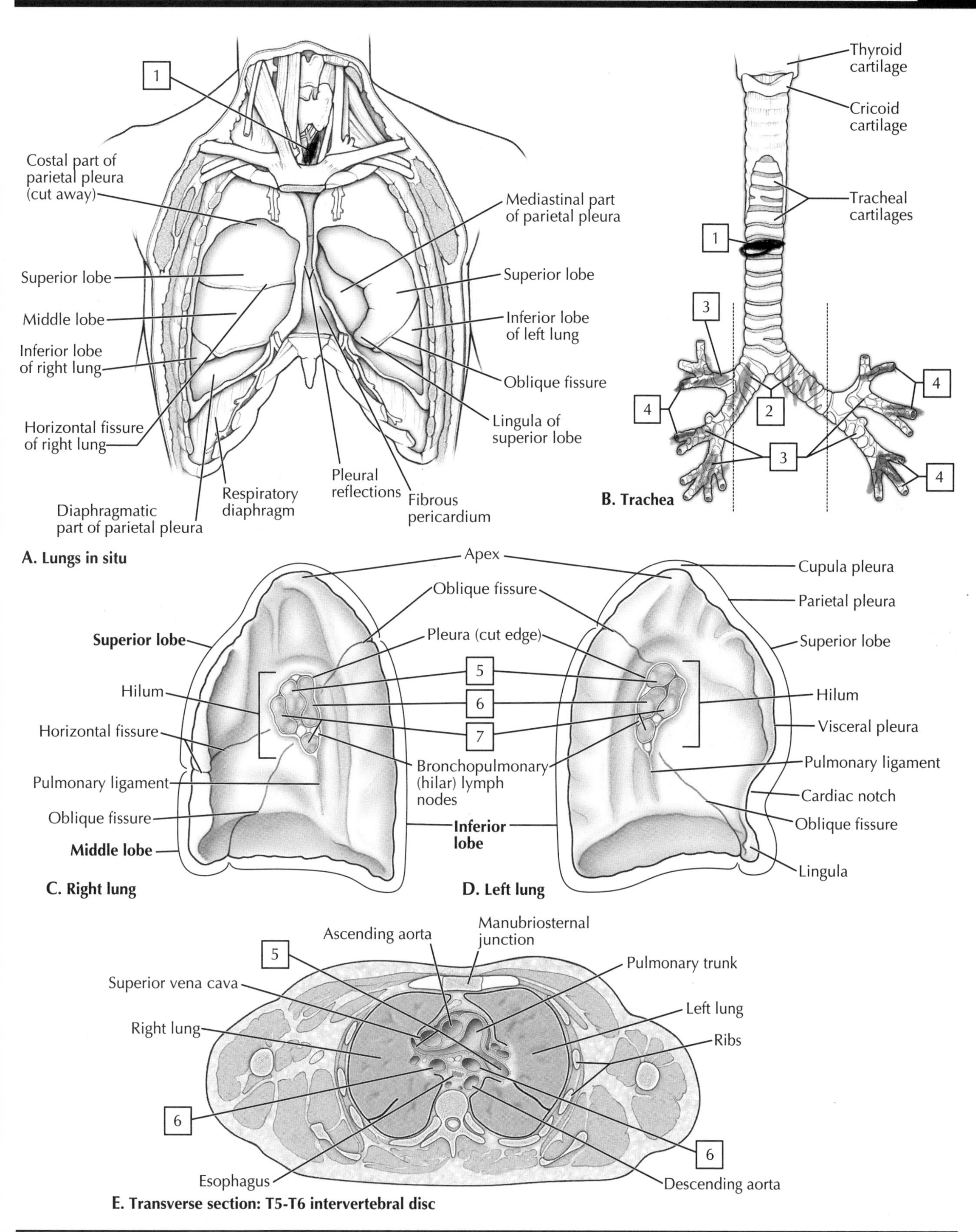


**A. Lungs in situ**

**B. Trachea**

**C. Right lung**

**D. Left lung**

**E. Transverse section: T5-T6 intervertebral disc**

# Respiratory Mechanisms

The mechanics of ventilation involve the dynamic interaction of the lungs, chest wall, and respiratory diaphragm.

During quiet respiration, contraction of the respiratory diaphragm alone accounts for about 75% of inspiration. The external intercostal muscles of the thoracic wall (see Plate 3-11) and selected muscles of the neck (scalenes) also can assist in inspiration, especially during active exercise. Expiration involves the elastic recoil of the lungs themselves and is assisted by relaxation of the respiratory diaphragm and contraction of some of the intercostal and abdominal muscles (rectus abdominis and abdominal oblique muscles).

Blood from the right ventricle of the heart perfuses the lungs (via the pulmonary arteries) at a resting rate of about 5 L/min under low pressure. Pulmonary capillary plexuses envelop the alveolar sacs, where most of the gas exchange occurs. Pulmonary veins collect the oxygenated blood and return it to the left side of the heart for distribution throughout the systemic circulation.

Gas exchange occurs at the level of the alveoli and capillaries and involves the following:

- Across the type I alveolar cells
- Across the fused basement membranes of the type I cells and endothelial cells
- Across the capillary endothelial cells

**COLOR** the following features of the intrapulmonary circulation, using the colors suggested:

- ☐ **1. Pulmonary artery (lower oxygen content) (blue)**
- ☐ **2. Pulmonary vein (saturated with oxygen) (red)**
- ☐ **3. Type II alveolar cell (secretes surfactant) (orange)**
- ☐ **4. Type I alveolar cell (yellow)**
- ☐ **5. Capillary endothelial cell (purple)**
- ☐ **6. Type I alveolar cell and endothelial cell fused basement membranes (light blue)**
- ☐ **7. Interstitial cells (green)**
- ☐ **8. Red blood cell (red)**
- ☐ **9. Alveolar macrophage (brown) (in alveolar air space)**

Type II alveolar cells secrete **surfactant,** which forms a thin film over the fluid that normally coats the surface of the alveolus, thus reducing the surface tension of the fluid-lined alveoli and helping to lower the pressure needed to inflate the alveoli.

As blood flows through the alveolar capillaries, oxygen diffuses from the alveolus into the red blood cell, where it binds to hemoglobin. At the same time, carbon dioxide diffuses out of the red blood cell and into the alveolus. Normally, blood traverses the entire capillary length in 0.75 second, and even faster when the cardiac output is increased. However, gas exchange is so efficient that it normally occurs in about 0.5 second. Almost all of the oxygen carried to the body's tissues by the blood is bound to hemoglobin; only a small fraction is dissolved and transported in the plasma. The interalveolar septum (separates the alveolar air space from the capillary lumen) is the **blood-air barrier;** it is very thin, and allows for gases to diffuse rapidly across it. The septum consists of three layers:

- Type I pneumocyte and its surfactant layer in the alveolar air space
- Fused basal lamina of the type I pneumocyte and the capillary endothelial cell
- Endothelium of the continuous capillaries

***Clinical Note:***

Failure to produce sufficient amounts of **surfactant**, as can occur in premature infants because of the underdevelopment of the type II alveolar cells, can result in an increase in the work of breathing and cause respiratory distress **(infant respiratory distress syndrome [IRDS])**. Because the lungs are not needed in utero, they are among one of the last systems to functionally develop in the fetus and often the limiting factor in survival of a premature infant.

Collapse of the lung, termed **atelectasis**, can occur when air enters the pleural cavity through a chest wound or rupture of the visceral pleura, sometimes seen in **pneumonia**. Air in the intrapleural space results in a **pneumothorax**. Removal of this air via a chest tube and repair of the hole will return the lung to its normal function.

In **emphysema**, the walls of adjacent alveoli break down and the alveolar chambers become larger. Commonly, this is caused by narrowing of the bronchioles and destruction of the alveolar wall, often from inflammation. Smoking is a major risk factor.

Chronic obstructive pulmonary disease (**COPD**) is a broad class of obstructive lung diseases such as chronic bronchitis, asthma, and emphysema. The elastic recoil of the lung decreases, causing collapse of the airways during expiration, leading to **dyspnea**, which is labored breathing (air hunger). This increases the work of expiration and can lead to a "barrel-chested" appearance caused by hypertrophy of the intercostal muscles.

Finally, **lung cancer** is the leading cause of cancer-related deaths worldwide, and cigarette smoking is the cause in about 85% to 90% of all cases.

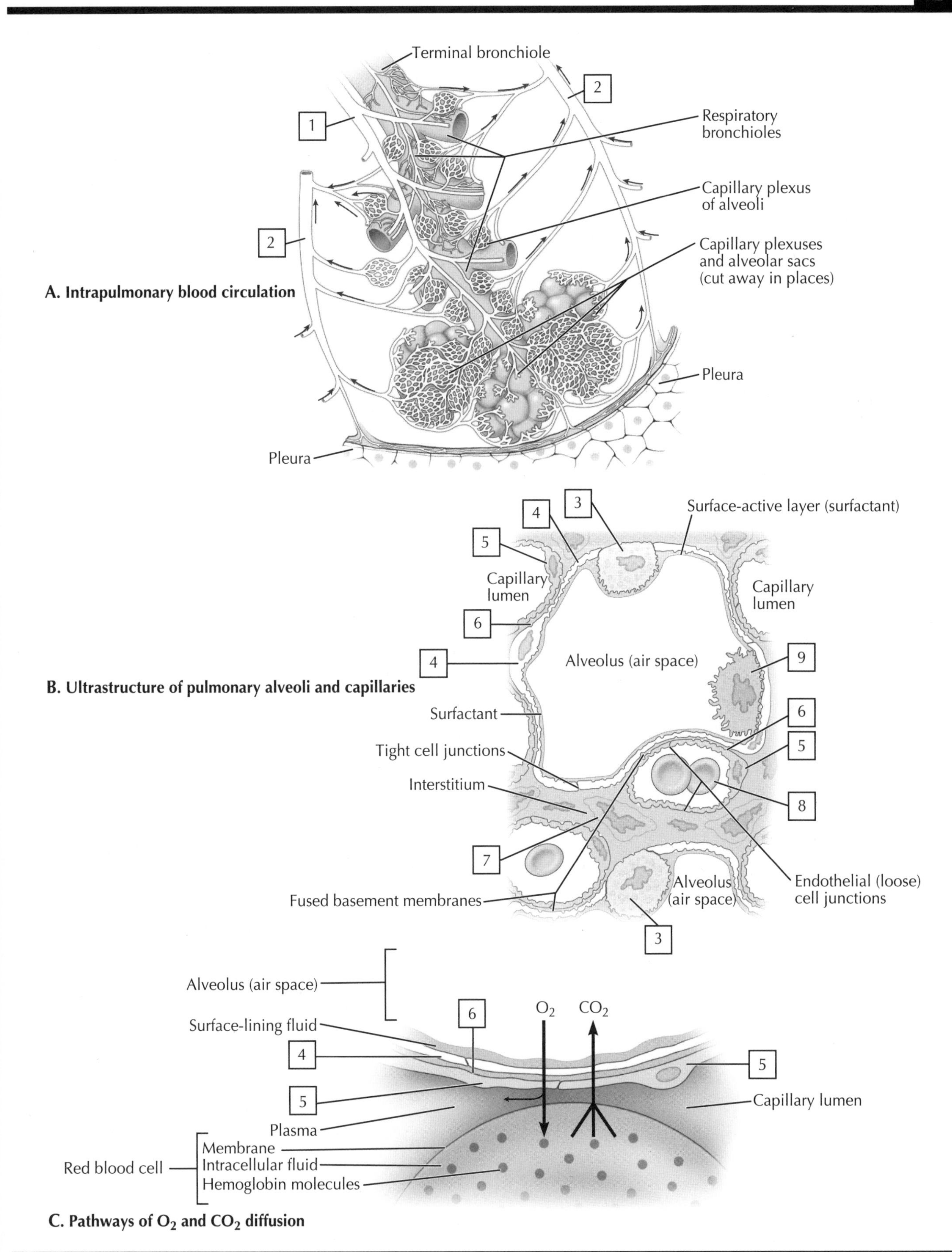
Terminal bronchiole
2
1
Respiratory bronchioles
Capillary plexus of alveoli
2
Capillary plexuses and alveolar sacs (cut away in places)
A. Intrapulmonary blood circulation
Pleura
Pleura
4
3
Surface-active layer (surfactant)
5
Capillary lumen
Capillary lumen
6
4
Alveolus (air space)
9
B. Ultrastructure of pulmonary alveoli and capillaries
Surfactant
6
Tight cell junctions
5
Interstitium
8
7
Fused basement membranes
Alveolus (air space)
Endothelial (loose) cell junctions
3
Alveolus (air space)
Surface-lining fluid
6
$O_2$
$CO_2$
4
5
5
Capillary lumen
Plasma
Red blood cell
Membrane
Intracellular fluid
Hemoglobin molecules
C. Pathways of $O_2$ and $CO_2$ diffusion

## REVIEW QUESTIONS

1. A premature infant is having great difficulty breathing because of an incomplete coating of surfactant on the alveolar epithelium. Which of the following cells secrete surfactant?
   A. Alveolar endothelial cells
   B. Alveolar macrophages
   C. Simple ciliated columnar cells
   D. Type I pneumocytes
   E. Type II pneumocytes

2. A small child aspirates a peanut into her lung. Where in the lung is that peanut most likely to be found?
   A. Lower lobe of the left lung
   B. Primary bronchus of the left lung
   C. Primary bronchus of the right lung
   D. Tertiary bronchus of the left lung
   E. Tertiary bronchus of the right lung

3. A patient's frontal sinus appears to be blocked and infected. Where does the frontal sinus normally drain?
   A. Inferior meatus
   B. Middle meatus
   C. Nasopharynx
   D. Sphenoethmoidal recess
   E. Superior meatus

4. A child bites into a very cold ice cream cone and immediately feels an intense pain, referred to as a "brain freeze." Which of the following regions is the most likely source of this pain?
   A. Hard palate
   B. Mandible
   C. Maxillary sinus
   D. Soft palate
   E. Sphenoidal sinus

**For each description below (5-7), color or highlight the relevant anatomical structure in the image.**

5. This cell possesses phagocytic characteristics and helps keep the alveolar sac free of debris.
6. This cell lines the alveolar sac but does not participate directly in gas exchange.
7. This cell does participate in gas exchange and is coated with surfactant.

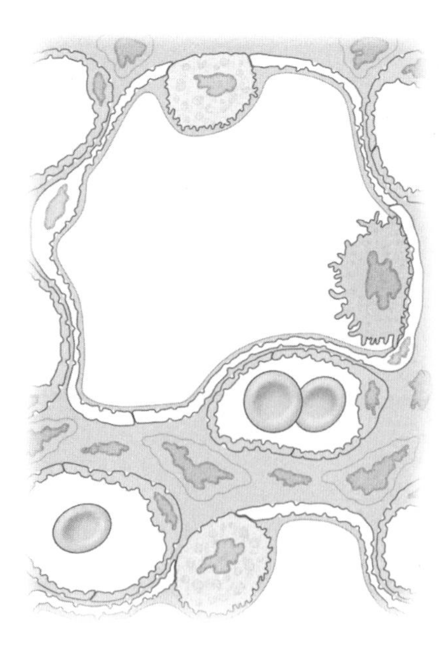

8. What type of epithelium normally lines the trachea? ______________________

9. Identify three important functions of the respiratory system. ______________________

10. The lungs are contained within a pleural sac. What two layers of connective tissue form these sacs? ______________________

## ANSWER KEY

1. E
2. C
3. B
4. C (nerves to the maxillary teeth run in the mucosa of the sinus walls and are sensitive to cold)
5. Alveolar macrophage
6. Type II pneumocyte (but it does secrete surfactant)
7. Type I pneumocyte

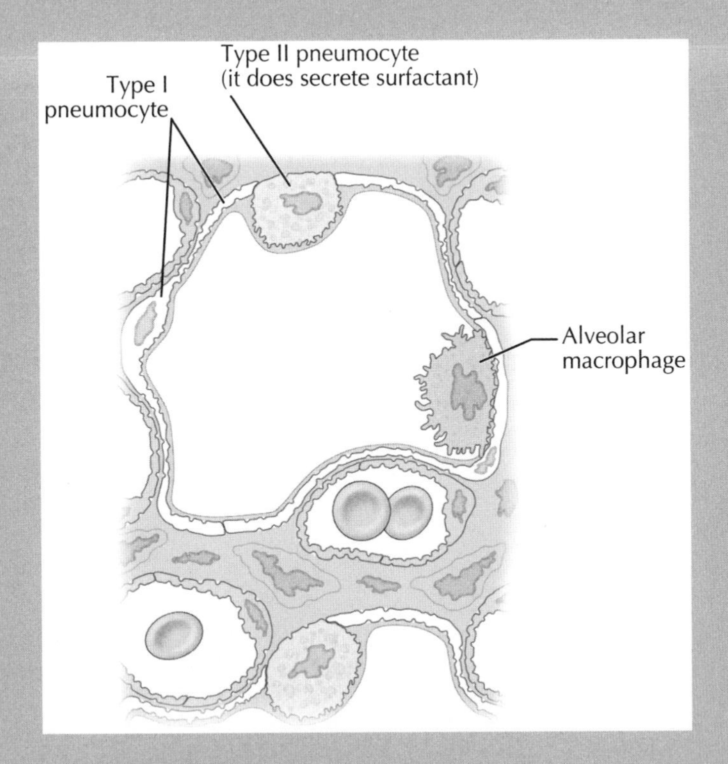


8. Ciliated pseudostratified columnar epithelium
9. Filters and humidifies the air, moves air into and out of the lungs, provides a large surface area for gas exchange, helps regulate the body's pH, participates in vocalization, and assists the olfactory system with the detection of smells
10. Visceral pleura (on the lung surface) and parietal pleura (lines the thoracic cavity)

# Chapter 8 Digestive System

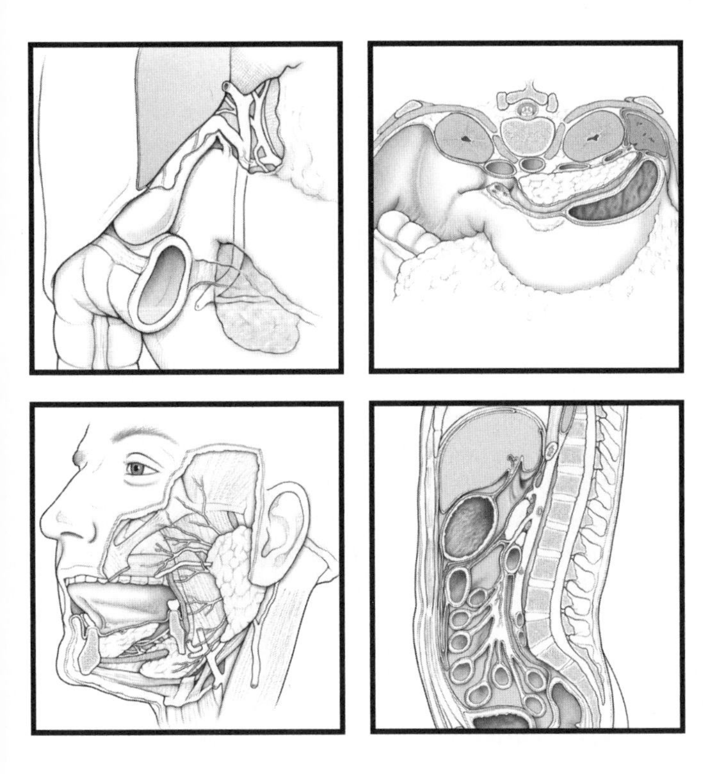

# 8 Overview

The digestive system consists of the epithelial-lined tube beginning with the **oral cavity** and extending to the **anal canal**, and also includes glands, such as the:

- **Salivary glands**: three major glands and thousands of microscopic minor salivary glands scattered throughout the oral mucosa
- **Liver**: the largest gland in the body
- **Gallbladder**: stores and concentrates bile needed for fat digestion
- **Pancreas**: an exocrine (digestive enzymes) and endocrine organ

The epithelial-lined tube that is the GI tract measures about 25 feet from mouth to anal canal in adults and includes the following cavities and visceral structures:

- **Oral cavity**: tongue, teeth, and salivary glands
- **Pharynx**: throat, subdivided into the nasopharynx, oropharynx, and laryngopharynx
- **Esophagus**
- **Stomach**
- **Small intestine**: subdivided into the duodenum, jejunum, and ileum
- **Large intestine**: subdivided into the cecum, ascending colon, transverse colon, descending colon, sigmoid colon, rectum, and anal canal

**COLOR** each of the following visceral components of the thoracic and abdominal GI tracts, using a different color for each component:

- ☐ 1. **Liver**
- ☐ 2. **Gallbladder**
- ☐ 3. **Duodenum (phantom in figure behind the transverse colon)**
- ☐ 4. **Ascending colon**
- ☐ 5. **Cecum**
- ☐ 6. **Ileum**
- ☐ 7. **Rectum**
- ☐ 8. **Anal canal**
- ☐ 9. **Sigmoid colon**
- ☐ 10. **Jejunum**
- ☐ 11. **Descending colon**
- ☐ 12. **Transverse colon**
- ☐ 13. **Stomach**
- ☐ 14. **Esophagus**
- ☐ 15. **Oral cavity**

Clinically, because of the structural complexity of the abdominal viscera, it is important for clinicians to know where underlying visceral structures lie in relationship to the surface of the abdominal wall. To facilitate this exercise, the abdomen can be divided into **four quadrants** or into **nine regions**, as shown in parts ***B*** and ***C***. Additionally, various reference planes are used clinically in the physical exam to divide the abdomen into regions, as summarized below.

| PLANES OF REFERENCE | DEFINITION |
|---|---|
| Median | Vertical plane from xiphoid process to pubic symphysis |
| Transumbilical | Horizontal plane across umbilicus (these two planes divide abdomen into quadrants) |
| Subcostal | Horizontal plane across inferior margin of 10th costal cartilage |
| Intertubercular | Horizontal plane across the tubercles of the ilium and the body of the L5 vertebra |
| Midclavicular | Two vertical planes through the midpoint of the clavicles (these two planes and the subcostal and intertubercular planes divide the abdomen into nine regions) |

Functionally, the digestive system begins with ingestion and mechanical and enzymatic digestion of food in the oral cavity, followed by propulsion or swallowing, and peristalsis, through the oropharynx, esophagus, stomach, and intestines. Churning, peristalsis, and chemical digestion occur from the stomach proximally down through the large intestine, and absorption occurs largely in the small intestine with the absorption of water in the large intestine. Compaction and defecation of feces occur in the rectum and distally via the anal canal. Accessory digestive organs include the teeth and tongue, the salivary glands, the liver and gallbladder, and the pancreas.

***Clinical Note:***

Many of the clinical problems associated with the digestive tract involve the sub-diaphragmatic organs (stomach to anal canal, and the accessory organs). These problems may include inflammation, malsecretion and malabsorption problems, physical obstructions, cancer, and dysfunctional innervation of bowel segments.

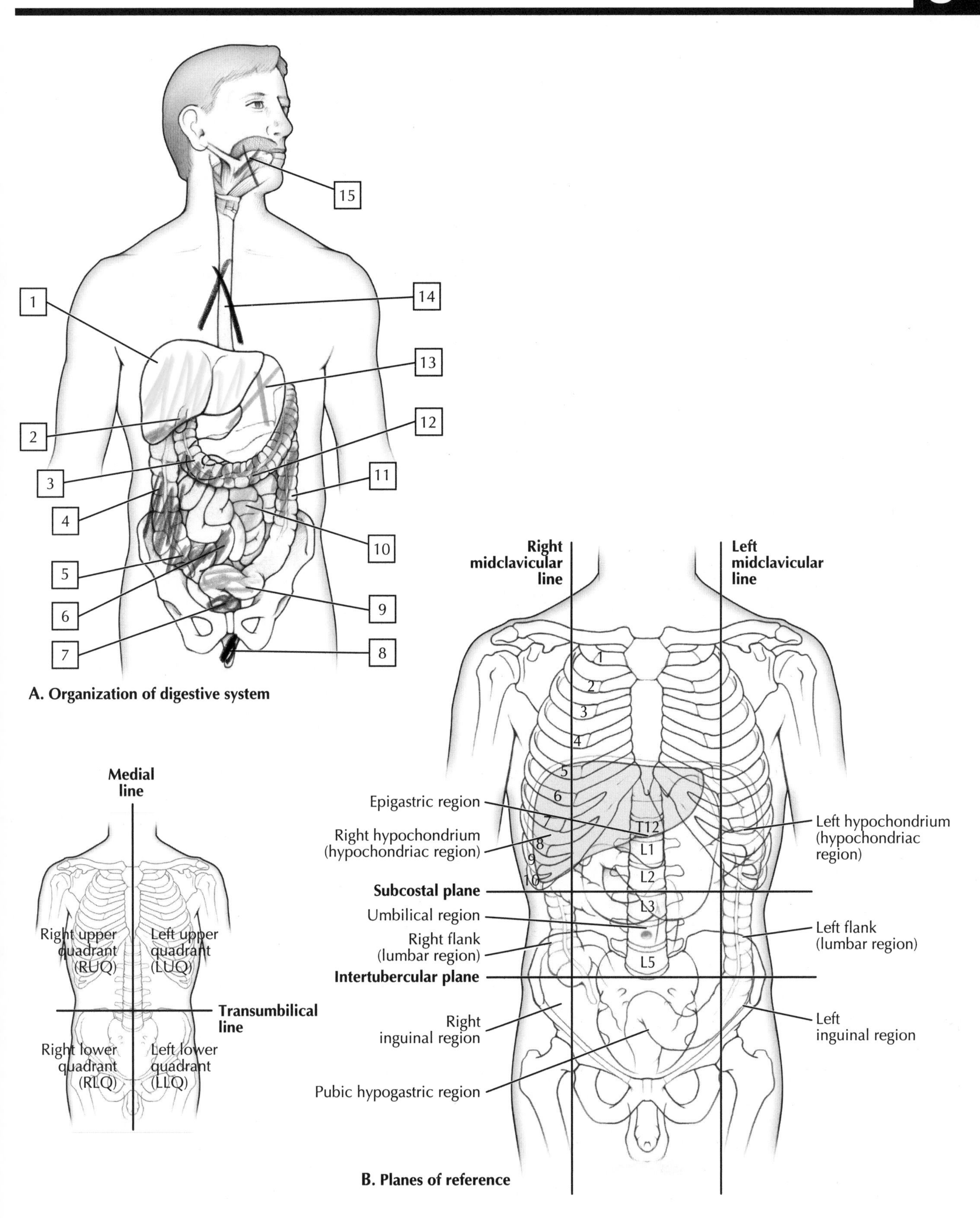


A. Organization of digestive system

B. Planes of reference

# Oral Cavity

The oral cavity is the first portion of the GI tract and consists of the following:

- Mouth (oral vestibule), which is the narrow space between the lips or cheeks, and the teeth and gums
- Oral cavity proper, which includes the palate (hard and soft), teeth, gums (gingivae), salivary glands, and tongue

The mucosae of the palate, cheeks, tongue, and lips contain thousands of **minor salivary glands** that secrete directly into the oral cavity. Additionally, three pairs of major salivary glands provide **saliva** to aid in the digestion, softening, and deglutition (swallowing) of food. Saliva is largely water (97% to 99%), hypo-osmotic, and slightly acidic (pH 6.75-7.00). Saliva keeps the mucosal surfaces moist and lubricated to protect against abrasion, controls oral bacteria by secreting lysozyme, secretes calcium and phosphate for tooth formation and maintenance, and secretes amylase to begin the digestion of starches. The **serous acinar cells** of the salivary glands secrete the proteinaceous and enzymatic components of saliva, whereas the **mucous acinar cells** secrete a watery mucus. Finally, **lingual lipase**, secreted by the serous glands of the tongue, mixes with saliva and begins the digestion of fats. The average output of saliva is about 1000-1500 ml per day. The major salivary glands are summarized in the table below.

| GLAND | GLAND TYPE AND INNERVATION |
|---|---|
| Parotid | Serous gland innervated by glossopharyngeal nerve (CN IX) parasympathetics that course to gland ultimately via the auriculotemporal nerve (branch of trigeminal nerve [CN $V_3$]); secretes via the parotid (Stensen's) duct |
| Submandibular | Seromucous gland innervated by facial nerve (CN VII) parasympathetics that course to gland via the chorda tympani branch of CN VII and join the lingual nerve (branch of trigeminal nerve [CN $V_3$]) to synapse in the submandibular ganglion (branch of CN $V_3$]); secretes via the submandibular (Wharton's) duct |
| Sublingual | Large mucous gland innervated by facial nerve (CN VII) parasympathetics that course to the gland in a similar fashion to the submandibular gland pathway listed above; secretes saliva via multiple small ducts in the sublingual fold |

See Plates 4-20 and 4-22 for the innervation of the salivary glands.

The parotid gland secretes saliva via its parotid (Stensen's) duct. The submandibular gland secretes saliva via its submandibular (Wharton's) duct, and the sublingual gland secretes saliva via numerous small ducts located at the base of the anterolateral tongue. As the saliva passes through the ducts, its electrolyte composition is modified such that the saliva entering the mouth is hypotonic to plasma and has a high bicarbonate concentration. The glycoprotein mucin secreted by the salivary glands dissolves in water and lubricates the oral cavity and hydrates the food entering the mouth. Saliva also protects against microorganisms by secreting IgA antibodies, lysozyme, a cyanide compound, and defensins (a mixture of antibiotic-like proteins).

**COLOR** the following features of the oral cavity, using a different color for each feature:

- ☐ 1. **Hard palate**
- ☐ 2. **Soft palate**
- ☐ 3. **Palatine tonsil**
- ☐ 4. **Tongue**
- ☐ 5. **Uvula**
- ☐ 6. **Sublingual glands**
- ☐ 7. **Submandibular gland**
- ☐ 8. **Parotid gland**

***Clinical Note:***

**Gingivitis** is an inflammation of the gums caused by bacterial accumulation in the crevices between the teeth and gums. Both plaque and tartar buildup can cause irritation of the gums that leads to bleeding and swelling and, if left untreated, can result in damage to the bone and loss of teeth.

Any disease process that inhibits saliva secretion will adversely affect the oral cavity, allowing decomposing food particles and bacteria to accumulate, resulting is **halitosis** (bad breath).

**Mumps** is a childhood disease that results in inflammation of the parotid salivary glands, caused by the myxovirus (it can spread to others in saliva). Mumps is accompanied by significant swelling of the parotid glands, a moderate fever, and painful swallowing. Most children now receive a vaccine to prevent mumps.

Touching the posterior aspect of the tongue elicits a "**gag reflex**" mediated by CN IX and CN X, which leads to muscular contraction in the walls of the oropharynx. CN IX provides the afferent limb of the reflex. This reflex can be tested by placing a tongue depressor on the back of the tongue.

**Injury to the CN XII**, such as might occur with a mandibular fracture, may result in paralysis and atrophy of one side of the tongue. When the patient is asked to "stick out their tongue," it deviates to the paralyzed side, due to the unopposed action of the unaffected genioglossus muscle on the normal side.

The sublingual veins beneath the thin mucosa under the tongue provide a nice route for the transmucosal absorption of a drug. This route of treatment is used for the absorption of nitroglycerin, a vasodilator to treat **angina pectoris** (chest pain from an ischemic heart).

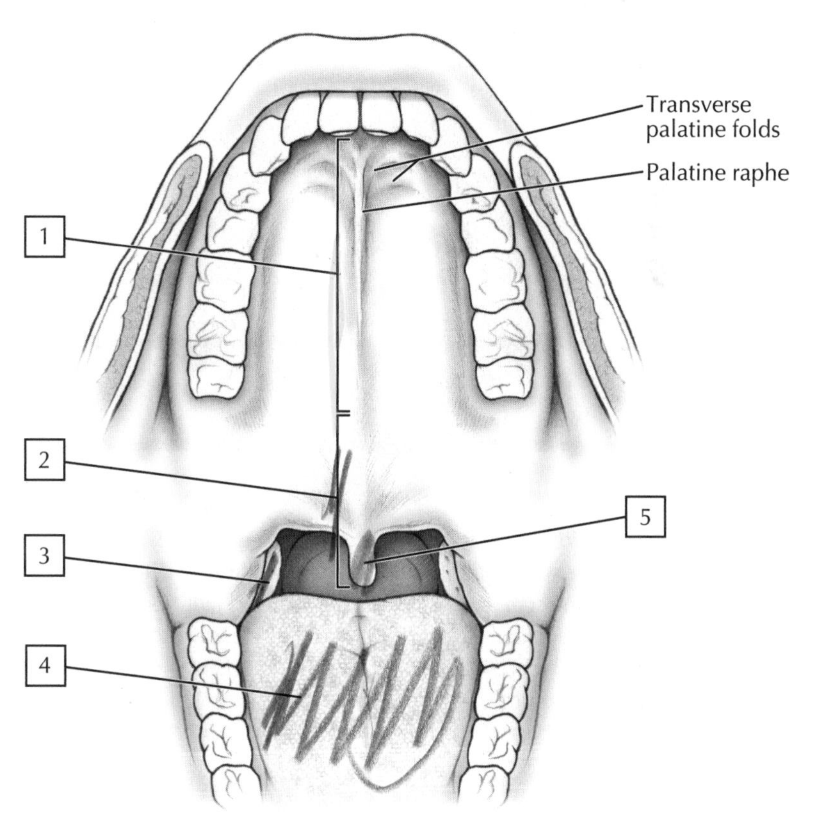


A. Anterior view

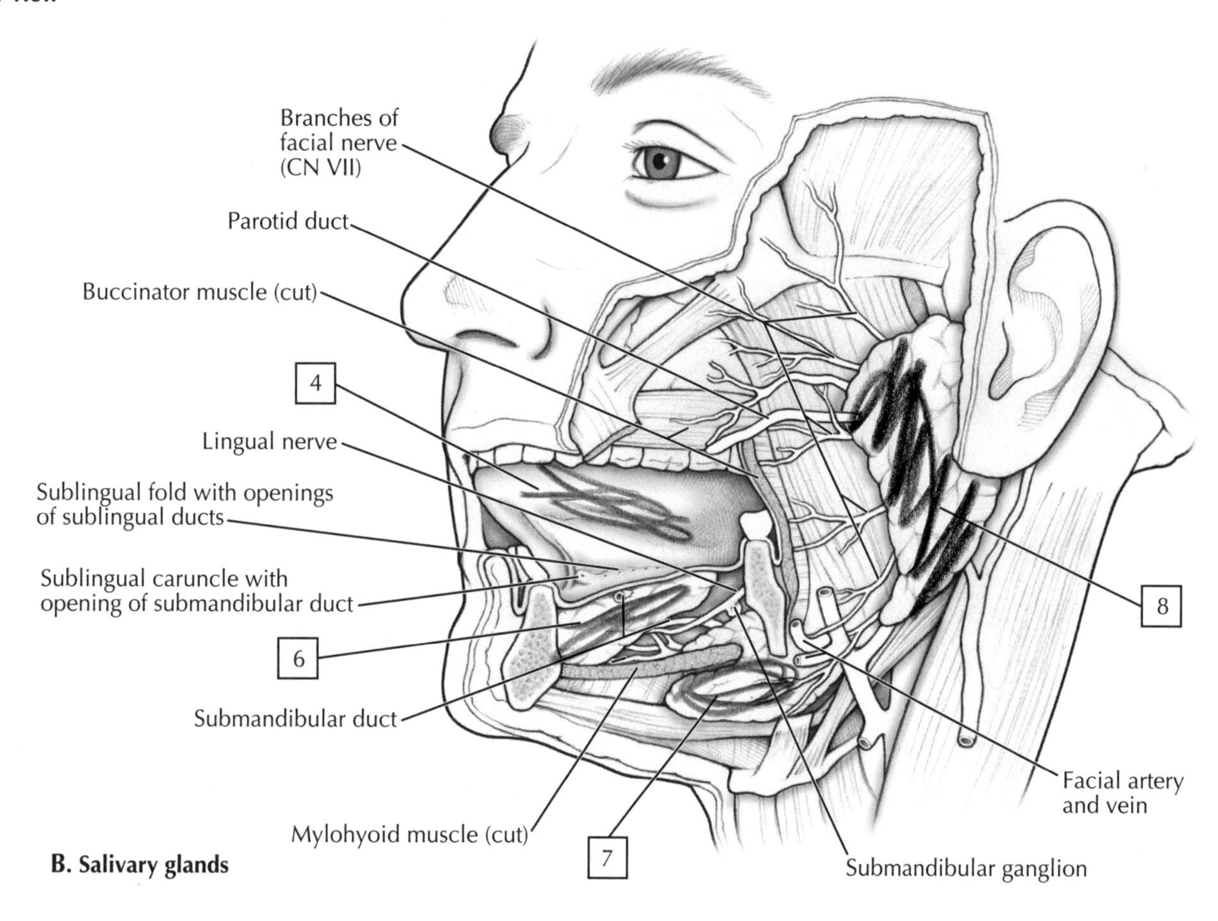


B. Salivary glands

# 8 Teeth

The teeth are hard structures set in the upper (maxilla) and lower (mandible) jaws in dental alveoli or sockets. The tooth has a crown, neck, and root; these formations, as well as other anatomical features of the tooth, are summarized in the following table.

| | |
|---|---|
| Crown | Anatomical crown: the portion of the tooth that has a surface of enamel |
| Root | Anatomical root: the portion of the tooth that has a surface of cementum |
| Apex of the root | The end tip of the root, which provides entrance of the neurovascular connective tissue into the pulp cavity |
| Enamel | The hard, shiny surface of the anatomical crown and the hardest part of the tooth |
| Cementum | A thin dull layer on the surface of the anatomical root |
| Dentin | The hard tissue that underlies both the enamel and cementum and constitutes the majority of the tooth |
| Pulp cavity | Contains the dental pulp (highly neurovascular connective tissue) |

Modified with permission from Norton N: *Netter's Head and Neck Anatomy for Dentistry,* 3rd Edition, Philadelphia, 2017, Elsevier, pp. 357-359.

Humans have two sets of teeth:

- **Deciduous teeth**: our primary dentition, which consists of 20 teeth that usually have all appeared by the age of 2.5-3 years (2 incisors, 1 canine, and 2 molars in each of the 4 quadrants of the jaws)
- **Permanent teeth**: our secondary dentition, which consists of 32 teeth that usually begin to appear around the age of 6 years (2 incisors, 1 canine, 2 premolars, and 3 molars in each quadrant), replacing the deciduous teeth. The third molars also are known as the "wisdom teeth" because they are normally the last teeth to erupt.

**COLOR** each of the following teeth and the features of a typical tooth, using a different color for each feature:

- ☐ **1. Incisors**
- ☐ **2. Canine**
- ☐ **3. Premolars**
- ☐ **4. Molars**
- ☐ **5. Enamel**
- ☐ **6. Dentin**
- ☐ **7. Gingival (gum) epithelium (stratified squamous)**
- ☐ **8. Cementum**
- ☐ **9. Root canals (containing vessels and nerves)**

***Clinical Note:***

**Tooth decay** (dental caries) can lead to cavities, which are caused by bacteria that convert food debris into acids that form plaque. The plaque adheres to the teeth and, if not removed in a timely fashion, can mineralize to form tartar. The acids in the plaque can erode into the enamel and form a cavity. Foods rich in sugars and starches promote cavity formation.

**Enamel** is an acellular material and is the hardest substance in the body. It is heavily mineralized with calcium salts and densely layered hydroxyapatite crystals that lie in a perpendicular orientation with relation to the tooth's surface. Unfortunately, once the tooth erupts, the enamel-producing cells degenerate. If cracked or lost because of tooth decay, the enamel will not recover and the tooth must be filled by the dentist.

If the nerve innervating a tooth dies, it will turn dark and its blood supply may be compromised, and the pulp cavity may become infected. In such cases, the diseased tooth will be removed by **root canal therapy**.

**Dental plaque** may result from bacterial metabolism, which produces acids that can dissolve the calcium salts in the tooth. Good oral hygiene with frequent brushing and flossing can remove this harmful plaque; failure to do this can result in calcification of the accumulating plaque, forming calculus (tartar) that disrupts the seal between the tooth and gum, leading to an infection called **gingivitis**. The gums become red, swollen, and sore and may bleed. Left untreated, the immune system attacks the bacteria and affected soft tissues around the tooth, which leads to a loss of bone and periodontitis.

**Periodontal disease** accounts for a significant loss of teeth as adults age, but can be prevented if diagnosed early enough.

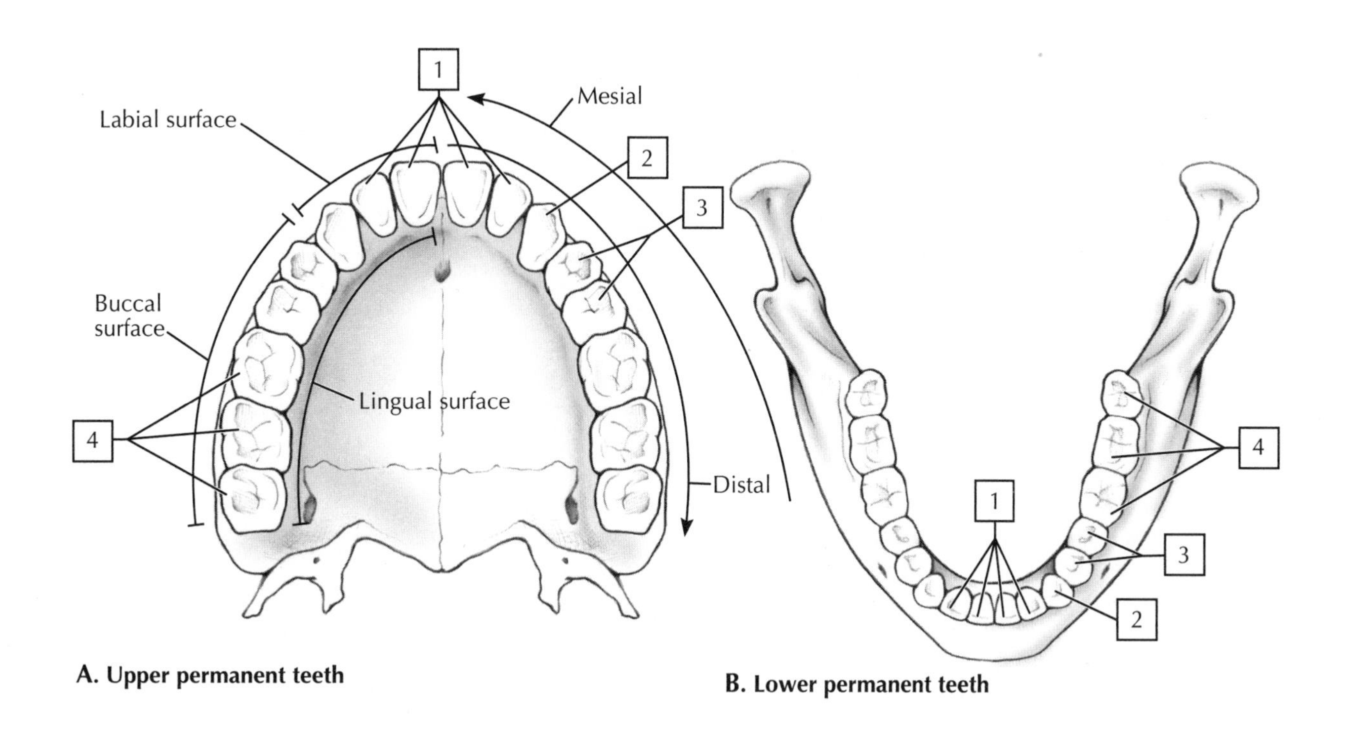


A. Upper permanent teeth

B. Lower permanent teeth

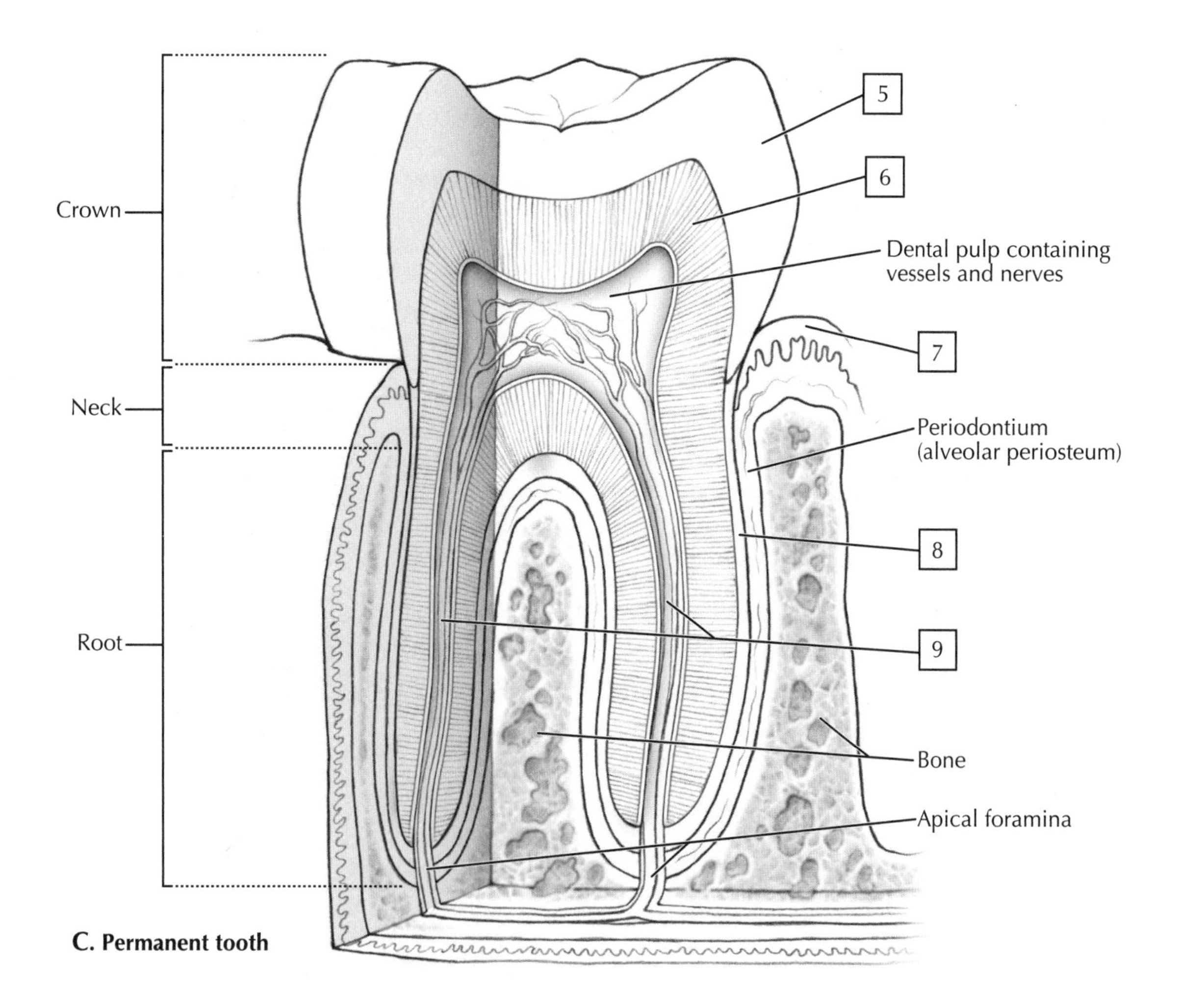


C. Permanent tooth

# 8 Pharynx and Esophagus

## Pharynx

The pharynx is subdivided into the **nasopharynx**, **oropharynx**, and **laryngopharynx**, and has been previously reviewed in the muscular and respiratory system sections (see Plate 7-1). The mucosa of the oropharynx and laryngopharynx is stratified squamous epithelium, providing protection during swallowing, and is interspersed with mucous glands to keep the epithelium moist with a thin mucus covering. The laryngopharynx opens anteriorly into the laryngeal inlet and posteriorly is continuous with the esophagus. Deep to the mucosa lie the pharyngeal constrictor muscles (see Plate 3-5) that help propel the food into the esophagus.

## Esophagus

The upper third of the esophagus contains skeletal muscle, the lower third contains smooth muscle, and the middle third contains mixed skeletal and smooth muscle. The muscular walls form an outer longitudinal layer and inner circular layer, and these layers participate in peristalsis, which moves food toward the stomach. As the esophagus approaches the stomach, the smooth muscle thickens and forms the **lower esophageal sphincter (LES)**. Normally the resting tone of the LES is high, which prevents the reflux of gastric contents into the esophagus. As the peristaltic wave carries a bolus of food to the stomach, release of nitric oxide and vasoactive intestinal peptide from the myenteric plexus, which is under vagal control, causes a relaxation of the LES, and food enters the stomach.

**Swallowing** (deglutition) involves the following sequence of coordinated movements:

**A.** The tongue pushes a bolus of food up against the hard palate
**B.** The soft palate elevates to close off the nasopharynx
**C.** The tongue pushes the bolus back into the oropharynx
**D.** When the bolus reaches the epiglottis, the larynx elevates and the tip of the epiglottis tips downward over the laryngeal opening (aditis); this prevents aspiration into the larynx
**E.** Contraction of the pharyngeal constrictor muscles squeeze the bolus into two streams that pass on either side of the epiglottis and into the upper esophagus
**F.** The soft palate pulls downward to assist in moving the bolus around the epiglottis
**G.** The laryngeal vestibular folds (the rima vestibuli) is the space between the vestibular folds and rima glottidis (space between the vocal folds) close to protect the larynx
**H.** Once the bolus is in the esophagus, all structures return to their starting positions

Solid foods pass from the oropharynx to the stomach in about 4-8 seconds; fluids pass in about 1-2 seconds.

**COLOR** the following features of the pharynx and esophagus, using a different color for each feature:

☐ 1. **Soft palate**
☐ 2. **Uvula**
☐ 3. **Epiglottis**
☐ 4. **Esophagus**
☐ 5. **Stomach**

***Clinical Note:***

**Gastroesophageal reflux disease (GERD)** is a relatively common problem caused by a decreased tone of the lower esophageal sphincter (LES) or a sliding hiatal (stomach herniation into the thorax) hernia. Reflux of the acidic gastric contents can cause abdominal pain, dyspepsia, gas, heartburn, dysphagia, and other problems. The chronic inflammation of the lower esophageal wall may result in esophagitis, ulceration, or stricture.

Talking and swallowing at the same time is generally prevented by a reflexive protective mechanism so that food will not enter the respiratory passageways. This reflex triggers a violent "**cough reflex**" in an attempt to expel the food.

**Heartburn** (gastroesophageal reflux disease, GERD) is a burning substernal pain that occurs when gastric acid is refluxed into the esophagus. It can happen from eating or drinking too much and/or too fast.

**Oral cancer** (specifically squamous cell carcinoma) accounts for more than 90% of cancers in this region, including cancer of the tongue, floor of the mouth, gingiva (the gums), lip (usually the lower lip), and oropharynx. Risk factors include alcohol use, tobacco use, and for lip cancer, ultraviolet (sun) exposure.

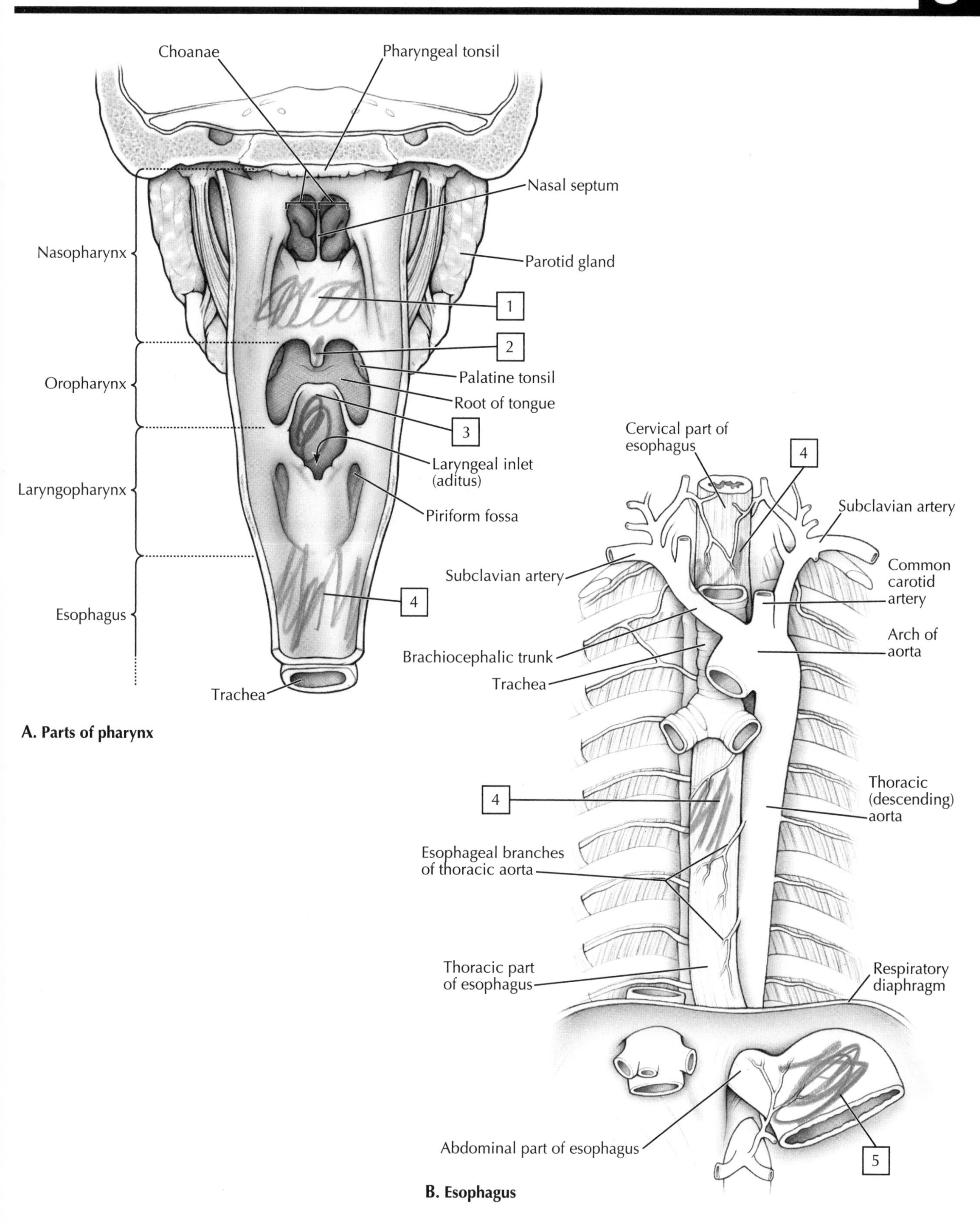


A. Parts of pharynx

B. Esophagus

# 8 Peritoneal Cavity and Mesenteries

The abdominal cavity is lined by muscles that assist in movements of the trunk, assist in respiration, and, by increasing intraabdominal pressure, facilitate micturition, defecation, and childbirth. The viscera of the abdominopelvic cavity lie within a potential space called the **peritoneal cavity** (not unlike the pleural and pericardial cavities), which has the following features:

- **Parietal peritoneum**: a serosal lining that covers the inner aspects of the walls of the abdominopelvic cavity
- **Visceral peritoneum:** a direct continuation of the parietal peritoneum, which reflects from the inner abdominal wall and covers the visceral structures of the abdomen
- **Mesenteries:** a double layer of visceral peritoneum that reflects from the inner wall of the abdomen and envelops portions of the abdominal viscera
- **Retroperitoneal viscera:** lie against the posterior abdominal wall deep to the parietal peritoneum and do not possess a suspending mesentery; retroperitoneal structures are considered to be either **primarily retroperitoneal** (never had a mesentery) or **secondarily retroperitoneal** (lost their mesentery during embryonic development when the bowel was rotated, pushed against the abdominal wall, and became fused to the posterior abdominal wall)
- **Intraperitoneal viscera:** are suspended from the abdominal walls by a mesentery
- **Serous fluid:** secreted in small amounts by the peritoneum and lubricates the viscera, thus reducing friction during peristalsis and other movements of the abdominal viscera when they rub against one another

These features and several others are depicted in part ***A*** in a midsagittal view and are summarized in the following table.

| FEATURE | DESCRIPTION |
|---|---|
| Greater omentum | An "apron" of peritoneum hanging from greater curvature of stomach, folding back on itself to attach to transverse colon |
| Lesser omentum | Double layer of peritoneum extending from lesser curvature of stomach and proximal duodenum to liver (hepatoduodenal and hepatogastric ligaments) |
| Mesenteries | Double fold of peritoneum suspending parts of bowel and conveying vessels, lymphatics, and nerves of bowel |
| Peritoneal ligaments | Double layer of peritoneum attaching viscera to walls or to other viscera |

The **omental bursa** is the cul-de-sac posterior to the stomach and anterior to the pancreas (see part ***B***). It also is known as the lesser sac, while the remainder of the abdominopelvic cavity is the greater sac.

As the simple gut tube of the embryo, which is suspended by a mesentery, begins to grow in length and breadth, it twists upon itself so that the significant length of bowel necessary for complete digestion can be accommodated in the confined space of the abdomen. As this twisting and growth occur, portions of the bowel and its accessory digestive glands are pushed to the posterior abdominal wall and fuse to the parietal peritoneum, thus losing their mesenteries and becoming **retroperitoneal viscera** (sometimes referred to as "secondarily retroperitoneal viscera" because at one time in human embryonic development they did have a mesentery). Other portions of the bowel retain their mesenteries and continue to be intraperitoneal. Summarized below are those portions of the bowel that are largely intraperitoneal (have a mesentery, which is listed) or retroperitoneal (have lost their mesentery).

| INTRAPERITONEAL | RETROPERITONEAL |
|---|---|
| Stomach (lesser omentum) | Duodenum (most of it) |
| Jejunum and ileum (mesentery of the small intestine) | Ascending colon |
| Transverse colon (transverse mesocolon) | Descending colon |
| Sigmoid colon (sigmoid mesocolon) | Rectum |

**COLOR** the following features of the peritoneal cavity, using a different color for each feature:

- ☐ 1. **Lesser omentum (mesentery suspending the stomach)**
- ☐ 2. **Transverse mesocolon (suspends the transverse colon)**
- ☐ 3. **Mesentery of the small intestine (suspends the jejunum and ileum)**
- ☐ 4. **Greater omentum (apron of peritoneum filled with fat)**

***Clinical Note:***

**Abdominal wall hernias** (ventral hernias) occur when there is a protrusion of peritoneal contents (mesentery, fat, and/or a portion of bowel) through the abdominal wall in the groin region (an inguinal hernia), or a herniation of the stomach through the diaphragm (hiatal hernia). Other abdominal wall hernias can include one at the site of the umbilicus (belly button), a midline hernia of the rectus sheath, and an incisional hernia that may occur at the site of a previous laparotomy (postoperative abdominal wall scar).

Injury to the anterior abdominal wall, e.g., in a motor vehicle accident, may rupture the respiratory diaphragm to such an extent that portions of the abdominal viscera can herniate into the thoracic cavity. Most of these hernias occur on the left side (herniation of the stomach, small intestine and mesentery, transverse colon, and/or the spleen) because the liver on the right side provides a physical barrier.

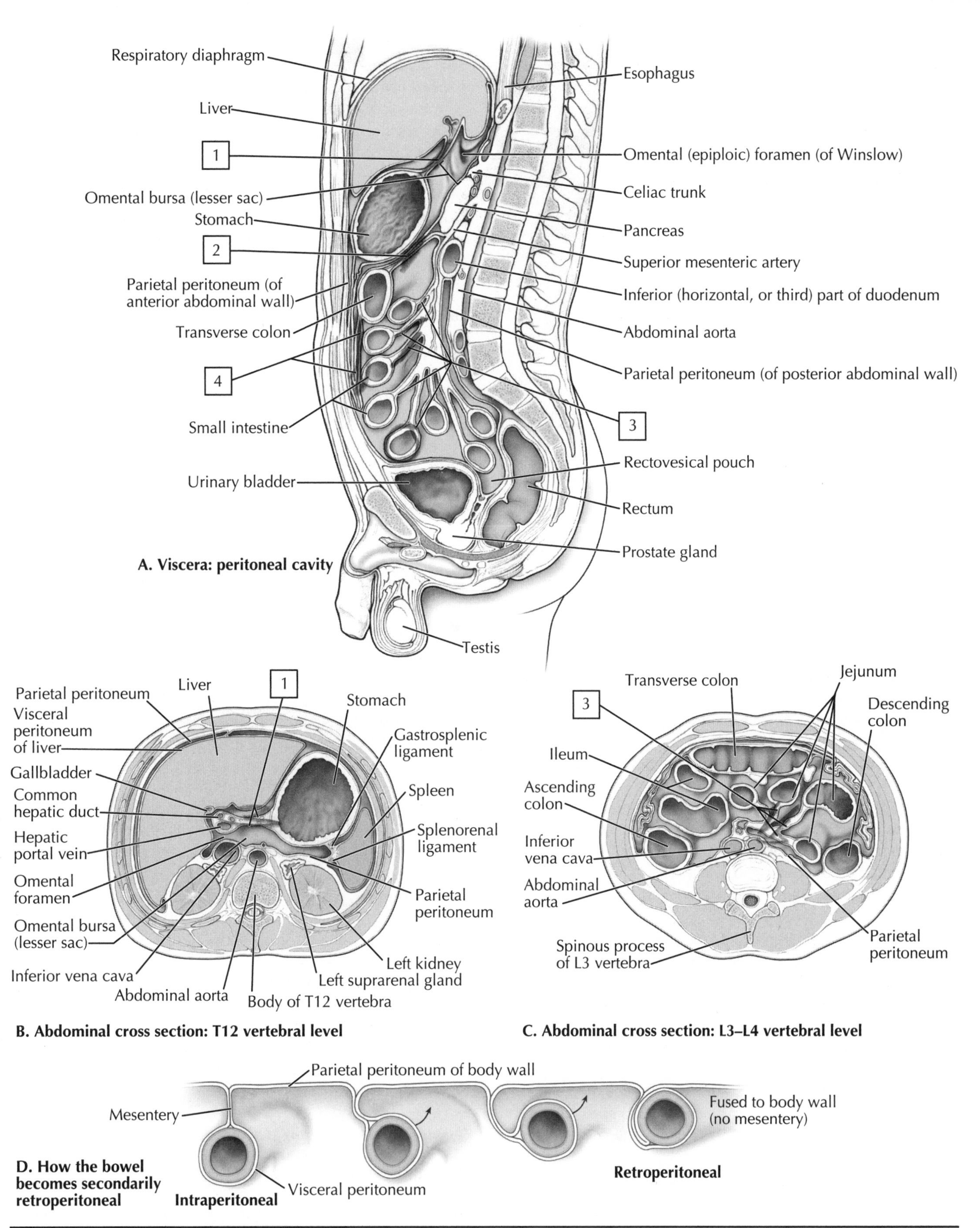


**A. Viscera: peritoneal cavity**

**B. Abdominal cross section: T12 vertebral level**

**C. Abdominal cross section: L3–L4 vertebral level**

**D. How the bowel becomes secondarily retroperitoneal**

# 8 Stomach

The stomach is a muscular bag, with its smooth muscle layers oriented in several different planes, an arrangement that causes macerated foods entering from the esophagus to become blended. The stomach begins the major enzymatic digestion of the food into a semiliquid mixture or slurry called **chyme,** which passes on to the duodenum. Features of the stomach are summarized in the table below.

| FEATURE | DESCRIPTION |
|---|---|
| Lesser curvature | Right border of stomach; lesser omentum attaches here and extends to liver (hepatogastric ligament) |
| Greater curvature | Convex border with greater omentum suspended from its margin |
| Cardiac part | Area of stomach that communicates with esophagus superiorly |
| Fundus | Superior part just under left dome of respiratory diaphragm |
| Body | Main part between fundus and pyloric antrum |
| Pyloric part | Portion that is divided into proximal antrum and distal canal |
| Pylorus | Site of pyloric sphincter muscle; joins first part of duodenum |

The stomach is flexible and can assume a variety of configurations during digestion, depending upon the contractions of its smooth muscle walls and how full and distended it is. Despite this flexibility, it still is tethered superiorly to the esophagus and distally to the first portion of the duodenum. Both the stomach and this proximal portion of the duodenum are suspended in a mesentery called the **lesser omentum** (hepatogastric and hepatoduodenal ligaments). Most of the duodenum is a retroperitoneal structure having lost its mesentery along most of its length. Behind the stomach is the **lesser sac**, or omental bursa, a space that communicates with the greater sac via the **epiploic foramen** (of Winslow). The greater sac is the entire rest of the peritoneal cavity. The omental bursa is a cul-de-sac that forms posterior to the stomach and anterior to the retroperitoneal pancreas; it is formed as a result of the twisting of the tethered stomach (attached to the esophagus proximally and distally to the duodenum) and its differential growth during embryonic life.

The mucosa of the stomach is thrown into large, longitudinal folds called **rugae** and into thousands of microscopic folds and gastric pits lined with a renewing (simple columnar) epithelium. At the base of the gastric pit are the gastric or fundic glands, which contain the following four cell types:

- **Mucous neck cells**: secrete mucus to protect the stomach lining
- **Chief cells**: situated deep in the glands, these cells secrete primarily **pepsinogen**, which is converted to pepsin once it contacts the gastric juice and aids in the digestion of proteins
- **Parietal cells**: mostly found in the neck of the gastric glands; they secrete **hydrochloric acid (HCl)** and **intrinsic factor** (which form complexes with vitamin $B_{12}$ so it can be absorbed in the ileum)
- **Enteroendocrine cells**: concentrated mostly near the base of the glands, they secrete a host of hormones or hormone-like substances (gastrin, histamine, endorphins, serotonin, cholecystokinin, and somatostatin) that help regulate digestion

**COLOR** the following features of the stomach and its mucosa, using a different color for each feature:

- ☐ 1. **Fundus of stomach**
- ☐ 2. **Body of stomach**
- ☐ 3. **Pyloric antrum**
- ☐ 4. **Pyloric canal (contains the pyloric smooth muscle sphincter that releases measured amounts of chyme into the duodenum during digestion)**
- ☐ 5. **Mucous neck cells (mucus)**
- ☐ 6. **Parietal cells (HCl and intrinsic factor)**
- ☐ 7. **Chief cells (pepsinogen)**
- ☐ 8. **Enteroendocrine cells (gastric hormones and regulatory peptides)**

***Clinical Note:***

**Hiatal hernias** are herniations of the stomach through the esophageal hiatus. Two anatomical types of hiatal hernias are recognized:

- Sliding, rolling, or axial hernia: makes up 95% of hiatal hernias
- Paraesophageal or nonaxial hernia: usually involves only the fundus of the stomach

**Peptic ulcers** are GI lesions that may extend through the muscularis mucosae and are remitting, relapsing lesions (can come and go). Gastric acid, aspirin, alcohol, and *Helicobacter pylori* infection (about 70% of gastric ulcers) are common aggravating factors.

Chronic stress also can produce ulcers by increasing the exposure to gastric acid and pepsin.

**Vomiting (emesis)** usually results from extreme stretching of the stomach or intestine, or by the presence of irritants (bacterial toxins, excessive alcohol, highly seasoned or spicy foods, and some drugs). Occasionally, disorienting maneuvers such as spinning or riding a roller coaster can cause nausea and may trigger vomiting.

**Gastroesophageal reflux disease** (GERD) can occur when the lower esophageal sphincter becomes compromised, causing inflammation of the lower esophagus.

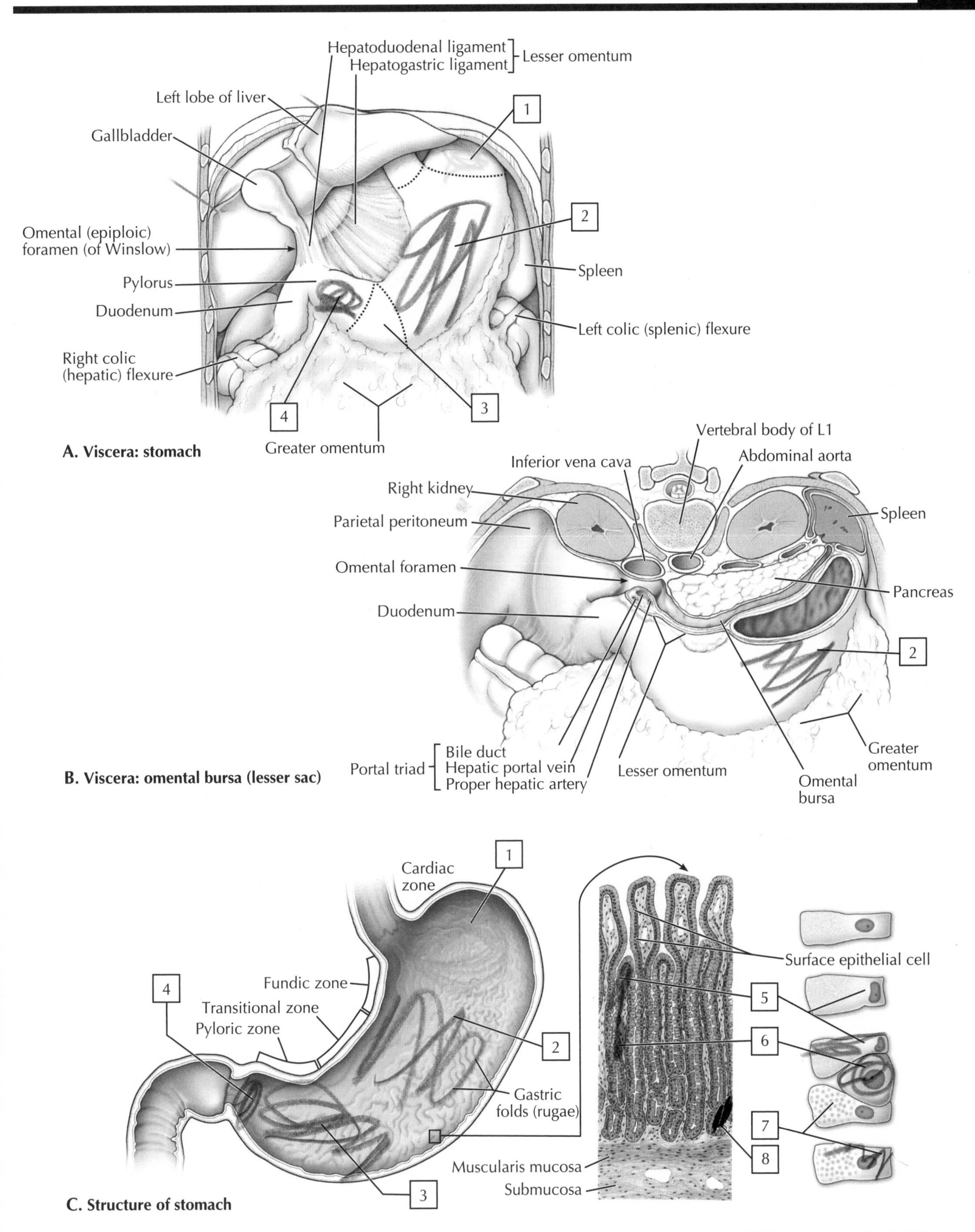


A. Viscera: stomach

B. Viscera: omental bursa (lesser sac)

C. Structure of stomach

# 8 Small Intestine

As an embryonic midgut structure, the small intestine is supplied with blood by the **superior mesenteric artery** and drained by the **hepatic portal system** (see Plate 5-19). The small intestine includes the:

- **Duodenum**: first part of the small intestine (about 25 cm long); it is largely retroperitoneal
- **Jejunum**: the proximal two-fifths of the mesenteric small intestine (about 2.5 m long); this is where most of the absorption takes place
- **Ileum**: the distal three-fifths of the mesenteric small intestine (about 3.5 m long), which then opens via the ileocecal valve into the cecum of the large intestine

## Duodenum

The duodenum is where bile and pancreatic enzymes are added to the chyme, which has just arrived from the stomach. The features of the duodenum are summarized below.

| PART OF DUODENUM | DESCRIPTION |
| --- | --- |
| Superior | First part; attachment site for hepatoduodenal ligament of lesser omentum |
| Descending | Second part; site where bile and pancreatic ducts empty |
| Inferior | Third part; part that crosses inferior vena cava (IVC) and aorta and is crossed anteriorly by superior mesenteric vessels |
| Ascending | Fourth part; portion tethered by suspensory ligament at duodenojejunal flexure |

## Jejunum and Ileum

The jejunum has a larger diameter, thicker walls, greater vascularity, less fat in its mesentery, fewer lymph nodules, and larger and taller circular folds (**plicae circulares**) than the ileum. Both the jejunum and ileum are suspended in an elaborate mesentery (two folds of peritoneum that convey vessels, lymphatics, and nerves) that originates from the midposterior abdominal wall and tethers the approximately 6 meters of small intestine.

The jejunum and ileum have a large surface area for secretion and absorption. The surface area is increased by the presence of **circular folds, villi**, and **microvilli** (brush border on the columnar epithelium). Simple columnar epithelium lines the bowel, and the lamina propria contains lymphatics, vessels, and connective tissue cells. Intestinal glands (crypts of Lieberkühn) extend into the lamina propria, and **aggregated lymphatic nodules** (Peyer's patches) increase in number as one moves distally toward the ileum.

The small intestine is engaged in **mechanical digestion and propulsion**, mixing the intestinal contents with digestive juices, and providing time for the digestive mucosa to absorb the breakdown products of digestion. **Chemical digestion** occurs simultaneously with mechanical digestion aided by digestive enzymes from the intestinal glands, pancreas, and by bile produced in the liver, and stored and concentrated in the gallbladder (see Plates 8-9 and 8-10). The intestinal glands secrete about 1 to 2 liters of "intestinal juice" daily, initiated largely by intestinal distention and irritation from the hypertonic and acidic chyme. This intestinal juice is mostly water and mucus, and has a pH of 7.4-7.8, i.e., it is slightly alkaline.

**COLOR** the following features of the small intestine, using a different color for each feature:

- ☐ 1. **First (superior) part of the duodenum (tethered by the hepatoduodenal ligament containing the bile duct, proper hepatic artery, and portal vein)**
- ☐ 2. **Second (descending) part of the duodenum**
- ☐ 3. **Third (horizontal) part of the duodenum**
- ☐ 4. **Fourth (ascending) part of the duodenum**
- ☐ 5. **Circular fold**
- ☐ 6. **Villi**
- ☐ 7. **Lymph nodule**

***Clinical Note:***

**Crohn's disease** is an idiopathic (thought to be an autoimmune disease with a genetic component), episodic, and chronic inflammatory bowel condition that usually involves the small intestine and colon. Often it occurs between the ages of 15 and 30 years and presents with abdominal pain, diarrhea, fever, and other signs and symptoms. The lumen of the bowel is narrowed, mucosal ulcerations are present, and the bowel wall is thick and rubbery; thus it affects the entire thickness of the bowel.

**Peptic ulcer disease** produces lesions the extend through the muscularis mucosae and may be acute (small, shallow lesions), whereas chronic lesions may erode into the muscularis externa or perforate the serosa. Ninety-eight percent of peptic ulcers occur in the duodenum (usually the first part of the duodenum) or stomach, but duodenal peptic ulcers account for about 80% of these lesions. Clinically, the two most serious complications of gastric or duodenal peptic ulcers are perforation and hemorrhage. Diet and/or stress may produce peptic ulcers.

**Volvulus** is the twisting of a bowel loop that may cause bowel obstruction and constriction of its vascular supply, which may lead to infarction. Volvulus most often affects the small intestine due to the mesenteric mobility of this portion of the bowel (the more mobile sigmoid colon is the most common site in the large intestine).

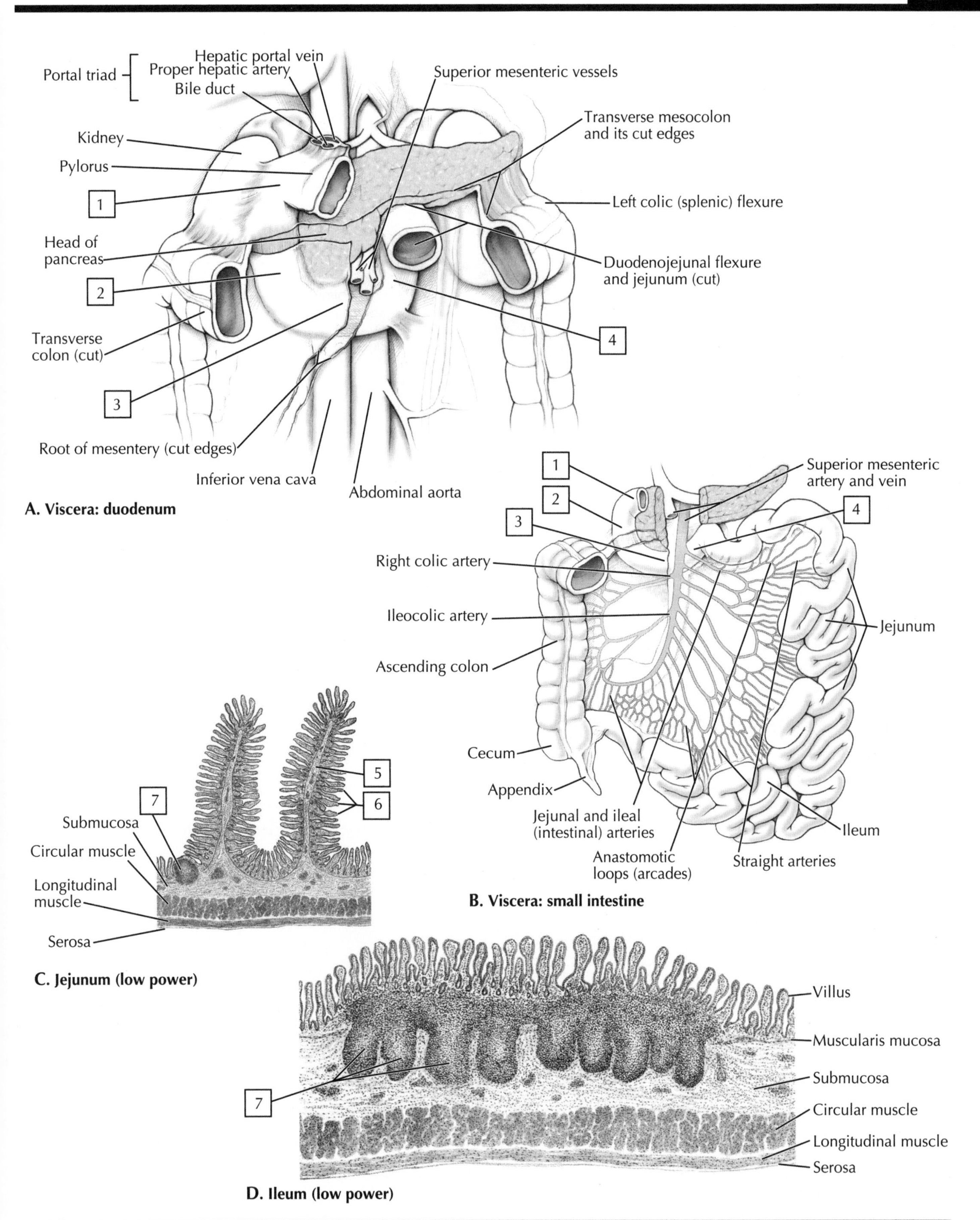


**A. Viscera: duodenum**

**B. Viscera: small intestine**

**C. Jejunum (low power)**

**D. Ileum (low power)**

# 8 Large Intestine

The large intestine is supplied by both the **superior** and **inferior mesenteric arteries**, because the proximal portion of the large bowel is derived from the embryonic midgut and the distal portion from the hindgut (distal transverse colon to rectum) (see Plate 5-15). The large intestine includes the:

- Cecum (and its vermiform appendix)
- Ascending colon (retroperitoneal)
- Transverse colon (has a transverse mesocolon)
- Descending colon (retroperitoneal)
- Sigmoid colon (has a sigmoid mesocolon)
- Rectum (retroperitoneal)
- Anal canal (lies below the pelvic diaphragm and ends at the anus)

The large intestine serves primarily to reabsorb water and electrolytes from the feces and to store feces until they are eliminated from the body. The large intestine has the same layers as the small intestine, but the mucosa is thicker, with deeper crypts, but does not have villi or circular folds and virtually no cells that secrete digestive enzymes; **lymphatic nodules** are common. **Goblet cells** also are common and secrete mucus, which lubricates the bowel lumen and facilitates the passage of feces. The mucosa has partial folds called **plicae semilunares**, and the outer longitudinal smooth muscle layer is organized into three thickened bands (**taeniae coli**) that run from the cecum to the rectum and help propel the feces along the length of the bowel. Contraction of the muscle layers produces sacculations called **haustra** that give the colon its typical appearance. Additionally, the colon is studded with small sacs of fat (**appendices epiploicae**).

The terminal end of the large intestine is the rectum and anal canal. Normally, the anal canal is closed because of the tonic contraction of the **internal** (smooth muscle) and **external** (skeletal muscle) **anal sphincters**. When the rectum is distended by fecal material, the internal sphincter relaxes, but defecation does not occur until the voluntary external sphincter is relaxed and the smooth muscles of the distal colon and rectum contract. Since the rectum must generate strong contractions to expel the feces, its muscle layers are well developed.

**COLOR** the following features of the large intestine, using a different color for each feature:

- ☐ 1. **Cecum and appendix**
- ☐ 2. **Ascending colon**
- ☐ 3. **Transverse colon**
- ☐ 4. **Descending colon**
- ☐ 5. **Sigmoid colon**
- ☐ 6. **Rectum**
- ☐ 7. **Anal canal**
- ☐ 8. **Internal anal sphincter (involuntary, smooth muscle; parasympathetic innervation)**
- ☐ 9. **External anal sphincter (voluntary, skeletal muscle; somatic innervation)**

Most of the bacteria entering the cecum of the large intestine are dead, killed by the action of lysozyme, defensins, HCl, and various enzymes. However, some bacteria still are present in the colon and, together with bacteria that enter the GI tract from the anus, they constitute the **bacterial flora** of the large intestine. They ferment some of the carbohydrates that are indigestible, e.g., cellulose, and in the process release irritating acids and a mixture of gases, some of which are quite odiferous. Normally, about 500 ml of gas (flatus) are produced each day and even more with certain carbohydrate-rich meals, such as beans! These bacterial flora also synthesize B complex vitamins and vitamin K, critical for our liver since it synthesizes some important clotting proteins. While over 30 different substances are involved in coagulation of the blood, at least four of them are vitamin K-dependent.

***Clinical Note:***

**Colonic diverticulosis** usually is an acquired herniation of colonic mucosa through the muscular wall, creating a diverticulum or little saccule that may contain a fecal deposit or concretion. This condition is most common in the distal colon and sigmoid colon and may be caused by exaggerated peristaltic contractions, increased intraluminal pressure, and/or an intrinsic weakness in the muscular wall.

**Colorectal cancer** is second only to lung cancer in the site-specific mortality rate and accounts for about 15% of cancer-related deaths in the United States. Risk factors include heredity, a diet high in fat, increasing age, inflammatory bowel disease, and the presence of polyps.

About 38% of colorectal cancers occur in the cecum and ascending colon, 38% in the transverse colon, 18% in the descending colon, and about 8% in the sigmoid colon. The incidence is highest in the USA, Canada, Australia, and New Zealand. Males are affected 20% more than females, and the peak incidence is between 60 and 79 years. The interior surface of the colon may be visualized and photographed in a procedure known as a **colonoscopy**. This procedure involves the use of a long fiberoptic endoscope (colonoscope) inserted through the anus and rectum, and passed through the full extent of the large intestine. A small amount of colonic tissue also may be biopsied for diagnosis.

**Appendicitis** is an acute inflammation of the appendix. Initially, the pain is diffuse and occurs around the periumbilical region. However, as the intensity of the inflammation increases, the pain localizes to the lower right quadrant.

**Ulcerative colitis** is an idiopathic inflammatory bowel disease that begins in the rectum and extends proximally. Usually, the inflammation is limited to the mucosal and submucosal layers of the bowel. Submucosal vessels, especially the rectal veins, may become enlarged and form **internal hemorrhoids**, usually as a result of an increased pressure in the portal venous circulation. Straining to "pass one's bowels" also may increase rectal venous pressure, irrespective of portal hypertension, and cause internal hemorrhoids.

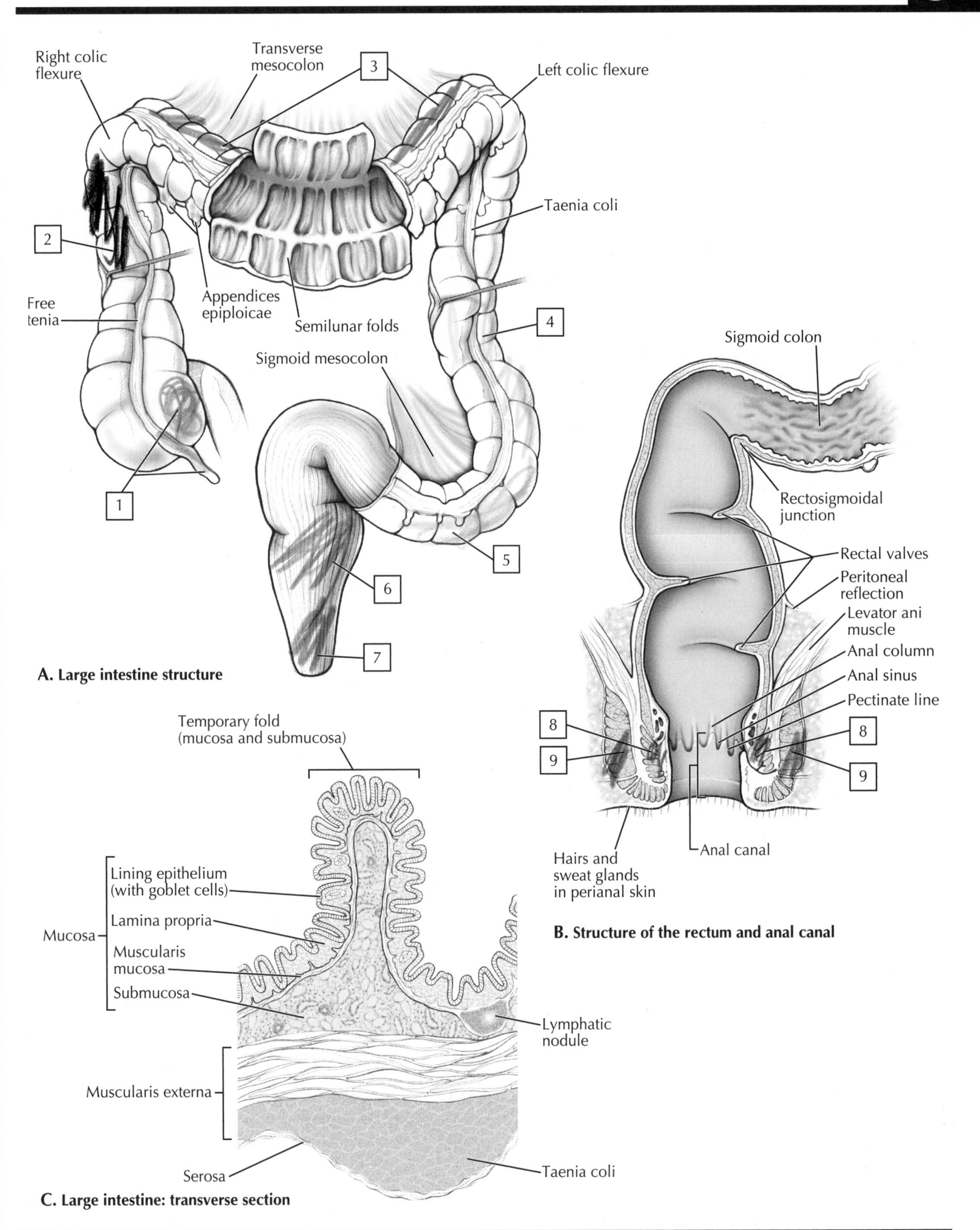
Right colic flexure
Transverse mesocolon
3
Left colic flexure
Taenia coli
2
Free tenia
Appendices epiploicae
Semilunar folds
4
Sigmoid mesocolon
1
5
6
7
A. Large intestine structure
Sigmoid colon
Rectosigmoidal junction
Rectal valves
Peritoneal reflection
Levator ani muscle
Anal column
Anal sinus
Pectinate line
8
9
8
9
Hairs and sweat glands in perianal skin
Anal canal
B. Structure of the rectum and anal canal
Temporary fold (mucosa and submucosa)
Lining epithelium (with goblet cells)
Lamina propria
Mucosa
Muscularis mucosa
Submucosa
Lymphatic nodule
Muscularis externa
Serosa
Taenia coli
C. Large intestine: transverse section

# 8 Liver

The liver is the largest solid organ in the body; anatomically, it is divided into four lobes:

- Right lobe (largest lobe)
- Left lobe
- Quadrate lobe (lies between the gallbladder and the round ligament of the liver)
- Caudate lobe (lies between the IVC, ligamentum venosum, and porta hepatis)

Functionally, the liver is divided into right and left lobes based upon its vasculature, with each lobe receiving a major branch of the hepatic artery, portal vein, hepatic vein (drains the liver's blood into the IVC), and biliary drainage.

| FEATURE | DESCRIPTION |
|---|---|
| Lobes | Divisions, in functional terms, into right and left lobes, with anatomical subdivisions of right lobe into quadrate and caudate lobes |
| Round ligament | Ligament that contains obliterated umbilical vein |
| Falciform ligament | Peritoneal reflection off anterior abdominal wall with round ligament in its margin |
| Ligamentum venosum | Ligamentous remnant of fetal ductus venosus, allowing fetal blood from placenta to bypass liver |
| Coronary ligaments | Reflections of peritoneum from liver to diaphragm |
| Bare area | Area of liver pressed against diaphragm, which lacks visceral peritoneum |
| Porta hepatis | Site at which vessels, ducts, lymphatics, and nerves enter or leave liver |

The liver is important because it receives the venous drainage from the GI tract, its accessory organs, and the spleen via the portal vein (see Plate 5-19). The liver serves a number of important functions:

- Storage of energy sources (glycogen, fat, protein, and vitamins)
- Production of cellular fuels (glucose, fatty acids, and keto acids)
- Production of plasma proteins and clotting factors
- Metabolism of toxins and drugs
- Modification of many hormones
- Production of bile acids
- Excretion of substances (bilirubin)
- Storage of iron and many vitamins
- Phagocytosis of foreign materials that enter the portal circulation from the bowel

Liver cells receive blood from the **portal vein** (about 75%) and from the **proper hepatic artery** (about 25%). Hepatocytes (liver cells) are arranged in plates of cells, which are separated from each other by **hepatic sinusoids**. The blood moves from the portal vein and hepatic arteriole branches through the sinusoid to the **central vein**. This arrangement forms **hepatic lobules** composed of hexagonal units of cells around the central vein. At the margin of the lobule is the **portal triad**, made up of a branch of the hepatic artery, a branch of the portal vein, and a bile duct. From the central vein, blood flows into the hepatic veins and IVC. The sinusoids contain **phagocytic cells** (Kupffer cells), which clear damaged red blood cells and foreign antigens. Bile is produced by the hepatocytes (about 900 ml/day) and drains into intralobular bile ductules and then the larger bile ducts (right and left). Ultimately, bile is collected into the **gallbladder**, where it is stored and concentrated.

**COLOR** the following features of the liver, using the suggested colors for each feature:

- ☐ 1. **IVC (blue)**
- ☐ 2. **Gallbladder (green)**
- ☐ 3. **Round ligament of the liver (yellow)**
- ☐ 4. **Hepatic artery branch (at portal triad) (red)**
- ☐ 5. **Portal vein branch (at portal triad) (blue)**
- ☐ 6. **Bile duct (at portal triad) (green)**
- ☐ 7. **Several hepatocytes (brown)**

***Clinical Note:***

**Cirrhosis of the liver** is largely an irreversible disease, characterized by diffuse fibrosis, parenchymal nodular regeneration, and disturbed hepatic architecture. Progressive fibrosis disrupts portal blood flow (leading to portal hypertension), beginning at the level of the sinusoids and central veins. Common causes of cirrhosis include:

- Alcoholic liver disease (60% to 70%)
- Viral hepatitis (10%)
- Biliary diseases (5% to 10%)
- Genetic causes (5%)
- Other (10% to 15%)

Inflammation of the liver is called **hepatitis**, usually caused by a viral infection. In the USA, about 40% of the hepatitis cases are caused by the HVB (hepatitis virus B), transmitted via blood transfusions, contaminated needles, or sexual contact.

**Portal hypertension** (see Plate 5-19) can occur via one of three mechanisms: prehepatic (obstructed blood flow to the liver); posthepatic (obstructed blood flow from the liver to the heart); and intrahepatic (cirrhosis or another liver disease that affects hepatic sinusoidal blood flow).

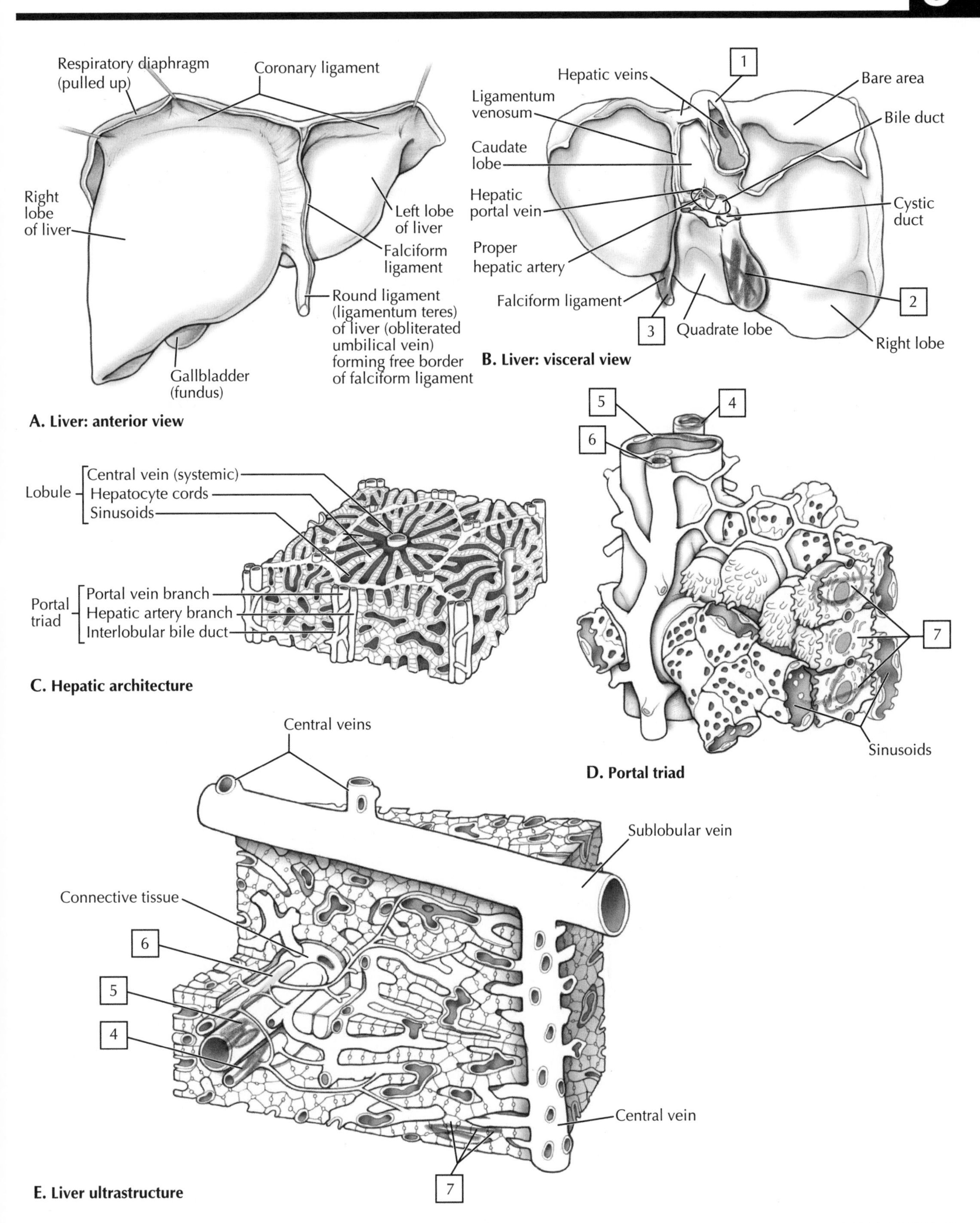


A. Liver: anterior view

B. Liver: visceral view

C. Hepatic architecture

D. Portal triad

E. Liver ultrastructure

## Gallbladder

The gallbladder stores and concentrates bile, which is secreted by the hepatocytes in the liver. Once the bile has been secreted by the hepatocyte, it takes the following journey (see Plate 8-9):

- Passes into a bile canaliculus (capillary)
- Passes from canaliculi into intralobular ductules
- Passes from intralobular ductules to bile ducts
- Collects in the right and left hepatic ducts
- Enters the common hepatic duct
- Enters the cystic duct and is stored and concentrated in the gallbladder
- Upon stimulation (largely by vagal efferents and cholecystokinin [CCK]), bile leaves the gallbladder and enters the cystic duct
- Passes inferiorly down the bile duct
- Enters the hepatopancreatic ampulla (of Vater)
- Empties into the second part of the duodenum

The liver produces about 900 ml of bile per day. Between meals, the bile is stored in the gallbladder (capacity of 30-50 ml), where it is also concentrated. Consequently, bile that reaches the duodenum is a mixture of the more dilute bile that flows directly from the liver and the concentrated bile that flows from the gallbladder. The mucosa of the gallbladder is specialized for electrolyte and water absorption, which allows the gallbladder to concentrate the bile. Vagal parasympathetic nerve fibers cause the gallbladder to contract, along with CCK, and thus release bile. Sympathetic innervation inhibits bile secretion.

## Exocrine Pancreas

The pancreas is both an **exocrine** and **endocrine** organ (see Plate 11-6). The pancreas lies posterior to the stomach in the floor of the **lesser sac** (omental bursa) and is a retroperitoneal organ except for the distal tail, which is in contact with the spleen. The pancreatic head is nestled within the C-shaped curve of the duodenum, with its uncinate process lying posterior to the superior mesenteric vessels.

The acinar cells of the exocrine pancreas (a compound tubuloacinar gland) secrete a number of enzymes that are necessary for digestion of proteins, starches, and fats. The pancreatic ductal cells secrete fluid with a high bicarbonate content that neutralizes the acid entering the duodenum from the stomach. Pancreatic secretion is under neural (vagus nerve [CN X]) and hormonal control (secretin and CCK), and the pancreatic exocrine secretions empty primarily into the main pancreatic duct (of Wirsung), which joins the bile duct at the hepatopancreatic ampulla (of Vater). Pancreatic secretions include a variety of proteases, amylase, lipases, and nucleases. A smaller, variable accessory pancreatic duct (of Santorini) also empties into the second part of the duodenum about 2 cm above the major duodenal ampulla (not shown in the illustrations).

**COLOR** the following features of the gallbladder and pancreas, using a different color for each feature:

- ☐ 1. **Gallbladder**
- ☐ 2. **Common hepatic duct**
- ☐ 3. **Cystic duct**
- ☐ 4. **Bile duct**
- ☐ 5. **Pancreas**
- ☐ 6. **Hepatopancreatic ampulla**
- ☐ 7. **Main pancreatic duct**

***Clinical Note:***

**Gallstones** occur in 10% to 15% of the population in developed countries and usually are precipitates of cholesterol (crystalline cholesterol monohydrate) or pigment stones (bilirubin calcium salts), or mixed stones. Risk factors include increasing age, obesity, female gender, rapid weight loss, estrogenic factors, and gallbladder stasis. The stone may pass through the duct system, collect in the gallbladder, or block the cystic or bile ducts, causing inflammation and obstruction to the flow of bile. Pain from this obstruction produces **biliary colic**, pain felt in the epigastric region. Possible treatments for gallstones include pulverizing them with ultrasound vibrations (lithothripsy), dissolving the stones with drugs, using laser therapy to vaporize the gallstones, or surgically removing the gallbladder.

**Jaundice** (yellow skin color) results when the bile salts and bile pigments, normally passed into the digestive tract, accumulate in the blood and end up being deposited in the skin.

**Pancreatic cancer** is the fifth leading cause of cancer deaths in the United States. Most of these cancers arise from the exocrine pancreas, and about 60% are found in the head of the pancreas (can cause obstructive jaundice). Most of these cancers are ductular adenocarcinomas that arise from the exocrine portion of the pancreas. Islet tumors of the endocrine pancreas are less common. Metastases via the lymphatics are common; additionally, because of the pancreas' strategic location, the duodenum, stomach, liver, colon, and spleen may be directly involved.

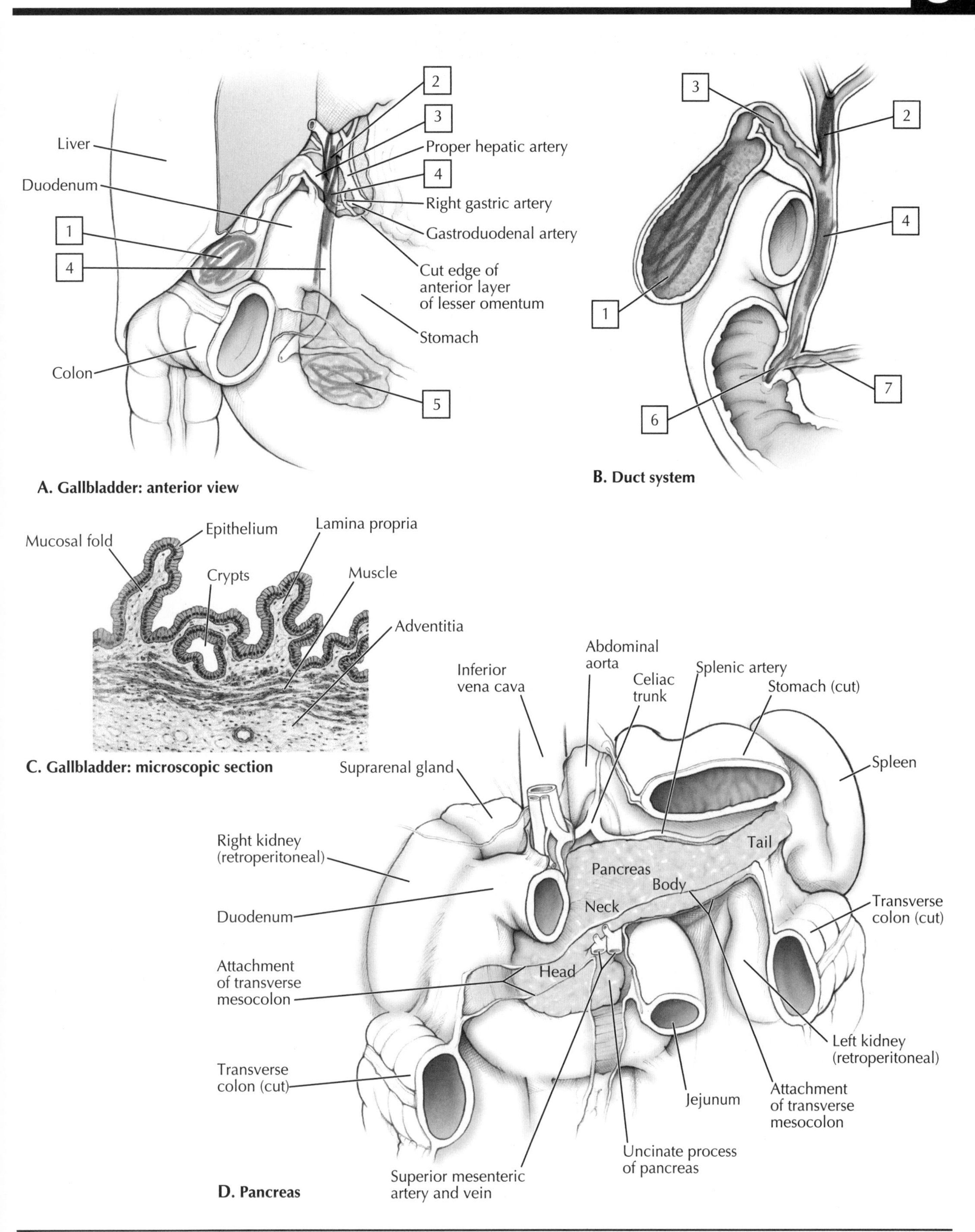


A. Gallbladder: anterior view

B. Duct system

C. Gallbladder: microscopic section

D. Pancreas

## REVIEW QUESTIONS

**For each description below (1-4), color or highlight the relevant structure on the image.**

1. This is the most extensive mesentery in the abdominopelvic cavity.
2. This organ is suspended from the liver by the hepatogastric ligament.
3. This portion of the small bowel is largely retroperitoneal.
4. This retroperitoneal structure is both an endocrine and exocrine organ.

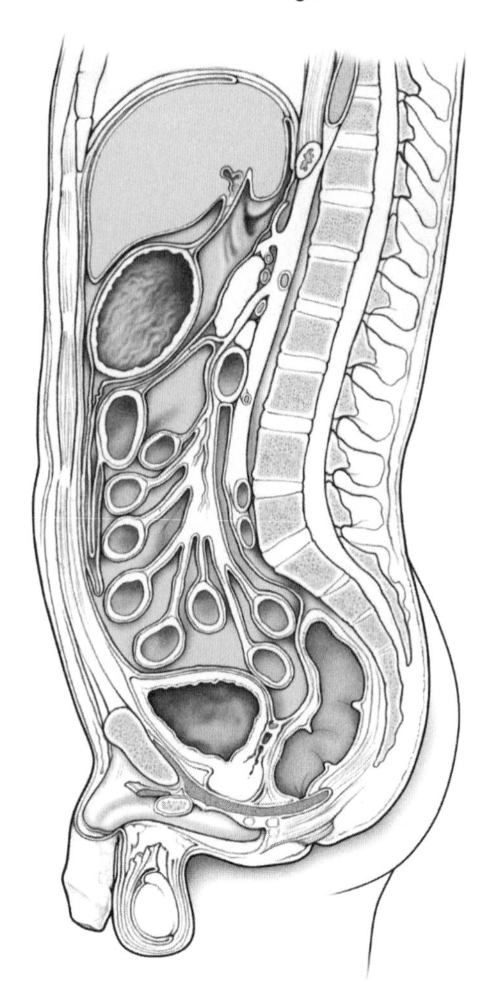

5. Which of the following structures is involved in a hiatal hernia?
   A. Duodenum
   B. Gallbladder
   C. Jejunum
   D. Sigmoid colon
   E. Stomach

6. Which of the following features is unique to the colon?
   A. Haustra
   B. Lymphatic nodules
   C. Mesentery
   D. Simple columnar epithelium
   E. Visceral peritoneum

7. Histologically, the portal triad refers to the presence of a branch of the portal vein, the hepatic artery, and which of the following structures?
   A. Bile duct
   B. Central vein
   C. Hepatic sinusoid
   D. Hepatocyte cords
   E. Kupffer cells
8. The cul-de-sac posterior to the stomach and anterior to the pancreas is known by this term. ___________________
9. Bile leaving the gallbladder passes down the bile duct and enters which portion of the GI tract? ___________________
10. As food enters the oral cavity and is mixed with saliva, what enzyme is secreted by the serous glands of the tongue to aid in digestion? ___________________

## ANSWER KEY

1. Mesentery of the small intestine (jejunum and ileum)
2. Stomach
3. Duodenum

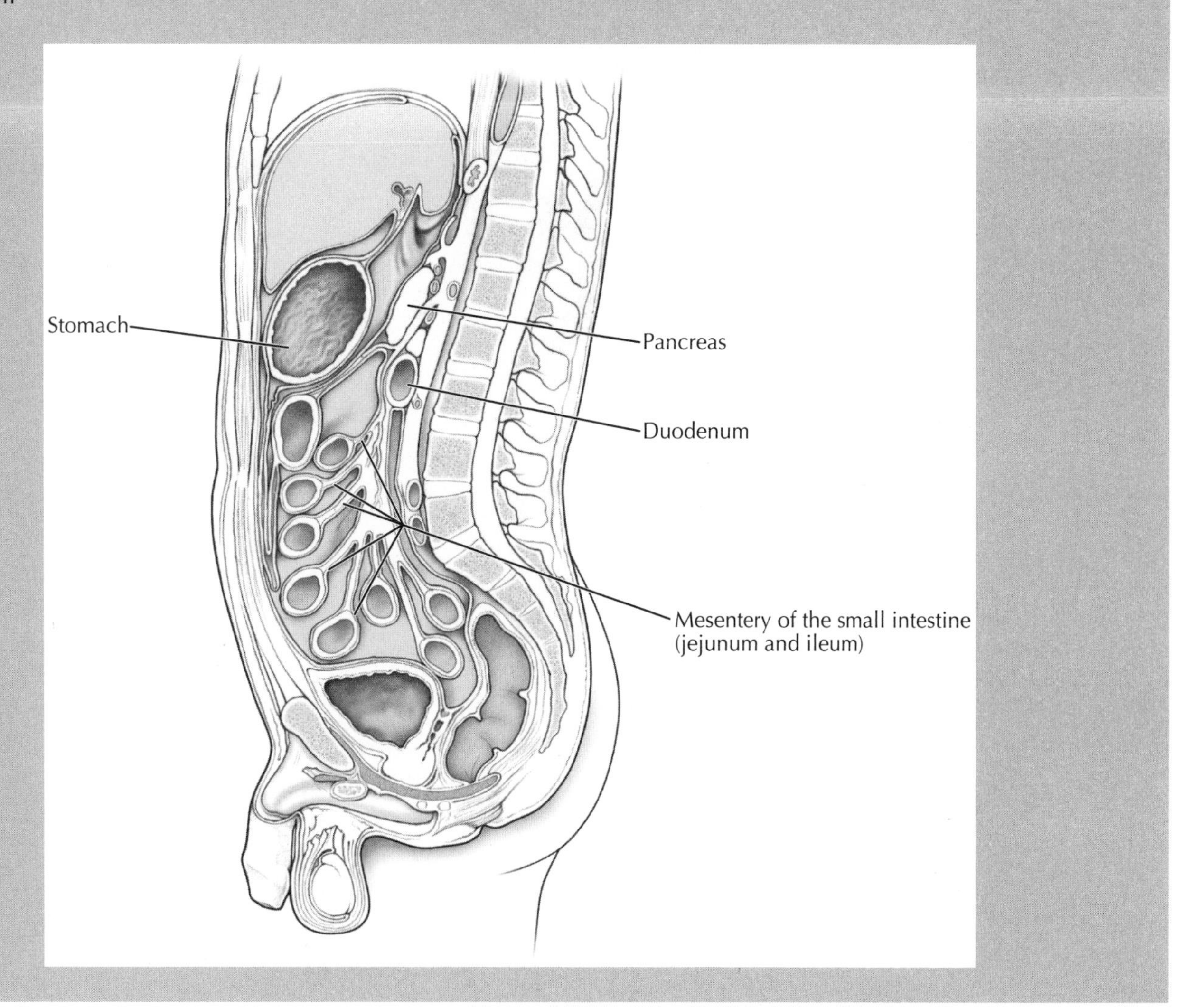

4. Pancreas
5. E
6. A
7. A
8. Lesser sac (omental bursa)
9. Second part of the duodenum
10. Lingual lipase

# Chapter 9 Urinary System

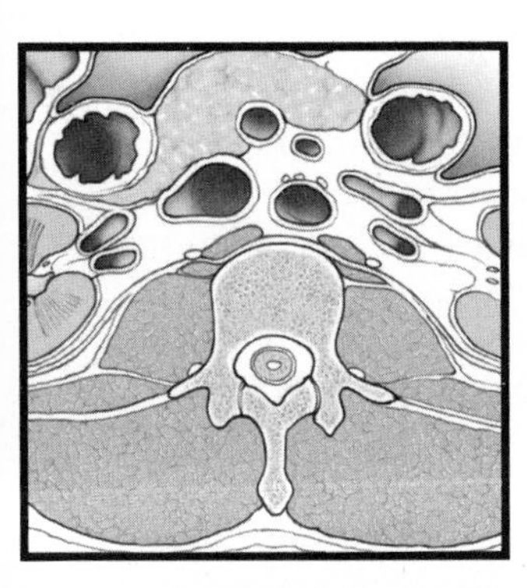

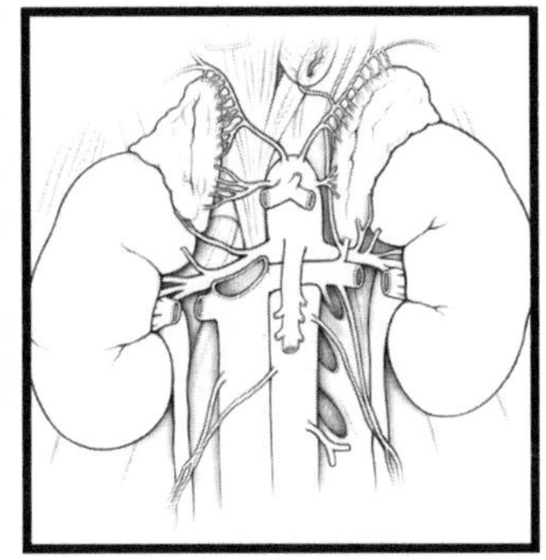

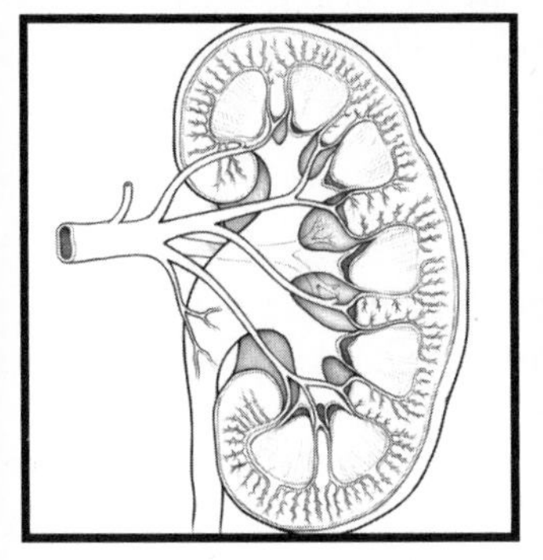

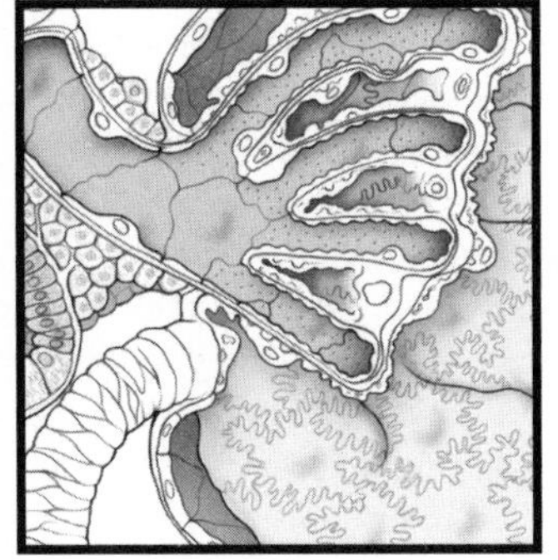

# 9 Overview of the Urinary System

The urinary system includes the following components:

- **Kidneys:** paired retroperitoneal organs that filter the plasma and produce urine; they are located high in the posterior abdominal wall just anterior to the muscles of the posterior wall, and each kidney is "capped" by an adrenal (suprarenal) endocrine gland (see Plate 11-5)
- **Ureters:** course retroperitoneally from the kidney to the pelvis and convey urine from the kidneys to the urinary bladder
- **Urinary bladder:** lies subperitoneally in the anterior pelvis, stores urine, and, when appropriate, discharges the urine via the urethra
- **Urethra:** courses from the urinary bladder to the exterior

The kidneys function to:

- Filter plasma and begin the process of urine formation
- Reabsorb important electrolytes, organic molecules, vitamins, and water from the filtrate
- Excrete metabolic wastes, metabolites, and foreign chemicals, such as drugs
- Regulate fluid volume, composition, and pH
- Secrete hormones that regulate blood pressure, erythropoiesis, and calcium metabolism
- Metabolize vitamin D to its active form
- Convey urine to the ureters, which then conduct the urine to the bladder

The kidneys filter about 180-200 L of fluid each day through a tuft of capillaries known as the **glomerulus**; the filtrate is delivered to a tubule and collecting duct system, which, together with the glomerulus, is called the **nephron**. Each kidney has about 1.25 million nephrons, which are the functional units of the kidney. Grossly, each kidney measures about 12 cm long × 6 cm wide × 3 cm thick and weighs about 150 g, although variability is common. Approximately 20% of the blood pumped by the heart passes to the kidney each minute for plasma filtration, although most of the fluid and important plasma constituents are returned to the blood as the filtrate courses down the tubules of the nephron. We generally just take our kidneys for granted, but they participate in a host of important functions (see above) and can even be donated (only one, of course!) assuming you have a good tissue match!

Each ureter is about 25 to 30 cm long, lies in a retroperitoneal position, and contains a thick, smooth muscle wall. The urinary bladder serves as a reservoir for the urine and is a muscular "bag" that expels the urine, when appropriate. The urethra in the female is short (3 to 5 cm) and in the male is long (about 20 cm). The male urethra runs through the prostate gland, the external urethral sphincter, and the corpus spongiosum of the penis (see Plate 10-8).

**COLOR** each of the following structures, using a different color for each structure:

- ☐ 1. **Kidney**
- ☐ 2. **Ureter**
- ☐ 3. **Urinary bladder**
- ☐ 4. **Urethra**

***Clinical Note:***

The fat surrounding each kidney plays an important role in maintaining their proper position within the posterior abdominal wall (note that the right kidney is slightly lower in position than the left kidney, due to the presence of the liver on the right side). If this fatty encasement is decreased due to significant emaciation and weight loss, the kidneys may descend somewhat in a condition called **renal ptosis**. This can cause the ureter to kink, thereby obstructing the normal flow of urine to the urinary bladder. This backup of urine into the kidney can lead to **hydronephrosis** and renal failure.

**Pyelitis** is an infection of the renal pelvis and calyces, while **pyelonephritis** is an infection or inflammation of the entire kidney, often due to fecal bacteria that spread from the anal region and gain access to the urethra, urinary bladder, and ureter. This condition is more common in females than males due to the proximity of the female urethra to the anus.

**Anuria** is an abnormally low urinary output (less than 50 ml/day) that may result from a low glomerular blood flow, compromising renal filtration; this may occur because of an infection (**nephritis**), an adverse reaction to a blood transfusion, or from traumatic injury directly to the kidney (automobile accident, combat injury).

**Diuretics** are chemicals that increase urinary output. For example, in diabetes mellitus, high glucose levels may act as an osmotic diuretic. Alcohol consumption also promotes diuresis (increased urine output) by inhibiting the release of ADH (antidiuretic hormone). Other diuretics inhibit $Na^+$ reabsorption and normal water reabsorption; examples include caffeine, prescribed drugs to treat hypertension, or congestive heart failure.

Of the **malignant tumors** of the kidney, 80% to 90% are adenocarcinomas that arise from the tubular epithelium. They account for about 2% of all adult cancers, often occur after the age of 50, and occur twice as often in men as in women. In children, **Wilms tumor** accounts for about 7% of all malignancies in children (usually infants) and is associated with congenital malformations related to chromosome 11.

Due to their development (kidneys ascend from the pelvis during embryonic development), kidneys may possess multiple renal arteries and/or veins (**accessory or polar vessels**). This occurs because some of the vessels may fail to degenerate (accessory renal vessels), a condition that occurs in about 25% of people.

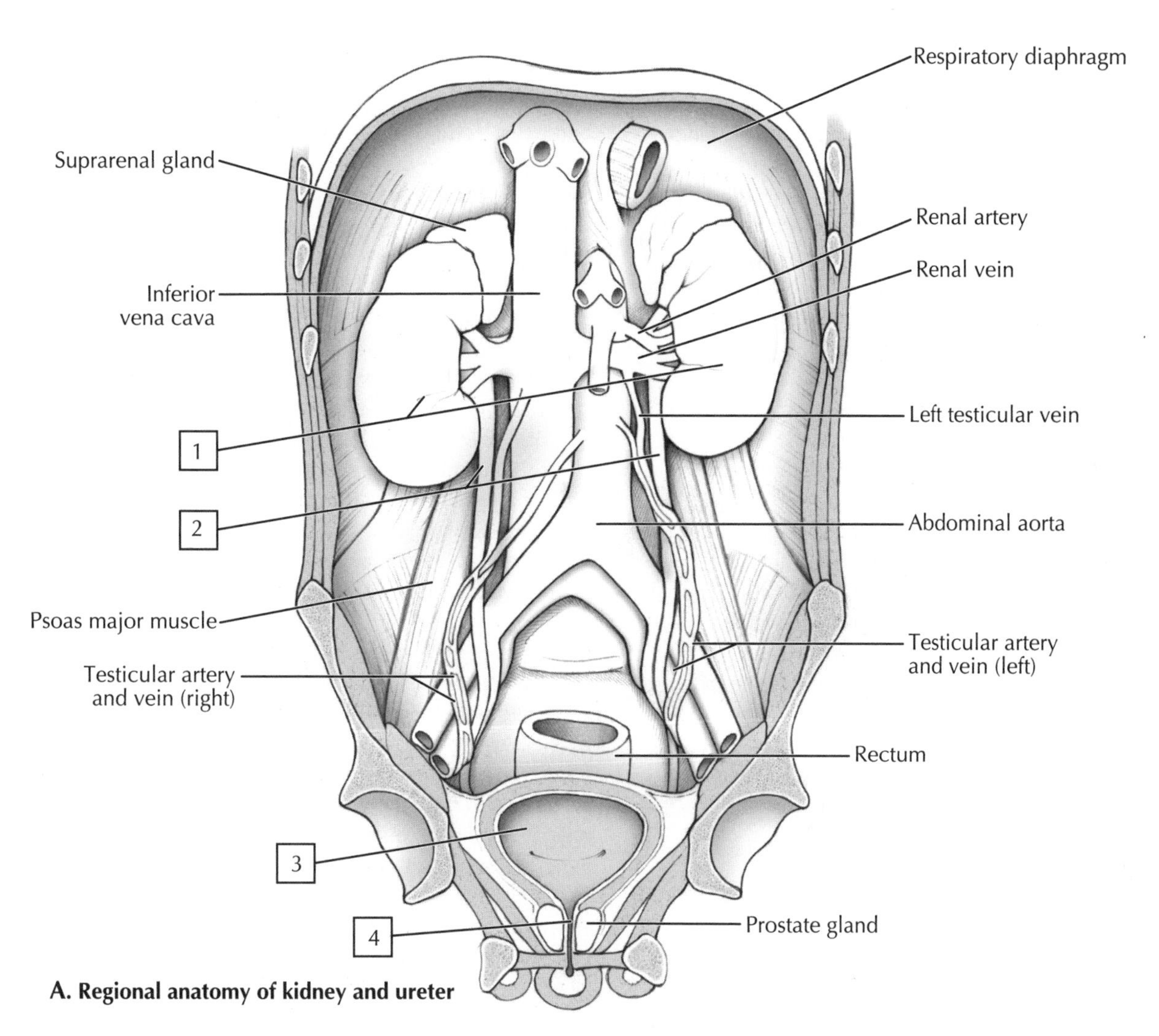


**A. Regional anatomy of kidney and ureter**

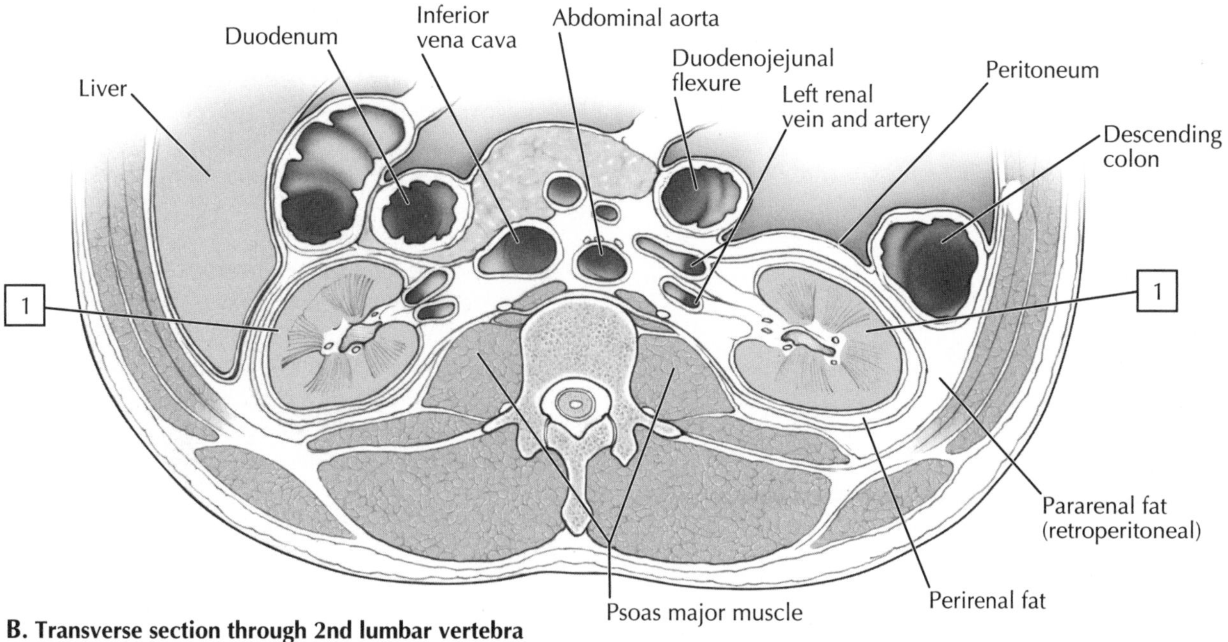


**B. Transverse section through 2nd lumbar vertebra**

# Kidney

Each kidney is enclosed in a **capsule** and, when viewed internally, displays a distinct **cortex** (outer layer) and **medulla** (inner layer). Nephrons are located in the outer cortex and in a juxtamedullary region, or the deepest part of the cortex. The **tubules** of the cortical nephrons extend only a short distance into the medulla, whereas the tubules of the juxtamedullary nephrons extend deep within the medulla. The renal medulla is characterized by the presence of 8 to 15 **pyramids** (collections of tubules), which taper down at their apex to form the papilla, where the urine drips into a **minor calyx**. Several minor calyces form a **major calyx**, and several major calyces empty into a single **renal pelvis** and the proximal **ureter**.

Each kidney is supplied by a large renal artery, which then divides into the following branches:

- **Segmental arteries:** one artery for each of about five segments
- **Interlobar arteries:** several arise from each segmental artery and course between the renal pyramids, ascending to the cortex and arching over the base of each pyramid
- **Arcuate arteries:** the arching terminal portions of the interlobar arteries at the base of each renal pyramid
- **Interlobular arteries:** arise from the arcuate arteries and ascend into the renal cortex (90% of the blood flow to the kidney perfuses the renal cortex)
- **Afferent arterioles:** arise from the interlobular arteries and pass (one each) to the nephron's glomerulus to form the glomerular capillary tuft
- **Efferent arterioles:** glomerular capillaries of the juxtamedullary nephrons reunite to form efferent arterioles that descend into the medulla and form the vasa recta countercurrent system and peritubular capillary network (maintains an osmotic gradient for tubular function; see Plate 9-3)

**COLOR** each of the following features of the kidney, using a different color for each feature:

- ☐ 1. **Kidney**
- ☐ 2. **Renal vein**
- ☐ 3. **Proximal ureter**
- ☐ 4. **Renal artery**
- ☐ 5. **Renal cortex**
- ☐ 6. **Renal pyramids (medulla)**
- ☐ 7. **Minor calyces**
- ☐ 8. **Major calyces**
- ☐ 9. **Renal pelvis**

***Clinical Note:***

Precipitates within the kidney can form **renal stones** (nephrolithiases), which can enter the urinary collecting system and cause renal colic (loin to groin pain) and potentially obstruct the flow of urine. About 12% of the US population will have renal stones, which are two to three times more common in males and relatively uncommon in African Americans and Asian Americans. The types of stones include:

- Calcium oxalate (phosphate): about 75% of stones
- Magnesium ammonium phosphate: about 15% of stones
- Uric acid or cystine: about 10% of stones

As the renal stone passes through the major calyx and renal pelvis to the ureter, it is more likely than not to obstruct flow in one of these three locations (or all three):

- At the junction between the renal pelvis and proximal ureter
- In the ureter, where it crosses the common iliac vessels (midureter)
- At the ureterovesical junction, where the ureter passes through the muscular wall of the urinary bladder

**Diuretics** are chemicals that increase urinary output. They may be osmotic diuretics that are not reabsorbed but carry water out with them. For example, alcohol promotes diuresis by inhibiting the release of ADH (antidiuretic hormone, which inhibits urine output). Some diuretics inhibit sodium reabsorption and, therefore, water normally follows in an obligatory fashion; such diuretics include coffee and tea, and a variety of drugs used to treat hypertension or edema.

**Renal failure** is a serious problem. It occurs from trauma, infections, and poisoning by heavy metals or organic solvents. However, it usually develops slowly where renal filtration gradually decreases and nitrogenous wastes accumulate in the blood, raising its pH. There are five stages of renal failure and in the fifth stage the kidneys are only about 10% to 15% effective! At this point one needs dialysis or a kidney transplant. Risk factors include increasing age, obesity, diabetes mellitus, hypertension, smoking, family history, and race (African Americans, Native Americans, and Asian Americans are especially vulnerable).

**Renal fusion** refers to various defects in which the two kidneys fuse to become one. Horseshoe kidney occurs when the two developing kidneys fuse (usually their lower lobes fuse) anterior to the aorta and often in the lower abdomen. Fused kidneys usually are close to the midline of the abdomen and have multiple renal arteries. Renal obstruction, stone formation, and infection are potential complications associated with renal fusion.

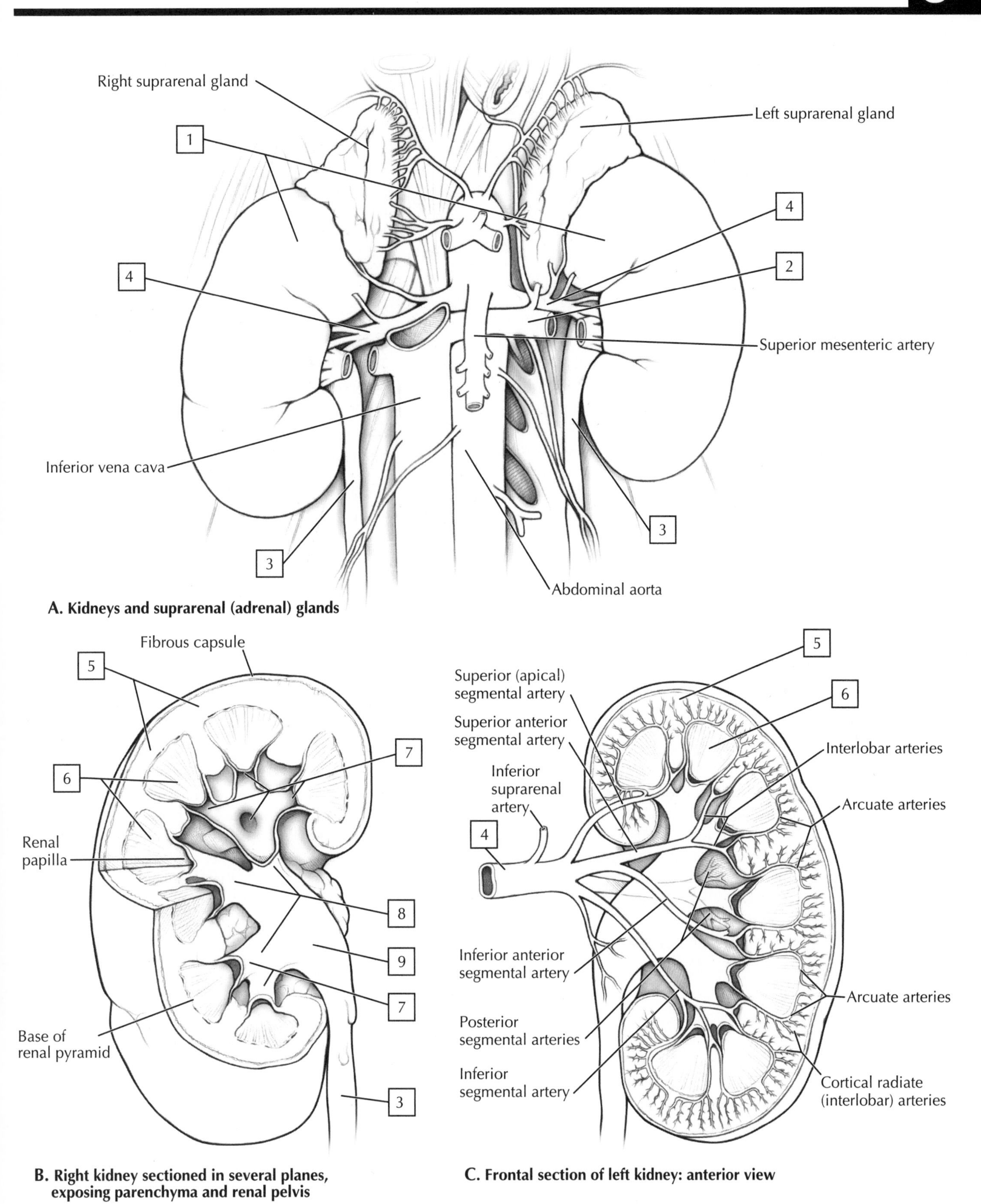


A. Kidneys and suprarenal (adrenal) glands

B. Right kidney sectioned in several planes, exposing parenchyma and renal pelvis

C. Frontal section of left kidney: anterior view

# 9 Nephron

The nephrons differ somewhat in structure, depending on their location. **Cortical nephrons** have their glomeruli in the upper cortex or midcortex and generally have short **loops of Henle** (tubules that dilute the urine but do not concentrate it), as opposed to **juxtamedullary nephrons,** which have long loops of Henle that extend deep into the inner medulla. Cortical nephrons account for about 85% of the nephrons, while the juxtamedullary nephrons account for only about 15% of all of the nephrons in the kidney. However, the juxtamedullary nephrons are important for concentrating the urine. Each kidney contains over 1 million nephrons!

Each nephron, which is the functional unit of the kidney that produces the ultrafiltrate of blood plasma and eventually forms the urine, consists of the following elements:

- **Glomerulus:** a capillary tuft formed by the afferent arteriole, which is encased in **Bowman's capsule** and is responsible for filtering the plasma
- **Proximal convoluted tubule (PCT):** connected to the glomerulus, it receives the plasma ultrafiltrate and conveys it down the loop of Henle
- **Loop of Henle:** consists of a single long tubule of varying thickness and lined with epithelial cells that are involved in reabsorption and secretion along the tubule's length
- **Distal convoluted tubule (DCT):** receives the remaining tubular fluid from the loop of Henle, monitors the fluid's osmolarity, and conveys the fluid to the collecting duct
- **Collecting duct:** terminal end of the nephron, where the final concentration of the urine is "fine-tuned" before it is conveyed to the minor calyces

The glomerulus filters the plasma. This ultrafiltrate is devoid of cells and virtually all proteins (unless they are smaller in size than albumin). The endothelium of the glomerulus is fenestrated but prevents the passage of blood cells. **Podocytes** envelop the fenestrated endothelium and keep proteins from being filtered. Adjacent to the afferent arteriole that delivers blood to the glomerulus is a specialization of the DCT wall called the **macula densa,** which monitors the NaCl in the fluid of the DCT. If the NaCl levels are low, the macula densa stimulates the release of **renin** from juxtaglomerular cells. The renin ultimately causes an increase in angiotensin II and aldosterone (renin-angiotensin-aldosterone [RAA] system). These hormones stimulate NaCl and water reabsorption by the nephron (angiotensin II acts on the proximal tubule and aldosterone acts on the collecting duct). The juxtaglomerular cells adjacent to the macula densa of the DCT also monitor the blood pressure in the afferent arteriole and, if the blood pressure is low, the cells release renin to elevate the blood pressure via the RAA system and sympathetic activity.

**COLOR** the following features of the nephron, using the colors suggested for each feature:

- ☐ 1. **Proximal tubule: convoluted and straight segments (blue)**
- ☐ 2. **Juxtamedullary glomerulus (purple)**
- ☐ 3. **Distal ascending loop of Henle (thick limb and DCT) (orange)**
- ☐ 4. **Thin descending and ascending loops of Henle (green)**
- ☐ 5. **Collecting duct (gray)**
- ☐ 6. **Cells lining the DCT (orange)**
- ☐ 7. **Afferent arteriole (red)**
- ☐ 8. **Juxtaglomerular cells (purple)**
- ☐ 9. **Endothelium of glomerular capillaries (yellow)**
- ☐ 10. **Podocytes (brown)**
- ☐ 11. **Bowman's capsule (green)**
- ☐ 12. **Epithelium of PCT (blue)**

***Clinical Note:***

**Obstructive uropathy** is a condition where the normal flow of urine is compromised, and it may occur anywhere from the level of the renal nephron to the urethral opening. In the kidney, it may be caused by infections, cancer, calculus (kidney stones); in the ureter it may be caused by stenosis, kinks, chronic infections, calculus, neoplasm, compression by nodes, abscess, appendicitis, trauma; in the urinary bladder it can be from a neoplasm, calculus, a diverticulum of the bladder wall, congenital bladder neck obstruction; in the urethra it can be from cysts, neoplasm, stricture, prostatitis (males only), and several other causes!

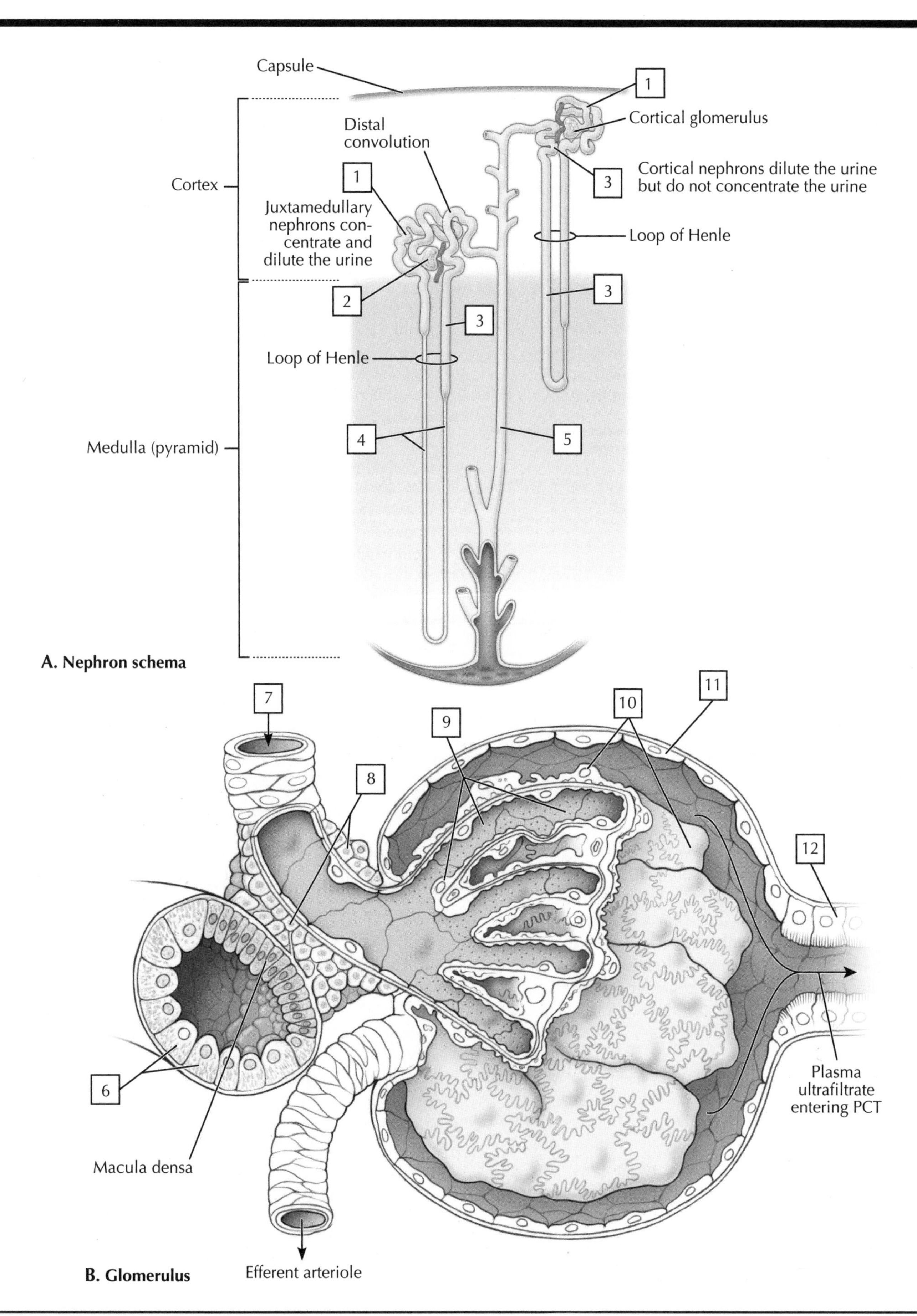
Capsule
1
Cortical glomerulus
Distal convolution
Cortex
1
Juxtamedullary nephrons concentrate and dilute the urine
3
Cortical nephrons dilute the urine but do not concentrate the urine
Loop of Henle
2
3
3
Loop of Henle
Medulla (pyramid)
4
5
A. Nephron schema
7
9
10
11
8
12
6
Plasma ultrafiltrate entering PCT
Macula densa
B. Glomerulus
Efferent arteriole

## Glomerular Filtration

The volume of fluid filtered by the renal glomeruli per unit of time is called the **glomerular filtration rate (GFR)**. Remember that in the average person, about 180 L of fluid is filtered per day (125 ml/min); because plasma accounts for about 3 L of our total blood volume, that means that the kidneys filter the blood plasma about 60 times per day! The amount of blood delivered to the glomerulus or leaving it is controlled by neural and hormonal mechanisms acting on the afferent and efferent arterioles.

## Tubular Reabsorption

Once the ultrafiltrate of the plasma enters the proximal convoluted tubule (PCT), it is modified by the renal tubules (PCT, Loop of Henle, distal convoluted tubule [DCT], and collecting duct), as summarized in the following table.

**REABSORPTION OF SEVERAL COMPONENTS FROM THE ULTRAFILTRATE**

| SUBSTANCE | AMOUNT FILTERED/DAY | PERCENTAGE REABSORBED |
|---|---|---|
| Water | 180 L | 99 |
| Sodium | 630 g | 99.5 |
| Glucose | 180 g | 100 |
| Urea | 54 g | 44 |

Reabsorption occurs both by diffusion and by mediated transport. For example, many substances are reabsorbed in combination with sodium (cotransported). Except in the descending limb of the loop of Henle, sodium is actively reabsorbed in all tubular regions, and water reabsorption is by diffusion and is dependent upon sodium reabsorption. About two-thirds of the sodium and water is reabsorbed in the proximal tubule; in fact, tubular reabsorption is generally high for nutrients, ions, and water but lower for waste products such as urea (see table above: 44% reabsorption).

## Tubular Secretion

Tubular secretion involves a process whereby substances in the capillaries that parallel the renal tubules diffuse or are actively transported into the tubular lumen. Important substances secreted include:

- Hydrogen ions
- Potassium
- Organic anions such as choline and creatinine (waste product of muscle)
- Foreign chemicals

## Renal Sodium and Water Regulation

Sodium filtration is regulated at the level of the glomerulus by the **baroreceptor reflex,** and its reabsorption is regulated at the tubular level by **aldosterone** (secreted by the adrenal cortex), which stimulates reabsorption. Other factors also play a role, but water reabsorption is linked to sodium movement until it reaches the collecting duct system, where water then comes under the control of **vasopressin** (antidiuretic hormone, ADH). ADH is synthesized primarily in the supraoptic nuclei (and also the paraventricular nuclei) of the hypothalamus, but is stored in and released from the posterior pituitary gland (see Plate 11-1). Low ADH levels result in a dilute urine (water excretion), whereas high ADH levels activate water channels (called **aquaporins**) that reabsorb water and create a concentrated urine.

The kidneys also play an important role in regulating the following:

- Water retention is facilitated by ADH and the countercurrent multiplier system (renal vasa recta), which creates a medullary interstitial fluid that is hyperosmotic
- Potassium levels, by both tubular reabsorption and secretion
- Calcium and vitamin D homeostasis, in concert with parathyroid hormone
- Homeostatic regulation of plasma hydrogen ion concentration (acid-base balance) in concert with the respiratory system
- Regulation of bicarbonate concentration and generation of new bicarbonate by the production and excretion of ammonium

**COLOR** each of the following dynamic features of tubular function, using the colors suggested for each feature:

- ☐ 1. **Water movement (blue)**
- ☐ 2. **Solute movement (yellow)**
- ☐ 3. **Filtrate (green)**
- ☐ 4. **PCT tubule cells (brown) (possess a high surface area for reabsorption)**
- ☐ 5. **Thin descending segment cells of the loop of Henle**
- ☐ 6. **DCT cells**
- ☐ 7. **Collecting duct cells**

***Clinical Note:***

A **variety of hormones** act on kidney function. Simplistically, parathyroid hormone (PTH) responds to a decrease in plasma calcium ($Ca^{2+}$) and increases reabsorption and a decrease in phosphate reabsorption. Antidiuretic hormone (ADH, vasopressin) responds to an increase in plasma osmolarity and a decrease in blood volume, thus increasing water permeability. Aldosterone responds to a decrease in blood volume (via the renin-angiotensin II system) and an increase in plasma $K^+$, and increases $Na^+$ reabsorption, $K^+$ secretion, and $H^+$ secretion. Atrial natriuretic peptide (ANP) responds to an increase in atrial pressure and increases the glomerular filtration rate (GFR), and decreases $Na^+$ reabsorption. Angiotensin II responds to a decrease in blood volume (via renin) and increases $Na^+$ - $H^+$ exchange and $HCO_3^-$ reabsorption (in the proximal tubule).

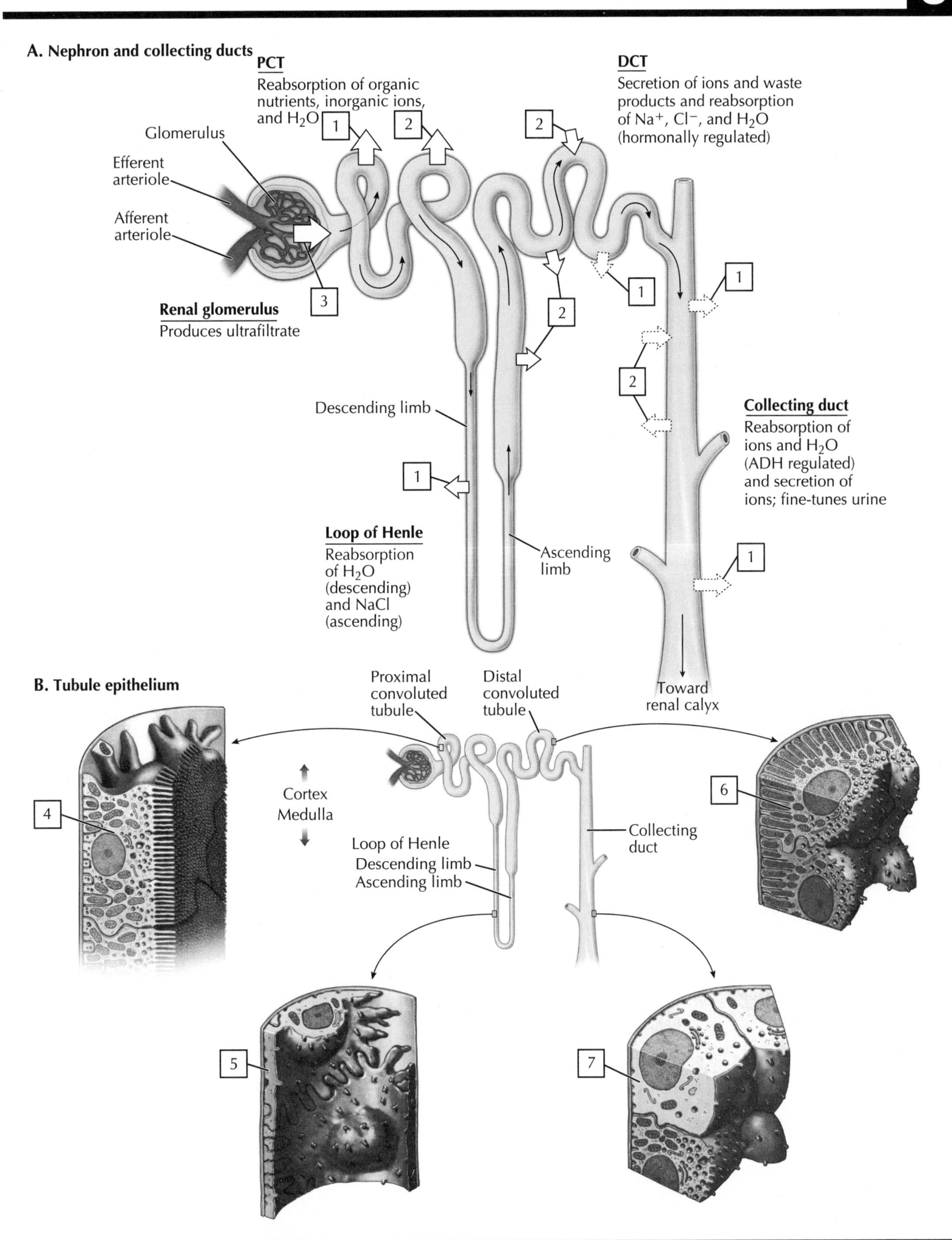
A. Nephron and collecting ducts
PCT
Reabsorption of organic nutrients, inorganic ions, and $H_2O$
DCT
Secretion of ions and waste products and reabsorption of $Na^+$, $Cl^-$, and $H_2O$ (hormonally regulated)
Glomerulus
Efferent arteriole
Afferent arteriole
Renal glomerulus
Produces ultrafiltrate
Descending limb
Collecting duct
Reabsorption of ions and $H_2O$ (ADH regulated) and secretion of ions; fine-tunes urine
Loop of Henle
Reabsorption of $H_2O$ (descending) and NaCl (ascending)
Ascending limb
Toward renal calyx
B. Tubule epithelium
Proximal convoluted tubule
Distal convoluted tubule
Cortex
Medulla
Loop of Henle
Descending limb
Ascending limb
Collecting duct
1
2
3
4
5
6
7

# 9 Urinary Bladder and Urethra

The renal calyces, pelvis, ureters, bladder, and proximal urethra are lined by **transitional epithelium** (urothelium), which has the unique ability to "unfold" or expand as the passageways and urinary bladder become distended. The ureters are enveloped in smooth muscle arranged in three layers, but the urinary bladder is enveloped with smooth muscle that is randomly mixed in its orientation and is known as the **detrusor** ("drive away") **muscle**. The proximal urethra in both sexes is lined with transitional epithelium, which then gives way to pseudostratified columnar and stratified squamous epithelium as the urethra opens to the exterior.

The urinary bladder lies **subperitoneally** behind the pubic symphysis. The urinary bladder stores the urine until it is appropriate to void (urination); it can hold up to 800 to 1000 ml of urine. The interior, posteroinferior wall of the urinary bladder demonstrates a smooth area called the **trigone**, demarcated by the two ureteric openings superiorly and the single urethral opening at the base of the urinary bladder.

Micturition (voiding or urination) involves several important steps:

- Normally sympathetic nerve fibers relax the urinary bladder wall, allowing for distention, and constrict the internal urethral sphincter (smooth muscle) located at the neck of the urinary bladder (females do not have this internal urethral sphincter)
- Micturition is initiated by the stimulation of stretch receptors in the detrusor muscle, sending afferent signals to the spinal cord levels S2-S4 via the pelvic splanchnic nerves
- Parasympathetic efferents (via the pelvic splanchnics) induce a reflex contraction of the detrusor muscle, relaxation of the internal sphincter in males, and enhance the "urge" to void
- When convenient (and sometimes not!), somatic efferents via the pudendal nerve (S2-S4) cause voluntary relaxation of the external urethral sphincter (in both sexes), and micturition occurs
- When empty, the external sphincter contracts (in males the bulbospongiosus muscle expels that last few drops of urine from the urethra), and the detrusor muscle once again relaxes under sympathetic control

The female urethra is short (3 to 5 cm), is encircled by the urethral sphincter (blends with another skeletal muscle called the sphincter urethrovaginalis; see Plate 3-16), and opens into the vestibule. The male urethra is longer (about 20 cm) and descriptively is divided into three parts:

- **Prostatic urethra:** proximal portion of the male urethra that runs through the prostate gland
- **Membranous urethra:** short, middle portion that is enveloped by the external urethral sphincter (skeletal muscle)
- **Spongy (penile, cavernous) urethra:** courses through the bulb of the penis, the pendulous portion of the penis, and the glans penis to open at the external urethral orifice

In both sexes, urethral glands open into the lumen and lubricate the urethral mucosa (see Plate 3-16, bulbourethral glands in males and greater vestibular glands in females).

**COLOR** the following features of the urinary bladder and urethra, using a different color for each feature:

- ☐ **1. Detrusor muscle of the female urinary bladder wall**
- ☐ **2. Trigone in the female and male urinary bladder**
- ☐ **3. Female urethra**
- ☐ **4. Sphincter urethrae muscle in the female**
- ☐ **5. Internal urethral sphincter in the male**
- ☐ **6. Membranous urethra**
- ☐ **7. External urethral sphincter in the male**
- ☐ **8. Spongy urethra**
- ☐ **9. Prostatic urethra**

***Clinical Note:***

**Stress incontinence** (involuntary release of urine) usually occurs with an increase in intraabdominal pressure caused by coughing, sneezing, defecation, or lifting. Normally, the sphincter mechanism (urethral sphincter) is strong enough to keep the urine from leaving the urinary bladder. However, weakening of the sphincter mechanism of the urinary bladder, vagina, and other support structures of the pelvic floor can lead to stress incontinence; predisposing factors include multiparity (multiple childbirths, leading to stretching of the sphincter during vaginal delivery), obesity, chronic cough, and heavy lifting.

Generally, urine is clear and a pale yellow in color. If concentrated, it will be a deep yellow color. This color is due to urochrome, a by-product of the body's destruction of hemoglobin. A cloudy urine may indicate a **urinary tract infection**.

Urine is slightly aromatic in odor but if left standing, it smells like ammonia, a result of bacteria metabolizing its urea solutes. In individuals with **diabetes mellitus**, the urine may smell fruity because of its acetone content.

Urine is generally slightly acidic (pH of about 6); an acidic diet of protein and whole wheat usually is the cause of this acidic pH. An alkaline diet, e.g., a vegetarian diet, chronic vomiting and/or a urinary tract infection, may cause the urine to become alkaline (pH of 8).

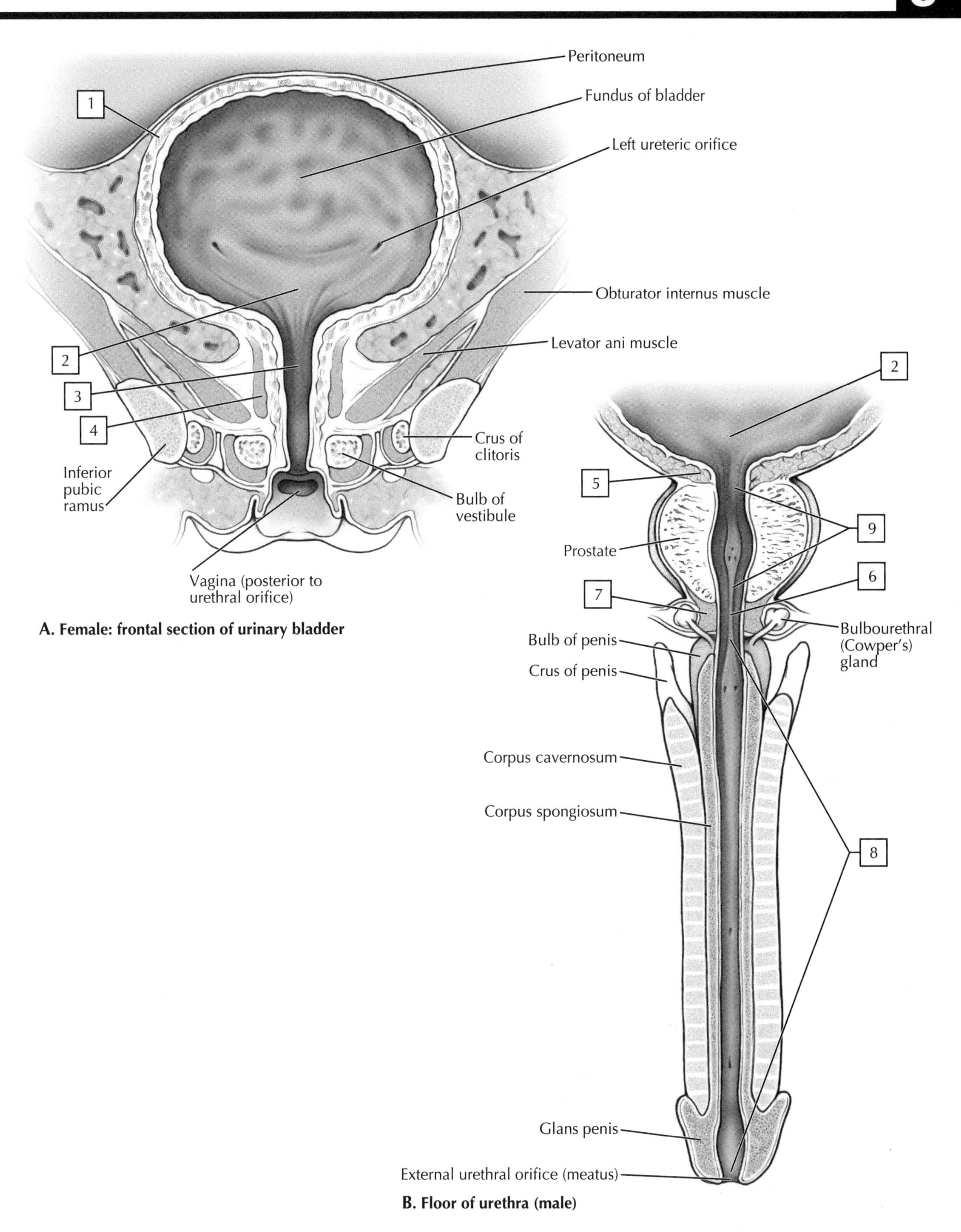


**A. Female: frontal section of urinary bladder**

**B. Floor of urethra (male)**

## REVIEW QUESTIONS

**For each description below (1-4), color the relevant structure or feature on the image.**

1. This region of the kidney contains the most nephrons and their glomeruli.
2. Most of the renal tubules and the vasa recta are found in this area.
3. These structures collect urine from each pyramid.
4. This structure conveys urine to the urinary bladder.

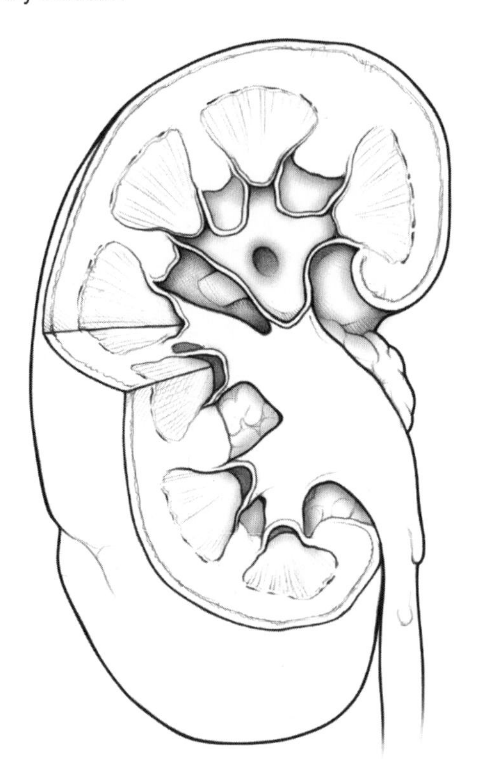

5. Descriptively, the kidneys do not reside within the abdominal peritoneal cavity, nor are they suspended in a mesentery. What terminology would a clinician use to describe the location of the kidneys? ____________________
6. Renal stones may be passed down the ureter to the bladder but can become lodged at three primary points along their journey to the bladder. Where are these three points? ____________________
7. At the level of the renal glomerulus, cells envelop the glomerulus to prevent the passage of cells and proteins from being filtered. What are these cells called? ____________________
8. High levels of this hormone result in the retention (reabsorption) of water in the collecting ducts. ____________________
9. Which of the following nerves is critical for maintaining the voluntary urethral sphincter (external sphincter) in males and must be spared, if possible, during pelvic or perineal surgery?
   A. Femoral
   B. Inferior gluteal
   C. Obturator
   D. Pelvic splanchnics
   E. Pudendal

10. Which portion of the nephron is critical for monitoring the osmolarity of the tubular fluid?
    A. Bowman's capsule
    B. Collecting duct
    C. Distal convoluted tubule
    D. Loop of Henle
    E. Proximal convoluted tubule

## ANSWER KEY

1. Renal cortex
2. Renal pyramids (medulla)
3. Minor calyces
4. Ureter

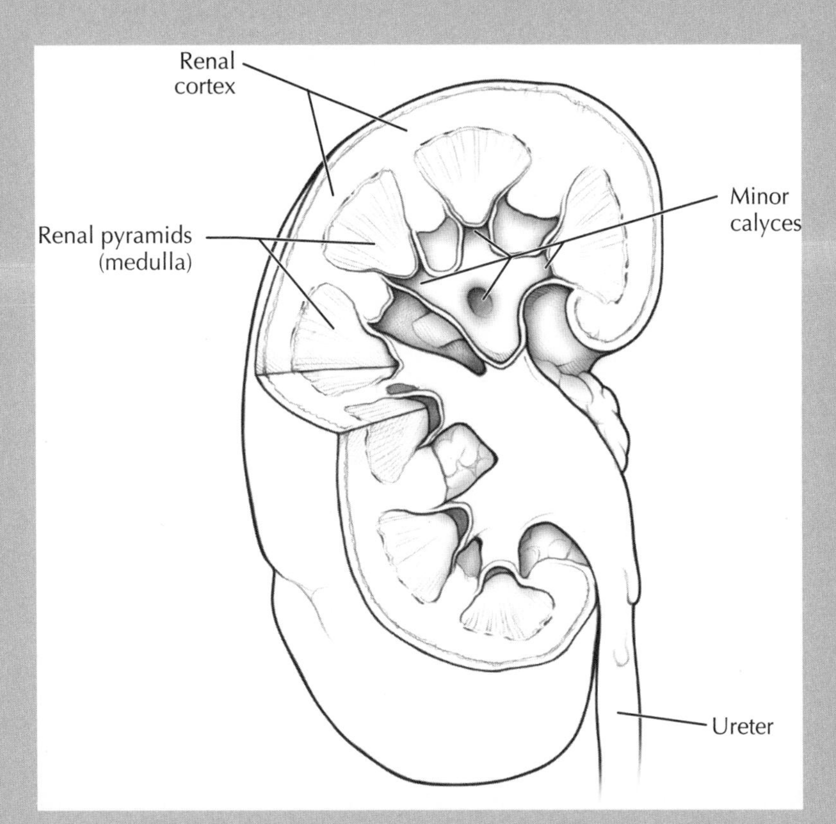


5. The kidneys are retroperitoneal organs.
6. At the junction of the renal pelvis and ureter, at the point where the ureter crosses the common iliac vessels, and at the ureterovesical junction as it passes through the muscular wall of the bladder
7. Podocytes
8. Antidiuretic hormone (ADH; also called vasopressin)
9. E
10. C

# Chapter 10 Reproductive Systems

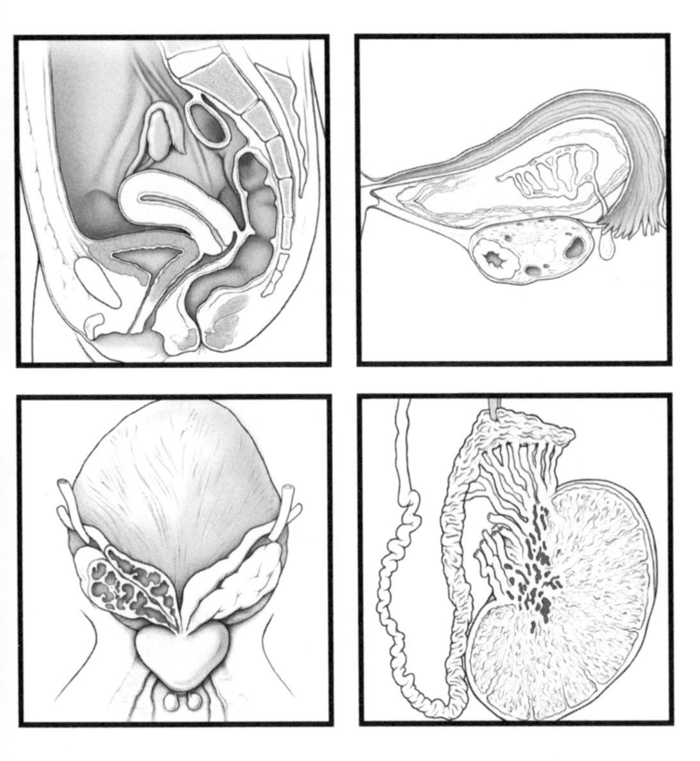

# Overview of the Female Reproductive System

The female reproductive system is composed of the following structures:

- **Ovaries**: the paired gonads of the female reproductive system, they produce the female germ cells called ova (oocytes, eggs) and secrete the hormones **estrogen** and **progesterone**
- **Uterine tubes (fallopian tubes or oviducts)**: paired tubes that extend from the superolateral walls of the uterus and open as fimbriated funnels into the pelvic cavity adjacent to the ovary (to "capture" the ovulated oocyte)
- **Uterus**: a pear-shaped, hollow muscular (smooth muscle) organ that protects and nourishes a developing fetus
- **Vagina**: a musculoelastic distensible tube (also referred to as the birth canal) approximately 8 to 9 cm long that extends from the uterine cervix (neck) to the vestibule

The female reproductive viscera are summarized in the following table.

| FEATURE | CHARACTERISTICS |
|---|---|
| Ovary | Is suspended between suspensory ligament of ovary (contains ovarian vessels, nerves, lymphatics) and ovarian ligament (tethered to uterus) |
| Uterine tube (fallopian tube, oviduct) | Runs in mesosalpinx of broad ligament, which suspends the tube and ovary and reflects off the uterus; subdivided into its fimbriated end, infundibulum, ampulla, isthmus, and intrauterine portions |
| Uterus | Consists of body (fundus and isthmus) and cervix; is supported by pelvic diaphragm and ligaments; is enveloped in the broad ligament |
| Vagina | Fibromuscular tube that includes the fornix (recess around protruding uterine cervix) |

The ovaries are suspended from the lateral pelvic walls by the **suspensory ligament of the ovary** (contains the ovarian neurovascular elements) and tethered to the uterus medially by the **ovarian ligament**. The uterus, uterine tubes, and ovaries also are supported by the **broad ligament**, a kind of "mesentery" that consists of peritoneum that reflects off of the pelvic walls and sweeps up to embrace these visceral structures, not unlike the mesenteries of the bowel. These features are summarized in the table below.

| FEATURE | CHARACTERISTICS |
|---|---|
| Broad ligament of uterus | Is a peritoneal fold the suspends the uterus and uterine tubes; includes mesovarium (enfolds ovary), mesosalpinx (enfolds uterine tube), and mesometrium (remainder of ligament) |
| Ovaries | Are suspended by suspensory ligament of ovary from lateral pelvic wall and tethered to uterus by ovarian ligament |
| Uterine tubes | Consist of fimbriated end (collects ovulated ova), infundibulum, ampulla, isthmus, and uterine parts; courses in the mesosalpinx of the broad ligament |
| Transverse cervical (cardinal or Mackenrodt's) ligaments | Are fibromuscular condensations of pelvic fascia that support uterus |
| Uterosacral ligaments | Extend from sides of cervix to sacrum, support uterus, and lie beneath peritoneum (form the uterosacral fold) |

The perineum is a diamond-shaped region extending from the pubic symphysis laterally to the two ischial tuberosities and then posteriorly to the tip of the coccyx. The anterior half of the diamond-shaped region is the **urogenital triangle** and it includes the vulva or external female genitalia. A labia (lip) majora, covering the erectile tissue of the bulb of the vestibule, surrounds the labia minor, which demarcates the **vulva** and the openings of the urethra and vagina. The erectile tissue of the clitoris (crus, body, and glans, similar to the male but smaller) demarcates the two lateral boundaries of the urogenital triangle that lie along the ischiopubic ramus and meet at the pubic symphysis anteriorly. This region is innervated by the **pudendal nerve** (somatic branches of S2-S4) and supplied by branches of the **internal pudendal artery** (see Plate 5-16).

**COLOR** the following features of the female reproductive system, using a different color for each feature:

☐ 1. **Uterine tube**

☐ 2. **Ovary**

☐ 3. **Uterus (fundus, body, and cervix)**

☐ 4. **Vagina**

☐ 5. **Clitoris (the erectile bodies: crus, body, and glans) (crus covered by the ischiocavernosus muscle) (see Plate 3-16)**

☐ 6. **Urethral opening**

☐ 7. **Labia minora**

☐ 8. **Labia majora**

☐ 9. **Vaginal opening**

☐ 10. **Bulb of the vestibule (large bilateral erectile tissue flanking the vagina and urethral openings, covered by the bulbospongiosus muscle, and extending anteriorly to form a connection with the glans of the clitoris; see Plate 3-16)**

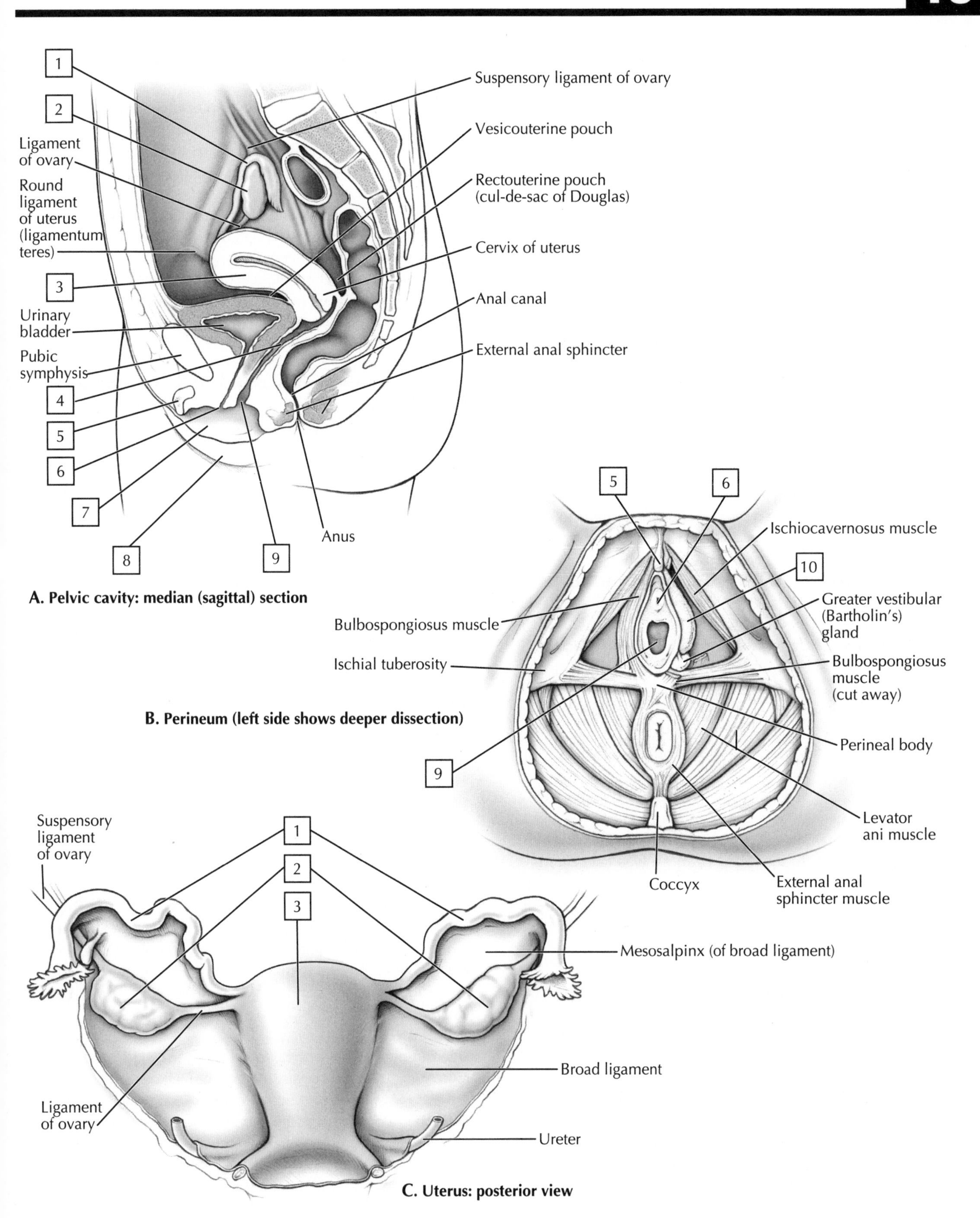


A. Pelvic cavity: median (sagittal) section

B. Perineum (left side shows deeper dissection)

C. Uterus: posterior view

# Ovaries and Uterine Tubes

The ovaries develop retroperitoneally high in the posterior abdominal wall and, like the testes, descend during fetal growth into the pelvic cavity, where they become enveloped within the **broad ligament** and are suspended between the lateral pelvic wall and the uterus medially. At birth, the two ovaries together contain about 600,000 to 800,000 primary oocytes (no new ones will be formed), but the vast majority will never fully mature; only about 400 will reach maturity and ultimately be ovulated, while the rest will degenerate.

The sequence of ovarian events culminating in the ovulation of a mature oocyte (ovum) include:

1. During fetal development, the oogonia (future eggs) become primary oocytes; they begin their first meiotic division but become arrested in this state until puberty.
2. At puberty, only the primordial follicles that ultimately will become mature complete their first meiotic division to form a secondary oocyte.
3. The secondary oocyte resides in a primary follicle, surrounded by a single layer of granulosa cells, and then it begins to grow to become a mature primary follicle.
4. As the oocyte grows in size, the granulosa cells proliferate (secrete estrogen and some progesterone), forming a secondary follicle with a fluid-filled space called the antrum.
5. About 10 to 20 such “preantral” follicles begin to mature at the beginning of each menstrual cycle, but usually only one becomes dominant and will mature, while the others degenerate.
6. The mature follicle, also known as a Graafian follicle, increases in size (about 10 mm in diameter) and begins to bulge out under the surface of the ovarian capsule. Ovulation is the process of the secondary oocyte breaking free from the Graafian follicle. The oocyte is expelled during the middle of the menstrual cycle (14th day of the 28-day cycle).
7. The secondary oocyte (contains the haploid number of chromosomes and is arrested in metaphase of the second meiotic division) is “captured” by the fimbriated end of the uterine tube, while the remaining granulosa cells on the ovarian surface enlarge and form a glandular-like structure called the corpus luteum (secretes estrogen, progesterone, and inhibin). The secondary oocyte remains viable for about 24 hours; if fertilization does not occur, it degenerates and passes through the uterine tube.
8. The corpus luteum lasts about 10 days and then degenerates, unless the egg is fertilized.
9. If fertilization occurs, the secondary oocyte completes its second meiotic division and forms a mature ovum (egg), with a maternal pronucleus containing a set of 23 chromosomes; the sperm that fertilizes the ovum carries its 23 chromosomes in a male pronucleus, and these two pronuclei become the zygote, with the diploid (2n) complement of 46 chromosomes. The zygote undergoes a mitotic division (its first cleavage), which begins the development of the embryo.
10. The conceptus then moves through the uterine tube and implants into the uterine endometrium around the fifth day following fertilization.
11. During early pregnancy, the corpus luteum maintains the pregnancy by secreting estrogen and progesterone; it regresses between the second and third months, as the placenta takes over the job of maintaining the pregnancy.

The uterine tubes are divided into the following segments:

- **Infundibulum** and its fimbriated end: envelops the ovary to capture the ovulated egg
- **Ampulla**: the next segment, where fertilization usually occurs
- **Isthmus**: a narrow, medial segment of the tube
- **Intramural portion**: lies within the uterine wall and opens into the uterine cavity

**COLOR** each of the following features of the ovary and uterine tube, using a different color for each feature:

- ☐ **1. Isthmus**
- ☐ **2. Ampulla**
- ☐ **3. Fimbriated end of the infundibulum**
- ☐ **4. Primary follicle**
- ☐ **5. Secondary follicles**
- ☐ **6. Mature Graafian follicle**
- ☐ **7. An ovulated ovum (secondary oocyte)**
- ☐ **8. Mature corpus luteum**

***Clinical Note:***

**Ovarian cysts** are fluid-filled sacs that usually arise from the Graafian follicles, and are often benign and asymptomatic.

**Polycystic ovarian syndrome** is a fairly common hormonal disorder. It is characterized by infrequent or prolonged menstrual periods. The abnormally enlarged ovaries have multiple subcortical follicular cysts.

**Ectopic pregnancy** is when the conceptus (fertilized ovum) implants in tissue outside of the uterus. The uterine (fallopian) tube is the most common site, although ectopic pregnancies may occur in the ovary, abdomen, or cervix. About 50% of the women with ectopic pregnancies have a history of pelvic inflammatory disease (PID).

**Ovarian cancer** is the most lethal cancer of the female reproductive tract. The vast majority of malignancies occur from the surface epithelium, with the cancerous cells often breaking through the ovarian capsule and seeding the peritoneal surface, thus invading the adjacent pelvic organs (omentum, mesentery, intestines). The cancer cells also may spread via the venous drainage to the liver and lungs.

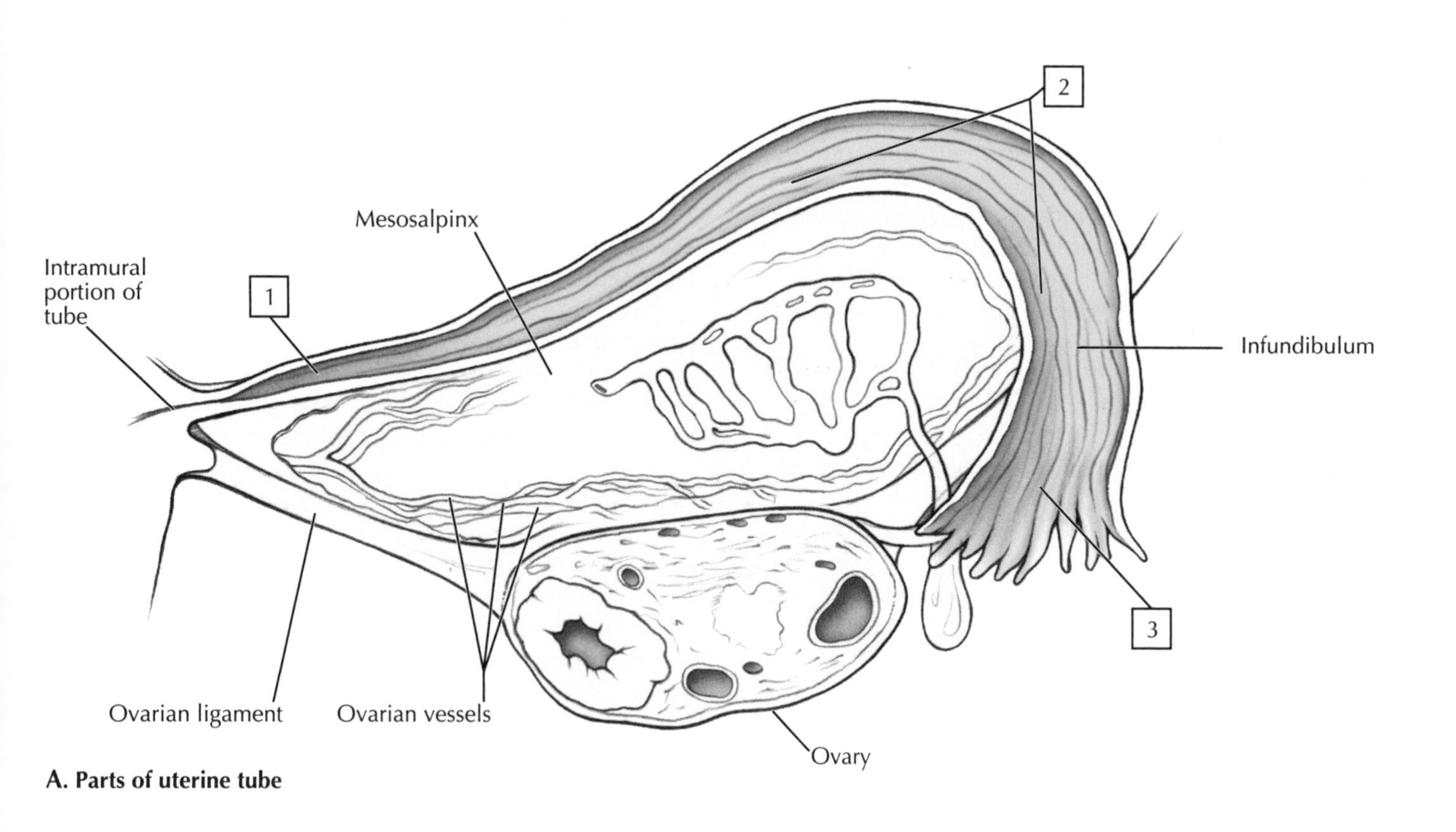


A. Parts of uterine tube

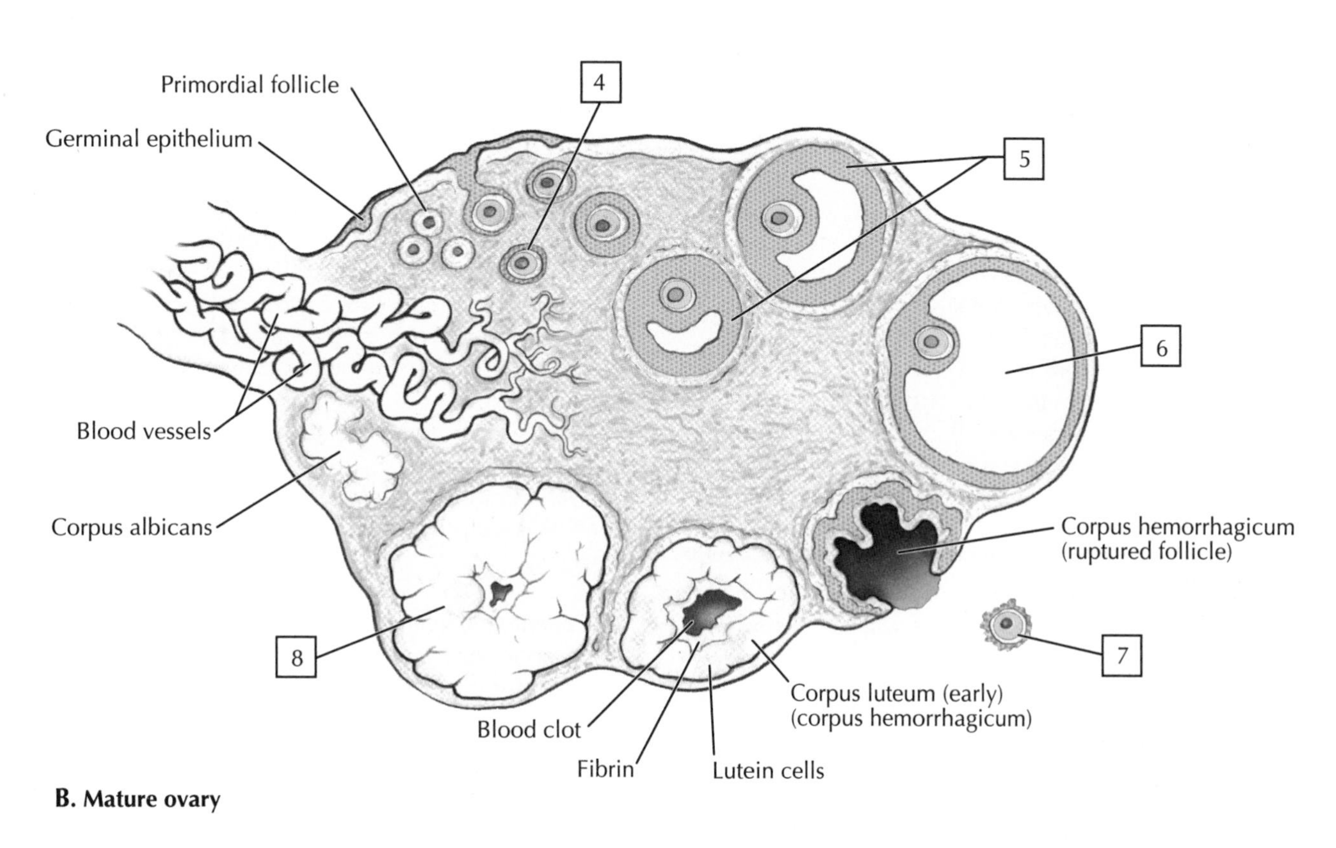


B. Mature ovary

# Uterus and Vagina

## Uterus

The uterus (womb) is a pear-shaped organ suspended in the **broad ligament** (mesometrium) and tethered laterally by its connections to the uterine tubes and by the ovarian ligament and its attachment to the ovary. Additionally, reflecting from its anterolateral aspect is the **round ligament of the uterus,** a distal remnant of the female gubernaculum (the proximal remnant is the ovarian ligament attached to the ovary), which pulls the ovary down from its developmental site in the posterior abdominal wall into the pelvis. The round ligament of the uterus passes through the inguinal canal and ends as a fibrous fatty band in the labia majora (homologue to the male scrotum).

The uterus has several parts:

- **Fundus**: that part lying superior to the attachments of the two uterine tubes
- **Body**: the middle portion of the uterus that tapers inferiorly into the cervix
- **Cervix**: the "neck" of the uterus, it lies subperitoneally, has a narrow endocervical canal, and opens into the superior part of the vagina

The uterine wall is lined internally by the **endometrium**, which proliferates significantly during the first half of the menstrual cycle in preparation for the possible implantation of a conceptus (also see Plate 10-4). If fertilization does not occur, the endometrium degenerates and is sloughed off during the 3 to 5 days of **menstruation** that mark the beginning of the next menstrual cycle. The middle layer of the uterine wall is the **myometrium**, a thick, smooth muscle layer, and the outer layer is the **perimetrium**, a serous layer (visceral peritoneal covering).

## Vagina

The vagina is about 8-9 cm long and is a musculoelastic tube extending from the uterine cervix to its opening in the vestibule (area enclosed by the labia minora). The lumen is lined by a stratified, squamous, nonkeratinized epithelium that is lubricated by mucus from cervical glands. The lamina propria of the vagina possesses an extensive nerve supply and a venous plexus that becomes engorged with blood during sexual stimulation. After menopause, a decrease in estrogen levels leads to an atrophy of the vaginal epithelium.

**COLOR** the following features of the uterus and vagina, using a different color for each feature (also see Plate 10-4):

- ☐ 1. **Fundus of the uterus**
- ☐ 2. **Body of the uterus**
- ☐ 3. **Cervix of the uterus**
- ☐ 4. **Vagina**
- ☐ 5. **Stratum basale (regenerates a new stratum functionale after menstruation) of the endometrium**
- ☐ 6. **Stratum functionale (thick surface layer that proliferates and is sloughed off during menstruation) of the endometrium**
- ☐ 7. **Uterine glands**

***Clinical Note:***

**Uterine prolapse** may occur when the support structures of the uterus, especially the cardinal (transverse cervical) ligaments, uterosacral ligaments, and levator ani muscle are weakened. Late reproductive and older age groups of women are most affected. Birth trauma, obesity, chronic cough, heavy lifting, and weakened support ligaments are risk factors.

**Cervical carcinoma** usually occurs near the external cervical os, where the epithelium changes from simple columnar to stratified squamous epithelium (the transformation zone). About 85% to 90% of cervical carcinomas are squamous cell carcinomas, whereas 10% to 15% are adenocarcinomas. Risk factors include early sexual activity, multiple sex partners, human papillomavirus (HPV), and smoking, with an age range usually between 40 and 60 years.

**Uterine fibroids (leiomyomas)** are benign tumors of smooth muscle and connective tissue cells of the myometrium of the uterus. The fibroids are firm and range in size from 1 to 20 cm. About 30% of all women may be affected, and 40% to 50% of women older than 50 years, making these benign tumors the most common in women.

**Endometrial carcinoma** is the most common malignancy of the female reproductive tract. It often occurs in women between the ages of 55 and 65 years, and risk factors include obesity (increased estrogen synthesis from fat cells without concomitant progesterone synthesis), estrogen replacement therapy without concomitant progestin, breast or colon cancer, early menarche or late menopause (prolonged estrogen stimulation), chronic anovulation, no prior pregnancies or breastfeeding, and diabetes.

**Hysterectomy** is an excision of the uterus and is performed either through the lower anterior abdominal wall or via the vagina.

A **cervical examination** and cytology (Papanicolaou [Pap] smear) is performed to obtain a sample of cervical cells for microscopic examination to determine if cervical cancer is present. Cervical cancer is the second most common cancer in women.

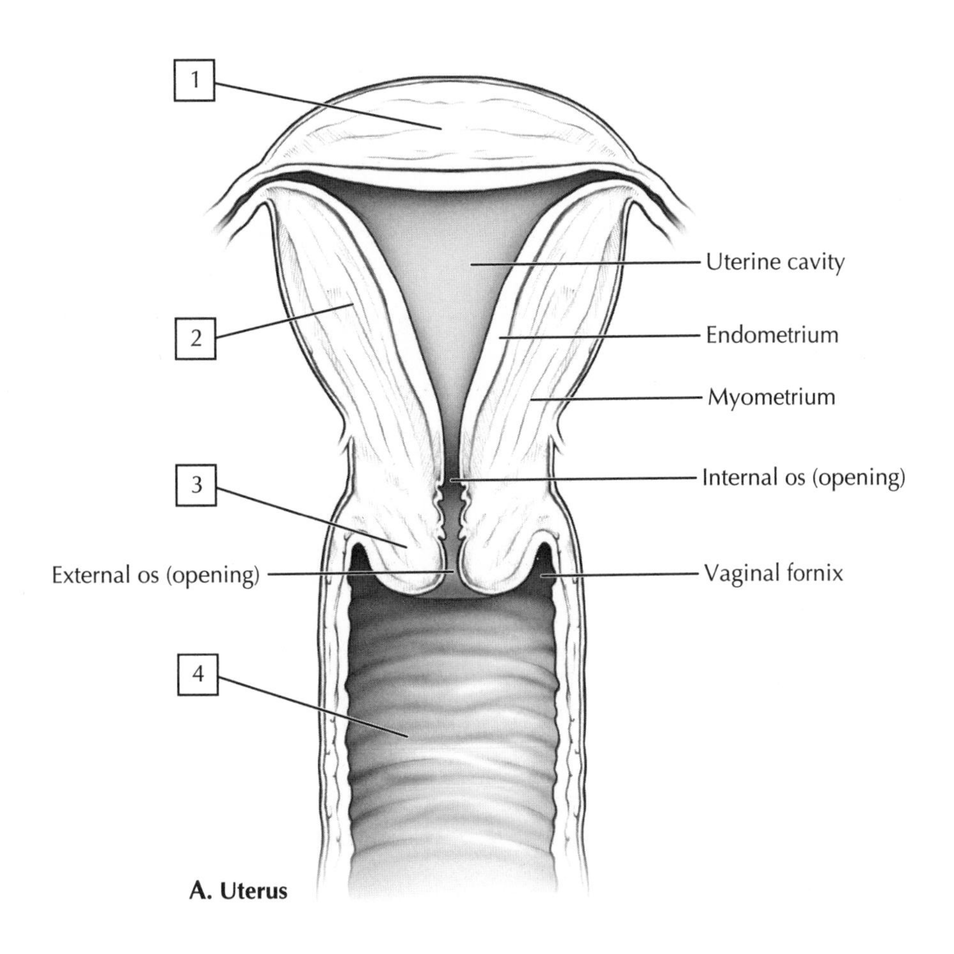


A. Uterus

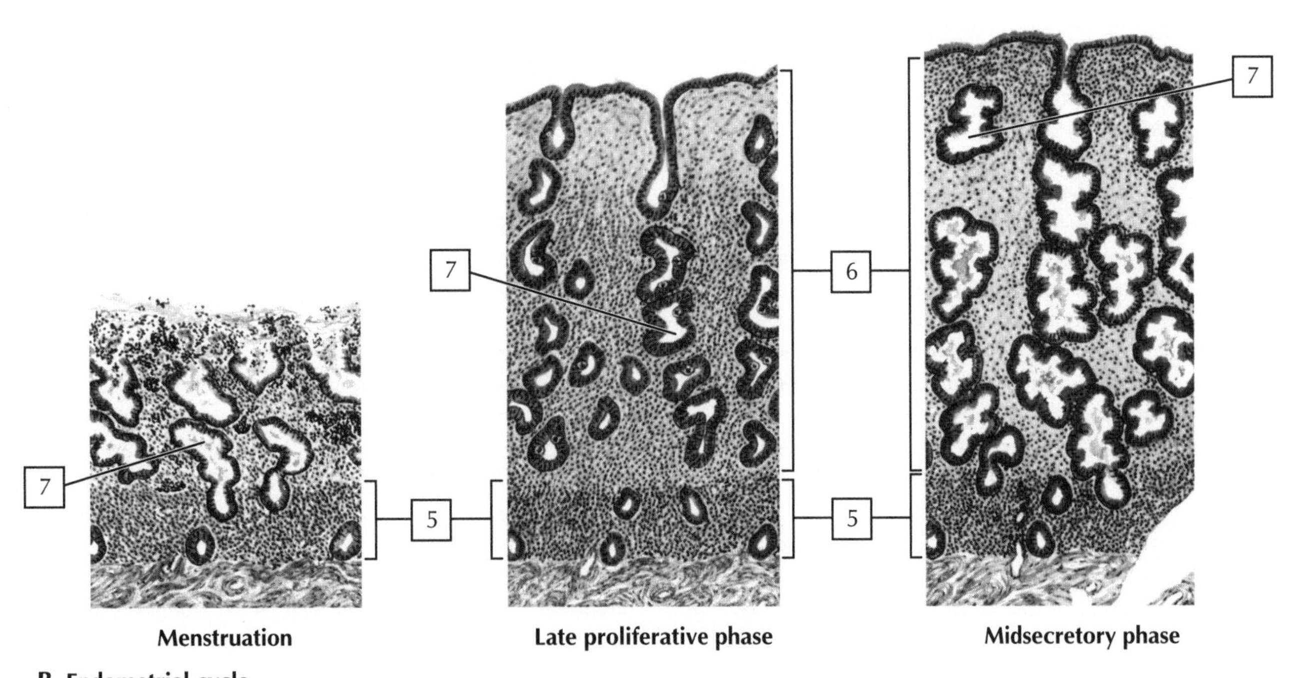


B. Endometrial cycle

# Menstrual Cycle

The menstrual cycle is a sequence of morphological and functional changes during a woman's 28-day cycle, assuming no pregnancy occurs. It is divided into four phases (slight variations in the days are common among women, and some textbooks do not include the Ischemic Phase):

1. **Menstrual Phase** (days 1-4 or 5): menstrual bleeding begins on day 1 with the necrosis and discharge of the functionalis layer of the uterine endometrium; menstrual blood loss averages about 35-50 ml
2. **Follicular or Proliferative or Estrogen Phase** (days 4-15): coincides with the proliferation of the granulosa cells (requires estrogen) in a selected follicle up to ovulation, and the rapid regeneration and repair of the endometrial lining of the uterus; the uterine endometrium thickens from about 0.5 mm to 2-3 mm in thickness
3. **Luteal or Secretory Phase** (days 15-26 or 27): happens midcycle around day 15 and coincides with surges of luteinizing hormone (LH) and follicle-stimulating hormone (FSH) that induce stimulation of ovulation of the oocyte by the mature Graafian follicle; the follicular cells transform into the corpus luteum and produce large amounts of progesterone, estrogen, and inhibin (negative feedback on the hypothalamus to inhibit gonadotropin-releasing hormone [GnRH]; LH and FSH also participate in this feedback)
4. **Ischemic Phase** (day 28): if fertilization does not occur, the corpus luteum degenerates beginning around the 25th day of the cycle and menses commence on the 28th day, and a new menstrual cycle begins again with the menstrual phase

During the follicular phase, the rising levels of estrogen feed back onto the hypothalamus and pituitary to increase the surge of GnRH that is followed by the LH and FSH peaks during the ovulatory phase. If fertilization does not occur, the corpus luteum degenerates beginning around the 25th day of the cycle and menses commence after the 28th day, as the new menstrual cycle begins again.

If fertilization and implantation occur, then the plasma levels of estrogen and progesterone continually increase, with estrogen stimulating the myometrium growth and progesterone inhibiting uterine contractility so the fetus can reach term (9 months) before birth. The **corpus luteum** is responsible for the secretion of these hormones during the first 2 months, under the stimulation of **human chorionic gonadotropin (hCG)** secreted by the trophoblast cells of the implant. After about 60 to 80 days, the placenta takes over and secretes the estrogen and progesterone necessary to maintain the pregnancy.

The menstrual cycle also results in changes in the uterine endometrium and includes the following phases:

- **Menstrual**: lasts about 4 to 5 days and marks the beginning of the cycle when the endometrium degenerates (because no implantation has occurred) and is sloughed off as the menstrual flow (this stage is seen in the Uterine Cycle image around the 28th day; labeled "Bleeding" and extending to the 4th day)
- **Proliferative**: from about day 5 to 14, when the endometrium thickens tremendously; this growth is stimulated by estrogen
- **Secretory**: after ovulation, the endometrium increases its secretory activity (nutrient-rich mucus) under the influence of progesterone ("promotes gestation"), becomes edematous, and thickens in anticipation of a possible implantation

**COLOR** the following features of the menstrual cycle, using the colors suggested for each feature:

- ☐ **1. Corpus luteum (yellow with a red center)**
- ☐ **2. Veins and venous lakes of the endometrium (blue)**
- ☐ **3. Spiral arteries of the endometrium during the cycle (red)**
- ☐ **4. LH levels (line in table) (orange)**
- ☐ **5. FSH levels (brown)**
- ☐ **6. Progesterone levels (blue)**
- ☐ **7. Estrogen levels (green)**
- ☐ **8. Inhibin levels (purple)**

***Clinical Note:***

Approximately 10% to 15% of infertile couples may benefit from various **assisted reproductive strategies,** including:

- Artificial insemination: use of a donor's sperm
- GIFT: gamete intrafallopian transfer
- IUI: intrauterine insemination (with a partner's sperm or a donor's sperm)
- IVF/ET: in vitro fertilization with embryo transfer into the uterine cavity
- ZIFT: in vitro fertilization with zygote transfer into the fallopian tube

**Menorrhagia** is the abnormal bleeding from the uterus resulting in a heavy menstrual flow. Causes may include a hormonal imbalance, fibroids, endometrial polyps, or uterine cancer.

**Endometriosis** is not uncommon and is characterized by the appearance of endometrial tissue in unusual locations in the lower abdomen. It affects females between puberty and menopause, but is most common between the ages of 20 and 30. Symptoms include pelvic pain and premenstrual bleeding. This condition often subsides after menopause, when estrogen levels decrease.

**Ovarian** cysts are fluid-filled sacs that usually arise from epithelial components of the ovary, primarily from Graafian follicles, and usually are benign and asymptomatic. Women in their reproductive-age period are most commonly affected.

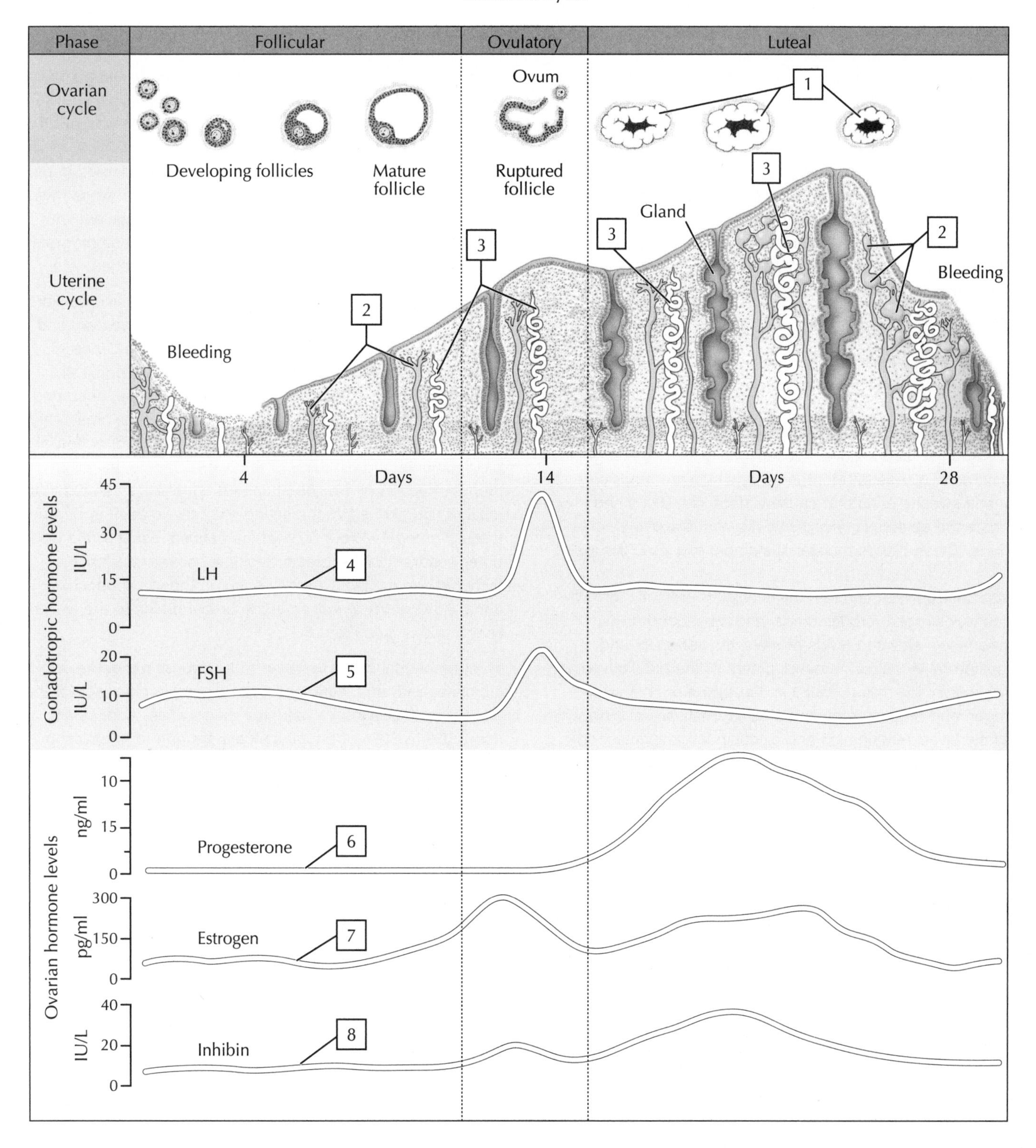
Menstrual Cycle
Phase
Follicular
Ovulatory
Luteal
Ovarian cycle
Ovum
1
Developing follicles
Mature follicle
Ruptured follicle
3
Gland
3
3
2
Bleeding
Uterine cycle
2
Bleeding
4
Days
14
Days
28
Gonadotropic hormone levels
45
30
IU/L
15
0
LH
4
20
FSH
5
IU/L
10
0
Ovarian hormone levels
10
ng/ml
15
0
Progesterone
6
300
pg/ml
150
0
Estrogen
7
40
IU/L
20
0
Inhibin
8

# Female Breast

The female breast extends from approximately the second to the sixth ribs and from the sternum medially to the midaxillary line laterally. The mammary gland tissue lies in the **superficial fascia**. Histologically, it is really a modified sweat gland that develops under hormonal influence. It is supported by strands of fibrous tissue called the **suspensory ligaments** (of Cooper). The nipple usually lies at approximately the fourth intercostal space and is surrounded by the pigmented **areola**. The glandular architecture includes the following features:

- **Secretory alveoli**: cells in the lobules of tubuloalveolar glands release "milk" via merocrine mechanisms (in which the protein secretory product is released by exocytosis) and apocrine mechanisms (in which the fatty component of the secretion is released in membrane-enclosed droplets)
- **Intralobular ducts**: collect the alveolar secretions and convey them along to interlobular ducts
- **Interlobular ducts**: coalesce into about 15 to 25 lactiferous ducts
- **Lactiferous ducts**: drain the milk toward the nipple and exhibit dilated segments just deep to the nipple called **lactiferous sinuses**, before opening on the nipple surface

The **areola** is the circular pigmented skin surrounding the nipple and contains modified sebaceous glands, sweat glands, small areolar glands (of Montgomery), which are modified apocrine sweat glands that cause surface elevations and lubricate the epidermis, along with numerous sensory nerve endings. These glands moisten the nipple and keep it supple.

Breast development is under the control of prolactin, growth hormone, estrogen, progesterone, and adrenocorticoids. In pregnancy, elevated levels of prolactin, estrogen, and progesterone increase the development of the tubuloalveolar glands but inhibit milk production. **Lactation** occurs when estrogen and progesterone levels fall dramatically at birth while prolactin levels remain high and oxytocin levels increase to stimulate milk release. In the absence of pregnancy or suckling (active nursing), the tubuloalveolar glands regress and become inactive. After menopause, the glandular tissue largely atrophies and is replaced by fat, although some of the lactiferous ducts may remain.

**COLOR** the following features of the female breast, using a different color for each feature:

- ☐ 1. **Areola**
- ☐ 2. **Nipple**
- ☐ 3. **Lactiferous ducts**
- ☐ 4. **Lactiferous sinuses**
- ☐ 5. **Fatty subcutaneous tissue**
- ☐ 6. **Gland lobules**

***Clinical Note:***

**Fibrocystic change (disease)** is a general term covering a large group of benign conditions that occur in about 80% of women and are often related to cyclic changes in the maturation and involution of the glandular tissue. **Fibroadenoma**, the second most common tumor of the breast after carcinoma, is a benign neoplasm of the glandular epithelium. It usually affects reproductive-age women 20-35 years of age, although most cases are in women younger than 30. These tumors are well-circumscribed palpable masses that occur as a single lump in one breast or multiple lumps in both breasts. Estrogen stimulation increases their size, while they regress after menopause. Both conditions can present with palpable masses, which also can be seen with mammography or ultrasonography, and warrant follow-up evaluation.

**Breast cancer** (usually ductal carcinoma or invasive lobular carcinoma) is the most common malignancy in women, and women in the United States have the highest incidence in the world. Approximately two-thirds of all cases occur in postmenopausal women. The most common type, occurring in about 75% of cases, is an infiltrating ductal carcinoma, which may involve the suspensory ligaments, causing retraction of the ligaments and dimpling of the overlying skin. Invasion and obstruction of the subcutaneous lymphatics can result in dilation and skin edema, creating an "orange peel" appearance (*peau d'orange*). About 60% of the cancers occur in the upper outer quadrant of the breast (region closest to the axilla). Lymphatic metastases usually occur in the axilla, because about 75% of the lymph from the breast drains to the axillary lymph nodes (see Plate 6-7).

Most breast cancers are linked to **hormonal exposure,** which increases with age, early onset of menarche, older age of the first full-term pregnancy, and late menopause. Additionally, about 5% to 10% of breast cancers are due to a familial or genetic link mutation in the autosomal dominant (BRCA1 and BRCA2) tumor suppressor genes.

Several options are available to **treat breast cancer**, including chemotherapy, hormonal therapy, immunotherapy, and "local" approaches such as radiation therapy or surgery. In a partial mastectomy, also called lumpectomy or quadrantectomy, the surgeon performs a breast-conserving surgery that removes the tumor along with a halo or margin of normal tissue.

**Radical mastectomy** involves removal of the whole breast (total) along with the axillary lymph nodes, fat, and chest wall muscles. A modified radical mastectomy is where the whole breast is removed along with most of the axillary and pectoral lymph nodes. A total (simple) mastectomy involves removal of the whole breast with or without some axillary lymph nodes.

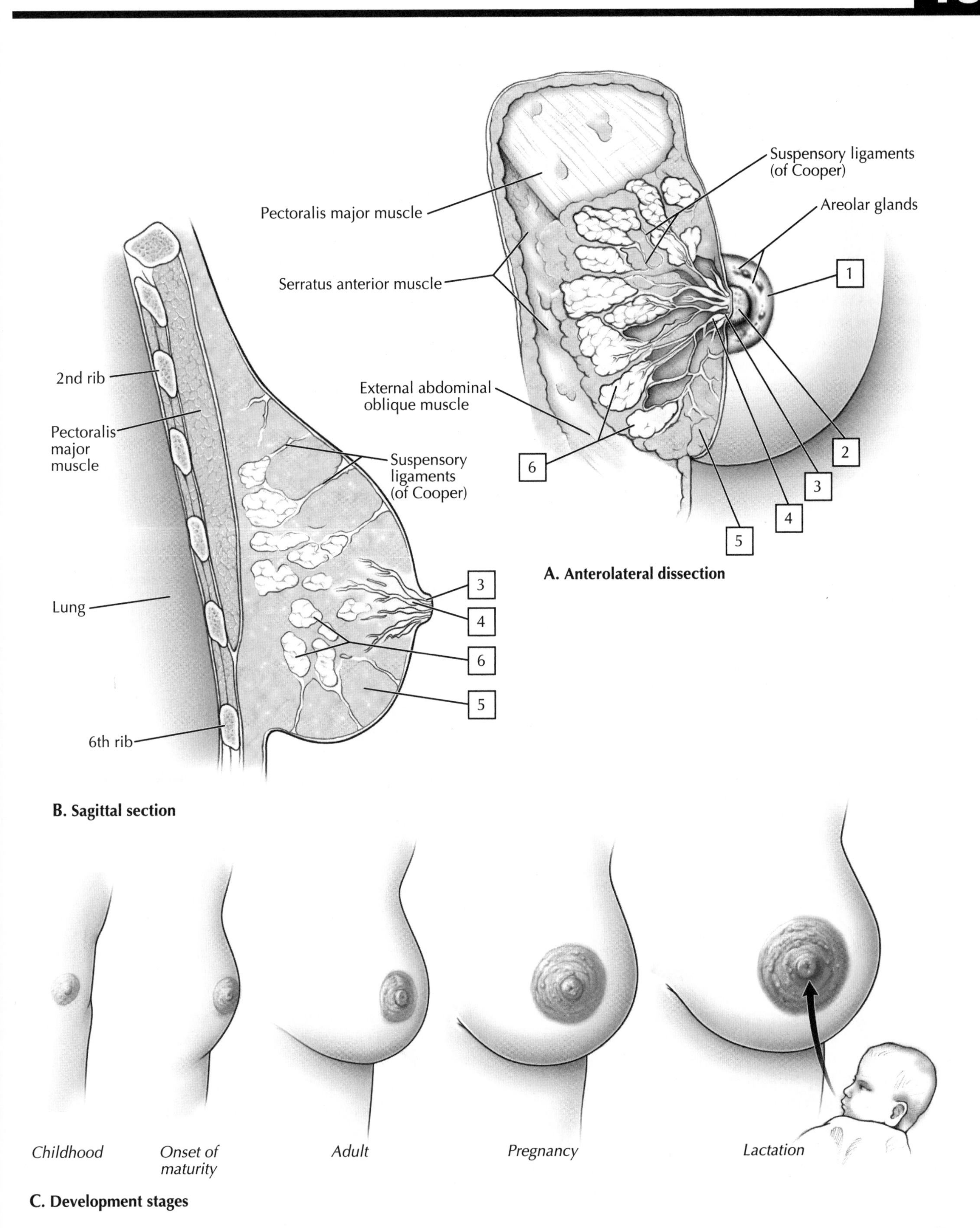


A. Anterolateral dissection

B. Sagittal section

C. Development stages

# Overview of the Male Reproductive System

The male reproductive system is composed of the following structures:

- **Testes**: the paired gonads of the male reproductive system; they are egg shaped and about the size of a chestnut, produce the male germ cells called **spermatozoa**, and reside in the scrotum (externalized from the abdominopelvic cavity)
- **Epididymis**: a convoluted tubule that receives the spermatozoa and stores them as they mature (the convoluted tubule stretched out in length is almost 23 feet long!)
- **Ductus (vas) deferens**: a muscular (smooth muscle) tube about 40 to 45 cm long that conveys sperm from the epididymis to the ejaculatory duct (seminal vesicle)
- **Seminal vesicles**: paired tubular glands that lie posterior to the prostate, are about 15 cm long, produce seminal fluid, and join the ductus deferens at the **ejaculatory duct**
- **Prostate gland**: a walnut-sized gland that surrounds the urethra as it leaves the urinary bladder and produces prostatic fluid that is added to semen (the sperm are suspended in glandular secretions)
- **Urethra**: a canal that passes through the prostate gland, enters the penis, and conveys the semen for expulsion from the body during ejaculation; also conveys urine from the urinary bladder to the outside

The male reproductive viscera are summarized in the following table (see Plate 10-7 for more detail).

| FEATURE | CHARACTERISTICS |
|---|---|
| Testes | Develop in retroperitoneal abdominal wall and descend into scrotum |
| Epididymis | Consists of head, body, and tail; functions in maturation and storage of sperm |
| Ductus (vas) deferens | Begins in the tail of the epididymis, ascends in the spermatic cord through inguinal canal to join duct of seminal vesicle to form the ejaculatory duct |
| Seminal vesicles | Secrete alkaline seminal fluid (see below for a list of contents) |
| Prostate gland | Surrounds prostatic urethra and secretes prostatic fluid (see below) |

The pelvic extent of the ductus deferens, the seminal vesicles, and the prostate gland lie deep to the peritoneum of the male pelvis. The peritoneum reflects off of the pelvic walls and passes over the superior aspect of the bladder and onto the anterior and lateral aspects of the lower rectum (see part *A*). The trough formed by this peritoneal reflection between the bladder anteriorly and the rectum posteriorly is called the **rectovesical pouch**, and it is the lowest extent of the abdominopelvic peritoneal cavity in the male (in the sitting or standing position). Fluids in the peritoneal cavity would eventually collect in this lowest point within the male peritoneal cavity, especially if sitting upright or standing.

The seminal vesicles produce a viscous, alkaline fluid (about 70% of the seminal fluid in semen) that helps to both nourish the spermatozoa and protect them from the acidic environment of the female vagina. Seminal fluid contains fructose as a metabolic substrate for the sperm, but in addition produces simple sugars, amino acids, ascorbic acid, and prostaglandins.

The prostate produces about 20% of the semen (glandular secretions plus the spermatozoa from the testes) and consists of a thin, milky, slightly alkaline (pH 7.29) secretion that helps to liquefy the rather coagulated semen after it is deposited in the female vagina. The prostatic secretion also contains citric acid, proteolytic enzymes, various ions (calcium, sodium, potassium, etc.), fibrolytic enzymes, prostatic acid phosphatase (PAP), and prostate-specific antigen (PSA). Each ejaculation contains about 2 to 5 ml of semen, has a pH of about 7 to 8, and normally contains about 100 million spermatozoa per ml. It is estimated that about 20% of sperm in the ejaculate are morphologically abnormal and about an equal number are immobile!

Small bulbourethral glands (Cowper's glands) are located posterolateral to the membranous urethra and secrete a mucous-like secretion that lubricates the urethra (see Plates 3-16 and 10-8).

**COLOR** the following features of the male reproductive system, using a different color for each feature:

- ☐ 1. **Ductus deferens**
- ☐ 2. **Testis**
- ☐ 3. **Epididymis**
- ☐ 4. **Prostate gland**
- ☐ 5. **Seminal vesicle**

***Clinical Note:***

**Benign prostatic hypertrophy** (BPH) is fairly common and usually occurs in aging males (90% of men over the age of 80 will have some BPH). It is caused by hyperplasia of the glandular and stromal cells of the prostate. This growth can lead to symptoms that may include urinary urgency, decreased stream force, frequency of urination, and nocturia (frequent nighttime urges to urinate).

**Prostatitis** is an inflammation of the prostate, frequently caused by various strains of bacteria that come in a reflux of urine from the urethra.

**Prostate cancer** is the second most common visceral cancer in males (lung cancer is first) and the second leading cause of death in men older than 50 years. Seventy percent of the cancers arise in the posterolateral portion of the gland (adenocarcinomas) and are palpable by digital rectal examination.

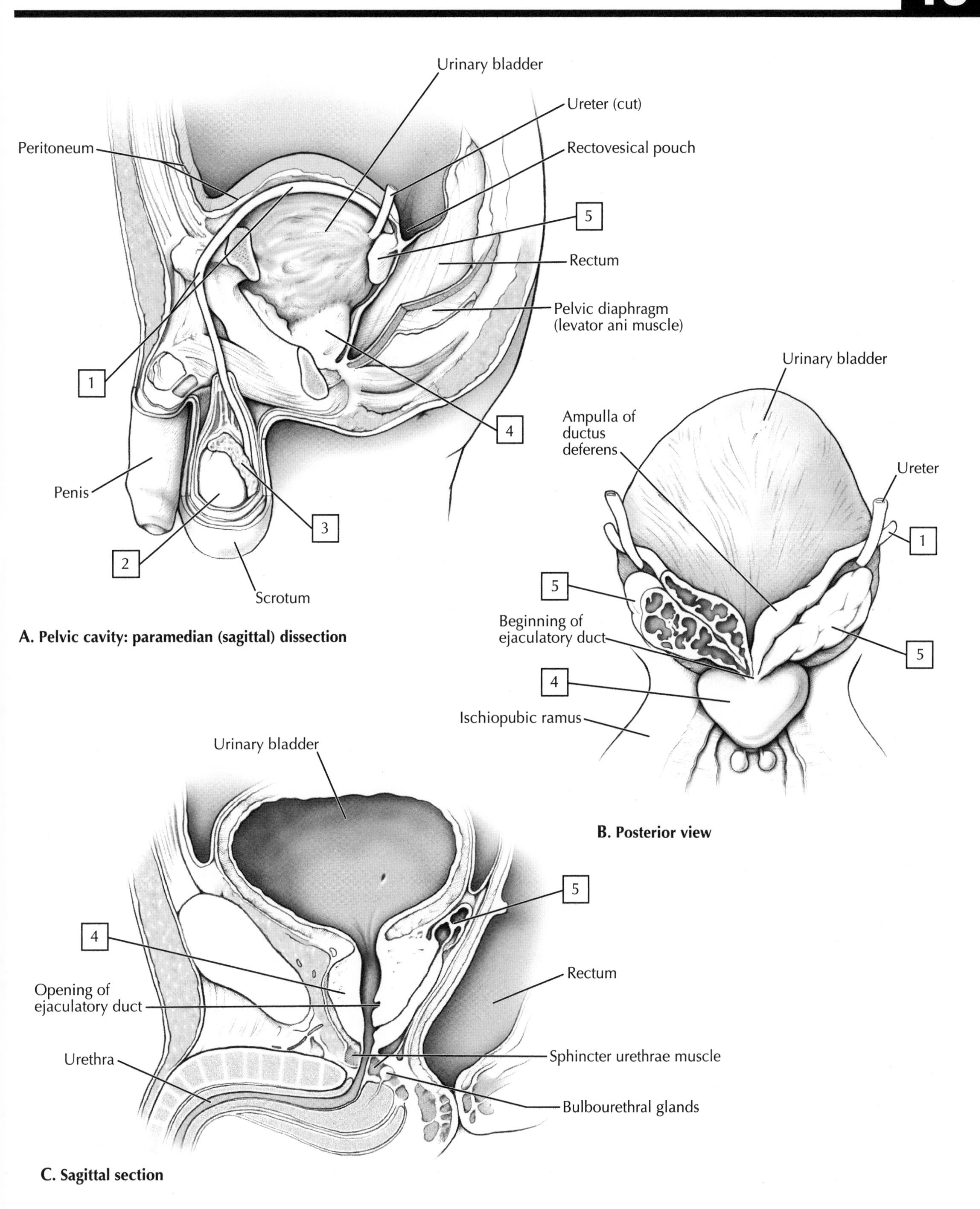


A. Pelvic cavity: paramedian (sagittal) dissection

B. Posterior view

C. Sagittal section

# Testis and Epididymis

The testes develop retroperitoneally, high in the posterior abdominal wall, and, like the ovaries, descend during fetal growth into the pelvic cavity. But, rather than remaining there, they continue their descent through the inguinal canal and into the scrotum. The testes are exteriorized because spermatogenesis (formation of spermatozoa) occurs optimally at a slightly lower temperature than the core body temperature (37° C). The testes also produce androgens (male hormones).

Each testis is encased within a thick capsule (tunica albuginea) and divided into lobules that contain **seminiferous tubules** and interstitial connective tissue that includes **Leydig cells,** which produce testosterone. During early fetal life, Leydig cells secrete testosterone and insulin-like protein 3 (INSL3) that stimulates descent of the testes during embryonic development. Adult Leydig cells continue to secrete testosterone to maintain spermatogenesis and also secrete oxytocin, which is necessary for transport of the spermatozoa toward the efferent ductules and the epididymis. The seminiferous tubules are lined with the germinal epithelium that gives rise to two types of cells: spermatogenic cells and Sertoli cells. **Spermatogenic cells** will ultimately form spermatozoa. **Sertoli cells** provide structural support, offer metabolic and nutritional support, and help form the **blood-testis barrier**, which prevents autoimmune responses from the lymphatic system from affecting the germ cells.

Spermatogenesis involves meiotic divisions that produce spermatids, according to the following sequence of differential events:

- **Spermatogonia**: stem cells that line the basal (outer) layer of the seminiferous tubule germinal epithelium undergo mitotic division to produce primary spermatocytes
- **Primary spermatocytes**: (destined to produce 4 sperm) are large germ cells that possess 46 chromosomes and undergo meiosis I to produce two secondary spermatocytes (1n) (which possess 23 chromosomes: 22 autosomes and either an X or a Y chromosome)
- **Secondary spermatocytes**: these cells are smaller than primary spermatocytes and undergo a second meiotic division very quickly to produce 4 spermatids (which contain 23 single chromosomes; the haploid number of chromosomes, [1n])
- **Spermatids**: these cells undergo a maturation process (called spermiogenesis) to form a head and tail and become spermatozoa, which then pass from the lumen of the seminiferous tubules to the epididymis for storage and maturation

**COLOR** the following germinal epithelial cells of the seminiferous tubule, using a different color for each cell:

- ☐ 1. **Leydig cells (interstitial cells that produce testosterone)**
- ☐ 2. **Spermatozoa**
- ☐ 3. **Spermatid**
- ☐ 4. **Secondary spermatocytes**
- ☐ 5. **Primary spermatocyte**
- ☐ 6. **Spermatogonia (basal stem cells)**
- ☐ 7. **Sertoli (support) cell**

The route of transfer of the immature spermatozoa from the testis to the epididymis includes the following pathway:

- **Tubulus rectus**: a straight tubule leading from the lobule's apex to the mediastinum (middle space) testis and its labyrinthine rete testis
- **Rete testis**: a network of anastomosing tubules that transfer the spermatozoa quickly to the efferent ductules
- **Efferent ductules**: about 10 or more tortuous ducts lined with ciliated epithelium that move the spermatozoa into the head of the epididymis and its single, highly convoluted duct, which is about 23 feet long and ultimately joins the proximal end of the ductus deferens

**COLOR** the following features of the testis and epididymis, using a different color for each feature:

- ☐ 8. **Ductus (vas) deferens**
- ☐ 9. **Epididymis (head, body, and tail)**
- ☐ 10. **Lobules (of seminiferous tubules)**
- ☐ 11. **Tunica albuginea (the thick "white" capsule of the testis)**
- ☐ 12. **Rete testis (in the mediastinum testis)**
- ☐ 13. **Efferent ductules**

***Clinical Note:***

**Testicular cancer** is characterized by a heterogeneous group of neoplasms. About 95% of them arise from the germ cells of the seminiferous tubules, and all of these are malignant. The peak age of incidence is 15 to 34 years. Sertoli cell and Leydig cell tumors are relatively uncommon and are more often benign.

**Orchitis** is inflammation of one or both testes, usually caused by a viral (mumps virus), bacterial (sexually transmitted pathogens, e.g., gonorrhea, chlamydia), or fungal infections.

**Spermatogenesis** can be compromised by a variety of factors including dietary deficiencies (vitamins), developmental disorders, systemic diseases or local infections, elevated testicular temperature, some medications, toxic agents (pesticides, chemicals in plastics), and ionizing radiation.

**Vasectomy** offers birth control with a low failure rate compared to the pill, condoms, intrauterine devices, and female tubal ligation. It can be performed as an office procedure with local anesthetic.

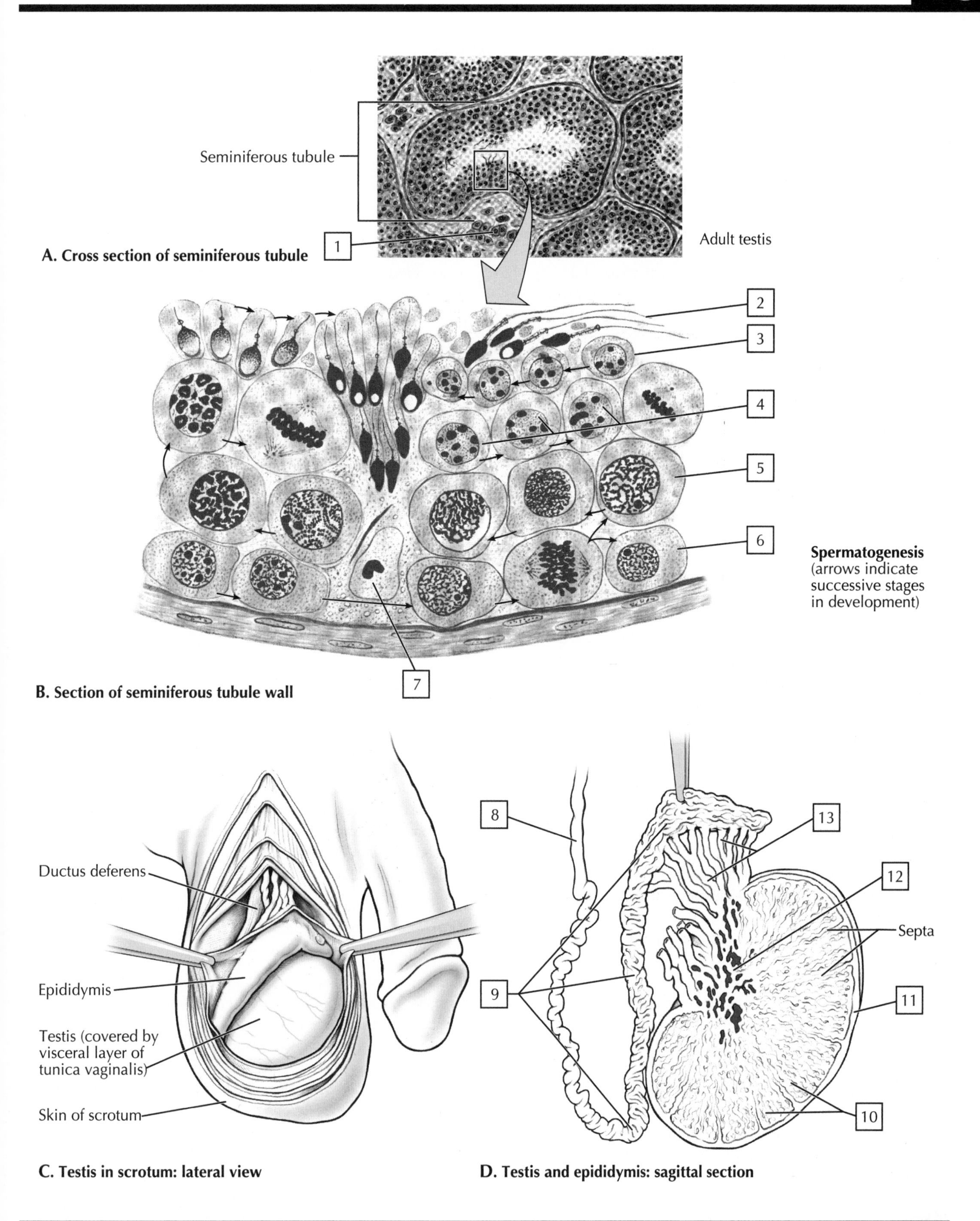


A. Cross section of seminiferous tubule

B. Section of seminiferous tubule wall

C. Testis in scrotum: lateral view

D. Testis and epididymis: sagittal section

# Male Urethra and Penis

## Urethra

The male urethra is about 20 cm long and descriptively is divided into three parts:

- **Prostatic urethra**: proximal portion of the male urethra that runs through the prostate gland
- **Membranous urethra**: short, middle portion that is enveloped by the external urethral sphincter (skeletal muscle)
- **Spongy (penile, cavernous) urethra**: courses through the bulb of the penis, the pendulous portion of the penis, and the glans penis to open at the external urethral orifice

As the prostatic urethra leaves the urinary bladder, it is surrounded by a sphincter of smooth muscle, the **internal urethral sphincter**. This sphincter is under sympathetic nervous control and closes off the urethra at the neck of the urinary bladder during ejaculation so that semen cannot pass superiorly into the bladder or urine in the bladder cannot gain access to the urethra. The membranous urethra also is surrounded by a sphincter, the **external urethral sphincter**, which is skeletal muscle and innervated by branches of the **pudendal nerve** (somatic control) (see Plate 3-16). Males have voluntary control of this sphincter.

The proximal portion of the spongy urethra receives the openings of two small glands, the **bulbourethral (Cowper's) glands**, which reside in the external urethral sphincter (deep transverse perineal muscle). These pea-sized glands secrete a clear, viscous alkaline mucus. Before ejaculation, these glands lubricate the lumen of the spongy urethra and neutralize its acidic environment, thus preparing the way for the semen.

## Penis

The penis provides a common outlet for urine and semen and is the copulatory organ in the male. It is composed of three bodies of erectile tissue:

- **Corpora cavernosa**: two lateral erectile bodies that begin along the ischiopubic ramus and meet at about the level of the pubic symphysis to form the posterior columns of the pendulous portion of the penis
- **Corpus spongiosum**: a single erectile body of tissue that begins in the midline of the perineum (bulb of the penis) and joins with the corpora cavernosa to form the anterior aspect of the pendulous portion of the penis (contains the spongy urethra)

The proximal portion of each of these cavernous bodies (the parts residing in the perineum) is covered by a thin layer of skeletal muscle (ischiocavernosus and bulbospongiosus muscle; see Plate 3-16), but the distal two-thirds of the three erectile bodies are wrapped in a dense connective tissue fascial sleeve **(Buck's fascia)**. The corpus spongiosum contains the spongy urethra and possesses less erectile tissue, so as not to obstruct the flow of semen during ejaculation by compressing the urethral lumen. Erection is achieved by parasympathetic stimulation, which relaxes the smooth muscle of the arterial walls supplying the erectile tissue and allows the flow of blood to engorge the erectile tissue sinuses. The erection compresses the veins, thus keeping the blood in the cavernous sinuses to maintain erection.

The spongy urethra passes into a dilated region called the navicular fossa within the glans penis and then terminates at the external urethral orifice. Along its length, the spongy urethra has openings for small urethral mucous glands (of Littré), which lubricate the urethral lumen. The glans penis is covered by an elastic prepuce (foreskin) that covers all or most of the glans penis (not shown in the image).

**COLOR** the following features of the male urethra and penis, using a different color for each feature:

- ☐ 1. **Prostatic urethra**
- ☐ 2. **Membranous urethra**
- ☐ 3. **Bulbourethral glands**
- ☐ 4. **Spongy urethra**
- ☐ 5. **Corpora cavernosa**
- ☐ 6. **Corpus spongiosum**
- ☐ 7. **Deep (Buck's) fascia of the penis (in cross section)**

***Clinical Note:***

**Erectile dysfunction** (ED) is an inability to achieve and/or maintain penile erection sufficient for sexual intercourse. Its occurrence increases with age and may be caused by a variety of factors, including:

- Depression, anxiety, and stress disorders
- Spinal cord lesions, multiple sclerosis, or prior pelvic surgery
- Vascular factors such as atherosclerosis, high cholesterol levels, hypertension, diabetes, smoking, and medications used to control these factors
- Hormonal factors

Available drugs to treat ED target the smooth muscle of the penile arteries, causing them to relax so that blood may pass easily into the cavernous sinuses.

**Circumcision** involves a surgical excision of the prepuce and exposes the glans penis. This is the most common minor surgical procedure done on males, usually at the request of the parents, but also is a religious practice in Islam and Judaism, and practiced by some aboriginal peoples in Africa and Australia. Medically, circumcision makes hygiene easier, reduces the incidence of urinary tract infections, reduces the risk of sexually transmitted infections, and decreases the risk of penile cancer (although this type of cancer is rare).


See Netter: *Atlas of Human Anatomy*, 7th Edition, Plate 363, 364, and 367.

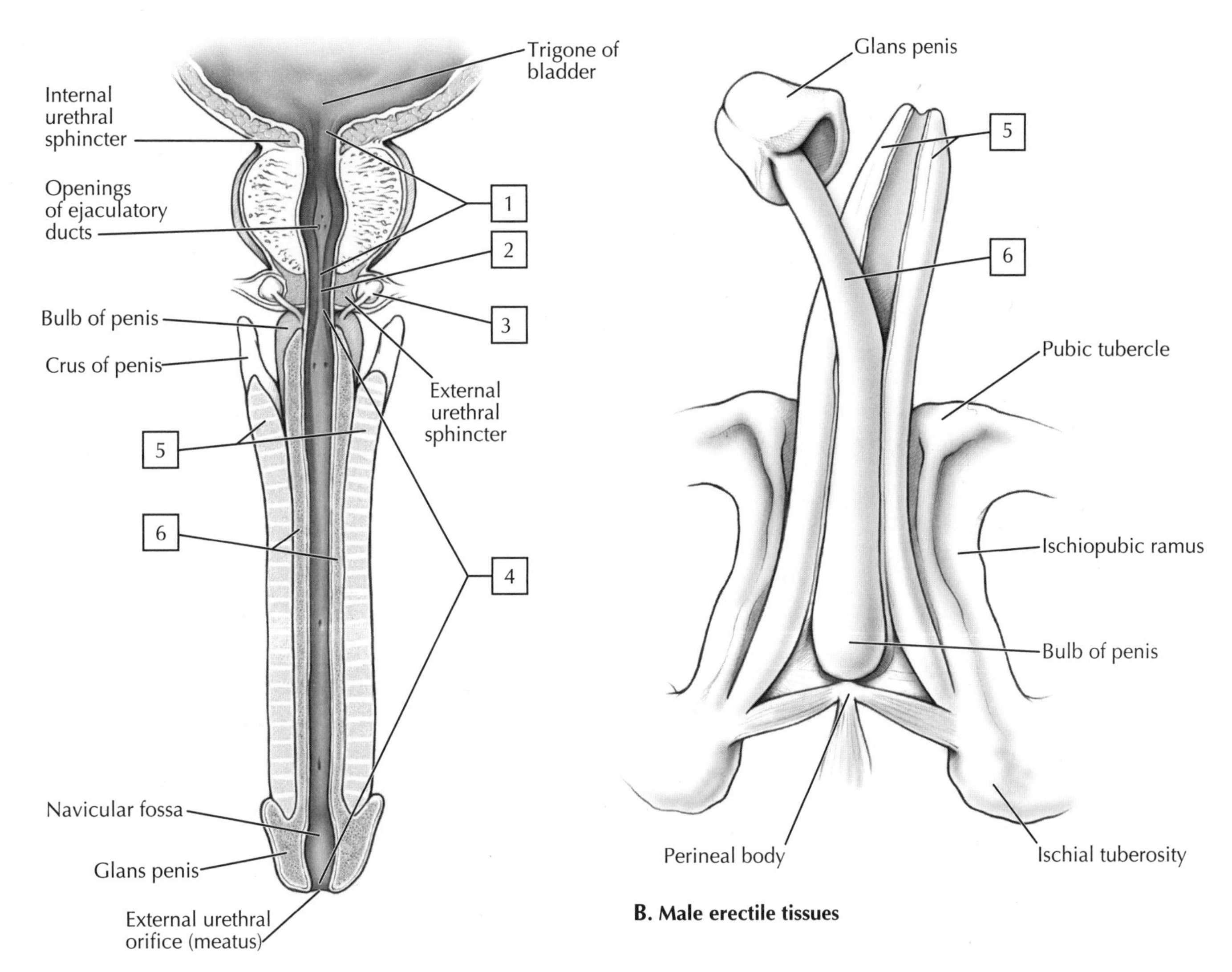


**A. Floor of urethra**

**B. Male erectile tissues**

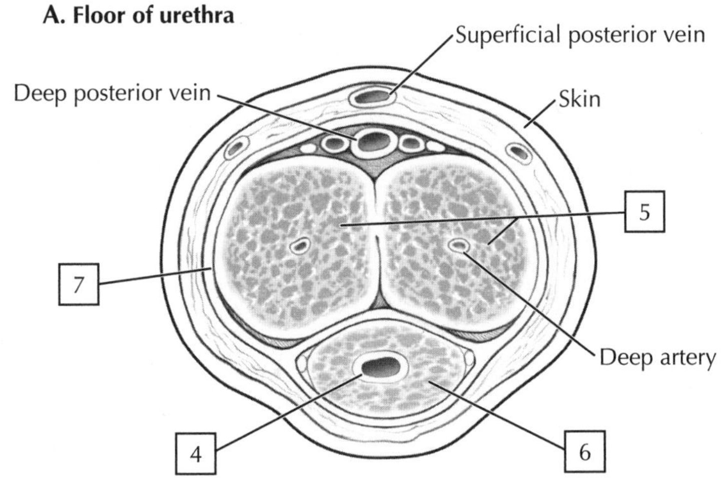


**C. Section through body of penis**

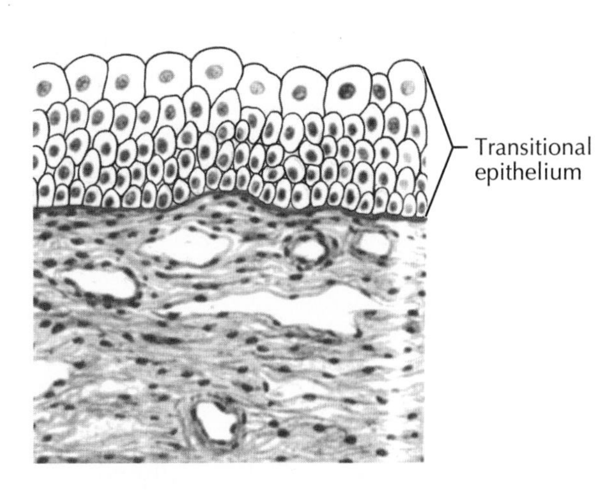


**D. Transitional epithelium in prostatic and membranous urethra**

## REVIEW QUESTIONS

1. Fertilization of the human ovum normally occurs in which of the following sites?
   A. Ampulla of the uterine tube
   B. Fimbriated portion of the uterine tube
   C. Fundus of the uterus
   D. Intramural portion of the uterine tube
   E. Isthmus of the uterine tube

2. While the interplay of all essential hormones is important in reproduction, which of the following is most important in maintaining a pregnancy?
   A. Estrogen
   B. FSH
   C. Inhibin
   D. LH
   E. Progesterone

3. Infertility in a 23-year-old man appears to be related to a lack of testosterone. Which of the following cells may be responsible for this condition?
   A. Leydig cells
   B. Seminiferous tubule cells
   C. Sertoli cells
   D. Spermatids
   E. Spermatogonial cells

**For each statement below (4-6), color the relevant feature or structure in the image.**

4. The ovaries are tethered to the uterus by this structure.
5. The vulva is demarcated by this hairless fold of tissue.
6. This portion of the uterus is often involved in cancer, and its epithelium can be easily assessed and monitored clinically by a routine Pap smear.

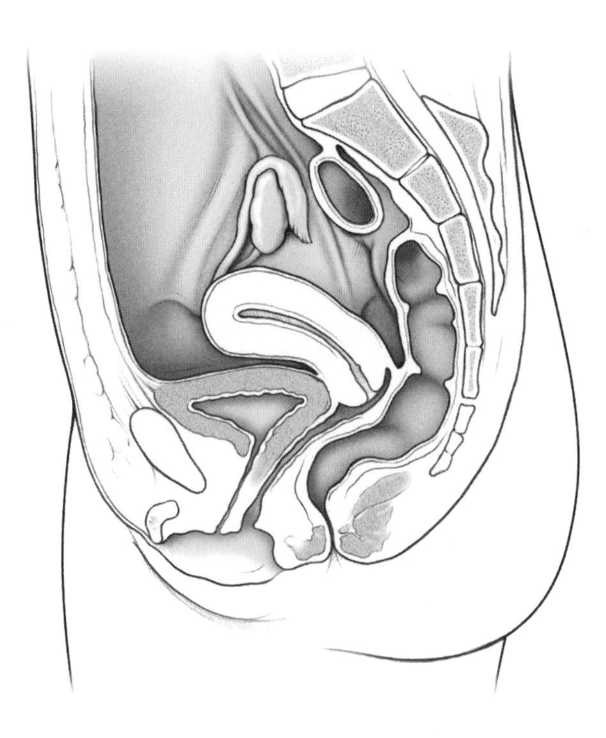

7. If the ovum is fertilized and implants in the uterine wall, it is maintained hormonally during the first 2 or 3 months by this structure. ___________________________
8. Sperm undergo their final maturation in this structure. ________________________
9. Which male structure accounts for about 70% of the volume of the ejaculate?_________________________________
10. The penile urethra is found in this erectile body of tissue. ____________________

## ANSWER KEY

1. A
2. E
3. A
4. Ligament of ovary
5. Labia minora
6. Cervix of uterus

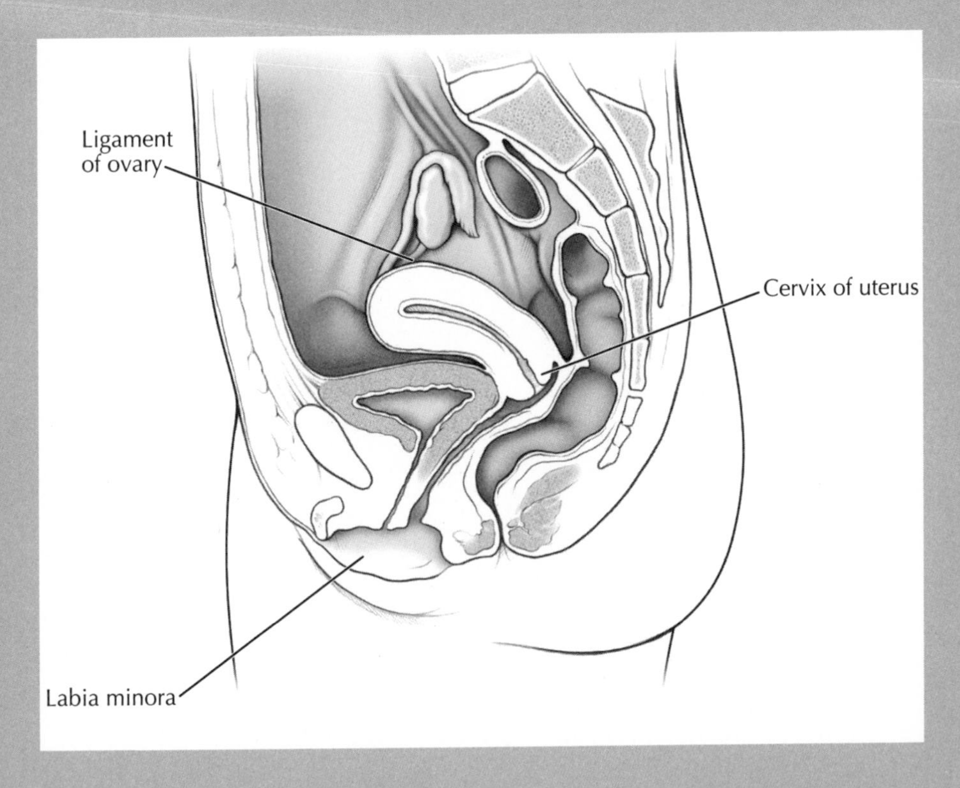


7. Corpus luteum in the ovary
8. Epididymis
9. Seminal vesicles
10. Corpus spongiosum

# Chapter 11 Endocrine System

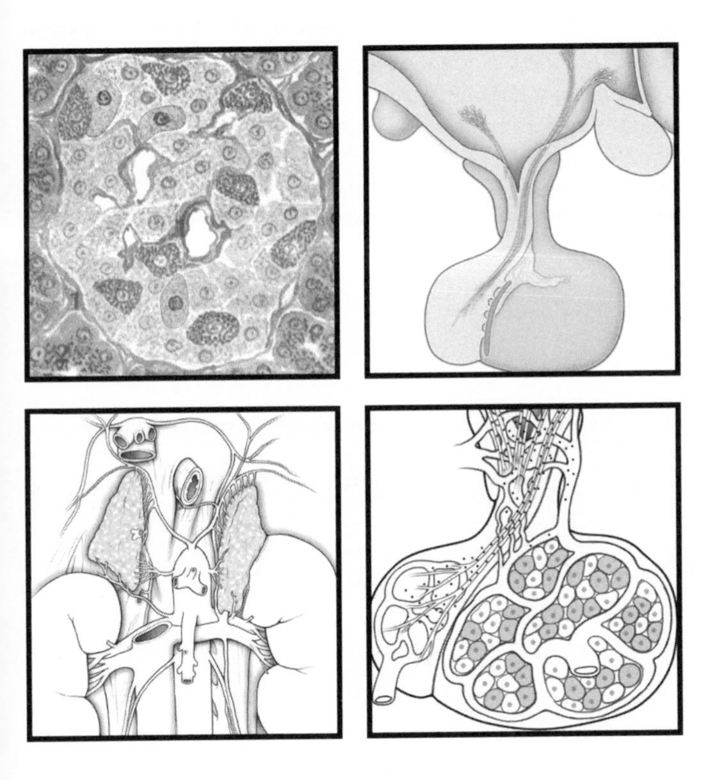

# 11 Overview

The endocrine system, along with the nervous and immune systems, facilitates communication, integration, and regulation of many of the body's functions. Specifically, the endocrine system interacts with target sites (cells and tissues), many a great distance away, by releasing hormones into the bloodstream. Generally speaking, the endocrine glands and hormones share several additional features:

- Secretion is controlled by **feedback mechanisms**
- Hormones bind target receptors on cell membranes or within cells (cytoplasmic or nuclear)
- Hormone action may be slow to appear but may have long-lasting effects
- Hormones are chemically diverse molecules (amines, peptides and proteins, steroids)

Hormones have a variety of ways of affecting cells or tissues (see part ***B***):

- **Autocrine**: a cell produces a hormone and is also affected by that hormone
- **Paracrine**: a hormone affects a cell directly adjacent to the cell that produced it or near the cell that produced it
- **Endocrine**: a hormone is secreted into the bloodstream or lymphatic system and affects a cell or tissues in another part of the body
- **Neurocrine**: a hormone affects nerves or is influenced by nerves

Major hormones and the tissues responsible for their release are summarized in the table below.

Additionally, the **placenta** releases human chorionic gonadotropin (HCG), estrogens, progesterone, and human placental lactogen (HPL), whereas other cells release a variety of growth factors. The endocrinology of the reproductive system is discussed in Chapter 10 (Plates 10-1 and 10-4 show hormones in women, and Plate 10-7 shows hormones in men).

Actually, there are many other hormones, and the table below covers only the major ones! As you can appreciate, the effects of the endocrine system are widespread and critically important in regulating bodily functions.

**COLOR** the major endocrine organs listed in the table (#1-14), using a different color for each organ/tissue and noting the major hormone(s) secreted by each organ or tissue. Trace the arrows in red at the bottom of the diagram to note the pathway that a hormone follows in affecting a cell.

**SUMMARY OF THE MAJOR HORMONES**

| | TISSUE/ORGAN | HORMONE |
|---|---|---|
| 1 | Hypothalamus | Antidiuretic hormone (ADH), oxytocin, thyrotropin-releasing hormone (TRH), corticotropin-releasing hormone (CRH), growth hormone–releasing hormone (GHRH), gonadotropin-releasing hormone (GnRH), somatostatin, prolactin-inhibiting factor (dopamine) |
| 2 | Pineal gland | Melatonin |
| 3 | Anterior pituitary | Adrenocorticotropic hormone (ACTH), thyroid-stimulating hormone (TSH), growth hormone (GH), prolactin, follicle-stimulating hormone (FSH), luteinizing hormone (LH), melanocyte-stimulating hormone (MSH) |
| 3 | Posterior pituitary | Oxytocin, vasopressin (antidiuretic hormone, ADH) |
| 4 | Thyroid gland | Thyroxine ($T_4$), triiodothyronine ($T_3$), calcitonin, interleukins, interferons |
| 5 | Parathyroid glands | Parathyroid hormone (PTH) |
| 6 | Thymus gland | Thymopoietin, thymulin, thymosin, thymic humeral factor |
| 7 | Heart | Atrial natriuretic peptide (ANP) |
| 8 | Digestive tract | Gastrin, secretin, cholecystokinin (CCK), motilin, gastric inhibitory peptide (GIP), glucagon, somatostatin, vasoactive intestinal peptide (VIP), ghrelin |
| | Liver | Insulin-like growth factors (IGFs), leptin, and many more |
| 9 | Adrenal glands | Cortisol, aldosterone, androgens, epinephrine, norepinephrine |
| 10 | Pancreatic islets | Insulin, glucagon, somatostatin, VIP, pancreatic polypeptide |
| 11 | Kidneys | Erythropoietin (EPO), calcitriol, renin, urodilatin |
| 12 | Fat | Leptin |
| 13 | Ovaries | Estrogens, progestins, inhibin, relaxin |
| 14 | Testes | Testosterone, inhibin |
| | White cells and some connective tissue cells | Various cytokines (interleukins, colony-stimulating factors, interferons, tumor necrosis factor [TNF]) |

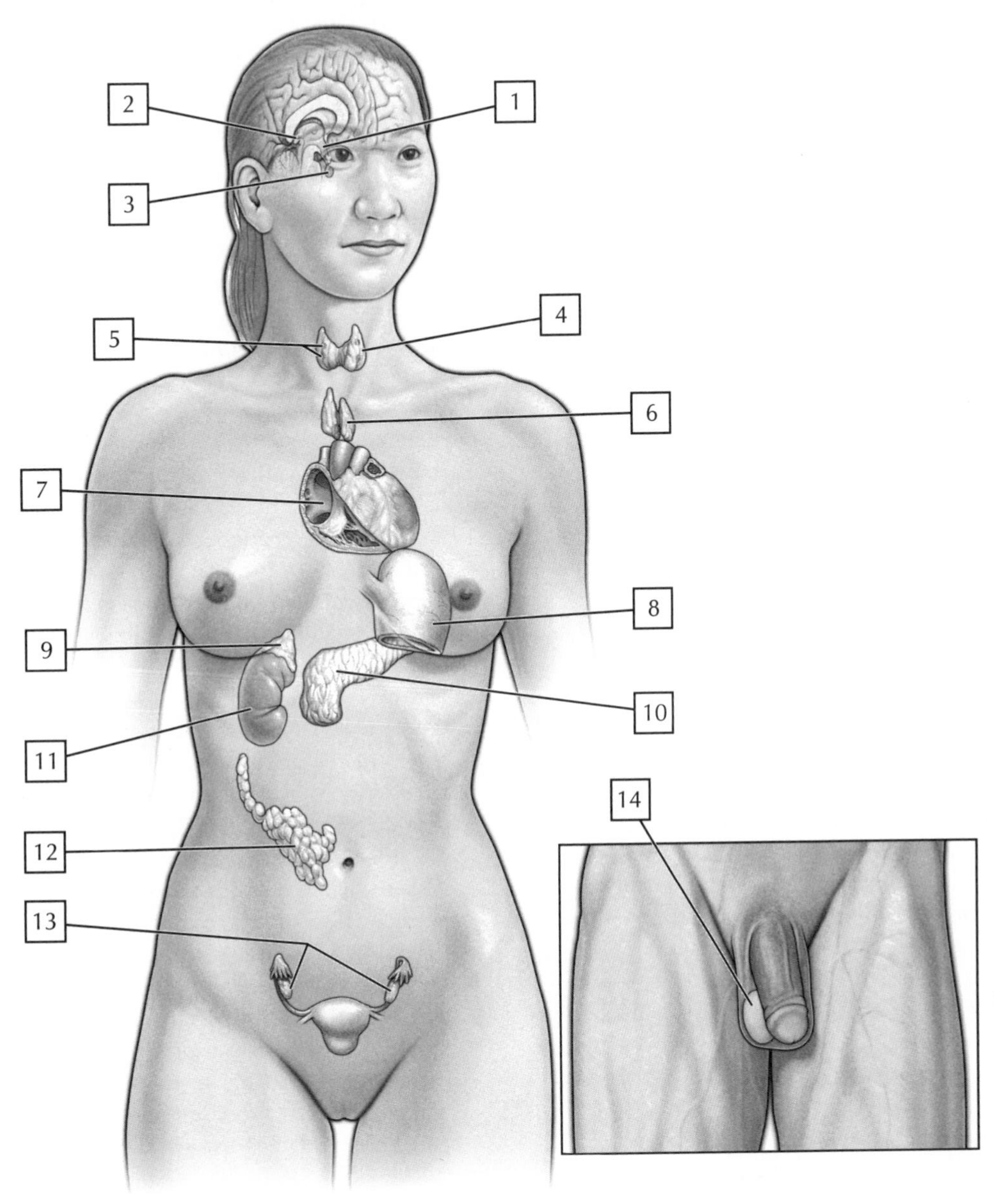


A. Overview of endocrine system

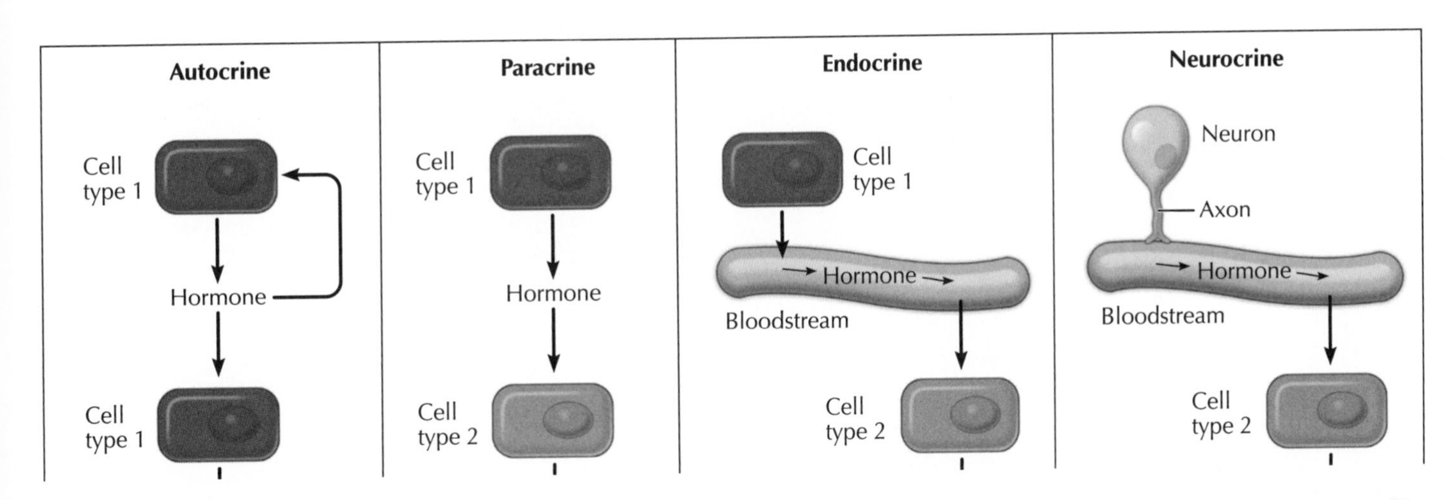

# 11 Hypothalamus and Pituitary Gland

## Hypothalamus

The hypothalamus (see Plate 4-11) is a portion of the diencephalon, along with the thalamus and epithalamus (pineal gland; reviewed previously in Plate 4-6). Functionally, the hypothalamus is very important in visceral control and homeostasis. Its neuroendocrine cells release hormones into the hypothalamic-hypophyseal portal system that stimulate or inhibit the secretory cells of the anterior pituitary gland. Neuroendocrine cells in the hypothalamus (paraventricular and supraoptic nucleus) also send axons into the posterior pituitary gland and median eminence, which really is a downgrowth of the brain's diencephalon. These axons release hormones into the systemic vasculature of the posterior pituitary, although it should be remembered that the hormones are synthesized and initially released from the hypothalamus.

## Pituitary Gland

The pituitary gland (hypophysis) lies within a bony seat or "Turkish saddle" called the sella turcica of the sphenoid bone and is connected to the overlying hypothalamus by a stalk called the **infundibulum** (see Plates 4-11 and 11-3). This pituitary stalk contains blood vessels and axons originating from several nuclei in the hypothalamus. The pituitary gland has three parts:

- **Anterior lobe**: also called the **adenohypophysis**, it is derived from an upward growth of the ectodermal tissue of the oropharynx (Rathke's pouch) and secretes seven different hormones (see table on Plate 11-1); unlike the posterior pituitary, the anterior pituitary is not directly connected to the hypothalamus but is connected to it by the vessels of the hypophyseal portal circulation
- **Posterior lobe**: also called the **neurohypophysis**, it is a neural extension of the hypothalamus that contains blood vessels and axonal terminals arising from the paraventricular and supraoptic nuclei of the hypothalamus; it releases two hormones, ADH and oxytocin
- **Intermediate lobe (pars intermedia)**: a very small intervening lobe between the anterior and posterior lobes that is poorly developed in humans; in humans its role is unclear but it does contain melanocyte-stimulating hormone (MSH) in frogs

**COLOR** the following features of the hypothalamus and pituitary gland, using a different color for each feature:

- ☐ **1. Cells and axons of the paraventricular nucleus of the hypothalamus**
- ☐ **2. Cells and axons of the supraoptic nucleus of the hypothalamus**
- ☐ **3. Cleft and connective tissue of the intermediate lobe**
- ☐ **4. Anterior pituitary**
- ☐ **5. Posterior pituitary**

***Clinical Note:***

**Endocrine diseases** are not uncommon and generally can be classified into four major categories:

1. **Hormone overproduction:** Usually this occurs from an increase in the number of endocrine cells. A good example of this is **hyperthyroidism** (Graves' disease), where abnormal antibodies mimic the action of thyroid-stimulating hormone (TSH), which significantly increases the number of thyroid cells.
2. **Hormone underproduction:** Diseases may destroy the endocrine organ, which can lead to underproduction of a hormone, e.g., **tuberculosis** of the adrenal glands, or **Hashimoto's disease,** where abnormal antibodies target and destroy thyroid hormone–secreting cells, or genetic abnormalities that affect normal development of an endocrine organ, as in **hypogonadism**.
3. **Altered tissue responses to hormones:** Genetic mutations in hormone receptors, as can occur in **diabetics,** where resistance to insulin in the muscles and liver is caused by altered signals originating from adipose (fat) tissue.
4. **Tumors of endocrine glands:** Most tumors of endocrine glands result in the overproduction of their hormone(s), such as occurs in **hyperthyroidism** (Graves' disease), where the excess synthesis and release of thyroid hormones can result in thyrotoxicosis and significantly increase tissue metabolism.

Fortunately, the release of hormones from the anterior lobe of the pituitary are carefully regulated by three different **regulatory mechanisms**:

(1) The pituitary is under significant control by the hypothalamus and its release of hypothalamic-regulating hormones into the **hypophyseal** portal veins
(2) **Paracrine** and **autocrine secretions** from cells within the pituitary gland itself
(3) **Feedback** from systemically circulating hormones provide negative feedback regulation of pituitary secretion

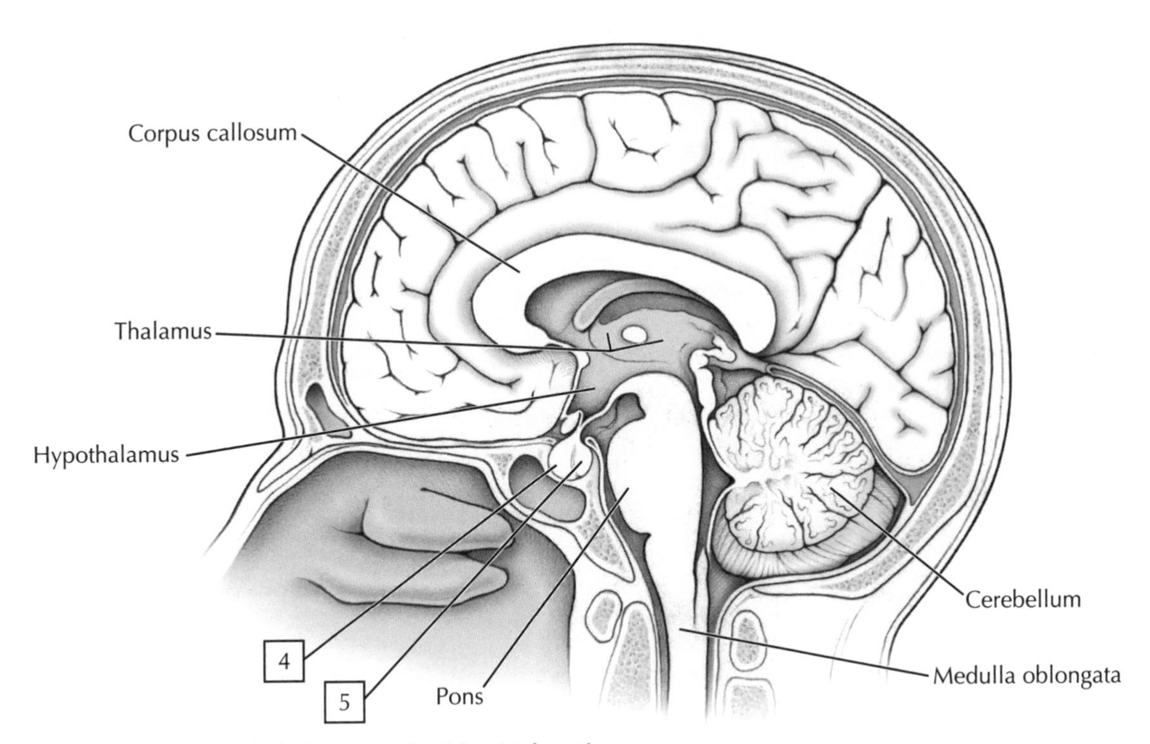


**A. Hypothalamus and pituitary gland: midsagittal section**

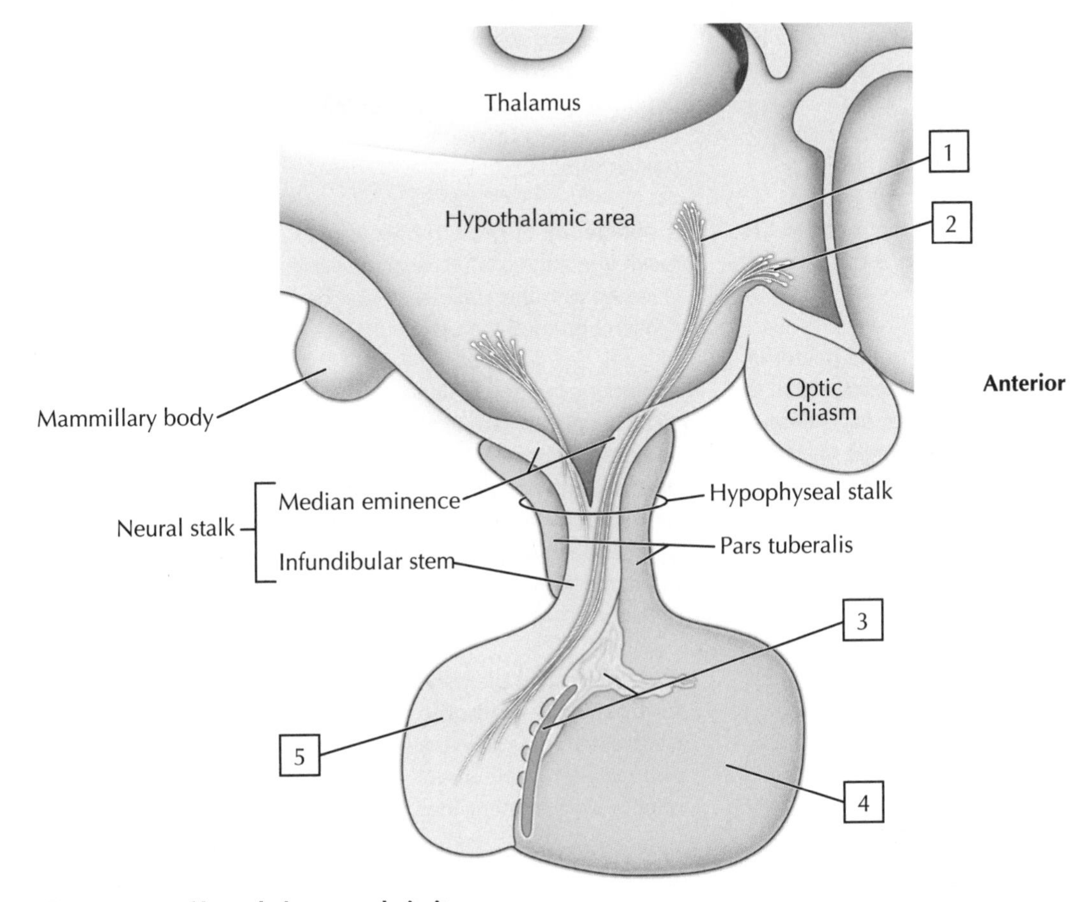


**B. Structure of hypothalamus and pituitary**

# Pituitary Hormones

The neuroendocrine cells of the hypothalamus release hormones into the hypothalamic-hypophyseal portal system that stimulate or inhibit the secretory cells of the anterior pituitary. These hormones include (abbreviations from table on Plate 11-1):

- **TRH (thyrotropin-releasing hormone)**: stimulates the release of TSH
- **CRH (corticotropin-releasing hormone)**: stimulates the release of ACTH
- **GHRH (growth hormone–releasing hormone)**: stimulates the release of GH
- **Somatostatin (somatotrophin release–inhibiting hormone)**: inhibits the release of GH
- **GnRH (gonadotropin-releasing hormone)**: stimulates the release of LH and FSH
- **Dopamine**: inhibits the release of prolactin

The cells of the anterior pituitary are of two primary types (based upon their histological staining characteristics, red or blue) and release the following hormones:

- **Somatotropes**: acidophilic cells (*stain red*) that secrete **GH**, which stimulates overall body growth, organ growth, increased lean body mass, and bone growth
- **Lactotropes (mammotropes)**: acidophilic cells (*stain red*) that secrete prolactin, which stimulates breast development and promotes milk production
- **Thyrotropes**: basophilic cells (*stain blue*) that secrete **TSH**, which stimulates the development and release of thyroxine from the thyroid gland
- **Corticotropes**: basophilic cells (*stain blue*) that secrete **ACTH**, which stimulates the adrenal cortex to release cortisol
- **Gonadotropes**: basophilic cells (*stain blue*) that secrete **LH** and **FSH**, which promote gamete production and hormone synthesis in the gonads

The axons that course from the hypothalamus to the posterior pituitary (neurohypophysis) can either store the hormones in the axon terminals until stimulated to release them or can release them immediately into the capillary system of the gland. Their release is controlled by neuronal and hormonal input on the hypothalamus. These hormones include:

- **Oxytocin**: stimulates milk expulsion ("let-down") from the breast (however, milk production is produced by prolactin) and uterine contractions during labor
- **ADH**: causes vasoconstriction and an increase in blood pressure (that is why ADH also is called **vasopressin**), and acts on the kidney to reabsorb water and help the body retain fluids

**COLOR** the following features of hormonal release from the pituitary gland, using the suggested colors for each feature:

- ☐ **1. Supraoptic and paraventricular neurons and their axons (purple)**
- ☐ **2. Acidophils of the anterior pituitary (red)**
- ☐ **3. Basophils of the anterior pituitary (blue)**
- ☐ **4. GH (arrow) targeting the liver (orange)**
- ☐ **5. TSH (arrow) targeting the thyroid gland (brown)**
- ☐ **6. ACTH (arrow) targeting the adrenal cortex (yellow)**
- ☐ **7. FSH (arrows) targeting the testis and ovary (blue)**
- ☐ **8. LH (arrows) targeting the testis and ovary (red)**
- ☐ **9. Prolactin (arrow) targeting the breast (green)**
- ☐ **10. Liver's release of insulin-like growth factors (IGFs) (pink)**

***Clinical Note:***

**Synthetic oxytocin** can be used clinically to artificially induce or stimulate the progression of labor (uterine contractions).

ADH (vasopressin) allows the kidneys to concentrate the urine, thus retaining water. An insufficiency of ADH secretion results in **diabetes insipidus**, a condition in which large volumes of hypotonic urine are excreted. Central diabetes insipidus is usually caused by trauma, disease, or surgery affecting the posterior pituitary.

A **deficiency of growth hormone** in prepubertal children can result in a short stature and may delay the onset of puberty. Children with this deficiency can be treated with recombinant growth hormone therapy.

**Prolactinoma** is a pituitary tumor (accounts for about 30% of all neoplastic pituitary tumors) that leads to amenorrhea, infertility, osteopenia, and galactorrhea in women, and erectile dysfunction and loss of libido in men. Smaller tumors can be treated with the dopamine agonist bromocriptine, which reduces the size of the tumor and inhibits prolactin secretion, but larger tumors must be surgically resected or treated with radiation.

ACTH-secreting pituitary adenomas are one of several causes for **Cushing's syndrome**, caused by elevated circulating corticosteroids. Clinical signs include abnormal adipose tissue deposition, muscle atrophy, hyperglycemia, and hypertension. It affects women more than men.

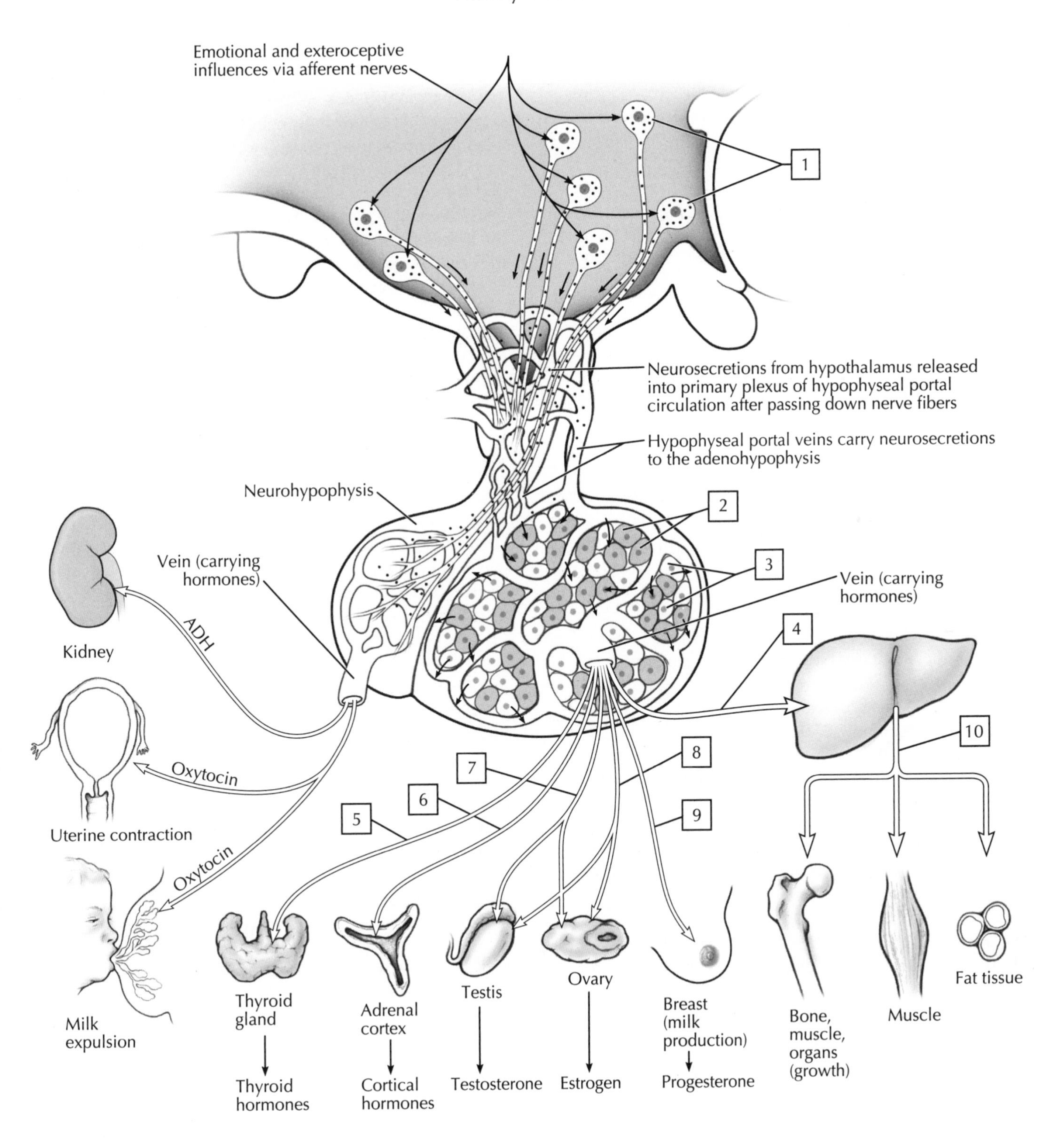
Pituitary Function
Emotional and exteroceptive influences via afferent nerves
1
Neurosecretions from hypothalamus released into primary plexus of hypophyseal portal circulation after passing down nerve fibers
Hypophyseal portal veins carry neurosecretions to the adenohypophysis
Neurohypophysis
2
3
Vein (carrying hormones)
Vein (carrying hormones)
4
ADH
Kidney
Oxytocin
Uterine contraction
Oxytocin
Milk expulsion
5
6
7
8
9
10
Thyroid gland
Thyroid hormones
Adrenal cortex
Cortical hormones
Testis
Testosterone
Ovary
Estrogen
Breast (milk production)
Progesterone
Bone, muscle, organs (growth)
Muscle
Fat tissue

# 11 Thyroid and Parathyroid Glands

## Thyroid

The thyroid gland is a **ductless endocrine gland** that weighs about 20 g and consists of right and left lobes joined by a midline isthmus. In about 50% of the population there is a small pyramidal lobe extending cranially from the gland. The thyroid lies anterior to the trachea and just inferior to the cricoid cartilage and, like most endocrine organs, has a rich vascular supply.

| FEATURE | CHARACTERISTICS |
|---|---|
| Lobes | Right and left, with a thin isthmus joining them |
| Blood supply | Superior and inferior thyroid arteries |
| Venous drainage | Superior, middle, and inferior thyroid veins |
| Pyramidal lobe | Variable (50% of time) superior extension of thyroid tissue |

The thyroid is composed of follicles formed by surrounding epithelial cells that synthesize, store, and secrete thyroxine ($T_4$, 90% of its secretion) and triiodothyronine ($T_3$). The follicular cells actively take up iodine to iodinate tyrosine molecules, forming $T_3$ and $T_4$, and storing them linked to thyroglobulin in the thyroid follicle (the thyroid gland is the only endocrine gland that stores its hormone to any significant degree). When stimulated by TSH from the anterior lobe of the pituitary gland, the thyroglobulin undergoes endocytosis and $T_3$ and $T_4$ are released into the bloodstream. $T_4$ is really a prehormone that is converted to the more active $T_3$ by the target tissues.

**Parafollicular cells** (C cells, labeled in part B) are located peripheral to the follicular cells and have no exposure to the follicle lumen. They secrete **calcitonin**, a hormone that regulates calcium metabolism and is a physiologic antagonist to **parathyroid hormone** (PTH).

Thyroxine ($T_4$) and $T_3$ have the following functions:

- Increase the metabolic rate of tissues
- Increase the consumption of oxygen
- Increase the heart rate, ventilatory rate, and renal function
- Are needed for growth hormone (GH) production and are especially important for central nervous system growth

**COLOR** the following features of the thyroid gland, using a different color for each feature:

☐ 1. **Superior thyroid arteries, from the external carotid artery, supplying the gland and inferior thyroid arteries from the subclavian artery**

☐ 2. **Internal jugular veins and their branches draining the thyroid gland**

☐ 3. **Common carotid arteries**

☐ 4. **Thyroid gland, the isthmus, and pyramidal lobe**

☐ 5. **Follicular cells surrounding a thyroglobulin-filled follicle**

## Parathyroid Glands

The parathyroid glands are paired superior and inferior glands located on the posterior aspect of the thyroid gland. Although there are usually four glands, their number and location can vary. The parathyroid glands secrete **PTH** in response to a decrease of calcium in the bloodstream. PTH acts on bone to cause resorption and release of calcium and phosphates into the bloodstream, and acts on the kidney to reabsorb calcium. PTH also alters vitamin D metabolism, which is critical for calcium absorption from the GI tract.

**COLOR** the following features of the parathyroid glands, using a different color for each feature:

☐ 6. **Parathyroids (superior and inferior pairs)**

☐ 7. **Target tissue sites (bone, kidney, small intestine)**

***Clinical Note:***

**Graves' disease**, an autoimmune disease, is the most common cause of hyperthyroidism in patients younger than 40 years, and affects women seven times more frequently than men. Excess synthesis and release of thyroid hormone results in thyrotoxicosis, which upregulates tissue metabolism.

**Hypothyroidism** is a disease in which the thyroid gland produces inadequate amounts of thyroid hormone to meet the body's needs Hashimoto's thyroiditis).

**Hyperparathyroidism** (about 85% are solitary benign adenomas) causes an increase in PTH and enhances calcium levels (hypercalcemia), leading to fatigue, constipation, polyuria, depression, skeletal pain, and nausea.

**Hypoparathyroidism** (PTH deficiency) usually results from direct trauma to the gland or resection of the gland during thyroid surgery. PTH also can be decreased if there is a dietary deficiency of magnesium, which is required for PTH secretion.

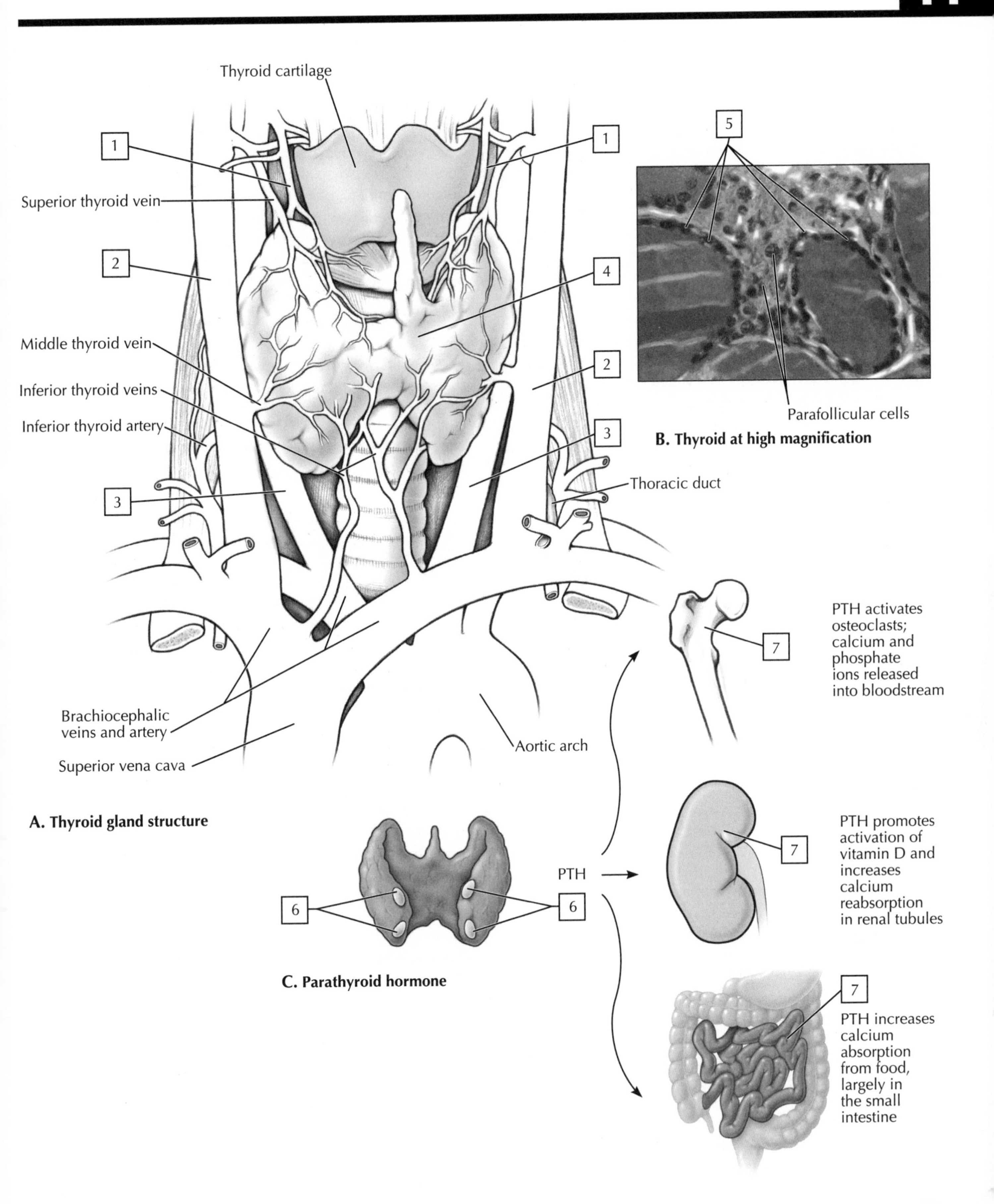

Thyroid cartilage
1
1
Superior thyroid vein
5
2
4
Middle thyroid vein
2
Inferior thyroid veins
Parafollicular cells
Inferior thyroid artery
3
B. Thyroid at high magnification
3
Thoracic duct
PTH activates osteoclasts; calcium and phosphate ions released into bloodstream
7
Brachiocephalic veins and artery
Aortic arch
Superior vena cava
A. Thyroid gland structure
PTH promotes activation of vitamin D and increases calcium reabsorption in renal tubules
7
PTH
6
6
C. Parathyroid hormone
7
PTH increases calcium absorption from food, largely in the small intestine

# Adrenal Glands

The paired adrenal (suprarenal) glands are **retroperitoneal ductless endocrine glands** that are nestled above the superior pole of each kidney, below the overlying diaphragm. Each gland normally weighs about 7 to 8 g, is highly vascularized, and consists of an outer **cortex** and an inner **medulla**. The right adrenal gland is often pyramidal in shape, and the left gland is semilunar in shape.

## Adrenal Cortex

Both the adrenal cortex and medulla are richly vascularized by a radially oriented plexus of vessels. The cortex produces more than two dozen steroid hormones and structurally is divided into three distinct histologic regions:

- **Zona glomerulosa**: the outer cortical region that lies just beneath the gland's capsule and produces mineralocorticoids, principally aldosterone
- **Zona fasciculata**: a middle region that produces glucocorticoids, principally cortisol (most important in humans), corticosterone, and cortisone
- **Zona reticularis**: the innermost cortical region that produces gonadocorticoids (adrenal androgens)

**Aldosterone** plays a critical role in regulating the extracellular fluid (ECF) compartment and blood volumes and in maintaining potassium balance. When the ECF compartment and blood volumes are reduced (e.g., from diarrhea or hemorrhage), renin is released from the kidney, which increases angiotensin II levels. Angiotensin II is a potent stimulator of aldosterone secretion, which then acts on sweat glands, salivary glands, the intestines, and kidneys to retain sodium and water in an effort to increase the ECF and blood volume.

**Cortisol** has both direct and indirect actions on a number of tissues, and is considered a hormone that is released during stress:

- Causes muscle wasting
- Fat deposition
- Hyperglycemia
- Insulin resistance
- Osteoporosis
- Immune suppression (antiinflammatory) and antiallergic actions
- Decreased connective tissue production, leading to poor wound healing
- Increased neural excitability
- Increased glomerular filtration rate (water diuresis), sodium retention, and potassium loss

**Adrenal androgens** play a role in puberty in both sexes and in females are the primary source of circulating androgens. They are responsible for the growth of pubic and axillary hair in women, whereas testicular testosterone does this in males. In general, the effects of androgens are anabolic, leading to increased muscle mass and bone formation. They also cause sebaceous gland hypertrophy (leading to acne), hairline recession, and growth of facial hair (think of the effects of anabolic steroid abuse by athletes).

## Adrenal Medulla

The medulla produces two hormones: epinephrine and norepinephrine. They have classically been thought of as neurotransmitters, but in this instance they are actually true hormones because they are released into the bloodstream. The cells of the adrenal medulla are actually the **postganglionic elements of the sympathetic division of the autonomic nervous system** (ANS) and produce the fight-or-flight response.

- **Epinephrine:** accounts for about 80% of the medullary secretions
- **Norepinephrine:** accounts for about 20% of the medullary secretions, but plays a larger role as a neurotransmitter in the ANS

**COLOR** the following features of the adrenal gland, using a different color for each feature:

- ☐ **1. Adrenal glands**
- ☐ **2. Capsule of the gland (inset)**
- ☐ **3. Zona glomerulosa (aldosterone) (inset)**
- ☐ **4. Zona fasciculata (cortisol) and its cells (inset)**
- ☐ **5. Zona reticularis (androgens) and its cells (inset)**
- ☐ **6. Medulla epinephrine and norepinephrine and its cells (inset)**

***Clinical Note:***

**Addison's disease**, or chronic adrenal cortical insufficiency, usually does not manifest itself until about 90% of the adrenal cortex has been destroyed. Manifestations include:

- Darkening of the hair
- Freckling of the skin; skin pigmentation
- Hypotension, hypoglycemia
- Loss of weight, fatigue, anorexia, vomiting, and diarrhea
- Muscular weakness

**Cushing's syndrome** is caused by any condition that results in high levels of blood cortisol. This may be due to a pituitary or ectopic ACTH-secreting tumor, hyperplasia, or a tumor of the adrenal gland (can also occur from the use of glucocorticoid drugs). Clinical features include:

- Red cheeks and a "moon" face
- Shoulder fat pads ("buffalo hump") and thin arms and legs
- Bruises and thin skin
- Osteoporosis, muscle wasting, hypertension
- Pendulous abdomen with red skin striae
- Poor wound healing

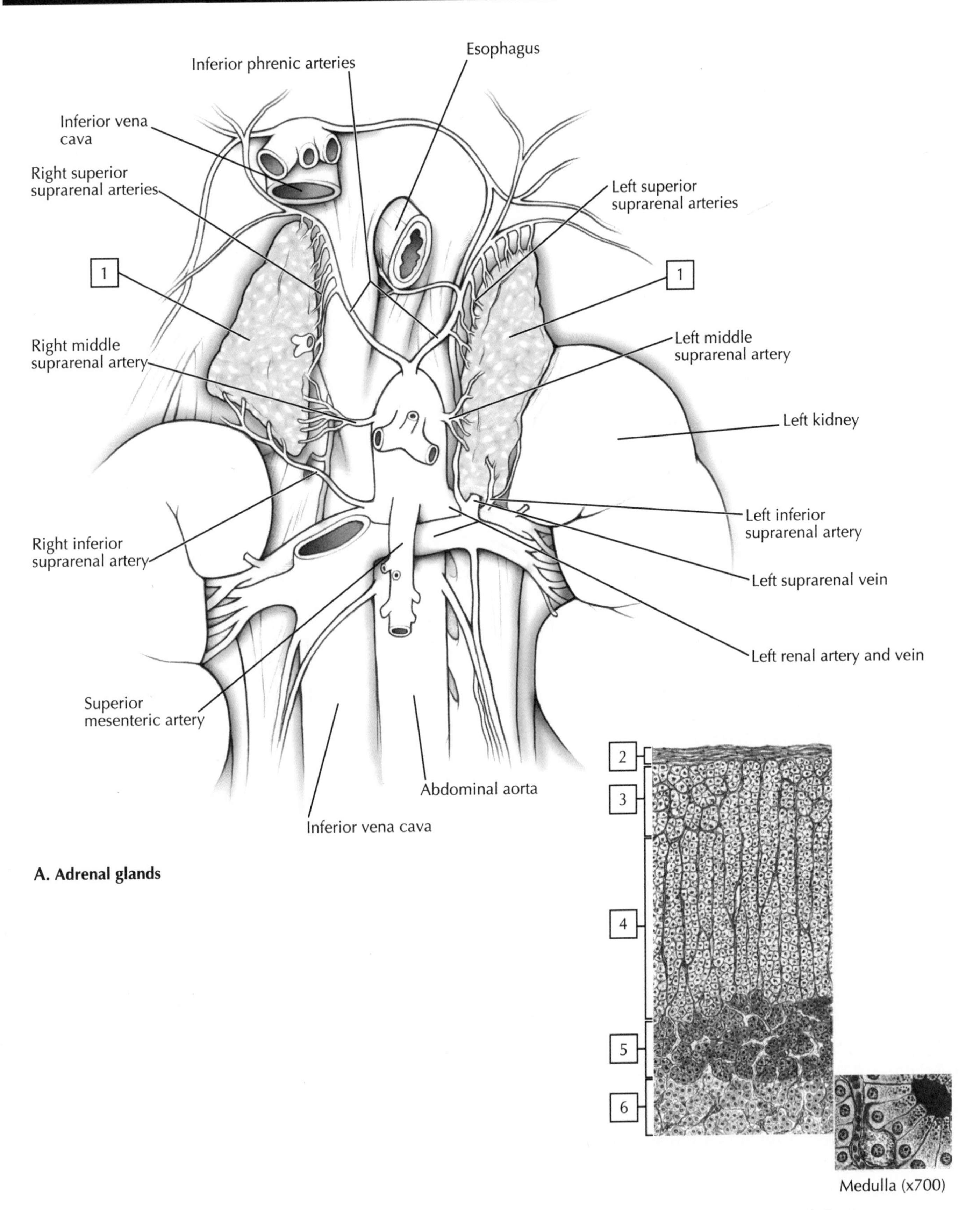


A. Adrenal glands

B. Normal human suprarenal gland

# Pancreas

The endocrine pancreas (the pancreas also is an important digestive exocrine gland) is represented by clusters of islet cells (of Langerhans), a heterogeneous population of cells responsible for the elaboration and secretion primarily of glucagon, insulin, and somatostatin. Several other hormones also are elaborated by the islets to a lesser extent, and include PP cells (pancreatic polypeptide, which stimulates gastric chief cells and inhibits bile secretion and intestinal motility) and epsilon cells (ghrelin, which stimulates appetite). Together these two hormones account for only about 5% of cells. The three main hormones are:

- **Glucagon**: secreted by the alpha cells (15% to 20% of cells)
- **Insulin**: secreted by the beta cells (65% to 70% of cells)
- **Somatostatin**: secreted by the delta cells (5% to 10% of cells)

**Glucagon** is a **fuel-mobilization hormone** that acts on the liver to break down glycogen and stimulates hepatic gluconeogenesis from amino acids. This results in an **increase in blood glucose concentration**. Glucagon also acts on adipose tissue to stimulate lipolysis and the release of fatty acids. The net effect of glucagon is that glucose, fatty acid, and keto acid levels in the bloodstream increase.

**Insulin** is a **fuel-storage hormone**. Insulin secretion increases in the presence of an increase in plasma glucose levels, especially after a meal. The major fuels of the body are glucose, fatty acids, and keto acids (derived from fatty acid metabolism). Insulin **stimulates the uptake of glucose** into cells, where it is stored in the form of glycogen (especially in the liver and muscle). Insulin also stimulates fat synthesis and inhibits lipolysis. Finally, insulin stimulates the uptake of amino acids into cells and their storage as protein. The net effect is that blood levels of glucose and keto acids are decreased.

Little is known about the role of **somatostatin** from the pancreas (secreted by delta cells). It may inhibit the release of insulin and glucagon secretion, and also suppresses the exocrine secretion of the pancreas.

**COLOR** the following features of the endocrine pancreas, using the colors suggested for each feature:

- ☐ **1. Pancreas (head, uncinate process, body, and tail) (green; see Plate 8-10)**
- ☐ **2. Delta cells (light blue) (somatostatin)**
- ☐ **3. Alpha cells (orange) (glucagon)**
- ☐ **4. Acini of the exocrine pancreas outside the islets (red)**
- ☐ **5. Beta cells (yellow) (insulin)**

***Clinical Note:***

**Diabetes mellitus (DM)** affects about 15 million people in the United States, and that percentage is probably an underestimate. There are two types of DM:

- **Type I:** insulin-dependent DM, in which insulin is absent or almost absent in the pancreatic islets because of the destruction of the islets by the body's immune system (autoimmune disease), thus requiring exogenous insulin administration; insulin is critical because most cells cannot take up glucose without it, but the exception is in the brain, liver, and exercising muscle, which have adequate normal glucose uptake due to the presence of GLUT2 transporters (and the upregulation of GLUT4 in exercising skeletal muscle).
- **Type II:** non–insulin-dependent DM, in which insulin is present in the plasma at normal or above normal levels but the target cells are hyporesponsive to the insulin because they have a reduction in insulin receptors; about 90% of DM is of the type II variety
  Uncontrolled hyperglycemia in Type II diabetes results in the following symptoms: polyuria, polydipsia (excessive thirst), and polyphagia (excessive eating).

Vascular complications account for about 80% of all deaths related to DM, and can include:

- **Retinopathy**: vascular microaneurysms and hemorrhages in vessels supplying the retina
- **Ischemic stroke**: cerebrovascular thrombosis, often from plaques that rupture in the carotid or cerebral vessels
- **Myocardial infarct**: occlusion of the coronary arterial branches supplying the heart
- **Kidney disease**: glomerulosclerosis of the renal glomerular vessels
- **Atherosclerosis**: plaque formation in the aorta and its major branches

**Pancreatic cancer** is about the fifth leading cause of cancer deaths in the United States. Pancreatic carcinomas, which are mostly adenocarcinomas, arise from the exocrine part of the organ (cells of the duct system).

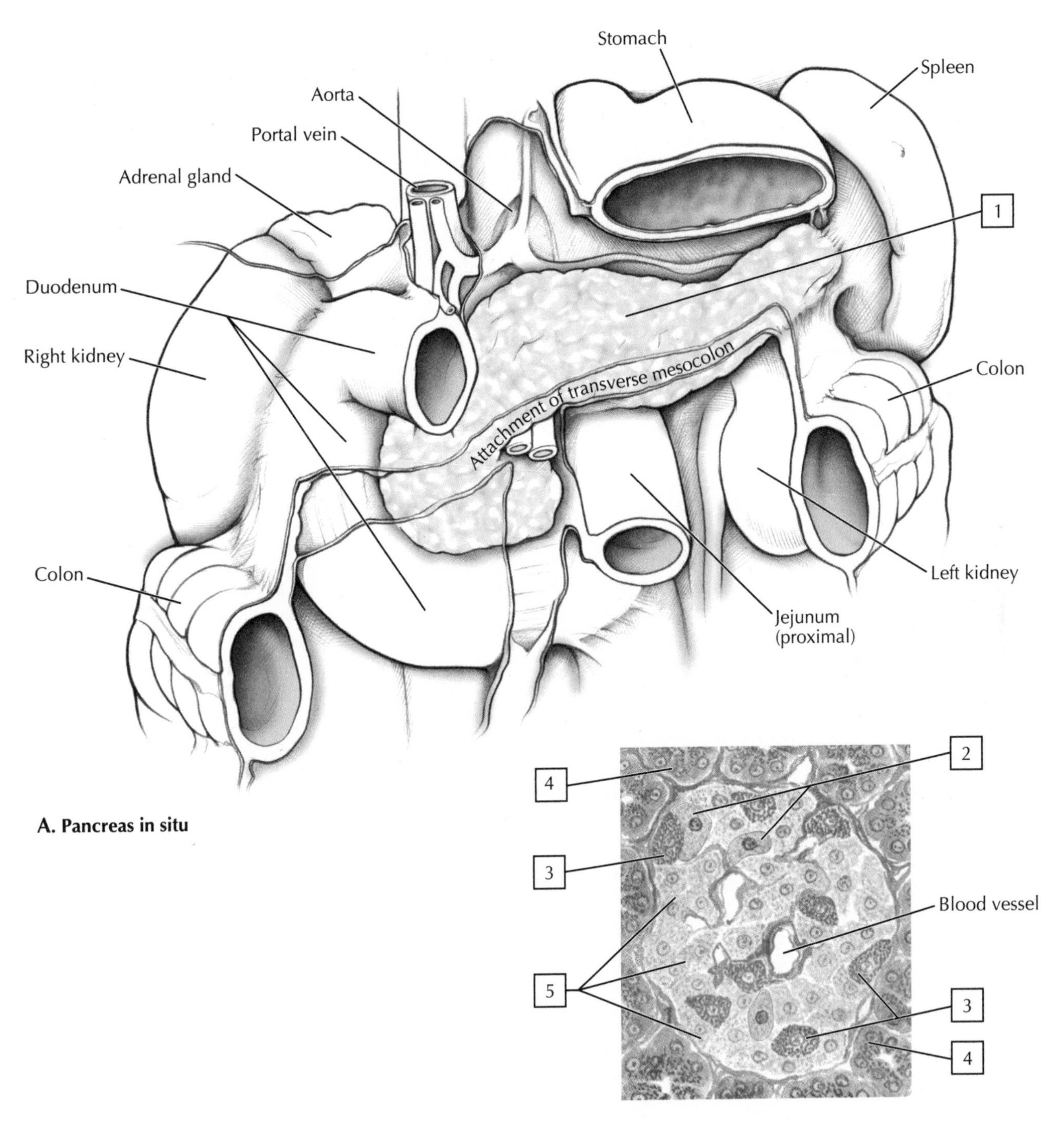


A. Pancreas in situ

B. Low-power section of pancreatic islet cells

# 11 Puberty

Puberty usually occurs between the ages of 10 and 14 years and marks the maturation of the reproductive organs in both sexes, as well as the development of the secondary sex characteristics. One to two years before puberty, adrenal androgen levels increase (adrenarche) and are responsible in both sexes for the early development of pubic, axillary, and facial hair and an increase in growth. Before full maturity, males may experience unexpected nocturnal emissions ("wet dreams") as their hormonal levels struggle to achieve their normal mature balance.

At puberty, the following events occur:

- Hypothalamus increases the release of gonadotropin-releasing hormone (GnRH)
- GnRH stimulates the release of luteinizing hormone (LH) and follicle-stimulating hormone (FSH) by the anterior pituitary
- In females, LH targets the ovary to produce androgens that are then converted to estrogens, LH also stimulates the production of progesterone, and FSH stimulates the production of estrogens from androgens. LH levels are greater than FSH at puberty and during the reproductive years; in senescence, the hormone levels are highest, and FSH is greater than LH
- Estrogen induces the changes in the accessory sex organs and secondary sex characteristics seen in puberty
- In males, LH acts on the testes to stimulate the production of testosterone, and testosterone and FSH together act on the testes to promote development of the spermatozoa
- Testosterone induces the changes in the accessory sex organs and secondary sex characteristics seen in puberty

The secondary sex characteristics commonly associated with puberty are illustrated and listed on the facing page. **Note:** the label GnRH is gonadotropin-releasing hormone from the hypothalamus. The female menstrual cycle is detailed in Plate 10-4, showing the effects of LH, FSH, progesterone, estrogen, and inhibin.

**COLOR** the features of puberty summarized in the illustration, using the colors suggested for each feature:

- ☐ **1. ACTH arrow (targeting the adrenal glands) (green)**
- ☐ **2. FSH arrow (targeting the ovaries and testes) (orange)**
- ☐ **3. LH arrow (targeting the ovaries and testes) (brown)**
- ☐ **4. Adrenal androgens (pink)**
- ☐ **5. Adrenal cortex (yellow)**
- ☐ **6. Ovaries (pink/light red)**
- ☐ **7. Testes (gray)**
- ☐ **8. Estrogen arrow (targets female sex characteristics) (red)**
- ☐ **9. Estrogen arrow (targets male sex characteristics) (blue)**
- ☐ **10. Progesterone arrow (targets female sex characteristics) (gold)**
- ☐ **11. Testosterone arrow (targets male sex characteristics) (purple)**

***Clinical Note:***

**Seminomas** are invasive germinal cell testicular tumors in men, usually in the range of 15-35 years old, that account for about 95% of solid tumors in men in this age group.

Some males show signs of **precocious virilization** that may be related to rather uncommon Leydig cell tumors that are hormonally active.

**Delayed puberty** may occur for several years in females. This condition may occur from poor nutrition from anorexia, extreme athleticism in young girls, or a systemic disease such as chronic renal failure, hypothyroidism, and Cushing's syndrome.

Approximately 1% of live births exhibit some degree of **sexual ambiguity**. Between 0.1% and 0.2% of live births are ambiguous enough to lead to surgical correction.

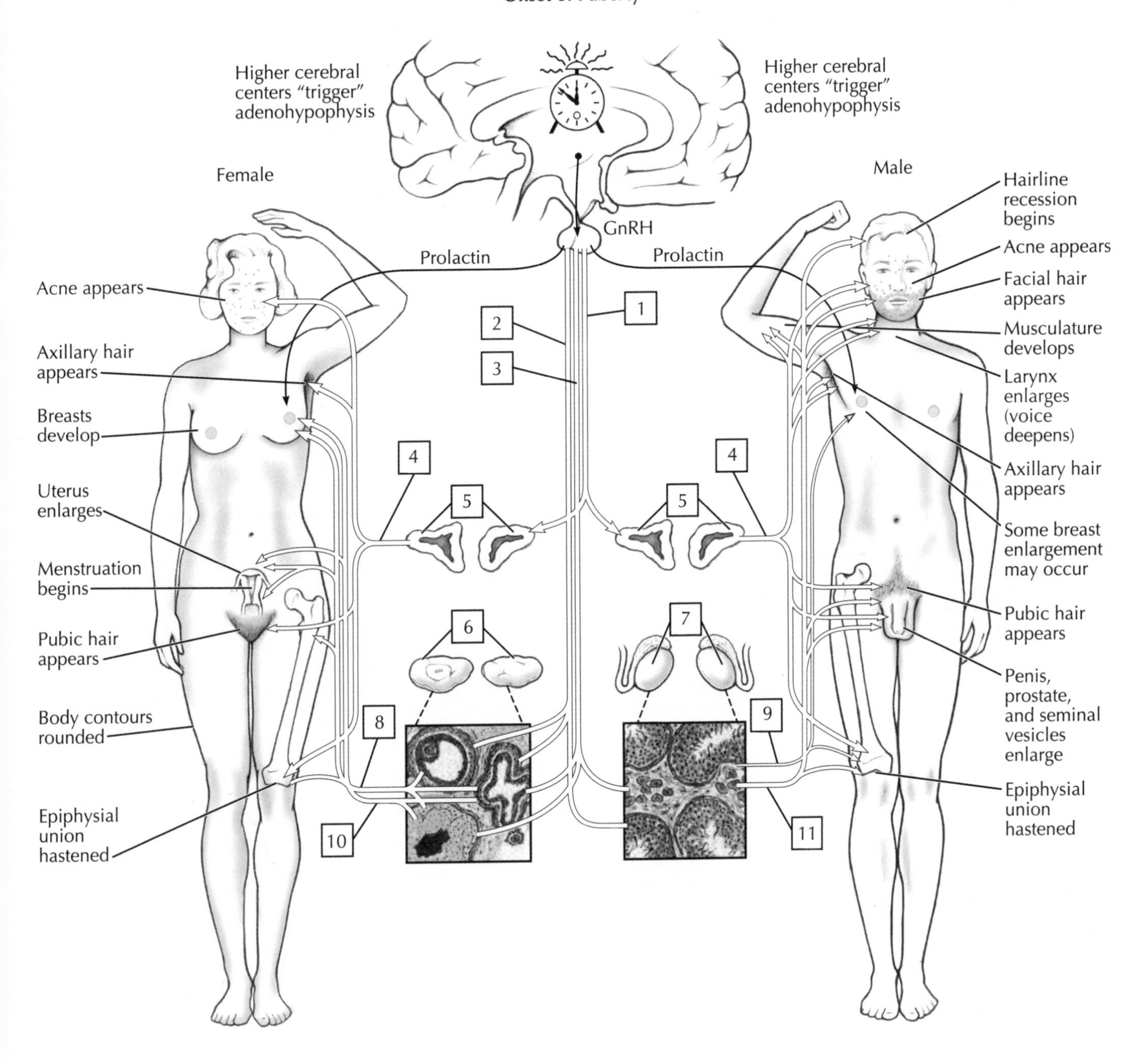
Onset of Puberty
Higher cerebral centers "trigger" adenohypophysis
Higher cerebral centers "trigger" adenohypophysis
Female
Male
GnRH
Prolactin
Prolactin
1
2
3
4
4
5
5
6
7
8
9
10
11
Acne appears
Axillary hair appears
Breasts develop
Uterus enlarges
Menstruation begins
Pubic hair appears
Body contours rounded
Epiphysial union hastened
Hairline recession begins
Acne appears
Facial hair appears
Musculature develops
Larynx enlarges (voice deepens)
Axillary hair appears
Some breast enlargement may occur
Pubic hair appears
Penis, prostate, and seminal vesicles enlarge
Epiphysial union hastened

# 11 Digestive System Hormones

It probably is fair to say that the largest endocrine organ in the human body is the gastrointestinal (GI) tract. The complex physiology of the GI tract, which involves digestion, absorption, peristalsis, metabolism, and storage, is regulated by the complex and integrated actions of the endocrine, neuroendocrine, nervous, and immune systems. The sheer number of different hormones involved is beyond the scope of this book, but some of the "major players" deserve to be introduced.

The composition of **saliva** is modified by the actions of ADH and aldosterone, whereas major GI hormones regulate the **secretory activity** of the stomach, pancreas, and liver. Likewise, **hormones** such as insulin, glucagon, cortisol, epinephrine, norepinephrine, and growth hormone play key roles in organic metabolism. The regulation of the body's energy stores, eating and fasting, obesity control, and thermoregulation all involve the integrated mechanisms of the endocrine and neuroendocrine systems.

When one is focusing principally on the abdominal GI tract, five major hormones play key roles. Dozens of other minor hormones and neuroendocrine molecules are necessary for optimal functioning, but these five are the primary ones and are summarized in the following table.

All of these hormones participate in a **feedback mechanism** that regulates the internal environment of the GI tract, and they also act on multiple target cells. Even between meals, hormones such as motilin initiate the **migrating myoelectric complex** (MMC), which consists of waves of peristalsis that clean the GI tract of residual food particles and move them into the colon. This essentially flushes the stomach and small intestine of bacteria that might otherwise flourish, multiply there, and cause disease.

Remember that the GI tract's enteric nervous system (see Plate 4-21) possesses over 20 different substances (neurotransmitters and neuromodulators) that, along with the hormones listed in the table below, play a key role in constriction, relaxation, and thus motility throughout the stomach, duodenum, and small and large intestines. Gastrin, gastric inhibitory peptide, glucose insulinotropic peptide, CCK, and secretin play key roles with the autonomic nerves in gastric emptying.

**COLOR** the following arrows demonstrating the target sites of major GI hormones, using the suggested color for each hormone's arrow:

- ☐ 1. **Gastrin (red)**
- ☐ 2. **Secretin (blue)**
- ☐ 3. **Cholecystokinin (green)**
- ☐ 4. **Gastric inhibitory peptide (yellow)**
- ☐ 5. **Motilin (orange)**

| HORMONE | NEUROENDOCRINE CELL TYPE AND LOCATION | STIMULUS FOR SECRETION | PRIMARY ACTION | OTHER ACTIONS |
|---|---|---|---|---|
| Gastrin | **G cell** Stomach, duodenum | Vagus nerve (CN X), organ distention, amino acids | Stimulates HC1 secretion | Inhibits gastric emptying |
| Secretin | S cell Duodenum | Acid | Stimulates pancreatic ductal cell $H_2O$ and $HCO_3^-$ secretion | Inhibits gastric secretion, inhibits gastric motility, and stimulates bile duct secretion of $H_2O$ and $HCO_3^-$ |
| Cholecystokinin | **I cell** Duodenum, jejunum | Fat, vagus nerve (CN X) | Stimulates enzyme secretion by pancreatic acinar cells and contracts the gallbladder | Inhibits gastric motility but increases lower gastric emptying |
| Gastric inhibitory peptide | **K cell** Duodenum, jejunum | Fat | Inhibits gastric secretion and motility | Stimulates insulin secretion |
| Motilin | **M cell** Duodenum, jejunum | Duodenal acidification and bile acids | Increased motility and initiates the migrating myoelectric complex (MMC) | Cleans the GI tract of residual food particles; flushes out bacteria |

***Clinical Note:***

**Vomiting** is a reflex action that is controlled by the vomiting center in the medulla oblongata. Irritant stimulation in the stomach or small intestine from enteric viruses or bacteria can initiate the reflex. Also, systemic irritants sensed by chemoreceptors in the fourth ventricular "trigger zone" (near the area postrema located at the inferior and posterior limit of the 4th ventricle, near the obex) can initiate vomiting. Injuries to the head and abnormal stimulation of the vestibular system, e.g., riding a roller coaster, also can initiate vomiting.

See Plate 11-6 for a more complete discussion of hormone effects on the gastrointestinal system.

Major GI Hormones

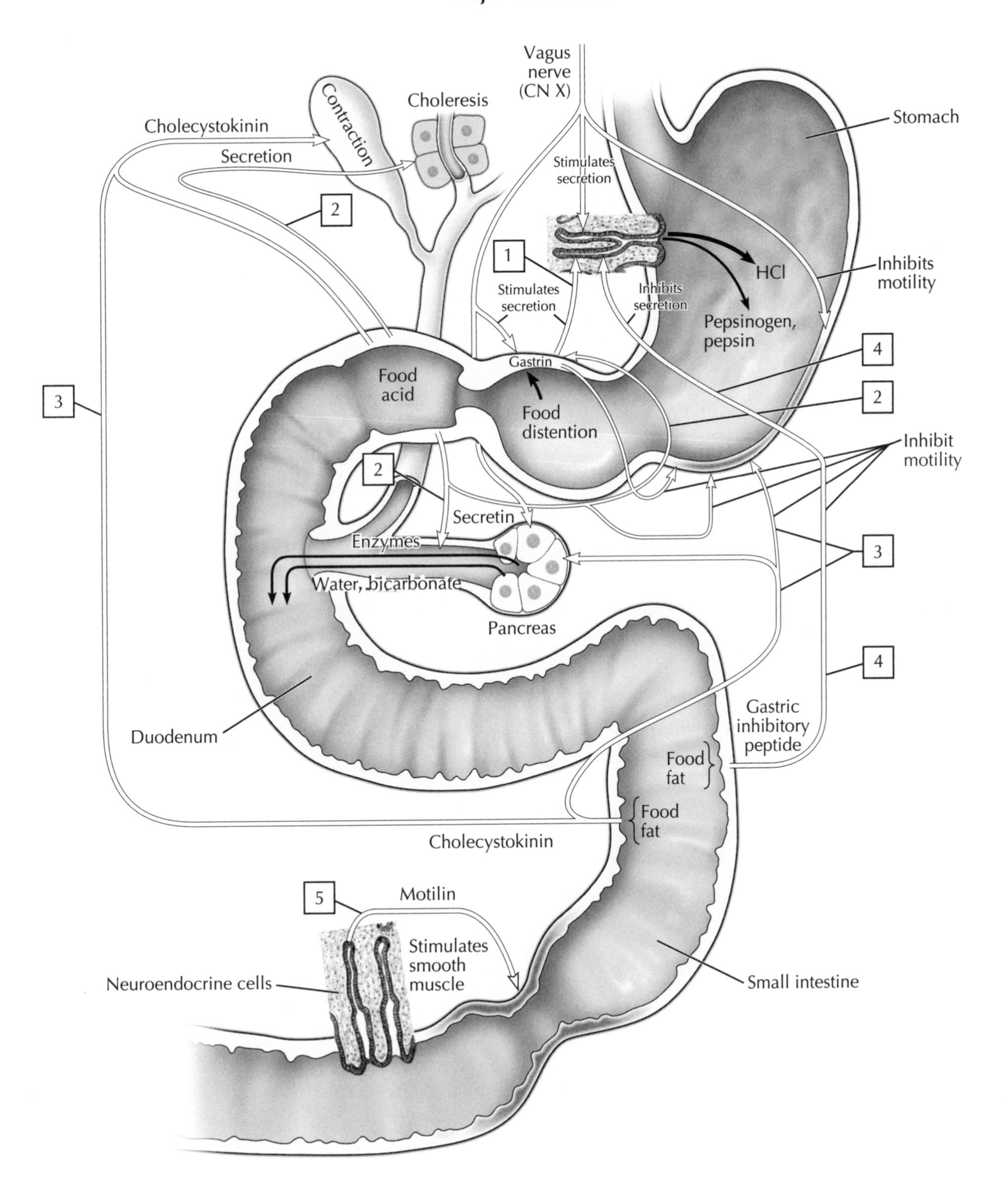

## REVIEW QUESTIONS

1. Which of the following endocrine organs is responsible for uterine contraction, milk expulsion, and concentration of the urine?
   A. Adrenal cortex
   B. Kidney
   C. Ovary
   D. Parathyroids
   E. Posterior pituitary
2. When a hormone is released into the bloodstream by an axon, which of the following types of cell-to-cell communication is occurring?
   A. Autocrine
   B. Endocrine
   C. Holocrine
   D. Neurocrine
   E. Paracrine
3. Graves' disease is an autoimmune disease caused by the excess synthesis and release of thyroid hormone. Which of the following symptoms is most likely to be observed in this condition?
   A. Coldness
   B. Dry skin
   C. Edema of the face
   D. Excitability
   E. Slow pulse
4. Cushing's syndrome is characterized by an increased secretion of what hormone from which gland? (Be specific.) ____________ ______________________________________________
5. One hormone in particular is known to be a "fuel-mobilization" hormone and another a "fuel-storage" hormone. Name these two hormones. ______________________________
6. Which endocrine organ (fat excluded) is probably the largest of the endocrine organs? ______________________________

**For each description below (7-10), color the appropriate endocrine organ in the image.**

7. This endocrine organ regulates the circulating amounts of calcium (an increase) in the bloodstream.
8. Somatostatin (SS), released from this structure, inhibits the release of growth hormone (GH).
9. This endocrine gland releases cortisol, aldosterone, androgens, epinephrine, and norepinephrine.
10. Gonadotropes are released by basophilic cells in this endocrine organ.

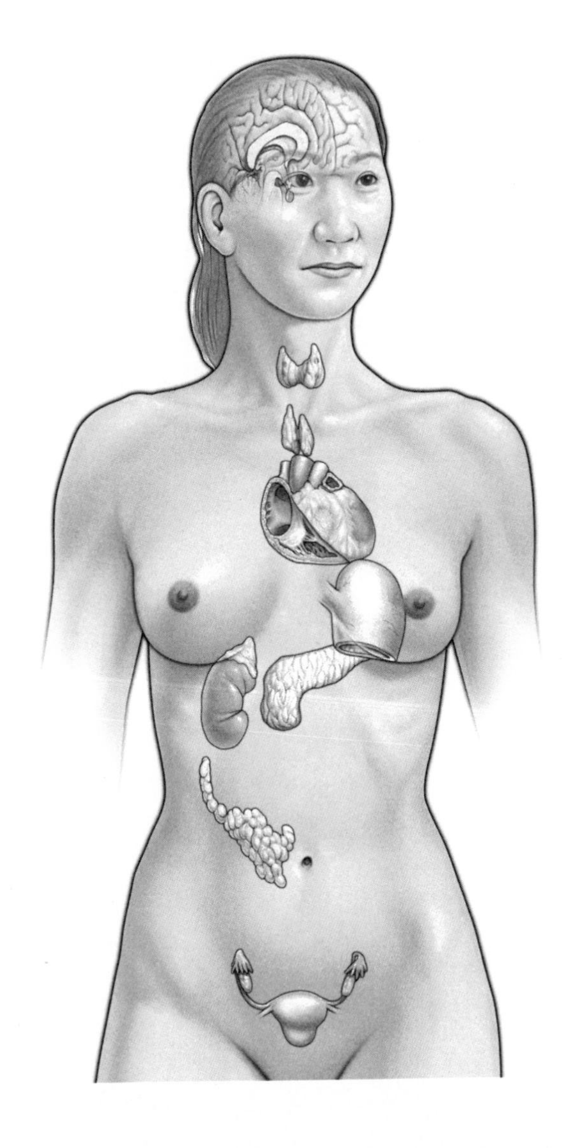

## ANSWER KEY

1. E
2. D
3. D
4. Glucocorticoids (principally cortisol) from the adrenal cortex (zona fasciculata)
5. Glucagon and insulin (from the pancreas)
6. Gastrointestinal tract
7. Parathyroid gland (PTH)
8. Hypothalamus
9. Adrenal gland
10. Anterior pituitary gland

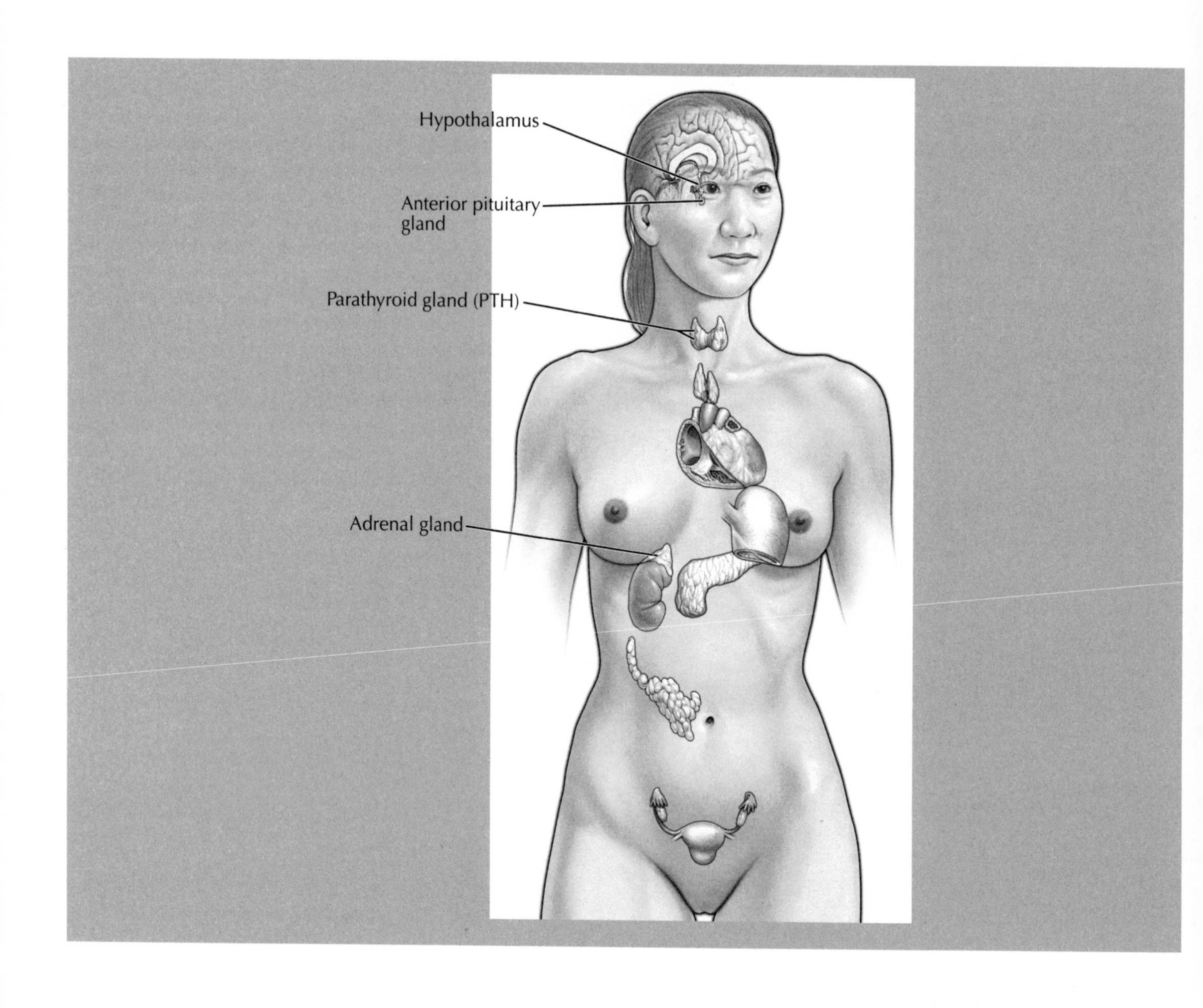
Hypothalamus
Anterior pituitary gland
Parathyroid gland (PTH)
Adrenal gland

# Index

Note: Locators cited are plate numbers. Numbers in regular type indicate the discussion; **boldface** numbers indicate the art in the plate.

## F

**J**



**K**



**L**

## U

**V**



**W**

**X**



**Z**